Lecture Notes in Physics

For information about Vols. 1–44, please contact your bookseller or Springer-Verlag.

Vol. 45: Dynamical Concepts on Scaling Violation and the New Resonances in e^+e^- Annihilation. Edited by B. Humpert. VII, 248 pages. 1976.

Vol. 46: E. J. Flaherty, Hermitian and Kählerian Geometry in Relativity. VIII, 365 pages. 1976.

Vol. 47: Padé Approximants Method and Its Applications to Mechanics. Edited by H. Cabannes. XV, 267 pages. 1976.

Vol. 48: Interplanetary Dust and Zodiacal Light. Proceedings 1975. Edited by H. Elsässer and H. Fechtig. XII, 496 pages. 1976.

Vol. 49: W. G. Harter and C. W. Patterson, A Unitary Calculus for Electronic Orbitals. XII, 144 pages. 1976.

Vol. 50: Group Theoretical Methods in Physics. 4th International Colloquium. Nijmegen 1975. Edited by A. Janner, T. Janssen, and M. Boon. XIII, 629 pages. 1976.

Vol. 51: W. Nörenberg and H. A. Weidenmüller, Introduction of the Theory of Heavy-Ion Collisions. 2nd enlarged edition. IX, 334 pages. 1980.

Vol. 52: M. Mladjenović, Development of Magnetic β-Ray Spectroscopy. X, 282 pages. 1976.

Vol. 53: D. J. Simms and N. M. J. Woodhouse, Lectures on Geometric Quantization. V, 166 pages. 1976.

Vol. 54: Critical Phenomena. Sitges International School on Statistical Mechanics, June 1976. Edited by J. Brey and R. B. Jones. XI, 383 pages. 1976.

Vol. 55: Nuclear Optical Model Potential. Proceedings 1976. Edited by S. Boffi and G. Passatore. VI, 221 pages. 1976.

Vol. 56: Current Induced Reactions. International Summer Institute, Hamburg 1975. Edited by J. G. Körner, G. Kramer, and D. Schildknecht. V, 553 pages. 1976.

Vol. 57: Physics of Highly Excited States in Solids. Proceedings 1975. Edited by M. Ueta and Y. Nishina. IX, 391 pages. 1976.

Vol. 58: Computing Methods in Applied Sciences. Proceedings 1975. Edited by R. Glowinski and J. L. Lions. VIII, 593 pages. 1976.

Vol. 59: Proceedings of the Fifth International Conference on Numerical Methods in Fluid Dynamics. 1976. Edited by A. I. van de Vooren and P. J. Zandbergen. VII, 459 pages. 1976.

Vol. 60: C. Gruber, A. Hintermann, and D. Merlini, Group Analysis of Classical Lattice Systems. XIV, 326 pages. 1977.

Vol. 61: International School on Electro and Photonuclear Reactions I. Edited by C. Schaerf. VIII, 650 pages. 1977.

Vol. 62: International School on Electro and Photonuclear Reactions II. Edited by C. Schaerf. VIII, 301 pages. 1977.

Vol. 63: V. K. Dobrev et al., Harmonic Analysis on the n-Dimensional Lorentz Group and Its Application to Conformal Quantum Field Theory. X, 280 pages. 1977.

Vol. 64: Waves on Water of Variable Depth. Edited by D. G. Provis and R. Radok. 231 pages. 1977.

Vol. 65: Organic Conductors and Semiconductors. Proceedings 1976. Edited by L. Pál, G. Grüner, A. Jánossy and J. Sólyom. 654 pages. 1977.

Vol. 66: A. H. Völkel, Fields, Particles and Currents. VI, 354 pages. 1977.

Vol. 67: W. Drechsler and M. E. Mayer, Fiber Bundle Techniques in Gauge Theories. X, 248 pages. 1977.

Vol. 68: Y. V. Venkatesh, Energy Methods in Time-Varying System Stability and Instability Analyses. XII, 256 pages. 1977.

Vol. 69: K. Rohlfs, Lectures on Density Wave Theory. VI, 184 pages. 1977.

Vol. 70: Wave Propagation and Underwater Acoustics. Edited by J. Keller and J. Papadakis. VIII. 287 pages. 1977.

Vol. 71: Problems of Stellar Convection. Proceedings 1976. Edited by E. A. Spiegel and J. P. Zahn. VIII, 363 pages. 1977.

Vol. 72: Les instabilités hydrodynamiques en convection libre forcée et mixte. Edité par J. C. Legros et J. K. Platten. X, 202 pages. 1978.

Vol. 73: Invariant Wave Equations. Proceedings 1977. Edited by G. Velo and A. S. Wightman. VI, 416 pages. 1978.

Vol. 74: P. Collet and J.-P. Eckmann, A Renormalization Group Analysis of the Hierarchical Model in Statistical Mechanics. IV, 199 pages. 1978.

Vol. 75: Structure and Mechanisms of Turbulence I. Proceedinas 1977. Edited by H. Fiedler. XX, 295 pages. 1978.

Vol. 76: Structure and Mechanisms of Turbulence II. Proceedings 1977. Edited by H. Fiedler. XX, 406 pages. 1978.

Vol. 77: Topics in Quantum Field Theory and Gauge Theories. Proceedings, Salamanca 1977. Edited by J. A. de Azcárraga. X, 378 pages 1978.

Vol. 78: Böhm, The Rigged Hilbert Space and Quantum Mechanics. IX, 70 pages. 1978.

Vol. 79: Group Theoretical Methods in Physics. Proceedings, 1977. Edited by P. Kramer and A. Rieckers. XVIII, 546 pages. 1978.

Vol. 80: Mathematical Problems in Theoretical Physics. Proceedings, 1977. Edited by G. Dell'Antonio, S. Doplicher and G. Jona-Lasinio. VI, 438 pages. 1978.

Vol. 81: MacGregor, The Nature of the Elementary Particle. XXII, 482 pages. 1978.

Vol. 82: Few Body Systems and Nuclear Forces I. Proceedings, 1978. Edited by H. Zingl, M. Haftel and H. Zankel. XIX, 442 pages. 1978.

Vol. 83: Experimental Methods in Heavy Ion Physics. Edited by K. Bethge. V, 251 pages. 1978.

Vol. 84: Stochastic Processes in Nonequilibrium Systems, Proceedings, 1978. Edited by L. Garrido, P. Seglar and P. J. Shepherd. XI, 355 pages. 1978

Vol. 85: Applied Inverse Problems. Edited by P. C. Sabatier. V, 425 pages. 1978.

Vol. 86: Few Body Systems and Electromagnetic Interaction. Proceedings 1978. Edited by C. Ciofi degli Atti and E. De Sanctis. VI, 352 pages. 1978.

Vol. 87: Few Body Systems and Nuclear Forces II, Proceedings, 1978. Edited by H. Zingl, M. Haftel, and H. Zankel. X, 545 pages. 1978.

Vol. 88: K. Hutter and A. A. F. van de Ven, Field Matter Interactions in Thermoelastic Solids. VIII, 231 pages. 1978.

Vol. 89: Microscopic Optical Potentials, Proceedings, 1978. Edited by H. V. von Geramb. XI, 481 pages. 1979.

Vol. 90: Sixth International Conference on Numerical Methods in Fluid Dynamics. Proceedings, 1978. Edited by H. Cabannes, M. Holt and V. Rusanov. VIII, 620 pages. 1979.

Lecture Notes in Physics

Edited by J. Ehlers, München, K. Hepp, Zürich
R. Kippenhahn, München, H. A. Weidenmüller, Heidelberg
and J. Zittartz, Köln
Managing Editor: W. Beiglböck, Heidelberg

130

Mathematical Methods and Applications of Scattering Theory

Proceedings of a Conference
Held at Catholic University
Washington, D. C., May 21 - 25, 1979

Edited by
J. A. DeSanto, A. W. Sáenz and W. W. Zachary

Springer-Verlag
Berlin Heidelberg GmbH 1980

Editors
John A. DeSanto
Albert W. Sáenz
Woodford W. Zachary
Naval Research Laboratory
Washington, D.C. 20375/USA

ISBN 978-3-540-10023-2 ISBN 978-3-540-38184-6 (eBook)
DOI 10.1007/978-3-540-38184-6

Originally published by Springer-Verlag Berlin Heidelberg New York in 1980

2153/3140-543210

EDITORS' PREFACE

The Conference on Mathematical Methods and Applications of Scattering Theory took place at Catholic University, Washington, D.C., from 21 May through 25 May 1979. The general area of the conference was mathematical methods used in scattering theory. The main motivation for holding it was to bring together mathematicians and physicists working in this area, and to give them an opportunity to interact in a wide variety of aspects of scattering theory and related fields.

The proceedings of the conference presented in this volume are dedicated in memoriam to our friend and colleague W. S. Ament of the Naval Research Laboratory, in recognition of his distinguished contributions to classical scattering theory.

The thirteen sessions of this five-day conference were mainly devoted to acoustic, electromagnetic, and quantum-mechanical scattering, and to inverse methods in scattering theory. Both mathematical foundations of the theory as well as physical applications were considered, with about equal time given to each.

There were fifty speakers at the conference, and they included some of the most distinguished names in foundational and applied scattering theory. The wide range of subjects discussed by these speakers, most of whom gave invited talks, can be appreciated by glancing at the table of contents of these proceedings. They ranged from inverse problems in speech and hearing to scattering by surface waves in acoustics and nuclear physics, and from multichannel quantum-mechanical scattering theory to nonlinear evolution equations.

We wish to thank Professor C. J. Nuesse, Executive Vice President and Provost of Catholic University, and Dr. A. Berman, Director of Research of the Naval Research Laboratory, for having given the Welcoming Address and the Opening Remarks at the conference, respectively. We are grateful to Professors J. D. Dollard, J. B. Keller, H. Moses, R. G. Newton, V. Twersky, H. Überall, and V. Weston for their willingness to chair conference sessions. We thank Professor Überall for his extensive organizational efforts and arrangements at Catholic University.

The financial support of the conference was provided by the Army Research Office, the Air Force Office of Scientific Research, the National Science Foundation, and the Office of Naval Research, and we acknowledge it gratefully. We are thankful to Dr. Berman for financially supporting the publication of these proceedings. The generosity of Catholic University in making available at no cost the Nursing Auditorium for the conference sessions is gratefully acknowledged.

We would like to thank Ms. R. Tosta, the Conference Secretary, for her untiring efforts, and Mr. J. Ertel, Ms. P. Burt, and Mr. M. McCord for their efficient handling of the audiovisual equipment and other contributions that assured the smooth functioning of the conference. It is also a pleasant duty to thank Ms. D. B. Wilbanks and her colleagues of the Computerized Technical Composition Section of NRL for their expert care in preparing a camera-ready copy of these proceedings, and Springer-Verlag for their efficient publication efforts.

J. A. DeSanto   A. W. Sáenz   W. W. Zachary

IN MEMORIAM

The editors of this book
dedicate it to the memory
of their friend
and colleague

W. S. Ament

(September 13, 1918 – March 12, 1980)

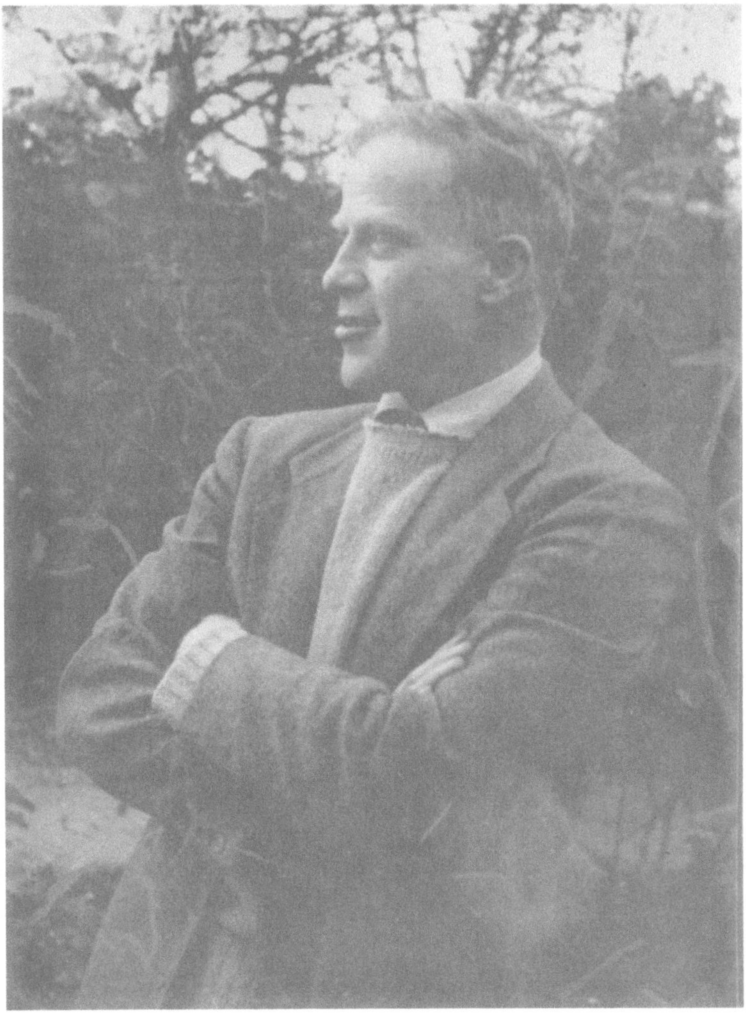

TABLE OF CONTENTS

CLASSICAL SCATTERING THEORY

MAINLY QUANTUM SCATTERING THEORY

INVERSE SCATTERING THEORY AND RELATED TOPICS

OPENING OF THE CONFERENCE ON MATHEMATICAL METHODS AND APPLICATIONS OF SCATTERING THEORY, CATHOLIC UNIVERSITY, 21-25 MAY 1979

A. Berman

Director of Research
Naval Research Laboratory
Washington, D.C. 20375

On behalf of the Naval Research Laboratory, I wish to welcome you to the Conference on Mathematical Methods and Applications of Scattering Theory.

We at NRL feel that scattering theory is an important aspect of our research program. As you can see from the representative NRL speakers at this conference, we are actively working in acoustic, electromagnetic, and quantum scattering theory, as well as on inverse scattering and remote sensing. For example, in the acoustic and electromagnetic areas, we are involved in problems of scattering from rough surfaces as models of the ocean surface, scattering from obstacles, and in inversion and remote sensing problems to identify oceanographic variability.

One of the main reasons for our co-sponsoring this conference was to bring together academic, government, and private industry scientists of different disciplines having strong research interests in some aspect of scattering theory.

Scientists who concentrate on developments in the foundations of their disciplines rarely see applied problems being discussed, and most often do not apply their work to practical problems. On the other hand, research scientists who are oriented toward applications do not usually inquire into the foundations of their subject. This conference will give these groups an opportunity to interact in many different facets of scattering theory. This has three obvious advantages. First, it is well known that heuristic insights gained by working on concrete practical problems may furnish very valuable clues for the development of complete mathematical theories of the physical phenomena involved. Second, it will enable foundational research scientists to see how some of their work is used in practice, to suggest additional uses, and to furnish constructive criticism of applications from a rigorous viewpoint. Third, it will enable research workers in applied scattering theory to see the progress in rigorous methods. These methods are important in isolating the essential elements of scattering phenomena, and in arriving at a qualitative understanding of the phenomena in question. This in turn furnishes a valuable guide for detailed analytical or numerical calculations, which are frequently necessary in applied scattering problems.

Because of the above reasons, we feel that the interactions between the conference participants will substantially benefit all disciplines of research concerned by demonstrating the formalism and progress in other areas, and will broaden the base of knowledge of both applied and basic researchers. Although it is well known that not enough interaction among scientists of different types takes place, very few conferences stress this interdisciplinarity. This is particularly true of conferences on scattering. We believe that this conference will make a significant contribution to interdisciplinary understanding and appreciation of the latter subject. Finally, we feel that it is timely because it will be an important contribution to knowledge in scattering theory and the related subjects covered by the conference.

I would like to thank the organizers of the conference, Drs. DeSanto, Sáenz, and Zachary from NRL, and Professor Überall from Catholic University, for doing a great deal of work and assembling an outstanding group of experts in scattering theory. This conference would not have been possible without the generous support of the Army Research Office, the Air Force Office of Scientific Research, the National Science Foundation, and the Office of Naval Research. Catholic University deserves our thanks for generously providing this Auditorium at no cost for the conference sessions and for their valuable contributions in effectively handling housing and other problems vital to the success of the conference.

Finally, we hope that you enjoy and benefit from the conference as much as we have enjoyed the opportunity to sponsor it.

LIST OF PARTICIPANTS

M. J. Ablowitz
Department of Mathematics
and Computer Science
Clarkson College

J. D. Achenbach
Department of Civil Engineering
Northwestern University

S. Ahn
Naval Research Laboratory

J. J. Ahner
Department of Mathematics
Vanderbilt University

S. T. Ali
Institut für Theoretische Physik
Technische Universität Clausthal

W. S. Ament
Naval Research Laboratory

R. H. Andreo
John Hopkins Applied Physics Laboratory

R. N. Baer
Naval Research Laboratory

G. Bencze
Department of Physics and Astronomy
University of Maryland

G. Blankenship
Department of Electrical Engineering
University of Maryland

N. Bleistein
Department of Mathematics
University of Denver

W-M. Boerner
Department of Information Engineering
University of Illinois
Chicago Circle

N. N. Bojarski
16 Pine Valley Lane
New Beach, California

D. Bollé
Instituuut voor Theoretische Fysica
Katholieke Universiteit Leuven

B. Bourgeois
3747 Duchess Trail
Dallas, Texas

W. Boyse
Research Laboratory
Cities Service Oil Company

S. L. Brodsky
Office of Naval Research

R. W. Buchal
Air Force Office of Scientific Research

R. Burridge
Courant Institute of Mathematical Sciences
New York University

C. Chandler
Department of Physics
University of New Mexico

J. Chandra
Army Research Office

Y. M. Chen
Department of Applied Mathematics and Statistics
State University of New York/Stony Brook

F. Coester
Physics Division
Argone National Laboratory

J. K. Cohen
Department of Mathematics
University of Denver

J. M. Cohen
Department of Physics
University of Pennsylvania

W. J. Condell
Office of Naval Research

J. Cooper
Department of Mathematics
University of Maryland

J. A. DeSanto
Naval Research Laboratory

J. A. Donaldson
Department of Mathematics
Howard University

J. D. Dollard
Department of Mathematics
University of Texas/Austin

J. P. Ertel
Department of Physics
Catholic University

R. Farrell
Johns Hopkins Applied Physics Laboratory

M. Feinstein
Johns Hopkins Applied Physics Laboratory

L. Flax
Naval Research Laboratory

G. V. Frisk
Woods Hole Oceanographic Institution

G. C. Gaunaurd
Naval Surface Weapons Center
White Oaks Laboratory

A. G. Gibson
Department of Mathematics and Statistics
University of New Mexico

C. I. Goldstein
Applied Matehmatics Department
Brookhaven National Laboratory

B. Gopinath
Bell Laboraties
Murray Hill

S. H. Gray
Naval Research Laboratory

J. Greenberg
National Science Foundation

I. W. Herbst
Department of Mathematics
University of Virginia

R. Hermann
Department of Mathematics
Rutgers University

J. Hietarinte
Department of Physics
University of Maryland

J. S. Howland
Department of Mathematics
University of Virginia

N. Hurt
Mechanics Research Institute, Inc.

A. Jacobs
Department of Geophysics
Stanford University

A. K. Jordan
Naval Research Laboratory

B. R. Junker
Office of Naval Research

D. J. Kaup
Department of Physics
Clarkson College

J. B. Keller
Courant Institute of Mathematical Sciences
New York University

Y. S. Kim
Department of Physics and Astronomy
University of Maryland

J. R. Klein
Department of Physics
Pennsylvania State University

R. E. Kleinman
Applied Mathematics Institute
University of Delaware

M. D. Kruskal
Applied Mathematics Program
Princeton University

R. H. Lang
Department of Electrical Engineering
George Washington University

O. W. LeVine
National Aeronatics and Space Agency
Goddard Space Flight Center

F. Lund
Facultad de Ciencias Físicas y Matemáticas

D. MacMillan
Department of Physics
University of Maryland

E. Marx
Harry Diamond Laboratories

R. Meneghini
National Aeronautics and Space Agency
Goddard Space Flight Center

H. E. Moses
Center for Atmospheric Research
University of Lowell

R. G. Newton
Department of Physics
Indiana University

T. A. Osborn
Department of Physics
University of Manitoba

W. Page
University of Maryland

H. C. Pao
Catholic University

W. Polyzou
Department of Physics and Astronomy
University of Maryland

R. T. Prosser
Department of Mathematics
Dartmouth College

E. F. Redish
Department of Physics and Astronomy
University of Maryland

J. C. W. Rogers
John Hopkins Applied Physics Laboratory

A. W. Sáenz
Naval Research Laboratory

H. Segur
Aeronautical Research Associates
of Princeton, Inc.

I. M. Sigal
Department of Mathematics
Princeton University

R. Simón
Georgetown University

I. Sloan
Department of Physics and Astronomy
University of Maryland

M. M. Sondhi
Bell Laboratories
Murray Hill

D. Stickler
Courant Institute of Mathematical Sciences

L. Thomas
Department of Mathematics
University of Virginia

E. Trubowitz
Courant Institute of Mathematical Sciences
New York University

V. Twersky
Department of Mathematics
University of Illinois
Chicago Circle

H. Überall
Department of Physics
Catholic University

V. K. Varadan
Department of Mechanical Engineering
Ohio State University

A. B. Weglein
Research Laboratory
Cities Service Oil Company

C. W. Werntz
Department of Physics
Catholic University

V.H. Weston
Department of Mathematics
Purdue University

C. H. Wilcox
Department of Mathematics
University of Utah

J. Willis
Naval Air Systems Command

W. W. Zachary
Naval Research Laboratory

CLASSICAL SCATTERING THEORY

MULTIPLE SCATTERING OF WAVES BY CORRELATED DISTRIBUTIONS

Victor Twersky*

Mathematics Department,
University of Illinois
Chicago, Illinois 60680

INTRODUCTION

We sketch several recent developments for scattering of waves by pair-correlated distributions of discrete obstacles, and indicate corresponding applications.

In order to introduce the representation theorems required by the dispersion equation for the coherent propagation coefficient (the bulk value), we start with key forms for the scattering of scalar waves by one obstacle, by a configuration of N obstacles [1] and by a statistical ensemble of configurations [2,3]. For simplicity, the scalar Helmholtz equation is considered explicity; analogous results are available for the vector and dyadic analogs [1]. We cite results or procedures of Foldy, Keller, Lax, Rayleigh, and Reiche in the course of the presentation, and compare wth several earlier explicit approximations [4-10]. See [3] and [10, (1962)] for detailed introductions, additional citations, and for related material on the incoherent scattering. We treat three, two, and one-dimensional scalar problems in parallel, but use three-dimensional terminology; the field corresponds to, e.g., the excess pressure in small amplitude acoustics. Applications to optics as well as acoustics are mentioned.

The basic equations are (6) and (8): the first specifies the field for a given configuration in terms of the multiple scattered amplitudes G of the obstacles, and the second relates G to the single scattered amplitudes g. Equation (8), essentially a reciprocity relation between the sets G and g, is regarded as a functional equation for G in terms of known g (the direct problem), but one could just as well regard G as known and seek g (the inverse problem). (See [1] for full discussion of (8) and of vector and tensor analogs. Special cases of (8) have been applied in detail to two obstacles [1], to gratings of parallel cylinders [11], and to singly and doubly periodic arrays of bounded obstacles [12].) Equations (9) and (10), the ensemble averages of (6) and (8), are the first two hierarchy integral relations for the statistical problem; we use a truncation procedure [7] discussed elsewhere in detail [3] and work with a simplified form of (8). The ensemble is specified by the average number ρ of scatterers in unit volume, and by $\rho f(\mathbf{R})$ with $f(\mathbf{R})$ as the distribution function for the separation $\mathbf{R}$ of pairs. For radially symmetric scatterers (or for more general shapes, aligned or averaged over alignment, and regarded as if inclosed within transparent spheres) $f(R)$ is taken as the liquid-state radial distribution function for impenetrable particles. More generally, we regard $f(\mathbf{R})$ as determined by the shape of the exclusion surface specified by the minimum separation of scattering centers. The general case we consider corresponds to differently aligned nonsimilar scatterer and exclusion surfaces.

1. REPRESENTATIONS

We consider a plane wave $\phi e^{-i\omega t} = e^{i\mathbf{k}\cdot\mathbf{r}-i\omega t}$ incident on an obstacle with center at $r = 0$, the center of its smallest circumscribing sphere (of radius a). In the external region (outside the scatterer's volume V bounded by S), the field $\psi = \phi + u$ satisfies $(\nabla^2 + k^2)\psi = 0$ with u as a radiative function:

$$u(\mathbf{r}) = c_o \int [h_o(k|\mathbf{r}-\mathbf{r}'|)\partial_m u(\mathbf{r}') - u\partial_m h_o] dS(\mathbf{r}') \equiv \{h_o, u\};$$

$$h_o(x) = h_o^{(1)}(x), H_o^{(1)}(x), e^{i|x|}; \quad c_o = \frac{k}{4\pi i}, \frac{1}{4i}, \frac{1}{2ik}, \tag{1}$$

where ∂_n is the outward normal derivative. The three forms of h_o and c_o correspond to three, two, and one-dimensional problems, i.e., to bounded obstacles, and to normal incidence on cylinders and slabs respectively. (For slabs, the brace operation represents the sum of the values at the entrance and exit faces.) For $r \sim \infty$ we have

$$u \sim h_0(kr)g(\hat{\mathbf{r}},\hat{\mathbf{k}}),\ \ g(\hat{\mathbf{r}},\hat{\mathbf{k}}) = \{e^{-i\mathbf{k}_r\cdot\mathbf{r}'},u\} \equiv g\{\mathbf{k}_r,\mathbf{k}\},\ \ \mathbf{k}_r = k\hat{\mathbf{r}},\ \mathbf{k} = k\hat{\mathbf{k}}, \tag{2}$$

with h as the asymptotic form of h_o, and $g(\hat{\mathbf{r}},\hat{\mathbf{k}})$ as the scattering amplitude. Using the complex spectral representation for h_o in (1), we obtain (at least for r greater than the scatterer's projection on $\hat{\mathbf{r}}$),

$$u(\mathbf{r}) = \int_c e^{i\mathbf{k}_c\cdot\mathbf{r}} g(\hat{\mathbf{r}}_c,\hat{\mathbf{k}}),\ \ \mathbf{k}_c = k\hat{\mathbf{r}}_c. \tag{3}$$

In three dimensions, $\hat{\mathbf{r}}_c = \hat{\mathbf{r}}(\theta_c,\phi_c)$ and $\int_c = (1/2\pi)\int d\Omega(\theta_c,\phi_c)$ with contours as for $h_o^{(1)}$; in two, $\hat{\mathbf{r}}_c = \hat{\mathbf{r}}(\theta_c)$ and $\int_c = (1/\pi)\int d\theta_c$ with contour as for $H_o^{(1)}$; in one, $\mathbf{k}_c = \pm k\hat{\mathbf{z}}$ and $\int_c$ selects the sign corresponding to $z = \pm|\hat{\mathbf{z}}|$. The associated integral operator for $j_o = \mathrm{Re}h_o$ represents the mean (M) over real directions of observation, i.e., $\mathrm{M}_3 = (1/4\pi)\int d\Omega(\theta,\phi)$, $\mathrm{M}_2 = (1/2\pi)\int d\theta$, and M_1 is one-half the forward and back scattered values. In terms of M, the energy theorem reads

$$-\sigma_o \mathrm{Re}\, g(\hat{\mathbf{k}},\hat{\mathbf{k}}) = \sigma_A + \sigma_s;\ \ \sigma_s = \sigma_o \mathrm{M}|g(\mathbf{r},\hat{\mathbf{k}})|^2;\ \ \sigma_o = \frac{4\pi}{k^2},\frac{4}{k},2, \tag{4}$$

where σ_A and σ_S are the absorption and scattering cross-sections.

For scatterers specified by $(\nabla^2 + K'^2)\psi' = 0$ in V, and $\psi = \psi'$ and $\partial_n\psi = B'\partial_n\psi'$ on S,

$$g(\hat{\mathbf{r}},\hat{\mathbf{k}}) = -c_o\int[(C'-1)k^2 e^{-i\mathbf{k}_r\cdot\mathbf{r}'}\psi'(\mathbf{r}') - (B'-1)\nabla e^{-i\mathbf{k}_r\cdot\mathbf{r}'}\cdot\nabla\psi']dV(\mathbf{r}')$$
$$\equiv [\![e^{-i\mathbf{k}_r\cdot\mathbf{r}'},\psi']\!] \equiv g[\![\mathbf{k}_r,\mathbf{k}]\!];C' = B'\eta'^2 = B'(K'^2/k^2). \tag{5}$$

If we identify ψ as the excess pressure, then for the simplest cases C' is the obstacle's relative compressibility and $1/B'$ its relative mass density. More generally the relative parameters are complex with $\mathrm{Im}\, C' > 0$ and $\mathrm{Im}\, B' < 0$ to account for energy losses. The relative index of refraction equals $\eta' = (C'/B')^{1/2}$; the relative acoustic impedance is given by $\zeta' = (C'B')^{-1/2}$.

Indicating the solutions corresponding to arbitrary directions of incidence $\hat{\mathbf{r}}_a,\hat{\mathbf{r}}_b$ by ψ_a,ψ_b (each subject to the same conditions on S and in V as $\psi = \phi + u$), the subsequent development applies for general obstacles satisfying $\{\psi_a,\psi_b\} = 0$. This condition gives $\{\phi_a,u_b\} = \{\phi_b,u_a\}$, i.e., the usual reciprocity relation $g(-\hat{\mathbf{r}}_a,\hat{\mathbf{r}}_b) = g(-\hat{\mathbf{r}}_b,\hat{\mathbf{r}}_a)$.

For a fixed configuration of N obstacles (with surfaces S_s, volumes V_s, centers $\mathbf{r}_s$, etc.) we write the external solution as

$$\Psi = \psi + \sum_{s=1}^{N} U_s(\mathbf{r}-\mathbf{r}_s;\mathbf{r}_1,\ \ldots,\ \mathbf{r}_N),\ \ U_s = \int_c e^{i\mathbf{k}_c\cdot(\mathbf{r}-\mathbf{r}_s)}G_s(\hat{\mathbf{r}}_c),\ \ G_s(\hat{\mathbf{r}}) = \{e^{-i\mathbf{k}_r\cdot\mathbf{r}'},U_s\}_s, \tag{6}$$

where the multiple scattering amplitude G_s is the form (2) over $S_s(\mathbf{r}')$, and $\mathbf{r}'$ is the local vector from r_s.

With reference to scatterer t,

$$\Psi = \Psi_t = \Phi_t + U_t,\ \ \Phi_t = \phi + \Sigma'_s U_s,\ \ \Sigma'_s = \sum_{s\neq t}, \tag{7}$$

such that Ψ_t,Φ_t,U_t satisfy the same relations at S_t and in V_t as ψ,ϕ,u for the corresponding scatterer in isolation. Thus $\{\psi_a,\Psi_t\}_t = 0$ gives $G_t(-\hat{\mathbf{r}}_a) = \{\phi_a,U_t\}_t = \{\Phi_t,u_a\}_t$; consequently, from the forms for g,G,U as in (2) and (6), we obtain a corresponding reciprocity relation

$$G_t(\hat{\mathbf{r}}) = g_t(\hat{\mathbf{r}},\hat{\mathbf{k}})e^{i\mathbf{k}\cdot\mathbf{r}_t} + \Sigma'\int_c g_t(\hat{\mathbf{r}},\hat{\mathbf{r}}_c)G_s(\hat{\mathbf{r}}_c)e^{i\mathbf{k}_c\cdot\mathbf{R}_{ts}},\ \ \mathbf{k}_c = k\hat{\mathbf{r}}_c,\ \mathbf{R}_{ts} = \mathbf{r}_t - \mathbf{r}_s = R_{ts}\hat{\mathbf{R}}_{ts}, \tag{8}$$

which holds at least if the scatterer's projections on $\hat{\mathbf{R}}_{ts}$ do not overlap. For a given configuration, this system of equations determines G in terms of g (the direct problem), or g in terms of G (the inverse).

From (6) and (7), the average of Ψ over a statistically homogeneous ensemble of configurations of N identical and aligned obstacles whose centers are uniformly distributed in V is

$$<\Psi> = \phi + \rho\int_{\mathrm{V}} <U_s(\mathbf{r}-\mathbf{r}_s)>_s d\mathbf{r}_s,\ \ \rho = N/\mathrm{V},\ \ <U_s>_s = <\Psi_s>_s - <\Phi_s>_s, \tag{9}$$

where $<\cdots>_s$ indicates the average over all variables but $\mathbf{r}_s$. If $\mathbf{r}$ is within $\mathrm{V}_s = \mathrm{V}$, we use $<\Psi_s(K')>_s - <\Phi_s(k)>_s$, and if not, the radiative form $<U_s(k)>_s$. We have $<U_s(k)>_s = \{h_o(k|\mathbf{r}-\mathbf{r}_s-\mathbf{r}'|), <U_s>_s\}$ over $S_s = S$, as well as $<U_s(k)>_s = \int_c <G_s(\hat{\mathbf{r}}_c)>_s e^{i(\mathbf{k}_c\cdot(\mathbf{r}-\mathbf{r}_s))}$. From (8), we express $<G_s>_s$ in terms of the pair distribution function $\rho f(\mathbf{R})$, such that $f(\mathbf{R}) \sim 1$ for $R \sim\infty$, and $f(\mathbf{R}) = 0$ for $R < b(\hat{\mathbf{R}})$. Here, $b(\hat{\mathbf{R}})$ is the minimum separation of centers, i.e., $\mathbf{R}_{ts} = \mathbf{b}(\hat{\mathbf{R}})$ specifies the exclusion surface S inclosing volume v containing only one center. Thus,

$$<G_t(\hat{\mathbf{r}})>_t = g_t(\hat{\mathbf{r}},\hat{\mathbf{k}})e^{i\mathbf{k}\cdot\mathbf{r}_t} + \rho\int_{V-v} d\mathbf{r}_s f(\mathbf{R}_{ts})\int_c g_t(\hat{\mathbf{r}},\hat{\mathbf{r}}_c) < G_s(\hat{\mathbf{r}}_c)>_{st} e^{i\mathbf{k}_c\cdot\mathbf{R}_{ts}}, \tag{10}$$

where $<G_s>_{st}$ is the average over all variables but $\mathbf{r}_s$ and $\mathbf{r}_t$.

If the scatterer centers are distributed in the slab region $0 \leqslant z \leqslant d$, then within the distribution exclusive of boundary transition layers (with thickness of order a), the internal field has the form

$$<\Psi> = A_1 e^{i\mathbf{K}_1\cdot\mathbf{r}} + A_2 e^{i\mathbf{K}_2\cdot\mathbf{r}} = \Sigma_i \Psi_i(\mathbf{r});\ \ \mathbf{K}_i = K_i\hat{\mathbf{K}}_i = k\eta_i\hat{\mathbf{K}}_i;\ \ \mathrm{Im}\,\eta_i > 0, \tag{11}$$

such that $\mathbf{K}_i \cdot \hat{\mathbf{t}} = \mathbf{k} \cdot \hat{\mathbf{t}}$ with $\hat{\mathbf{t}} \cdot \hat{\mathbf{z}} = 0$ (Snell's law); for the simpler cases, $\mathbf{K}_1$ and $\mathbf{K}_2$ are images in $z = 0$. The corresponding terms of $<\Psi_s(\mathbf{r}_s + \mathbf{r}')>_s$, $<\Phi_s>_s$, $<U_s>_s$ have the translational property indicated by $\psi^i(\mathbf{r}')\Psi_i(\mathbf{r}_s)$, $\phi^i\Psi_i$, $u^i\Psi_i$ where the fields ψ^i, ϕ^i, u^i are related on S and in V as for a single obstacle. We write $u^i \sim h(kr)g(\mathbf{k}_r|\mathbf{K}_i) \sim hg^i$, with g^i in the form (2) or (5) in terms of u^i or ψ^i. The average transmitted and reflected fields have the forms $T\phi = Te^{i\mathbf{k}\cdot\mathbf{r}}$ and $R\phi' = \mathrm{R}e^{i\mathbf{k}'\cdot\mathbf{r}}$, with $\mathbf{k}'$ as the image of $\mathbf{k}$ in $z = 0$.

From (9), we obtain transmission (T) and reflection (R) coefficients and extinction (of ϕ) and cancellation (of ϕ') relations [2] which reduce to earlier forms [4] in terms of the present g^i if boundary transition layers are negligible). We also obtain the boundary independent relation

$$k^2 - k^2 = -(\rho/c_o)g\,[\![\mathbf{K}|\mathbf{K}]\!] = -i\rho k\sigma_o g\,[\![\mathbf{K}|\mathbf{K}]\!],\ \ g\,[\![\mathbf{K}_i|\mathbf{K}_i]\!] \equiv [\![e^{-i\mathbf{K}_i\cdot\mathbf{r}'}, \psi^i]\!], \tag{12}$$

a form obtained originally by REICHE [5] for spherical dipoles and by FOLDY [6] for monopoles. The inclosure of the argument of $g\,[\![\mathbf{K}|\mathbf{K}]\!]$ indicates explicit restriction to the volume integral form (5). The analog for $B' = 1$ with $K'^2 - k^2$ as the scattering potential was derived originally by LAX [7], who interpreted $g\,[\![\mathbf{K}|\mathbf{K}]\!]$ as proportional to the result for one scatterer in K-space. See [2] for alternative representations, and for corresponding forms for the bulk parameters.

From (10) with $<G_s>_{st} \approx <G_s>_s$, analogous to $<\Phi_s>_{st} \approx <\Phi_s>_s$ as introduced by LAX [7], in terms of the radiative function

$$\mathrm{U} = \int_c g(\hat{\mathbf{r}},\hat{\mathbf{r}}_c)g(\mathbf{k}_c|\mathbf{K})e^{i\mathbf{k}_c\cdot\mathbf{R}} \equiv \int_c F(\mathbf{k}_r,\mathbf{k}_c|\mathbf{K})e^{i\mathbf{k}_c\cdot\mathbf{R}}, \tag{13}$$

we obtained [2] the dispersion equation

$$g(\mathbf{k}_r|\mathbf{K}) = -\frac{\rho}{c_o(K^2 - k^2)}\{e^{-iK\cdot R}, \mathrm{U}\}_S + \rho\int_{V-v}[f(\mathrm{R}) - 1]e^{-i\mathbf{K}\cdot\mathbf{R}}\mathrm{U}d\mathbf{R}. \tag{14}$$

Here $\mathrm{S} = \mathrm{S}(\mathbf{b})$ is the exclusion surface, and $\mathrm{V} - \mathrm{v}(\mathbf{b})$ now represents the corresponding depleted volume of all space (as appropriate for $f - 1 \approx 0$ for even moderate values of R). More complete results may be based on KELLER's procedure [8], or on alternative approximation [3] of $<G_s>_{st}$ or $<\Phi_s>_{st}$. See [3] for limitations on forms such as (14) obtained by truncating the hierarchy integrals.

In terms of

$$F\{\mathbf{k}_r, \mathbf{K}|\mathbf{K}\} \equiv \{e^{-i\mathbf{k}\cdot\mathbf{R}}, \mathrm{U}\}_S,\ \ M(\mathbf{k}_r, \mathbf{K}) = \rho\int_{V-v}[f(\mathbf{R}) - 1]e^{i(\mathbf{k}_r - \mathbf{K})\cdot\mathbf{R}}d\mathbf{R}, \tag{15}$$

we rewrite (14) as

$$g(\mathbf{k}_r|\mathbf{K}) = -\frac{\rho F\{\mathbf{k}_r, \mathbf{K}|\mathbf{K}\}}{c_o(K^2 - k^2)} + \int_c F(\mathbf{k}_r, \mathbf{k}_c|\mathbf{K})M(\mathbf{k}_c, \mathbf{K}). \tag{16}$$

2. APPLICATIONS

We include key elementary approximations for slight ($B' \approx 1, C' \approx 1$) scatterers small compared to wavelength ($ka \approx 0$) for comparison with more general results. For sparse uncorrelated distributions, we use $F\{\} \approx g(\hat{\mathbf{r}},\hat{\mathbf{k}})g(\mathbf{k}|\mathbf{K})$ in (16) for $\hat{\mathbf{r}} = \hat{\mathbf{k}}$ to obtain

$$K^2 - k^2 \approx -i\rho k\sigma_o g(\hat{\mathbf{k}},\hat{\mathbf{k}}) \equiv 2k(K_R - k),\ \ \mathrm{Re}\,(K_R/k) = 1 + (\rho\sigma_o/2k)\mathrm{Im}\,g(\hat{\mathbf{k}},\hat{\mathbf{k}}),$$

and

$$2\mathrm{Im}\,K_R = -\rho\sigma_o\mathrm{Re}\,g(\hat{\mathbf{k}},\hat{\mathbf{k}}) = \rho(\sigma_A + \sigma_S), \tag{17}$$

where K_R is essentially Rayleigh's result [9] generalized to arbitrary scatterers.

To modify Im K_R and include correlations, we approximate the complex Hankel-type integral $\int_c FM$ of (16) by the real Bessel-type, and replace $g(\mathbf{k}_r|\mathbf{K})$ by $g(\hat{\mathbf{r}},\hat{\mathbf{k}})$. For Im $g >>$ Re g, we obtain

$$2\mathrm{Im}\,K \approx \rho\sigma_A + \rho\sigma_o M[|g(\hat{\mathbf{r}},\hat{\mathbf{k}})|^2 W(\hat{\mathbf{r}},\hat{\mathbf{k}})], \quad W(\hat{\mathbf{r}},\hat{\mathbf{k}}) = 1 + \rho\int [f(\mathbf{R}) - 1]e^{ik(\hat{\mathbf{r}}-\hat{\mathbf{k}})\cdot\mathbf{R}}d\mathbf{R}, \tag{18}$$

where if $f(\mathbf{R}) = f(R)$ is the radial distribution function, then $W(\hat{\mathbf{r}},\hat{\mathbf{k}})$ is a standard form in x-ray diffraction by liquids. For small-spaced scatterers (spacing between closest centers small compared to λ), to lowest order in $k = 2\pi/\lambda$,

$$2\mathrm{Im}\,K \approx \rho(\sigma_A + \sigma_S W), \quad W = 1 + \rho\int [f(\mathbf{R}) - 1]d\mathbf{R}. \tag{19}$$

Alternatively, if in (16) for $\hat{\mathbf{r}} = \hat{\mathbf{K}}$ we introduce $F\{\mathbf{k}_r, \mathbf{K}|\mathbf{K}\} \approx g(\hat{\mathbf{r}},\hat{\mathbf{K}})g(\mathbf{k}_K|\mathbf{K})$, then

$$K^2 - k^2 \approx -\frac{\rho}{c_o}g(\hat{\mathbf{K}},\hat{\mathbf{K}})[1 - \frac{1}{g(\mathbf{k}_K|\mathbf{K})}\int_c g(\hat{\mathbf{K}},\hat{\mathbf{r}}_c)g(\mathbf{k}_c|\hat{\mathbf{K}})M(\mathbf{k}_c,\mathbf{K})]^{-1}, \quad \mathbf{k}_K = k\hat{\mathbf{K}}. \tag{20}$$

Essentially this form in terms of one g-function with $k\hat{\mathbf{K}}$ replaced by $\mathbf{K}$ is given by LAX [7]. The leading terms

$$K^2 - k^2 \approx -\frac{\rho}{c_o}[g(\hat{\mathbf{K}},\hat{\mathbf{K}}) + \int_c g(\hat{\mathbf{K}},\hat{\mathbf{r}}_c)g(\mathbf{r}_c\hat{\mathbf{K}})M(\mathbf{k}_c,\mathbf{K})], \tag{21}$$

reproduce (18) under the same restrictions.

We may obtain closed-form approximations for W without considering f explicitly, by applying statistical-mechanics theorems that relate W (proportional to the fluctuations in the number of particles) to a derivative of the equation of state (E) for the corresponding statistical mechanical fluid. Thus, if we use the scaled-particle approximations [13] for impenetrable particles that occupy the fraction w (volume fraction) of space, then from $(\partial E/\partial\rho)^{-1} = W$, we obtain [14,15] simple rational functions of w:

$$W_n = \frac{(1 - w^{n+1})}{[1 + (n - 1)w]^{n-1}}; \quad n = 3, 2, 1; \quad W_o = 1 - w. \tag{22}$$

Here, w corresponds to spheres, cylinders, or slabs with diameter b for $n = 3,2,1$. The case W_o (which arose in an earlier development [10]) corresponds to a random lattice gas, i.e., to uncorrelated space-occupying particles.

From (19), we write the extinction coefficient as

$$2\,\mathrm{Im}\,K = \tau = \mathrm{C}w + \mathrm{B}wW = \mathrm{C}w + \mathrm{B}S(w), \tag{23}$$

where C and B are proportional to the absorption and scattering cross sections of an isolated particle. The same form arises in electromagnetics and optics. We applied (23) in terms of W of (22) for lossless scatterers (C = 0) to account [14] for the transparency of biological structures having globular or cylindrical inclusions in ranges of the parameters where uncorrelated scattering theory suggests that the distributions are opaque; the opacity that results with swelling could then be interpreted as arising from a decrease in local order. (See [14] for detailed applications to the cornea.) We also considered $S(w) = wW(w)$ as a function of w for the cases in (22), and determined the maximum value $S(w_\partial)$ and the corresponding value $w = w_\partial$. For nonvanishing, small C/B, the maximum of τ is shifted to $w_\Delta > w_\partial$. To first order in C,

$$\tau(w_\Delta) \approx \mathrm{C}w_\partial + \mathrm{B}S(w_\partial), \quad w_\Delta \approx w_\partial + \mathrm{C/B}|S''(w_\partial)|, \tag{24}$$

where $S'' = \partial^2 S/\partial^2 w$.

For detailed consideration, we analyzed (15) and (16) by expanding the g functions in terms of angle functions appropriate for the symmetry of the exclusion surface. In particular, for radially symmetric scatterers (intrinsically symmetric or symmetrized by averaging over orientation) and radially symmetric pair correlations ($f = 0$ for $R < b$) we use the standard Legendre or Fourier expansions for $g(\hat{\mathbf{r}},\hat{\mathbf{k}})$ in terms of the known isolated scattering coefficients a_n. We rewrite the results [2] collectively as

$$\eta^2 - 1 = -i\Gamma\sum_{n=0}^{\infty} A_n\eta^{2n}, \quad A_n = a_n[1 + \sum_{\nu=0}^{\infty} A_\nu\eta^{-n+\nu}H_{n\nu}(\eta)], \quad \Gamma = \rho\sigma_o/k. \tag{25}$$

The A_n are determined by a_n and $H_{n\nu}$ (analogs of lattice sums), and the resulting series in $A_n(\eta)$ is a functional equation for η.

For spherical symmetry

$$H_{n\nu} = \sum_{|n-\nu|}^{|n+\nu|} d_m\begin{pmatrix}0; & 0\\ n; & \nu\end{pmatrix}\left[i\Gamma\frac{\eta^{\nu+n} - \eta^m}{\eta^2 - 1} + H_m\right], \quad \Gamma = \rho 4\pi/k^3,$$

$$H_m = 4\pi\rho\int_0^\infty [f(R) - 1]j_n(KR)h_n^{(1)}(kR)R^2dR, \tag{26}$$

where the coefficients d_m are known from the expansion $P_n(x)P_\nu(x) = \Sigma d_m \begin{pmatrix} 0, & 0 \\ n, & \nu \end{pmatrix} P_m(x)$ for the Legendre polynomials, and where $(\eta^{\nu+n} - \eta^m)/(\eta^2 - 1) = \eta^m + \eta^{m+2} + \ldots + \eta^{\nu+n-2}$ is nonsingular. For circular symmetry,

$$2H_{n\nu} = i\Gamma\eta^{\nu-n}\sum_{m=o}^{n-1}\eta^{2m} + H_{\nu-n} + H_{\nu+n}, \quad \Gamma = 4\rho/k^2,$$

and

$$H_n = H_{-n} = 2\pi\rho\int_o^\infty [f(R) - 1]J_n(KR)H_n^{(1)}(kR)RdR. \tag{27}$$

For slabs, only $n = 0$ and $n = 1$ arise:

$$H_o = 2\rho\int_o^\infty [f(R) - 1]\cos(KR)e^{ikR}dR,$$

$$H_1 = -i2\rho\int_o^\infty [f(R) - 1]\sin(KR)e^{ikR}dR, \quad \Gamma = 2\rho/k; \; H_{11} = i\Gamma + H_o. \tag{28}$$

If we retain only monopoles (a_o) and dipoles (a_1), then from (25),

$$A_o = a_o(1 + A_oH_o + A_1H_1\eta), \quad A_1 = a_1(1 + A_oH_1/\eta + A_1H_{11}); \; H_{11} = [i\Gamma + H_o + (n-1)H_1]/n. \tag{29}$$

Consequently,

$$\eta^2 = 1 - i\Gamma\,\frac{a_o + a_1\eta^2 - a_oa_1(H_{11} + \eta^2H_o - 2\eta H_1^2)}{1 - a_oH_o - a_1H_{11} + a_oa_1(H_oH_{11} - H_1^2)}. \tag{30}$$

An analogous result for circular cylinders was obtained by BOSE and MAI [16] by separating variables in circular coordinates, and applied for detailed considerations to an exponential correlation function $f - 1 = e^{-R/L}$ for $R > 2a$. Similarly, an analog for spheres [17] obtained by Fikioris and Waterman by separating variables in spherical coordinates corresponds to $f - 1{=}0$ for $R > 2a$. (The generalization to include octupoles, i.e., $\eta^2(a_o,a_1,a_2)$ is given in [2, (1977)] We used the virial expansion of f for spheres, as well as the closed forms W of (22) to apply (30) and corresponding expressions for the bulk parameters B and $C = B\eta^2$ to low frequency acoustics [15].

The same functions H_n arise for aligned nonradially symmetric scatterers, provided the exclusion surface is radially symmetric. For aligned elliptic cylinders and ellipsoids, instead of a single system of algebraic equations for the coefficients A_n in (25), we dealt with two coupled systems involving two sets of As. For nonradially symmetric statistics, we expand the g's in terms of special functions of angles appropriate to the symmetry of the exclusion surface, e.g., Mathieu functions for elliptic cylinders [2]. Low frequency results can be obtained from such series, or directly from (16).

The low frequency limit of (30) for pure dipoles specified by $B' = 1/\eta'^2$ equals

$$\frac{1}{\eta^2} - 1 = B - 1 = \frac{i\Gamma a_1}{1 - i\Gamma a_1/n} \Rightarrow B - 1 = \frac{w(B' - 1)}{1 + (B' - 1)(1 - w)/n} \tag{31}$$

with w as the volume fraction for dipoles. The result for B for $n = 3$ was obtained originally by MAXWELL [18] in the form

$$\frac{B - 1}{B + 2} = w\frac{B' - 1}{B' + 2}, \tag{32}$$

i.e., the form attributed to Clausius, Mossotti, Lorenz and Lorentz. Maxwell also gave the result for $n = 1$, and RAYLEIGH [19] derived the values for $n = 3,2$ for cubic and square lattices.

For the analogous problem of aligned anisotropic dipoles and similarly aligned ellipsoidal exclusion regions specified by the elliptic integral depolarization factors Q_i, for incidence along a principal axis, we obtain

$$\frac{1}{\eta_i^2} - 1 = B_i - 1 = \frac{i\Gamma a_{1i}}{1 - i\Gamma a_{1i}Q_i}, \quad \Sigma Q_i = 1, \quad i = 1,2,3. \tag{33}$$

If the dipoles are ellipsoids specified by B'_i and q_i, then (33) reduces to

$$\Delta_i = \frac{w\delta_i}{1 + \delta_iD_i}; \quad \Delta_i = B_i - 1, \quad \delta_i = B'_i - 1, \quad D_i = q_i - wQ_i, \tag{34}$$

where the compound depolarization factor D_i is always positive because of an implicit constraint that bounds w below the densest realizable packing of the ellipsoidal dipoles [2]. Form (34) also arises in electromagnetics for **E** parallel to an axis of dielectric particles specified by $B'_i = \epsilon_{si}/\epsilon_{oi} = \eta'^2$, where s and o indicate values for the particle and embedding medium: for this case $B_i = \epsilon_{bi}/\epsilon_{oi} = \eta_{bi}^2/\eta_{oi}^2$ with b indicating the bulk value; similarly for **H** parallel to an axis of particles specified by $B'_i = \mu_{si}/\mu_{oi}$. We write all cases collectively as

$$\frac{\Delta_i}{1+Q_i\Delta_i} = w\frac{\delta_i}{1+q_i\delta_i}, \quad \Delta_i = \frac{P_i}{P_{oi}} - 1, \quad \delta_i = \frac{p_i}{p_{oi}} - 1, \tag{35}$$

where P_i and and p_i correspond to bulk and particle parameters, respectively.

For the more general case of noncoincident axes of the corresponding dyadics $\tilde{\mathbf{q}}, \tilde{\mathbf{p}}, \tilde{\mathbf{Q}}, \tilde{\mathbf{p}}_o$, we have

$$\tilde{\mathbf{\Delta}} \cdot (\tilde{\mathbf{I}} + \tilde{\mathbf{Q}} \cdot \tilde{\mathbf{\Delta}})^{-1} = w\tilde{\boldsymbol{\delta}} \cdot (\tilde{\mathbf{I}} + \tilde{\mathbf{q}} \cdot \tilde{\boldsymbol{\delta}})^{-1}, \quad \tilde{\mathbf{\Delta}} = \tilde{\mathbf{P}} \cdot \tilde{\mathbf{p}}_o^{-1} - \tilde{\mathbf{I}}, \quad \tilde{\boldsymbol{\delta}} = \tilde{\mathbf{p}} \cdot \tilde{\mathbf{p}}_o^{-1} - \tilde{\mathbf{I}} \tag{36}$$

where $\tilde{\mathbf{I}}$ is the identity dyadic. This form $\tilde{\mathbf{T}} = w\tilde{\mathbf{T}}'$ generalizes Maxwell's original construction for spheres in that it relates the distant dyadic potentials of N ellipsoids to that of an equivalent (non-existing) ellipsoid. From (36),

$$\tilde{\mathbf{\Delta}} = w(\tilde{\mathbf{I}} + \tilde{\boldsymbol{\delta}} \cdot \tilde{\mathbf{D}})^{-1} \cdot \tilde{\boldsymbol{\delta}} = w\tilde{\boldsymbol{\delta}} \cdot (\tilde{\mathbf{I}} + \tilde{\mathbf{D}} \cdot \tilde{\boldsymbol{\delta}})^{-1}, \quad \tilde{\mathbf{D}} = \tilde{\mathbf{q}} - w\tilde{\mathbf{Q}}, \tag{37}$$

with $\tilde{\mathbf{D}}$ as the compound depolarization dyadic.

Thus, there are four distinct bases of anisotropy for a general composite medium: two are intrinsic (form independent) and are determined by the parameters of the inclusions ($\tilde{\mathbf{p}}$) and the embedding medium ($\tilde{\mathbf{p}}_o$); two depend on form, the form of the particle ($\tilde{\mathbf{q}}$) and of the exclusion region ($\tilde{\mathbf{Q}}$). The roles of $\tilde{\mathbf{p}}, \tilde{\mathbf{p}}_o$ and $\tilde{\mathbf{q}}$ (particle shape) are the same as for the analogous problem of a regular lattice; the shape of the exclusion region $\tilde{\mathbf{Q}}$ (correlation, configurational factor) is an analog of the shape of the unit cell. See [20] for detailed applications to bipolarization and birefringence studies.

FOOTNOTES AND REFERENCES

*Work supported in part by National Science Foundation Grants MCS 75-07391 and 78-01167.

[1] V. Twersky: "Scattering of waves by two objects," pp. 361-389 of *Electromagnetic Waves*, ed. by R.E. Langer (U. Wisconsin Press, 1962); "Multiple scattering by arbitrary configurations in three dimensions," J. Math. Phys. **3**, 83-91 (1962); "Multiple scattering of electromagnetic waves by arbitrary configurations," J. Math. Phys. **8**, 589-610 (1967); J.E. Burke, D. Censor, V. Twersky: "Exact inverse-separation series for multiple scattering in two-dimensions," J. Acoust. Soc. Am. **37**, 5-13 (1965)

[2] V. Twersky: "Propagation parameters in random distributions of scatterers," J. d' Anal. Math. **30**, 498-511 (1976); "Coherent scalar field in pair-correlated random distributions of aligned scatterers," J. Math. Phys. **18**, 2468-2486 (1977); "Coherent electromagnetic waves in pair-correlated distributions of aligned scatterers," J. Math. Phys. **19**, 215-230 (1978); "Constraint on the compound depolarization factor of aligned ellipsoids," J. Math. Phys. **19**, 2576-2578 (1978)

[3] V. Twersky: "On propagation in random media of discrete scatterers," Proc. Sympos. Appl. Math. **16**, 84-116 (Am. Math. Soc., Providence, Rhode Island, 1964)

[4] V. Twersky: "On scattering of waves by random distributions, II; two-space scatterer formalism," J. Math. Phys. **3**, 724-734 (1962)

[5] F. Reiche: "Zur Theorie der Dispersion in Gasen und Dampfen," Ann. Phys. **50**, 1, 121 (1916)

[6] L.L. Foldy: "The multiple scattering of waves," Phys. Rev. **67**, 107-119 (1945)

[7] M. Lax: "Multiple scattering of waves," Rev. Mod. Phys. **23**, 287-310 (1951); "The effective field in dense systems," Phys. Rev. **88**, 621-629 (1952)

[8] J. B. Keller: "Wave propagation in random media," Proc. Sympos. Appl. Math **13**, 227-246 (Am. Math. Soc., Providence, Rhode Island, 1962); "Stochastic equations and wave propagation in random media," *ibid.* **16**, 145-170 (1964); D.J. Vezzetti, J.B. Keller: "Refractive index, dielectric constant, and permeability for waves in a polarizable medium," J. Math. Phys. **8**, 1861-1870 (1967)

[9] Lord Rayleigh: "On the transmission of light through the atmosphere containing small particles in suspension, and on the origin of the color of the sky," Phil. Mag. **47**, 375-383 (1899)

[10] V. Twersky: "Multiple scattering of waves and optical phenomena," J. Opt. Soc. Am. **52**, 145-171 (1962); "Interface effects in multiple scattering by large, low-refracting, absorbing particles," J. Opt. Soc. Am. **60**, 908-914 (1970); "Absorption and multiple scattering by biological suspensions," J. Opt. Soc. Am. **60**, 1084-1093 (1970)

[11] V. Twersky: "On the scattering of waves by an infinite grating," IRE Trans. **AP-4**, 330-345 (1956); "On scattering of waves by the infinite grating of circular cylinders," **AP-10**, 737-765 (1962); E.J. Burke, V. Twersky: "On scattering of waves by the infinite grating of elliptic cylinders," IEEE Trans. **AP-14**, 465-480 (1966)

[12] V. Twersky: "Multiple scattering of sound by a periodic line of obstacles," J. Acoust. Soc. Am. **53**, 96-112 (1973); "Multiple scattering of waves by the doubly periodic planar array of obstacles," "Lattice sums and scattering coefficients for the rectangular planar array," "Low frequency coupling in the planar rectangular lattice," J. Math. Phys. **16**, 633-666 (1975)

[13] H. Reiss, H.L. Frisch, J.L. Lebowitz: "Statistical mechanics of rigid spheres," J. Chem. Phys. **31**, 369-380 (1959); E. Helfand, H.L. Frisch, J.L. Lebowitz: "The theory of the two-and one-dimensional rigid sphere fluids," J. Chem. Phys. **34**, 1037-1042 (1961)

[14] V. Twersky: "Transparency of pair-correlated, random distribution of small scatterers with applications to the cornea," J. Opt. Soc. Am. **65**, 524-530 (1975)

[15] V. Twersky: "Acoustic bulk parameters in distributions of pair-correlated scatterers," J. Acoust. Soc. Am. **64**, 1710-1719 (1978)

[16] S.K. Bose, A.K. Mai: "Longitudinal shear waves in a fiber-reinforced composite," Int. J. Solids Structures **9**, 1975-1985 (1973)

[17] J.G. Fikioris, P.C. Waterman: "Multiple scattering of waves II; hole corrections in the scalar case," J. Math. Phys. **5**, 1413-1420 (1964)

[18] J.C. Maxwell: *A Treatise on Electricity and Magnetism* (Cambridge, 1873; Dover, New York, 1954)

[19] Lord Rayleigh: "On the influence of obstacles arranged in rectangular order upon the properties of a medium," Phil. Mag. **34**, 481-501 (1892)

[20] V. Twersky: "Form and intrinsic birefringence," J. Opt. Soc. Am. **65**, 239-245 (1975); "Intrinsic, shape, and configurational birefringence," to appear in J. Opt. Soc. Am. (1979)

THE S-MATRIX AND SONAR ECHO STRUCTURE

Calvin H. Wilcox

Department of Mathematics
University of Utah
Salt Lake City, Utah 84112

INTRODUCTION

This paper deals with pulse mode sonar echo prediction; that is, the calculation of sonar echoes when the characteristics of the transmitter, scattering objects and ambient medium are known. The physical hypotheses are

- The medium is a stationary homogeneous unlimited fluid.
- Both the sonar system and the scattering objects are stationary.
- The scattering objects are rigid bodies.

The analysis is based on the theory of scattering for the wave equation developed in the author's monograph [1]. The principal result of this paper is an asymptotic calculation of the echo waveform, valid when both transmitter and receiver are in the far-field of the scatterers. The results show that, in this approximation, the dependence of the echo waveform on the scatterers is determined by the S-matrix of scattering theory. The work is a sequel to the author's article [2] where similar results are derived for plane-wave signals.

1. THE BOUNDARY VALUE PROBLEM.

In what follows, $x = (x_1,x_2,x_3) \in \mathbf{R}^3$ denotes the coordinates of a Cartesian system fixed in the medium and $t \in \mathbf{R}$ denotes a time coordinate. The acoustic field is characterized by a real-valued acoustic potential $u(t,x)$ which is a solution of the wave equation

$$\frac{\partial^2 u}{\partial t^2} - \nabla^2 u = f(t,x). \tag{1.1}$$

The function f, which characterizes the transmitter, will be called the source function. The scattering of a single pulse of duration T, emitted by a transmitter localized near a point x_0, will be analyzed. Hence, the space-time support of f is assumed to satisfy

$$\operatorname{supp} f \subset \{(t,x) \mid t_0 \leqslant t \leqslant t_0 + T \text{ and } |x - x_0| \leqslant \delta_0\}, \tag{1.2}$$

where δ_0 and t_0 are constants. The scatterers are represented by a closed bounded set $\Gamma \subset \mathbf{R}^3$ with complement $\Omega = \mathbf{R}^3 - \Gamma$. The common boundary $\partial\Gamma = \partial\Omega$ is assumed to be a smooth surface. It will be convenient to let the origin of coordinates lie in Γ and

$$\Gamma \subset \{x \mid |x| \leqslant \delta\}. \tag{1.3}$$

It is also assumed that

$$\delta + \delta_0 < |x_0|, \tag{1.4}$$

(the transmitter and scatterers are disjoint) and

$$T < |x_0| - \delta - \delta_0, \tag{1.5}$$

(the sources cease acting before the signal reaches the scatterers).

The total acoustic field produced by the transmitter is the presence of the scatterers Γ is the solution $u(t,x)$ of (1.1) in $\mathbf{R} \times \Omega$ that satisfies the boundary condition

$$\frac{\partial u}{\partial \nu} = \vec{\nu} \cdot \nabla u = 0, \tag{1.6}$$

for $(t,x) \in \mathbf{R} \times \partial\Omega$, where $\vec{\nu}$ is a normal to $\partial\Omega$, and the initial condition

$$u(t,x) = 0 \text{ for } t < t_0 \text{ and } x \in \Omega. \tag{1.7}$$

The corresponding signal field $u_0(t,x)$, generated by $f(t,x)$ when no scatterers are present, is given by the retarded potential

$$u_0(t,x) = \frac{1}{4\pi}\int_{\mathbf{R}^3} \frac{f(t - |x - x'|,x')}{|x - x'|}\, dx', \tag{1.8}$$

where $dx' = dx_1'\, dx_2'\, dx_3'$. The sonar echo $u_s(t,x)$ produced by the source function f and the scatterers Γ is defined by

$$u_s(t,x) = u(t,x) - u_0(t,x),\ t \in \mathbf{R},\ x \in \Omega. \tag{1.9}$$

Both supp $u_0(t,\cdot)$ and supp $u(t,\cdot)$ are contained in $\{x \,|\, |x - x_0| \leqslant t - t_0 + \delta_0\}$ which is disjoint from Γ for $t - t_0 + \delta_0 \leqslant |x_0| - \delta$. It follows that

$$u_s(t,x) = 0 \text{ for } t \leqslant t_0 + |x_0| - \delta - \delta_0 \text{ and } x \in \Omega. \tag{1.10}$$

The goal of this paper is to calculate $u_s(t,x)$, especially in the far field ($|x| >> 1$) and to analyze its dependence on the source function f and the scatterers Γ.

2. THE WAVE OPERATORS AND PULSE MODE SONAR ECHO STRUCTURE

The starting point for the calculation of $u_s(t,x)$ below is a construction of $u(t,x)$ in the Hilbert space $L_2(\Omega)$. To describe it let

$$L_2^m(\Omega) = \{u(x) \,|\, D^\alpha u(x) \in L_2(\Omega) \text{ for } 0 \leqslant |\alpha| \leqslant m\}, \tag{2.1}$$

where $\alpha = (\alpha_1,\alpha_2,\alpha_3)$, $|\alpha| = \alpha_1 + \alpha_2 + \alpha_3$ and $D^\alpha = \partial^{|\alpha|}/\partial x_1^{\alpha_1}\, \partial x_2^{\alpha_2}\, \partial x_3^{\alpha_3}$. Then the operator $A: L_2(\Omega) \rightarrow L_2(\Omega)$ defined by

$$D(A) = L_2^2(\Omega) \cap \{u \,|\, \partial u/\partial\nu = 0 \text{ on } \partial\Omega\}, \tag{2.2}$$

$$Au = -\nabla^2 u \text{ for all } u \in D(A), \tag{2.3}$$

is selfadjoint and non-negative (see [1] for details). The solution of the initial-boundary value problem (1.6), (1.2), (1.7) is given by Duhamel's integral [2]:

$$u(t,\cdot) = \int_{t_0}^{t} \{A^{-1/2} \sin(t-\tau)A^{1/2}\} f(\tau,\cdot)\, d\tau,\ t \geqslant t_0. \tag{2.4}$$

In particular, for $t \geqslant t_0 + T$

$$u(t,\cdot) = \int_{t_0}^{t_0+T} \{A^{-1/2} \sin(t-\tau)A^{1/2}\} f(\tau,\cdot)\, d\tau = \mathrm{Re}\{\mathrm{v}(t,\cdot)\}, \tag{2.5}$$

where

$$\mathrm{v}(t,\cdot) = i\int_{t_0}^{t_0+T} A^{-1/2} \exp\{-i(t-\tau)A^{1/2}\} f(\tau,\cdot)\, d\tau = \exp\{-itA^{1/2}\}h, \tag{2.6}$$

and

$$h = i\int_{t_0}^{t_0+T} A^{-1/2} \exp\{i\tau A^{1/2}\} f(\tau,\cdot)\, d\tau. \tag{2.7}$$

In the special case where $\Omega = \mathbf{R}^3$ (no scatterer), the operator A will be denoted by A_0. Thus $A_0: L_2(\mathbf{R}^3) \rightarrow L_2(\mathbf{R}^3)$, defined by $D(A_0) = L_2^2(\mathbf{R}^3)$ and $A_0 u = -\nabla^2 u$, is self-adjoint in $L_2(\mathbf{R}^3)$ and the signal $u_0(t,x)$ is given by

$$u_0(t,\cdot) = \mathrm{Re}\{\mathrm{v}_0(t,\cdot)\},\ t \geqslant t_0 + T, \tag{2.8}$$

where

$$\mathrm{v}_0(t,\cdot) = \exp\{-itA_0^{1/2}\} h_0, \tag{2.9}$$

and

$$h_0 = i\int_{t_0}^{t_0+T} A_0^{-1/2} \exp\{i\tau A_0^{1/2}\} f(\tau,\cdot)\, d\tau. \tag{2.10}$$

To compare $u(t,x)$ and $u_0(t,x)$ introduce the operator $J: L_2(\Omega) \rightarrow L_2(\mathbf{R}^3)$ defined by

$$Jh(x) = \begin{cases} j(x)h(x) & \text{for} \quad x \in \Omega, \\ 0 & \text{for} \quad x \in \mathbb{R}^3 - \Omega, \end{cases} \tag{2.11}$$

where $j \in C^\infty(\mathbb{R}^3)$, $0 \leqslant j(x) \leqslant 1$, $j(x) = 1$ for $|x| \geqslant \delta$, and $j(x) = 0$ in a neighborhood of Γ. J is a bounded operator with bound $||J|| = 1$. It will be convenient to extend the definition (1.9) of u_s by defining $u_s(t,x) = \mathrm{Re}\{v_s(t,x)\}$, where

$$v_s(t,x) = Jv(t,x) - v_0(t,x), \quad t \in \mathbb{R}, \ x \in \mathbb{R}^3. \tag{2.12}$$

The calculation of the far field form of $u_s(t,x)$ will be based on the theory of wave operators as developed in [1]. The wave operators W_+ and W_- are defined by the strong limits

$$W_\pm = s - \lim_{t \to \pm\infty} \exp\{itA_0^{1/2}\} \, J \exp\{-itA^{1/2}\}. \tag{2.13}$$

It is shown in [1] that these limits exist and define unitary operators $W_\pm : L_2(\Omega) \to L_2(\mathbb{R}^3)$. It follows that, for each $h \in L_2(\Omega)$,

$$Jv(t,\cdot) = J \exp\{-itA^{1/2}\}h = \exp\{-itA_0^{1/2}\} \, W_+ h + o_t(1), \quad t \to +\infty, \tag{2.14}$$

where $o_t(1)$ denotes an $L_2(\mathbb{R}^3)$-valued function of t that tends to zero in $L_2(\mathbb{R}^3)$ when $t \to \infty$. Equations (2.14) and (2.9) imply that

$$v_s(t,\cdot) = \exp\{-itA_0^{1/2}\}(W_+h - h_0) + o_t(1), \quad t \to \infty. \tag{2.15}$$

This result is used below to calculate the far field form of $u_s(t,x) = \mathrm{Re}\{v_s(t,x)\}$.

3. THE FAR FIELD APPROXIMATION AND THE SCATTERING OPERATOR

The scattering operator for the scatterer Γ is the unitary operator S in $L_2(\mathbb{R}^3)$ defined by

$$S = W_+ W_-^*, \tag{3.1}$$

where W_-^* denotes the adjoint of W_-. A connection between S and the approximation (2.15) will be derived by calculating the relationship between h and h_0. Equation (1.10) implies that $v_s(t,x) = 0$ for $t_0 + T \leqslant t \leqslant t_0 + |x_0| - \delta - \delta_0$ and $x \in \mathbb{R}^3$. It will be convenient to choose

$$t_0 = -|x_0| + \delta_0 + \delta, \tag{3.2}$$

so that the arrival time of the signal at Γ is non-negative (see (1.10)). With this convention,

$$J \exp\{-itA^{1/2}\}h = \exp\{-itA_0^{1/2}\}h_0 \text{ for } t_1 \leqslant t \leqslant 0, \tag{3.3}$$

where

$$t_1 = t_0 + T = -|x_0| + \delta_0 + \delta + T. \tag{3.4}$$

Taking $t = 0$ in (3.3) gives $Jh = h_0$, while taking $t = t_1$ gives

$$\exp\{it_1A_0^{1/2}\}J \exp\{-it_1A^{1/2}\}h = h_0. \tag{3.5}$$

The scatterer Γ is in the far field of the transmitter if $|x_0| >> 1$ or, by (3.4), if $t_1 << -1$. Combining this with (3.5) and the definition (2.13) gives

$$h_0 = W_-h + o_{x_0}(1), \quad |x_0| \to \infty, \tag{3.6}$$

where $o_{x_0}(1)$ is an $L_2(\mathbb{R}^3)$-valued function of x_0 that tends to zero in $L_2(\mathbb{R}^3)$ when $|x_0| \to \infty$. Multiplying (3.6) by S gives

$$W_+h = Sh_0 + o_{x_0}(1), \quad |x_0| \to \infty, \tag{3.7}$$

because $W_-^* \, W_- = 1$ (W_- is unitary). Combining (3.7) and (2.15) gives

$$v_s(t,\cdot) = \exp\{-itA_0^{1/2}\} \, (S-1)h_0 + o_t(1) + o_{x_0}(1). \tag{3.8}$$

Note that the term $o_{x_0}(1)$ tends to zero in $L_2(\mathbb{R}^3)$ when $|x_0| \to \infty$ *uniformly in t* because $\exp\{-itA_0^{1/2}\}$ is unitary. Equation (3.8) shows that, in the far field approximation, the dependence of the echo waveform on the scatterer is determined by the scattering operator.

The approximation (3.8) is used in Sec. 6 to derive an explicit integral formula for the far field echo waveform. The derivation is based on a known integral representation for S, formulated in Sec. 4, and the theory of asymptotic wave functions of [1] which is summarized in Sec. 5.

4. THE STRUCTURE OF THE SCATTERING OPERATOR

The steady-state theory of scattering and associated eigenfunction expansions for A are reviewed briefly in this section and applied to the construction of the scattering operator for Γ.

A_0 is a selfadjoint operator in $L_2(\mathbb{R}^3)$ with a purely continuous spectrum and the plane waves

$$w_0(x,p) = (2\pi)^{-3/2} \exp\{ix\cdot p\}, \ p \in \mathbb{R}^3, \tag{4.1}$$

are a complete family of generalized eigenfunctions. The corresponding eigenfunction expansion is the well-known Plancherel theory of the Fourier transform (see [1], Chap. 6).

Generalizations of the Plancherel theory to acoustic scattering by bounded objects were first given by N. A. SHENK [3] and Y. SHIZUTA [4]. In this work the generalized eigenfunctions are the distorted plane waves

$$w_\pm(x,p) = w_0(x,p) + w_\pm^s(x,p), \ x \in \Omega, \ p \in \mathbb{R}^3, \tag{4.2}$$

which are characterized by the properties that $w_\pm(x,p)$ is locally in $D(A)$ (i.e., $\phi w_\pm(\cdot,p) \in D(A)$ for all $\phi \in C_0^\infty(\mathbb{R}^3)$) ,

$$(\nabla^2 + |p|^2) w_\pm(x,p) = 0 \text{ for } x \in \Omega, \tag{4.3}$$

$$\frac{\partial w_\pm^s}{\partial |x|} \mp i|p| w_\pm^s = 0\left(\frac{1}{|x|^2}\right), \ |x| \to \infty. \tag{4.4}$$

For the existence, uniqueness and construction of $w_\pm(x,p)$ see [1-5]. Physically, $w_\pm^s(x,p)$ is the steady-state scattered field produced when the plane wave (4.1) is scattered by Γ. It has the far field form [1,2]

$$w_\pm^s(x,p) = \frac{e^{\pm i|p||x|}}{4\pi|x|} T_\pm(|p|\theta,p) + 0\left(\frac{1}{|x|^2}\right), \ |x| \to \infty, \tag{4.5}$$

where $\theta = x/|x|$. $T_\pm(p,p')$, the *scattering amplitude* or *differential scattering cross section* of Γ, is defined for all p and p' in $\mathbb{R}^3$ such that $|p| = |p'|$ and has the symmetry properties

$$T_\pm(p,p') = T_\pm(-p',-p) = \overline{T_\mp(p,-p')} = \overline{T_\pm(-p,-p')}, \tag{4.6}$$

where the bar denotes the complex conjugate.

The connection between S and $T_\pm(p,p')$ is based on the eigenfunction expansion theorem for A. The latter states that, for all $h \in L_2(\Omega)$, the limits

$$\hat{h}_\pm(p) = (\phi_\pm h)(p) = L_2(\mathbb{R}^3)\text{-}\lim_{M\to\infty} \int_{\Omega_M} \overline{w_\pm(x,p)}\, h(x) dx, \tag{4.7}$$

exist, where $\Omega_M = \Omega \cap \{x \mid |x| < M\}$, and

$$h(x) = L_2(\Omega)\text{-}\lim_{M\to\infty} \int_{|p|<M} w_\pm(x,p)\hat{h}_\pm(p) dp. \tag{4.8}$$

Moreover, the operators $\Phi_\pm : L_2(\Omega) \to L_2(\mathbb{R}^3)$ are unitary and for each bounded measurable function $\Psi(\lambda)$ on $\lambda \geqslant 0$,

$$(\Phi_\pm \Psi(A) h)(p) = \Psi(|p|^2)\Phi_\pm h(p). \tag{4.9}$$

The Fourier transform will be denoted by

$$\hat{h}(p) = (\Phi h)(p) = L_2(\mathbb{R}^3)\text{-}\lim_{M\to\infty} \int_{|x|<M} \overline{w_0(x,p)}\, h(x) dx. \tag{4.10}$$

An exposition of the eigenfunction expansion theory is given in [1]. In what follows the relations (4.7), (4.8) are written

$$\hat{h}_\pm(p) = \int_\Omega \overline{w_\pm(x,p)}\, h(x) dx, \tag{4.11}$$

$$h(x) = \int_{\mathbb{R}^3} w_\pm(x,p)\hat{h}_\pm(p) dp, \tag{4.12}$$

for brevity. However, the integrals are not convergent, in general, and (4.11), (4.12) must be interpreted in the sense of (4.7), (4.8).

The wave operators $W_\pm$ defined by (2.13) are known to have the representations [1]

$$W_+ = \Phi^*\Phi_-, \ W_- = \Phi^*\Phi_+. \tag{4.13}$$

Combining (4.13) and (3.1) gives the representation

$$S = W_+W_-^* = \Phi^*\hat{S}\Phi, \tag{4.14}$$

where the operator

$$\hat{S} = \Phi_-\Phi_+^*, \tag{4.15}$$

is called the S-matrix for the scatterer Γ. The operator $\hat{S} - 1$ has the integral representation

$$(\hat{S} - 1)\hat{h}(p) = \frac{i|p|}{2(2\pi)^{1/2}} \int_{S^2} T_+(p, |p|\theta')\hat{h}(|p|\theta')d\theta', \tag{4.16}$$

whose kernel is the differential scattering cross section of Γ. The integration in (4.16) is over the points θ' of the unit sphere S^2 in $\mathbf{R}^3$. The first proof of (4.16) for acoustic scattering is due to SHENK [3].

5. PULSE MODE SONAR SIGNALS IN THE FAR FIELD

The signals $u_0(t,x)$ originate in the region $|x - x_0| \leqslant \delta_0$ and reach points x in the far field, characterized by $|x - x_0| >> 1$, after a time interval of magnitude comparable with $|x - x_0|$. Hence, the far field form of $u_0(t,x)$ coincides with its asymptotic form for large t. The latter is provided by the theory of asymptotic wave functions developed in [1]. The theory is applied here to determine the far field form of $u_0(t,x)$.

The complex wave function $v_0(t,x)$ defined by (2.9), (2.10) has the Fourier representation

$$v_0(t,x) = (2\pi)^{-3/2} \int_{\mathbf{R}^3} \exp\{i(x\cdot p - t|p|)\}\hat{h}_0(p)dp. \tag{5.1}$$

Equations (2.10) and (4.9) imply that

$$\hat{h}_0(p) = (2\pi)^{1/2} i|p|^{-1}\hat{f}(-|p|,p), \tag{5.2}$$

where

$$\hat{f}(\omega,p) = (2\pi)^{-2}\int_{\mathbf{R}^4} \exp\{-i(t\omega + x\cdot p)\}f(t,x)dtdx, \tag{5.3}$$

is the 4-dimensional Fourier transform of f. Note that (5.2) suggests the concept of a *nonradiating source function*. f is said to be nonradiating if

$$f(t,x) = \partial^2 u_0/\partial t^2 - \nabla^2 u_0\,, \quad \mathrm{supp}\,u_0 \text{ bounded in } \mathbf{R}^4. \tag{5.4}$$

In this case $\hat{u}_0(\omega,p)$ exists, and (5.4) is equivalent to

$$\hat{f}(\omega,p) = (|p|^2 - \omega^2)\hat{u}_0(\omega,p). \tag{5.5}$$

Equations (5.2) and (5.5) imply $\hat{h}_0(p) = 0$ and, hence, $u_0(t,x) = 0$ for $t \geqslant t_0 + T$.

The asymptotic wave function associated with $u_0(t,x) = \mathrm{Re}\{v_0(t,x)\}$ is defined by

$$u_0^\infty(t,x) = s(|x|-t,\theta)/|x|, \quad x = |x|\theta, \tag{5.6}$$

where $s \in L_2(\mathbf{R} \times S^2)$ is defined by (see [1], Ch. 2)

$$s(\tau,\theta) = \mathrm{Re}\left\{(2\pi)^{-1/2}\int_0^\infty \exp\{i\tau\omega\}(-i\omega)\hat{h}_0(\omega\theta)d\omega\right\} = \mathrm{Re}\left\{\int_0^\infty \exp\{i\tau\omega\}\hat{f}(-\omega,\omega\theta)d\omega\right\}. \tag{5.7}$$

Direct calculation of $s(\tau,\theta)$ from (1.8) yields the alternative representation

$$s(\tau,\theta) = \frac{1}{4\pi}\int_{\mathbf{R}^3} f(-\tau+\theta\cdot x,x)dx. \tag{5.8}$$

It was shown in [1] that u_0^∞ describes the asymptotic behavior of u_0 in $L_2(\mathbf{R}^3)$for $t \to \infty$:

$$u_0(t,\cdot) = u_0^\infty(t,\cdot) + o_t(1), \quad t \to \infty. \tag{5.9}$$

The integral

$$E(u,K,t) = \frac{1}{2}\int_K \{|\nabla u(t,x)|^2 + (\partial u(t,x)/\partial t)^2\}dx, \tag{5.10}$$

may be interpreted as the acoustic energy in the set $K \subset \mathbf{R}^3$ at time t. It was shown in [1] that, if $h_0 \in L_2^1(\mathbf{R}^3)$, then $u_0^\infty(t,x)$ converges to $u_0(t,x)$ in energy when $t \to \infty$. More precisely,

$$\partial u_0(t,x)/\partial x_j = u_{0,j}^\infty(t,x) + o_t(1),\ t \to \infty,\ j = 0,1,2,3, \tag{5.11}$$

where $x_0 = t$,

$$u_{0,j}^\infty(t,x) = s_j(|x| - t,\theta)/|x|, \tag{5.12}$$

$$s_0(\tau,\theta) = -\partial s(\tau,\theta)/\partial\tau, \tag{5.13}$$

and

$$s_j(\tau,\theta) = -\theta_j s_0(\tau,\theta) \text{ for } j = 1,2,3. \tag{5.14}$$

The result (5.11) was used in [5] to calculate the asymptotic distribution of energy in cones

$$C = \{x = r\theta \,|\, r > 0, \theta \in C_0 \subset S^2\}. \tag{5.15}$$

Applying the results to the signal $u_0(t,x)$ generated by $f(t,x)$ gives

$$E(u_0,C,\infty) = \lim_{t\to\infty} E(u_0,C,t) = \pi\int_C |\hat{f}(-|p|,p)|^2 dp. \tag{5.16}$$

In particular, the total signal energy introduced by the source function f is

$$E(u_0,\mathbf{R}^3,\infty) = \pi\int_{\mathbf{R}^3} |\hat{f}(-|p|,p)|^2 dp. \tag{5.17}$$

6. PULSE MODE SONAR ECHOES IN THE FAR FIELD

Equation (3.8) implies that the echo $u_s(t,x)$ satisfies

$$u_s(t,\cdot) = \mathrm{Re}\{\exp(-itA_0^{1/2})(S-1)h_0\} + o_t(1) + o_{x_0}(1). \tag{6.1}$$

The first term on the right has the same form as the signal but with h_0 replaced by $(S-1)h_0$. It follows from the results of Sec. 5 that

$$u_s(t,x) = u_s^\infty(t,x) + o_t(1) + o_{x_0}(1), \tag{6.2}$$

where

$$u_s^\infty(t,x) = e(|x| - t,\theta)/|x|,\ x = |x|\theta, \tag{6.3}$$

and

$$e(\tau,\theta) = \mathrm{Re}\{(2\pi)^{-1/2}\int_0^\infty \exp(i\tau\omega)(-i\omega)((S-1)h_0)\hat{}(\omega\theta)d\omega\}. \tag{6.4}$$

Now by (4.14)

$$((S-1)h_0)\hat{} = \Phi(S-1)\Phi^*\Phi h_0 = (\hat{S}-1)\hat{h}_0, \tag{6.5}$$

and hence, by (4.16),

$$((S-1)h_0)\hat{}(\omega\theta) = \frac{i\omega}{2(2\pi)^{1/2}}\int_{S^2} T_+(\omega\theta,\omega\theta')\hat{h}_0(\omega\theta')d\theta'. \tag{6.6}$$

Combining (6.4) and (6.6) gives

$$e(\tau,\theta) = \frac{1}{4\pi}\,\mathrm{Re}\left\{\int_0^\infty \exp(i\tau\omega)\omega^2\int_{S^2} T_+(\omega\theta,\omega\theta')\hat{h}_0(\omega\theta')d\theta' d\omega\right\}. \tag{6.7}$$

Finally, by (5.2),

$$e(\tau,\theta) = \frac{1}{2(2\pi)^{1/2}}\,\mathrm{Re}\left\{i\int_0^\infty \exp(i\tau\omega)\omega\int_{S^2} T_+(\omega\theta,\omega\theta')\hat{f}(-\omega,\omega\theta')d\theta' d\omega\right\}. \tag{6.8}$$

Equation (6.2) implies that u_s^∞ gives the far field form of u_s. For $|x_0| >> 1, \Gamma$ is in the far field of the transmitter and the term $o_{x_0}(1)$ is small, uniformly for all t. For receivers in the far field region $|x| >> 1$ for Γ, the echo $u_s(t,x)$ arrives at times $t >> 1$ and hence the term $o_t(1)$ is small.

7. CONCLUDING REMARKS

Actual sonar transmitters do not, of course, generate signals by means of a source function $f(t,x)$. However, the purpose of a well designed transmitter is to generate a signal with a prescribed waveform $s(\tau,\theta)$. Now,

$$\hat{s}(\omega,\theta) = \frac{1}{(2\pi)^{1/2}} \int_{-\infty}^{\infty} \exp(-i\omega\tau) s(\tau,\theta) d\tau = (\pi/2)^{1/2} \hat{f}(-\omega,\omega\theta), \tag{7.1}$$

(see (5.7)) and, hence,

$$e(\tau,\theta) = \mathrm{Re}\left\{ \frac{i}{2\pi} \int_0^{\infty} \exp(i\tau\omega)\omega \int_{S^2} T_+(\omega\theta,\omega\theta')\hat{s}(\omega,\theta')d\theta' d\omega \right\}. \tag{7.2}$$

In particular, the transmitter characteristics influence the echo waveform only through $s(\tau,\theta)$. Hence, (7.2) is applicable to real transmitters with known waveforms $s(\tau,\theta)$.

It is known that $T_+(\omega\theta,\omega\theta')$ is a meromorphic function of ω with poles in the lower half plane [5]. The other functions in the integrand of (7.2) are entire holomorphic functions. Hence, the integral in (7.2) can be transformed by deforming the contour of integration in the complex ω-plane. This leads to an expansion of the echo waveform of the type occurring in the singularity expansion method [6].

REFERENCES

[1] C. H. Wilcox: *Scattering Theory for the d'Alembert Equation in Exterior Domains,* Lecture Notes in Mathematics, Vol. 442 (Springer, New York, 1975)

[2] C. H. Wilcox: "Sonar echo analysis," Math. Meth. Appl. Sci. **1**, 70-88 (1979)

[3] N. A. Shenk: "Eigenfunction expansions and scattering theory for the wave equation in an exterior domain," Arch. Rational Mech. Anal. **21**, 120-150 (1966)

[4] Y. Shizuta: "Eigenfunction expansion associated with the operator -Δ in the exterior domain," Proc. Japan Acad. **39**, 656-660 (1963)

[5] N. A. Shenk, D. Thoe: "Eigenfunction expansions and scattering theory for perturbations of - Δ ", Rocky Mt. J. Math. **1**, 89-125 (1971)

[6] C. E. Baum: "The singularity expansion method," in *Transient Electromagnetic Waves,* ed. by L. B. Felsen, Topics in Applied Physics, Vol. 10 (Springer, New York, 1976)

SOME APPLICATIONS OF FUNCTIONAL ANALYSIS IN CLASSICAL SCATTERING*

R.E. Kleinman

Department of Mathematical Sciences
University of Delaware
Newark, Delaware 19711

ABSTRACT

Problems in classical scattering, the Dirichlet, Neumann, and transmission problems for the Helmholtz equation, and the problem of a perfect reflector for Maxwell's equations are formulated as problems in integral equations. The spectrum of the integral operators is shown to vary in each case. However, a result from functional analysis on the perturbation of spectra is used not only to establish existence and uniqueness, but also to provide an iterative method for actually constructing the solution in each example.

INTRODUCTION

Integral equation methods in classical scattering have proven to be very powerful in establishing existence and uniqueness of solutions. Moreover, integral equation formulations underlie most constructive solution methods, exact and approximate, with the notable though restricted exception of separation of variables.

For closed bounded obstacles which perturb an incident time harmonic acoustic or electromagnetic field, one often reduces the problem of determining the scattered field to that of solving a Fredholm equation of the second kind. Symbolically this equation takes the form

$$(I-K_k)u = f$$

where u is the desired field quantity, f is a known function usually given in terms of the incident field, K_k is a bounded, linear, operator-valued function of the parameter k, and I is the identity operator. More suggestively, the equation may be written as $(\lambda I - K_k)u = f$ and the solution is sought when $\lambda = 1$. The operator K_k can often be shown to map some Banach space B into itself and the existence of $(I-K_k)^{-1}$, that is whether $\lambda = 1$ is a resolvent point of K_k, is to be established. Recall that the spectrum of K_k, $\sigma(K_k)$, is the set of all (possibly complex) numbers λ for which either $(\lambda I - K_k)^{-1}$ does not exist, is unbounded, or the range of $\lambda I - K_k$ is not dense in the Banach space B. If K_k is compact then $\sigma(K_k) = \{\lambda \mid K_k u = \lambda u \text{ for some } u \in B\}$. The resolvent set is the complement of the spectrum, and the spectral radius, $r_o(K_k)$, is the radius of the smallest circle in the complex λ-plane containing the spectrum i.e.

$$r_\sigma(K_k) = \sup_{\lambda \in \sigma(K_k)} \{|\lambda|\}.$$

Two results from functional analysis allow for an explicit construction of $(I-K_k)^{-1}$ in the form of a Neumann series, (see e.g. [1], §5)

Theorem 1. *If* $r_\sigma\ (K_k) < 1$ *then* $(I-K_k)^{-1} = \sum_{n=0}^{\infty} K_k^n$.

Theorem 2. *If* $r_\sigma\ (K_0) < 1$ *and* K_k *is continuous in* k *at* $k = 0$, *i.e.* $\lim_{k\to 0} \|K_k - K_0\| = 0$, *then* $(I-K_k)^{-1} = \sum_{n=0} K_k^n$ *for* $|k|$ *sufficiently small. Here* $\|\cdot\|$ *is the norm of the Banach space B.*

We illustrate how these results may be used to construct inverse operators for small k and thus solve the equations for some well known problems in classical scattering, viz., the Dirichlet, Neumann, and transmission problems in acoustic scattering, and the problem of a perfect conductor in electromagnetic scattering. In each of these cases, the reduction to an integral equation is well known. However, the standard integral operators have different spectral properties. In the electromagnetic case the spectral radius is less than 1 so that Theorem 2 may be applied directly; in the Neumann problem the spectral radius is 1 but $\lambda = 1$ is in the resolvent set; whereas in the Dirichlet case the spectral radius is again 1 but this time $\lambda = 1$ is in the spectrum. The transmission problem is shown to be a generalization of the Neumann problem. In each of these cases it is shown how Theorems 1 and 2 may be used to construct the solution.

All of the problems will involve a scattering surface S which is closed and bounded and divides Euclidian space $\mathbb{R}^n$ into an interior S_{in} and an exterior S_{ex} (see Fig. 1). In the Dirichlet problem the dimension n is 2, in the Neumann and electromagnetic cases n is 3, while the transmission problem is treated for arbitrary n. The surface S will be assumed to be Lyapunoff (see, e.g. [2]), in which case it has a Hölder-continuous normal everywhere, or piecewise Lyapunoff, consisting of a finite number of segments of Lyapunoff surfaces. On each segment the normal exists and is Hölder-continuous but the normal may not be continuous from one segment to another; hence, the entire surface may only have a normal almost everywhere. An additional requirement is needed in the piecewise Lyapunoff case, namely a two-sided cone condition. That is, there must be positive constants α and h such that each point of S is the vertex of two cones of half-vertex angle α and height h, one lying in S_{in} and one in S_{ex}. Points will be denoted by x, y, etc. and these represent vectors with appropriate dimension, e.g., $x = (x_1, x_2, x_3)$. The unit normal will always point from S into S_{ex} and will be denoted by $\hat{n}_y$ where the subscript indicates the point on S at the base of the normal.

1. ELECTROMAGNETIC SCATTERING BY A PERFECT CONDUCTOR IN $\mathbb{R}^3$

As our first example, we treat what is perhaps physically the most complicated of all the cases we shall consider, but curiously is the simplest from a functional analysis viewpoint. The problem of perfect reflection is one of finding electric and magnetic fields $E(x)$ and $H(x)$ which are vector valued functions in $\mathbb{R}^3$ of points in S_{ex}. Specifically we seek

$$E = E^S + E^i, \; H = H^S + H^i, \tag{1.1}$$

where E^i, H^i is a known incident field and the scattered field satisfies

$$\nabla \times H^S = -i\omega\epsilon E^S \;, \quad \nabla \times E^S = i\omega\mu H^S, \quad x \in S_{ex}, \tag{1.2}$$

$$\hat{r} \times \nabla \times E^S + ikE^S = \hat{r} \times \nabla \times H^S + ikH^S = o\left(\frac{1}{r}\right), \text{ uniformly in } \hat{r}, \tag{1.3}$$

and

$$\hat{n} \times E^S = -\hat{n} \times E^i, \quad \hat{n} \cdot H^S = -\hat{n} \cdot H^i, \quad x \in S, \tag{1.4}$$

where $\hat{r} = \dfrac{x}{|x|}$, $k = \omega\sqrt{\epsilon\mu}$, and ω, ϵ, μ are the frequency, permittivity, and permeability respectively.

By standard analysis [3,4], this problem may be reduced to solving a boundary integral equation for the unknown components of H (i.e., $\hat{n} \times H$) as follows:

$$\hat{n}_x \times H(x) - 2\hat{n}_x \times \int_S \nabla_y G_3(x,y,k) \times [\hat{n}_y \times H(y)] dS_y = 2\hat{n}_x \times H^i(x), \quad x \in S, \tag{1.5}$$

where $G_3(x,y,k) = -\dfrac{e^{ik|x-y|}}{4\pi|x-y|}$. Actually, (1.5) holds only for S Lyapunoff (not piecewise Lyapunoff) and although corresponding integral equations may be derived for nonsmooth boundaries the subsequent analysis does not easily generalize. Therefore, in this example we require S to be Lyapunoff. Equation (1.5) may be written in operator form as

$$J - K_k J = 2\hat{n}_x \times H^i, \tag{1.6}$$

where $J(x) = \hat{n}_x \times H(x)$ and K_k is defined through (1.5). The spectral properties of K_o have been studied by MÜLLER and NIEMEYER [5], KRESS [6,7], and WERNER [8], with more stringent smoothness requirements on S. GRAY [9] has extended these properties to the case when S is Lyapunoff. Taking the Banach space to be $C_T(S)$, continuous vector valued functions on S with vanishing normal component with the usual supremum norm, the following facts have been established.

K_k is a compact mapping of $C_T(S)$ into itself and $\sigma(K_o) \subset (-1,1)$, and thus $r_\sigma(K_o) < 1$. Moreover, $||K_k - K_o|| = O(k^2)$. Thus, all of the requirements of Theorem 2 are satisfied and the solution of (1.6) is

$$J = \sum_{n=0}^{\infty} K_k^n (2\hat{n} \times H^i) \tag{1.7}$$

for $|k|$ sufficiently small.

2. EXTERIOR NEUMANN PROBLEM IN $\mathbb{R}^3$

Now we consider an example where $r_\sigma(K_o) = 1$ but $1 \notin \sigma(K_o)$. This arises in the well-known Neumann problem in acoustic scattering. Here we wish to determine the velocity potential

$$u(x) = u^i + u^S, \tag{2.1}$$

where u^i is a known incident field and the scattered field satisfies

$$(\nabla^2 + k^2)u^S = 0, \quad x \in S_{ex}, \tag{2.2}$$

$$u^S = e^{ikr}\left[\frac{f(\hat{r})}{r} + o\left(\frac{1}{r}\right)\right], \tag{2.3}$$

$$\text{and } \frac{\partial u^S}{\partial n} = -\frac{\partial u^i}{\partial n}, \quad x \in S, \tag{2.4}$$

where $r = |x|$ and $\hat{r} = \frac{x}{r}$.

When S is Lyapunoff, this problem may be reduced to the following boundary integral equation [10]:

$$u(x) + 2\int_S \frac{\partial}{\partial n_y} G_3(x,y,k) u(y) dS_y = 2u^i(x), \quad x \in S, \tag{2.5}$$

where $G_3(x,y,k)$ is the free space Green's function defined in (1.5).

In operator form, with (2.5) providing the definition of K_k,

$$K_k u = 2u^i. \tag{2.6}$$

The fact that S is Lyapunoff enables one to show that K_k is a compact mapping of $C(S)$ into $C(S)$ with the usual sup norm (see e.g. [2]). We could write (2.6) in the form

$$(\lambda I - K_k)u = 2u^i, \tag{2.7}$$

where the solution is sought for $\lambda = -1$. However even for $k = 0$ this equation cannot be solved directly by iteration because $\lambda = 1$ is an eigenvalue. This is part of Plemelj's theorem (see e.g. [11], Sec. 9) which establishes that the eigenvalues of K_o are real, $\lambda = 1$ is a simple eigenvalue and $|\lambda| < 1$ for all other eigenvalues. Thus $\lambda = -1$ is not in the spectrum of K_o. However, the spectral radius of K_o is equal to 1; hence, direct iteration of (2.6) is not possible.

To overcome this difficulty in potential theory ($k = 0$), Neumann proposed a method subsequently shown (see e.g. [12,13]) to shift the unwanted eigenvalue into the interior of the unit circle. The method essentially consists of defining a new operator

$$L_o = \frac{I - K_o}{2}, \tag{2.8}$$

which transforms the eigenvalue equation for K_o,

$$K_o u = \lambda u, \tag{2.9}$$

into

$$L_o u = \frac{1-\lambda}{2} u = \mu u. \tag{2.10}$$

This has the effect of transforming $\lambda = 1$ into $\mu = 0$ and $\lambda = -1$ into $\mu = 1$ as shown in Fig. 2. Thus, under this transformation $r_\sigma(L_o) < 1$. KRESS and ROACH [14] have shown that (2.8) is not the optimal transformation in the sense of achieving the minimum spectral radius of the transformed operator. They propose a different transformation which was shown by KLEINMAN [15] to be optimal when all the eigenvalues of K_o have the same sign.

These results have been applied to the scattering problem ($k \neq 0$) by AHNER and KLEINMAN [16] and KLEINMAN and WENDLAND [17], who generalized Neumann's method as embodied in (2.8) in the obvious manner, defining a new operator to be

$$L_k = \frac{I - K_k}{2}, \tag{2.11}$$

in which case the integral equation to be solved, (2.6), becomes

$$(I - L_k)u = u^i. \tag{2.12}$$

They also showed that

$$||L_k - L_o|| = O(k^2), \tag{2.13}$$

which, together with the fact that $r_\sigma(L_o) < 1$, allows us to apply Theorem 2 to (2.12), which then has the solution

$$u = \sum_{n=0}^{\infty} L^n u^i \tag{2.14}$$

for $|k|$ sufficiently small.

If S is not smooth the argument is slightly more complicated because the operators are no longer compact. Here we must employ the idea of the Fredholm radius (see e.g. [18], Sec. 85) which identifies values of λ for which the essentials of compactness, namely discrete spectra and applicability of Fredholm's alternative, still apply. Here we use Radon's theorem generalized to three dimensions [19], which ensures that for piecewise Lyapunoff surfaces with a two-sided cone condition with height h and half-vertex angle α, the Fredholm radius of K_o is $\omega = \cos\alpha < 1$. There is still an eigenvalue of K_o at $\lambda = 1$, but the spectral picture, as shown in Fig. 3, is a little more complicated. There are two cases, depending on whether or not the smallest eigenvalue, λ_1, is larger or smaller than the negative of the Fredholm radius, but the essential feature is that the spectrum and the non-Fredholm points of L_o are all contained in a circle of radius less than 1 and $\mu = 1$ is the value for which the solution is sought. Despite the presence of corners, the integral equation to be solved is still given by (2.12), and the estimate (2.13) is still valid [17]. Thus, the conditions for applying Theorem 2 remain satisfied and the solution is still given by the Neumann series

$$u = \sum_{n=0}^{\infty} L_k^n u^i$$

for $|k|$ sufficiently small.

3. TRANSMISSION PROBLEMS IN $\mathbb{R}^n$

The preceding example may be subsumed under a more general setting scattering of an acoustic wave from a penetrable object in which both the density and propagation velocity differ from the exterior medium. Mathematically, the problem may be formulated in $\mathbb{R}^n$ for any $n \geq 2$. As before, we seek a function

$$u(x) = u^i(x) + u^S(x), \quad x \in \mathbb{R}^n, \tag{3.1}$$

such that

$$(\nabla^2 + k_e^2)u^S = 0, \quad x \in S_{ex}, \tag{3.2}$$

$$(\nabla^2 + k_i^2(x))u = 0, \quad x \in S_{in}, \tag{3.3}$$

$$u^S = e^{ik_e r}\left[\frac{f(\hat{r})}{r^{(n-1)/2}} + o\left(\frac{1}{r^{(n-1)/2}}\right)\right], \tag{3.4}$$

$$u^+(x) = u^-(x), \quad x \in S, \tag{3.5}$$

$$\frac{1}{\rho_e}\frac{\partial u^+}{\partial n} = \frac{1}{\rho_i}\frac{\partial u^-}{\partial n}, \quad x \in S, \tag{3.6}$$

where $+$ denotes the limit from S_{ex}, $-$ from S_{in}, k_e is constant with Im $k_e \geq 0$, $k_i(x)$ is a bounded function of x in $S_{in} \cup S$, and S is a piecewise Lyapunoff surface with a two-sided cone condition.

It has been shown [20] that the solution to this problem admits of the representation

$$u(x) - \left(1-\frac{\rho_e}{\rho_i}\right)\int_S [u(x)-u(y)]\,\frac{\partial G_n}{\partial n_y}(x,y,0)\,dS_y$$
$$-\left(1-\frac{\rho_e}{\rho_i}\right)\int_S u(y)\frac{\partial}{\partial n_y}[G_n(x,y,0)-G_n(x,y,k_e)]\,dS_y$$
$$-\frac{\rho_e}{\rho_i}\int_{S_{in}} G_n(x,y,k_e)[k_e^2-k_i^2(y)]u(y)\,d\tau_y = u^i(x), \quad x\in\mathbb{R}^n, \tag{3.7}$$

where $G_n(x,y,k) = -\frac{i}{4}\left(\frac{k}{2\pi|x-y|}\right)^{\frac{n-2}{2}} H^{(1)}_{\frac{n-2}{2}}(k|x-y|)$. $H_\nu^{(1)}$ is the Hankel function of the first kind, and when $n = 2$ or 3 we have $G_2(x,y,k) = -\frac{i}{4}H_o^{(1)}(k|x-y|)$ and $G_3(x,y,k) = -\frac{e^{ik|x-y|}}{4\pi|x-y|}$, respectively. In operator form (3.7) may be written

$$u - \left(1-\frac{\rho_e}{\rho_i}\right)L_{k_e}u - \frac{\rho_e}{\rho_i}Mu = u^i. \tag{3.8}$$

While (3.8) holds for all $x\in\mathbb{R}^n$, observe that if the interior density ρ_i becomes infinite and x is restricted to S, (3.8) reduces to the n-dimensional form of the Neumann problem (compare with (2.12)). If $\rho_e = \rho_i$ and x is restricted to S_{in}, then (3.8) is the n-dimensional form of the equation which is the basis for the Born approximation. Even if S has corners, both L_{k_e} and M map $C(S\cup S_{in})$ into itself, where $C(S\cup S_{in})$ is the space of continuous functions on the closure of S_{in}, equipped with the usual sup norm. Restricting x to lie in $S_{in}\cup S$ in (3.8), yields an integral equation of the second kind.

The extension of the Plemelj and Radon results to higher dimensions is straightforward if S is smooth [11], but quite complicated if S has corners [17,19,21,22]. Assuming the validity of this extension, then the spectral radius of L_o, restricted to functions defined on S (the Neumann case), is less than one as before. Moreover, the spectral radius of L_o cannot increase when L_o acts on functions defined on $S\cup S_{in}$ (it may decrease but not increase). Thus, for the operator defined in (3.7) and (3.8),

$$r_\sigma(L_o) < 1. \tag{3.9}$$

Hence,

$$r_\sigma\left(\left(1-\frac{\rho_e}{\rho_i}\right)L_o\right) < 1, \tag{3.10}$$

provided

$$\left|1-\frac{\rho_e}{\rho_i}\right| < \frac{1}{r_\sigma(L_o)}. \tag{3.11}$$

With this restriction on $\frac{\rho_e}{\rho_i}$ it is easy to establish that

$$\left\|\left(1-\frac{\rho_e}{\rho_i}\right)L_{k_e} + \frac{\rho_e}{\rho_i}M - \left(1-\frac{\rho_e}{\rho_i}\right)L_o\right\| < \epsilon \tag{3.12}$$

for any ϵ, provided $\left|\left(1-\frac{\rho_e}{\rho_i}\right)k_e\right|$ and $\left|\frac{\rho_e}{\rho_i}\right|\,||k_e^2-k_i^2||$ are sufficiently small. Again, the conditions of Theorem 2 may, therefore, be satisfied and (3.8) has the solution

$$u = \sum_{n=0}^{\infty}\left[\left(1-\frac{\rho_e}{\rho_i}\right)L_{k_e} + \frac{\rho_e}{\rho_i}M\right]^n u^i, \tag{3.13}$$

provided $\left|\left(1-\frac{\rho_e}{\rho_i}\right)k_e\right|$, $\left|1-\frac{\rho_e}{\rho_i}\right|$, and $\left|\frac{\rho_e}{\rho_i}\right|\,||k_e^2-k_i^2||$ are sufficiently small.

4. EXTERIOR DIRICHLET PROBLEM IN $\mathbb{R}^2$

As a final example, we consider a problem which involves solving an integral equation of the second kind at an eigenvalue. The nonuniqueness which is implied by Fredholm's alternative in this case is removed by utilizing additional properties of the solution. The boundary value problem in this case, the Dirichlet problem for the Helmholtz equation, is again that of determining a velocity potential

$$u = u^i + u^s, \tag{4.1}$$

where u^i is a given incident field and the scattered field satisfies

$$(\nabla^2 + k^2)u^S = 0, \quad x \in S_{ex}, \tag{4.2}$$

$$u^S = e^{ikr}\left[\frac{f(\hat{r})}{\sqrt{r}} + o\left(\frac{1}{\sqrt{r}}\right)\right], \tag{4.3}$$

$$u^S = -u^i, \quad x \in S. \tag{4.4}$$

We will consider only the case when S is smooth (Lyapunoff). A standard application of Green's theorem yields the representation

$$u(x) - \int_S G_2(x,y,k)\frac{\partial u(y)}{\partial n_y}dS_y = u^i(x), \quad x \in S_{ex}, \tag{4.5}$$

and, with the jump condition for the normal derivative of a single layer distribution, we obtain

$$\frac{\partial u}{\partial n_x} - 2\int_S \frac{\partial G_2(x,y,k)}{\partial n_x}\frac{\partial u(y)}{\partial n_y}dS_y = 2\frac{\partial u^i}{\partial n_x}, \quad x \in S. \tag{4.6}$$

Letting $v(x) = \dfrac{\partial u(x)}{\partial n_x}$, we may rewrite (4.6) as

$$(I - K_k^*)v = 2\frac{\partial u^i(x)}{\partial n_x}, \tag{4.7}$$

or $(\lambda I - K_k^*)v = 2\dfrac{\partial u^i}{\partial n_x}$ for $\lambda = 1$.

The operator K_k^* is the adjoint of the two-dimensional form of K_k introduced in (2.6). When $k = 0$, K_o (in two dimensions) is the operator which Plemelj originally proved has $\lambda = 1$ as an eigenvalue. Hence, $\lambda = 1$ is also an eigenvalue of K_o^*. But $\lambda = 1$ is precisely the value for which a solution of (4.7) is sought. This dilemma was partially resolved by AHNER [23], who showed how to solve (4.7) by projecting onto the orthogonal complement of the eigenspace correspond to $\lambda = 1$. This, however, requires the explicit determination of the appropriate eigenfunction, which is not always trivial, depending on the geometry of S. A treatment which accomplishes the projection, but avoids the necessity for finding the eigenfunction of K_o^* has recently been developed by COLTON and KLEINMAN [24]. Using the properties of u specified in (4.1)-(4.4), (4.6) is modified (apparently complicated) to be

$$\frac{\partial u(x)}{\partial n_x} - 2\int_S \frac{\partial}{\partial n_x}\left[G_2(x,y,k) - \pi\frac{G_2(0,y,k)G_2(x,0,k)}{\log k}\right]\frac{\partial u(y)}{\partial n_y}dS_y$$
$$= 2\frac{\partial u^i(x)}{\partial n_x} - 2\pi\frac{u^i(0)}{\log k}\frac{\partial}{\partial n_x}G_2(x,0,k). \tag{4.8}$$

Denoting the right hand side by $v^i(x)$, again letting $\dfrac{\partial u(x)}{\partial n_x}$ be $v(x)$, and using (4.8) to implicitly define an operator L_k, (4.8) may be written in operator form as

$$(I - L_k)v = v^i. \tag{4.9}$$

L_k maps the space of continuous functions into itself (provided S is Lyapunoff). Hence, again we use the Banach space $C(S)$ equipped with the sup norm. The operator L_o, where

$$L_o v = (K_o^* v)(x) - \frac{1}{2\pi}\log|x|\int_S v(y)\,dS_y, \tag{4.10}$$

no longer has $\lambda = 1$ as an eigenvalue. This is shown by COLTON and KLEINMAN [24], following the ingenious work of KRESS [25]. They show that

$$r_\sigma(L_o) < 1 \tag{4.11}$$

and also that

$$||L_k - L_o|| = O\left(\frac{1}{\log k}\right). \tag{4.12}$$

Thus, once again the requirements of Theorem 2 are fulfilled and (4.9) has the unique solution

$$v = \sum_{n=0}^{\infty} L^k v^i \tag{4.13}$$

for $|k|$ sufficiently small.

FOOTNOTES AND REFERENCES

* Research supported under AFOSR Grant 79-0085

[1] A. Taylor: *Introduction to Functional Analysis* (Wiley, New York, 1958), Chap. 5
[2] S.G. Mikhlin: *Mathematical Physics: An Advanced Course* (American-Elsevier, New York, 1970)
[3] A.W. Maue: "Zur Formulierung eines allgemeinen Beugungsproblems durch eine Integralgleichung," Z. Phys. **126**, 601-618 (1949)
[4] A.M. Poggio, E.K. Miller: "Integral equation solutions of three-dimensional scattering problems," in *Computer Techniques for Electromagnetics*, ed. by R. Mittra (Pergamon, Oxford, 1973)
[5] C. Müller, H. Niemeyer: "Greensche Tensoren und asymptotische Gesetze der elektromagnetischen Hohlraumschwingungen," Arch. Rational Mech. Anal. **1**, 305-358 (1961)
[6] R. Kress: "Über die Integralgleichung des Pragerschen Problems," Arch. Rational Mech. Anal. **30**, 381-400 (1968)
[7] R. Kress: "Grundzüge einer Theorie der verallgemeinerten harmonischen Vektorfelder," Meth. Verf. Math. Phys. **2**, 49-83 (1969)
[8] P. Werner: "On an integral equation in electromagnetic diffraction theory," J. Math. Anal. Appl. **14**, 445-462 (1966)
[9] G. Gray: "Low frequency iterative solution of integral equations in electromagnetic scattering theory," University of Delaware Applied Mathematics Institute Technical Report 35A (1978)
[10] R.E. Kleinman, G.F. Roach: "Boundary integral equations for the three dimensional Helmholtz equation," SIAM Review **16**, 214-236 (1974)
[11] K. Jörgens: *Lineare Integraloperatoren* (Teubner, Stuttgart, 1970)
[12] E. Goursat: *A Course in Mathematical Analysis* (Dover, New York, 1964), Vol. III, Part 2
[13] L.V. Kantorovich, V.I. Krylov: *Approximate Methods of Higher Analysis* (Noordhoff, Groningen, 1964)
[14] R. Kress, G.F. Roach: "On the convergence of successive approximations for an integral equation in a Green's function approach to the Dirichlet problem" J. Math. Anal. Appl. **55**, 102-111 (1976)
[15] R.E. Kleinman: "Iterative solutions of boundary value problems," in *Function Theoretic Methods for Partial Differential Equations*, Lecture Notes in Mathematics, Vol. 561 (Springer, New York, 1976)
[16] J.F. Ahner, R.E. Kleinman: "The exterior Neumann problem for the Helmholtz equation," Arch. Rational Mech. Anal. **52**, 26-43 (1973)
[17] R.E. Kleinman, W.L. Wendland: "On Neumann's method for the exterior Neumann problem for the Helmholtz equation," J. Math. Anal. Appl., **57**, 170-202 (1977)
[18] F. Riesz, B. Sz.-Nagy: *Functional Analysis* (Ungar, New York, 1955)
[19] W.L. Wendland: "Die Behandlung von Randwertaufgaben im R_3 mit Hilfe von Einfach- und Doppelschicht-Potentialen," Numer. Math. **11**, 380-404, (1968)
[20] R. Kittappa, R.E. Kleinman: "Transmission problems for the Helmholtz equation," to be published (1979)
[21] J. Král: "The Fredholm method in potential theory," Trans. Amer. Math. Soc. **125**, 511-547 (1966)
[22] Yu.D. Burago, V.G. Maz'ja, V.D. Saposhnikova: "On the theory of simple and double-layer potentials for domains with irregular boundaries," in *Problems in Mathematical Analysis,* Vol. 1, (Boundary Value Problems and Integral Equations), ed. by V.I. Smirnov (Consultants Bureau, New York, 1968), pp. 1-30
[23] J.F. Ahner: "The exterior Dirichlet problem for the Helmholtz equation," J. Math. Anal. Appl. **52**, 415-429 (1975)
[24] D. Colton, R. Kleinman: "The direct and inverse scattering problems for an arbitrary cylinder: Dirichlet boundary conditions," University of Delaware Applied Mathematics Institute Technical Report 55A (1979)
[25] R. Kress: *Integral Equations,* Lecture Notes (University of Strathclyde, Glasgow, 1977)

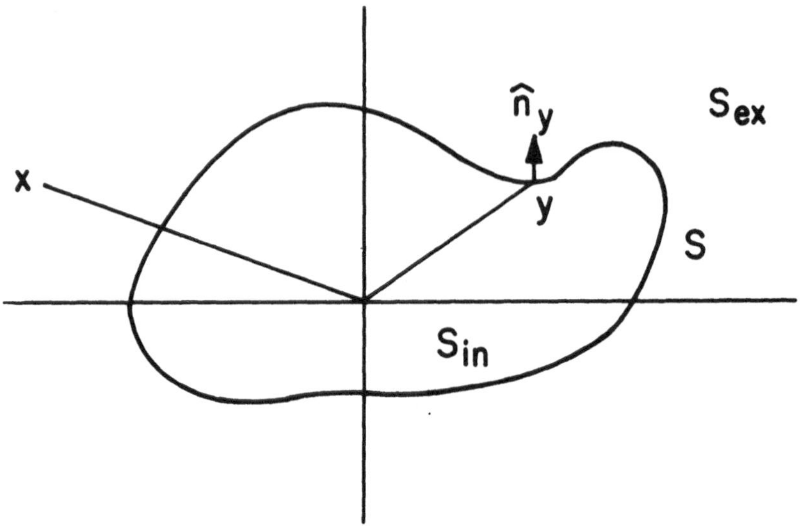

Fig. 1. Scattering geometry.

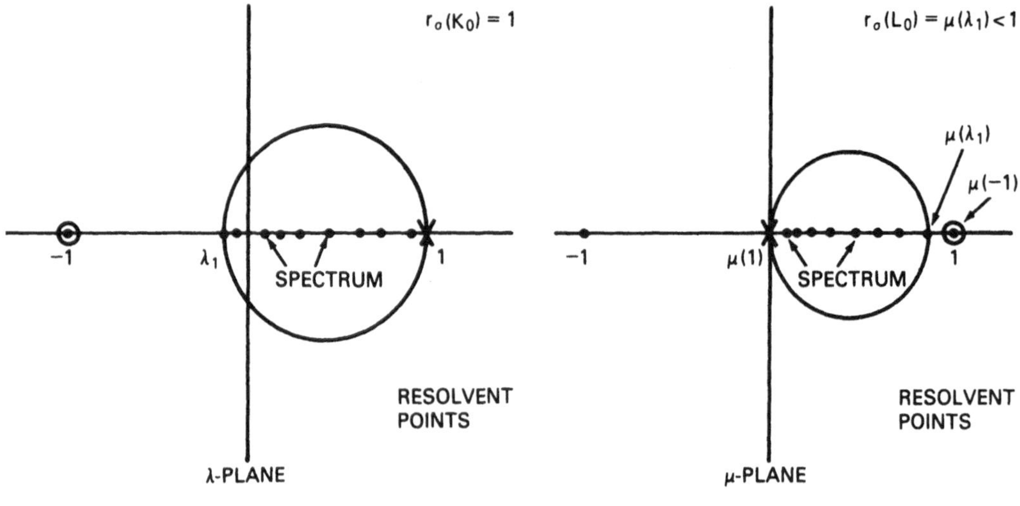

Fig. 2. Transformation of the spectrum.

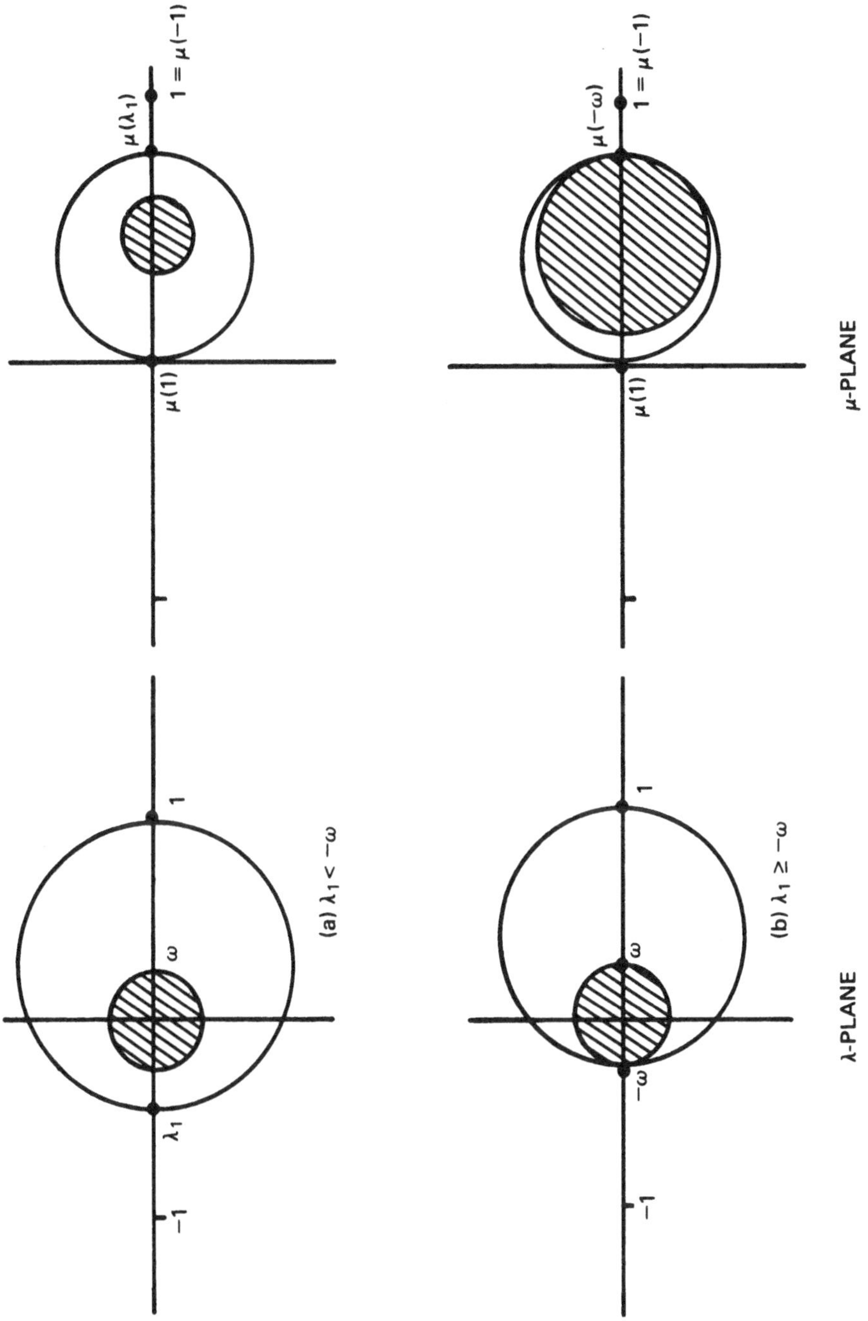

Fig. 3. Spectrum.

NUMERICAL METHODS FOR HELMHOLTZ-TYPE EQUATIONS IN UNBOUNDED REGIONS

Charles I. Goldstein*

Applied Mathematics Department
Brookhaven National Laboratory
Upton, New York 11973

INTRODUCTION

It is the purpose of this paper to describe certain numerical methods for approximately solving Helmholtz type equations in unbounded regions in two and three-dimensional space, subject to appropriate radiation conditions at infinity.

Mathematical problems of this kind arise from the propagation of linear periodic waves through an unbounded medium. For example, if we assume harmonic time dependence, the scattering of acoustic waves by a bounded obstacle may be mathematically formulated as an exterior Dirichlet or Neumann problem for the Helmholtz equation, $(-\Delta-k^2)u = 0$ (see Sec. 1.1 for a precise formulation of the problem). In addition, various problems relating to underwater acoustics and the scattering of electromagnetic waves in a waveguide may be formulated in terms of the Helmholtz equation in regions with infinite boundaries, as described in Sec. 1.5.

There are various difficulties in the numerical treatment of problems of this type. These include the unboundedness of the region under consideration, the oscillatory nature of the solutions and the fact that the resulting matrix equations are indefinite when the frequency is large. The last two questions will be briefly discussed in Sec. 3. This paper will be mainly concerned with techniques for replacing the given problem by an approximate or equivalent problem on a bounded region and then discretizing the resulting problem.

We shall refer to the numerical techniques discussed in Sec. 1 as direct discretization methods. These methods consist of first introducing an artificial boundary, Γ_∞, at a sufficiently large distance from the origin. Appropriate radiation boundary conditions are then imposed on Γ_∞ in order to approximately incorporate the asymptotic behavior of the solution at infinity in the numerical method. Two recently developed techniques for constructing highly absorbing radiation boundary conditions will be discussed in Sec. 1.2. Finally, the resulting boundary value problem is discretized using either finite difference or finite element methods. Finite element methods take account of the boundary condition on Γ_∞ in a natural way and are described in Sec. 1.

In Sec. 2, we shall describe an alternative approach based on reformulating the exterior problem as an integral equation on a bounded surface. A few different integral equation formulations will be described in Sec. 2.1. In Sec. 2.2, we discuss numerical techniques for solving the resulting integral equations. In Sec. 2.3, we consider techniques for coupling direct discretization and integral equation methods.

Finally, we remark that analytic and asymptotic methods have been extensively employed for solving problems in linear wave propagation. For a survey of these methods as well as the kinds of physical problems for which they are applicable, see [1] and the references cited there.

1. DIRECT DISCRETIZATION METHODS

1.1 The Exterior Problem

We begin by describing the exterior Dirichlet problem for the Helmholtz equation in $\mathbb{R}^3$, three dimensional Euclidean space. Let B denote a bounded region in $\mathbb{R}^3$ with smooth boundary, Γ, and let Ω denote the complement of $B \cup \Gamma$ in $\mathbb{R}^3$. Suppose that the complex-valued function $u(x)$ satisfies the following problem:

$$(-\Delta - k^2)u(x) = 0 \text{ in } \Omega, \tag{1.1}$$

$$\lim_{r\to\infty} r\left|\frac{\partial u(x)}{\partial r} - iku(x)\right| = 0, \tag{1.2}$$

and

$$u(x) = g(x) \text{ on } \Gamma, \tag{1.3}$$

where $x = (r,\theta,\phi)$ is an arbitrary point in $\overline{\Omega} = \Omega \cup \Gamma$, (r,θ,ϕ) denotes spherical polar coordinates, Δ denotes the Laplacian, $g(x)$ is a smooth function defined on Γ, and $k \neq 0$ is a real number. The exterior Neumann problem is defined by (1.1), (1.2), and

$$\frac{\partial u(x)}{\partial n} = g(x) \text{ on } \Gamma, \tag{1.4}$$

where n denotes the outward directed normal to Γ.

Condition (1.2) is known as a radiation condition and states that u represents an outgoing wave. It is well-known that when Γ and g are sufficiently smooth, there exists a unique solution of the exterior Dirichlet and Neumann problems. Futhermore, the free space Green's function for the Helmholtz equation is given by

$$F_o(x,y;k) = \frac{e^{ik|x-y|}}{4\pi|x-y|}, \quad x,y \in \mathbb{R}^3, \quad x \neq y. \tag{1.5}$$

Finally, we observe that there exists an $R_o > 0$ such that the solution, $u(x)$, of the exterior Dirichlet or Neumann problem has a uniformly convergent series expansion given by

$$u(x) = e^{ikr} \sum_{n=1}^{\infty} a_n(\theta,\phi,k) r^{-n} \text{ for } r \geqslant R_o. \tag{1.6}$$

A proof of this expansion was given by Atkinson. For a reference to this work as well as various other useful results, see [2].

Results analogous to those given above are equally valid for exterior Helmholtz problems in R^2. In particular, the free space Greens function for $\mathbb{R}^2$ is given by

$$F_o(x,y;k) = \frac{i}{4} H_o^{(1)}(k|x-y|), \quad x,y \in \mathbb{R}^2, \quad x \neq y, \tag{1.7}$$

where $H_o^{(1)}$ denotes the Hankel function of the first kind of order zero. The radiation condition in $\mathbb{R}^2$ analogous to (1.2) is given by

$$\lim_{r\to\infty} r^{1/2}\left|\frac{\partial u}{\partial r} - iku\right| = 0. \tag{1.8}$$

We also note that the above results as well as the numerical methods to be described in this section apply to other boundary conditions, including Robin boundary conditions and mixed boundary conditions (e.g., the Dirichlet condition may be imposed on one portion of the boundary and the Neumann condition may be imposed on the remainder of the boundary). Finally, we remark that the numerical methods to be described in this section are applicable even if the differential operator on the left side of (1.1) is perturbed so as to include variable coefficients with bounded support. However, for the sake of simplicity we shall confine our discussion to the exterior Dirichlet or Neumann problem for the Helmholtz equation in $\mathbb{R}^2$ or $\mathbb{R}^3$.

1.2 Radiation Boundary Conditions

In order to apply a direct discretization [3] method, we shall introduce an artificial boundary Γ_∞ enclosing the obstacle as well as an appropriate boundary condition of the form $\frac{\partial u}{\partial n} = T(u)$ on Γ_∞. In order to approximate the outgoing solution satisfying (1.2) or (1.8), it is desirable that our boundary condition be highly absorbing. For computational reasons, it is also desirable that the boundary condition be local. (Nonlocal means that in order to compute $\frac{\partial u}{\partial n}$ at a point of Γ_∞ it is necessary to compute u at each point of Γ_∞). We shall briefly describe two methods for constructing a hierachy of local, highly absorbing boundary conditions of increasingly better accuracy.

The first method was developed and applied by ENGQUIST and MAJDA [4,5] to deal with hyperbolic mixed initial boundary value problems. Consider, e.g., the wave equation in two dimensions using polar coordinates. A perfectly absorbing boundary condition that is nonlocal in both space and time was constructed for this equation in [4] using the theory of pseudodifferential operators and the reflection of singularities for solutions of differential equations. This nonlocal condition was then employed to obtain a hierarchy of local, highly absorbing boundary conditions for which the resulting initial boundary value problem is well-posed. We shall give the first two boundary conditions obtained in this manner, with Γ_∞ assumed to be a circle of radius R,

$$\left[\frac{\partial}{\partial r} + \frac{\partial}{\partial t} + \frac{1}{2R}\right] u|_{r=R} = 0 \tag{1.9}$$

and

$$\left[\frac{\partial^3}{\partial r \partial t^2} + \frac{\partial^3}{\partial t^3} - \frac{1}{2R^2}\frac{\partial^3}{\partial t \partial \theta^2} + \frac{1}{2R}\frac{\partial^2}{\partial t^2} + \frac{1}{2R^3}\frac{\partial^2}{\partial \theta^2}\right] u|_{r=R} = 0. \tag{1.10}$$

To apply these boundary conditions to the Helmholtz equation, it suffices to replace $\frac{\partial}{\partial t}$ by $-ik$. Observe that the boundary operator in (1.9) annihilates spherical waves (i.e., waves of the form $r^{-1/2}f(t-r)a(\theta)$). Higher order boundary operators such as that in (1.10) also compensate for angular dependence when the wave is not quite spherical.

A second method for constructing a sequence of highly absorbing boundary conditions was developed by BAYLISS and TURKEL [6], in connection with hyperbolic problems such as the wave equation. Their method is based on the following asymptotic expansion for solutions of the wave equation in $\mathbb{R}^3$:

$$u(t,r,\theta,\phi) \sim \sum_{j=1}^{\infty} \frac{f_j(t-r,\theta,\phi)}{r^j} \quad \text{for } r \geq R, \tag{1.11}$$

where R is sufficiently large. Expansion (1.11) generalizes (1.6) and was proved in [6] under suitable assumptions. Their boundary conditions are then obtained by treating (1.11) as a formal expansion, valid for $r \geq R$. The first two boundary conditions obtained in this manner are

$$\left[L + \frac{1}{r}\right] u|_{r=R} = 0, \tag{1.12}$$

and

$$\left[L + \frac{3}{r}\right]\left[L + \frac{1}{r}\right] u|_{r=R} = 0, \tag{1.13}$$

where $L = \frac{\partial}{\partial r} + \frac{\partial}{\partial t}$. The sequence of boundary operators constructed in [6] annihilate successively more terms on the right side of (1.11). Thus, the operator in (1.13) annihilates $f_1(t-r,\theta,\phi)r^{-1}$ and $f_2(t-r,\theta,\phi)r^{-2}$. Finally we note that in $\mathbb{R}^2$, the boundary condition obtained in this way corresponding to (1.12) is identical to (1.9). However, the higher order condition corresponding to (1.13) differs from (1.10).

1.3. Variational Formulation

We now introduce an artificial boundary, Γ_∞, enclosing the obstacle and employ one of the radiation boundary conditions discussed above. Specifically, we consider the exterior Dirichlet problem, (1.1)-(1.3), in $\mathbb{R}^3$. We choose Γ_∞ to be a sphere of radius R and (1.12) as our boundary condition on Γ_∞ with $\frac{\partial}{\partial t}$ replaced by $-ik$. Hence we wish to solve the following boundary value problem on Ω_R, the annular region bounded by Γ

$$(-\Delta - k^2)u_0 = 0 \text{ in } \Omega_R,\ u_0 = g \text{ on } \Gamma, \text{ and } \partial u_0/\partial r - iku_0 + r^{-1}u_0 = 0 \text{ on } \Gamma_\infty. \tag{1.14}$$

It may be shown that (1.14) is a well-posed problem (see, e.g., [7]). Furthermore, it is readily seen that (1.14) may be converted into the following problem:

$$(-\Delta - k^2)v_0 = F \text{ in } \Omega_R, v_0 = 0 \text{ on } \Gamma \text{ and } \partial v_0/\partial r - iKv_0 + R^{-1}v_0 = 0 \text{ on } \Gamma_\infty, \tag{1.15}$$

where F is a smooth function with bounded support in Ω_R.

In order to solve (1.15) by the finite element method, we reformulate it as a variational problem. Set $H^E = \{u : \|u\|_E < \infty \text{ and } u = 0 \text{ on } \Gamma\}$, where $\|u\|_E^2 = \|u\|_{H^1(\Omega_R)}^2 = \int_{\Omega_R} (|u|^2 + |\nabla u|^2)$. We now define

$$a(u,\mathrm{v}) = \int_{\Omega_R} (\nabla u \cdot \nabla \mathrm{v}^* - k^2 u \mathrm{v}^*) - \oint_{\Omega_\infty} \left\{ iku\mathrm{v}^* - R^{-1} u \mathrm{v}^* \right\}, \ \forall\, u, \mathrm{v} \in H^E, \tag{1.16}$$

where v^* denotes the complex conjugate of v. The following result may be proved using the argument of [8] (Sec. 5.3.2):

Theorem 1.1. *(a) There exists a constant C_1 such that*

$$\inf_{u \in H^E} \sup_{\mathrm{v} \in H^E} |a(u,\mathrm{v})| \geqslant C_1 \| u \|_E \| \mathrm{v} \|_E.$$

(b) There exists a unique function, $\mathrm{v}_0 \in H^E$, such that

$$a(\mathrm{v}_0, \mathrm{v}) = (F, \mathrm{v}) \text{ for each } \mathrm{v} \in H^E. \tag{1.17}$$

Furthermore, v_0 is the unique solution of (1.15).

Theorem 1.1 (b) follows from (a) as in [8]. Theorem 1.1 (a) will also be crucial in showing that the finite element discretization described below is well-posed. We remark that the exterior Neumann problem as well as the other exterior problems previously mentioned may be converted into variational problems in the same way. In the case of natural boundary conditions on Γ such as the Neumann condition, it is not necessary to impose any boundary condition on functions in H^E. Finally, we observe that

$$\| u - u_0 \|_{H^1(D)} \to 0 \text{ as } R \to \infty, \tag{1.18}$$

where D is a fixed, bounded subset of Ω and $u - u_0$ denotes the error between the solution of problem (1.1)-(1.3) and problem (1.14). Equation (1.18) may be proved using the argument in [7].

1.4. The Finite Element Method

In order to approximate the solution u of (1.1)-(1.3) we shall first very briefly describe a finite element method for approximately solving problem (1.15) using the variational formulation (1.17). We begin by replacing H^E by a one parameter family of finite dimensional subspaces S^h, defined for $h \in (0,1]$. We wish to approximate the solution v_o of (1.17) by the solution v_o^h of the following discretized version of (1.17):

$$a(\mathrm{v}_o^h, \mathrm{v}) = (F, \mathrm{v}), \ \forall\, \mathrm{v} \in S^h. \tag{1.19}$$

The finite element subspaces, S^h, are typically obtained by subdividing Ω_R into simple subsets, t_j^h, of diameter $O(h)$. S^h may then be defined as the subspace of H^E consisting of all continuous functions, v^h, vanishing on Γ, such that the restriction of v^h to each t_j^h is a polynomial of degree less than K for some integer $K \geqslant 2$. By expressing v_o^h as a linear combination of basis functions for S^h with unknown coefficients and then applying (1.19), we obtain a finite number of linear equations (resembling finite difference equations) for the desired approximate solution, v_o^h.

We observe that in general for curved boundaries Γ, it is not feasible to construct finite element subspaces of functions vanishing identically on Γ. See [8]-[10] for methods of overcoming this difficulty as well as detailed descriptions of the finite element method. For natural boundary conditions such as the Neumann condition, it is not necessary to impose any boundary condition on functions in S^h. (Note that the radiation boundary condition on Γ_∞ is a natural boundary condition.)

The following approximation property holds for a typical finite element space S^h described above and is crucial in proving that v_o^h converges to v_o as $h \to 0$:

$$\inf_{\chi \in S^h} \| \mathrm{v}_0 - \chi \|_E \leqslant C h^{K-1}, \tag{1.20}$$

where $K \geqslant 2$ and C is independent of h.

Theorem 1.2. *There exists a unique function $\mathrm{v}_0^h \in S^h$, satisfying (1.19) for h sufficiently small. Furthermore, there exists a constant C independent of h, such that*

$$\| \mathrm{v}_o - \mathrm{v}_o^h \|_E \leqslant C \inf_{\chi \in S^h} \| \mathrm{v}_0 - \chi \|_E \text{ for } h \text{ sufficiently small.}$$

The proof of Theorem 1.2 is based on Theorem 1.1 (a) and (1.20), using the argument of [8] (Sec. 6.3.2). Note that Theorem 1.2 combined with (1.20) gives the rate of convergence of v_o^h to v_o. L^2 and L^∞

error estimates may also be obtained. We remark that analogous results may be proved for more general radiation boundary conditions of the type described earlier.

It may be shown, using the argument in [7], that

$$\|u - u_0\|_{H^1(D)} \leqslant \frac{C}{R^\sigma}, \tag{1.21}$$

where u_o satisfies (1.14), σ depends on the radiation boundary condition, D is a fixed, bounded subset of Ω, and C is independent of R. Since u_o may be readily expressed in terms of v_o, we may employ v_o^h to obtain an approximation, u_o^h, to u_0. We thus obtain

$$\|u - u_o^h\|_{H^1(D)} \leqslant \|u - u_0\|_{H^1(D)} + \|u_o - u_o^h\|_{H^1(D)}. \tag{1.22}$$

The last term in (1.22) is a discretization error and may be estimated using Theorem 1.2. The first term on the right side of (1.22) may be estimated using (1.21). For this term to be small, we require R to be large, thus increasing the number of linear equations. For further details concerning this method, see [7].

1.5. Regions With Infinite Boundaries

There are various wave propagation problems in which it is desired to solve a Helmholtz type equation in a region with infinite boundary. In many cases, the radiation condition is expressed in terms of a modal expansion for the outgoing solution (sufficiently far away from the origin), instead of condition (1.2) or (1.8). A variety of such problems were treated in [11] and [12], where the modal expansion was obtained by separation of variables and the given problem was proved to be well-posed by means of the limiting absorption principle. For a detailed account of these results as well as some applications, see [13]. For extensions of these results to more general classes of domains, see [14] and [15].

In [16], the finite element method was applied to a problem of this type occurring in underwater sound propagation. For such problems, the formulation of the finite element method is analogous to that described above for the exterior problem. The main difference occurs in connection with the radiation boundary condition given by a modal expansion on the artificial boundary, Γ_∞.

It is often the case that instead of computing the solution u one is more interested in calculating certain functionals of u. For example, in various waveguide problems, it is the scattering coefficients that are of interest. See [17] (Chap. 5) for a description of several methods for calculating scattering coefficients for problems involving electromagnetic wave propagation in waveguides with discontinuities. To cite another example, a method for calculating the dispersion relation in periodic iris-loaded waveguides based on a mode-matching procedure is described in [18].

2. INTEGRAL EQUATION METHOD

2.1. Integral Equation Formulations

The exterior Dirichlet or Neumann problem for the Helmholtz equation may be reformulated as an integral equation on the boundary Γ in various ways. For example, consider the exterior Neumann problem given by (1.1), (1.2), and (1.4) in $\mathbb{R}^3$. (Note that the methods of this section are applicable to problems in $\mathbb{R}^2$ in an analogous fashion.) We express the solution $u(x)$ as a single-layer potential as follows:

$$u(x) = \oint_\Gamma \sigma(y) F_o(x,y;k)\, ds_y, \quad x \in \Omega, \tag{2.1}$$

where $F_o(x,y;k)$ is given by (1.5) and $\sigma(y)$ is to be determined. Employing standard jump relations from potential theory (see, e.g., [19]), it may be shown that

$$\frac{\sigma(x)}{2} - \oint_\Gamma \sigma(y) \frac{\partial F_o(x,y;k)}{\partial n_x} ds_y = g(x), \quad x \in \Gamma, \tag{2.2}$$

where n_x denotes the outward directed unit normal to Γ at the point x. Note that (2.2) is a Fredholm equation of the second kind for σ. If (2.2) is uniquely solvable for a given k, then u may be obtained from (2.1).

An integral equation analogous to (2.2) may be obtained for the exterior Dirichlet problem by expressing $u(x)$ as a double-layer potential,

$$u(x) = \oint_\Gamma \frac{\partial F_o(x,y;k)}{\partial n_y} \sigma(y)\, ds_y.$$

A difficulty associated with this method is that (2.2) (or its analogue for the Dirichlet problem) need not have a unique solution for certain values of k, denoted by Λ_D. The sequence of real numbers Λ_D consists of those values of k for which the following problem has a non-trivial solution in the interior region, B:

$$\Delta v = -k^2 v \ \text{ in } B, \quad v = 0 \text{ on } \Gamma. \tag{2.3}$$

In order to circumvent this difficulty, other (more complicated) integral equation formulations have been developed. For more detailed discussions of these and other questions, see [20] and [21], as well as the references cited there.

The exterior Dirichlet problem may also be represented by means of a single layer potential,

$$u(x) = \oint_\Gamma \sigma(y) F_o(x,y;k)\, ds_y, \quad x \in \Omega, \tag{2.4}$$

yielding a Fredholm equation of the first kind for σ,

$$g(x) = \oint_\Gamma \sigma(y) F_o(x,y;k)\, ds_y, \quad x \in \Gamma. \tag{2.5}$$

Again, there need not be a unique solution of (2.5) if $k \in \Lambda_D$. There is no simple analogue of this formulation for the exterior Neumann problem. Finally, we remark that in the case of a mixed boundary condition, a system of Fredholm equations results. For a discussion of these methods and appropriate references, see [22].

2.2. Numerical Methods

We first consider an integral equation formulation such as that given by (2.1) and (2.2). A standard approach for approximately solving (2.2) is to apply a suitable numerical quadrature formula in order to replace the integral by a finite sum. This reduces the integral equation to a finite linear system of equations. The appropriate solution is then obtained from (2.1) by again applying numerical quadrature. For further details of the method as well as numerical results, see [20], [23], and [24].

We next consider an integral equation of the first kind, such as that given by (2.4) and (2.5). The finite-element method has been employed to approximately solve the integral equation (2.5). The approximate solution was then computed from (2.4) using quadrature. This method has been analyzed and optimal error estimates proved for the Helmholtz equation in [25].

Finally, we point out that when $k \in \Lambda_D$ the system of linear equations approximating the integral equation (2.2) or (2.5) will be nearly singular. In fact, the resulting linear system will be ill-conditioned for frequencies in a neighborhood of these critical values. If one of the alternative formulations is employed to circumvent this difficutly, the resulting integral equation will be more complicated and the singularity may be more difficult to treat numerically. These matters are discussed further in [20].

2.3 Coupling of Integral Equation and Direct Discretization Methods

The integral equation method and direct discretization method each possess advantages and disadvantages. The integral equation method incorporates the exact radiation condition and results in relatively few linear equations to be solved. However, this method requires knowledge of the fundamental solution, so that problems with variable coefficients cannot usually be treated using this method. (In general, direct discretization methods are more flexible in treating variable coefficients and more general boundary conditions on Γ.) Furthermore, there may be numerical problems near interior eigenvalues, as previously mentioned. (Interior eigenvalues do not cause computational difficulties when direct discretization methods are applied to the exterior problems.) In addition, the singularity of the fundamental solution may cause problems, e.g., in computing the solution near the boundary Γ (as pointed out in [23]).

For direct discretization methods, the resulting system of linear equations is larger. (However, the matrix of this system is sparse and banded. The matrix resulting from the integral equation method is full.) Furthermore, the radiation condition is only approximately incorporated using a direct discretization method. As previously pointed out, however, recent developments in the theory of highly absorbing boundary conditions have proved useful in treating this difficulty.

Attempts have been made to combine the advantages of both of these methods by coupling them. In [22], an integral equation approach was employed to obtain a non-local, perfectly absorbing boundary condition on Γ_∞ for the solution of the exterior Dirichlet problem in $\mathbb{R}^2$. This boundary condition was then incorporated in the finite element method to approximately solve the problem. Optimal error estimates were established and the method was implemented and tested using piecewise linear finite elements. In [23], a finite difference method was combined with the integral equation method to solve the exterior Dirichlet problem. For references in the engineering literature concerning applications of this type of method, see [26].

3. ADDITIONAL COMMENTS

We remark that the numerical methods discussed above are also applicable to the exterior problem for the Laplace equation, subject to a different boundary condition at infinity. There is another method that has proved successful in treating this problem. This method consists of mapping the exterior region onto a bounded one and then discretizing the resulting problem. For the Helmholtz equation, however, the method fails, since the solution has an essential singularity at infinity. See [27] for a more detailed discussion of this method.

Another numerical problem inherent in treating the Helmholtz equation numerically is the "resolution problem". This means that in order to approximate the solution accurately for large k, one must decrease the grid size. (This is due to the oscillatory nature of the solution). Hence, there is a practical limitation on the size of k in applying numerical techniques. For large values of k, asymptotic methods have proved to be successful. For an extensive discussion of these methods, see [28] and the references cited there. In [24], an integral equation formulation is combined with an appropriate choice of coordinate system in order to increase the range of frequencies that can be treated numerically. In [29], a direct discretization method based on the use of piecewise exponentals is employed to treat the resolution problem.

Finally, we observe that the system of linear equations resulting from a direct discretization of Helmholtz type equations is generally sparse and banded, but indefinite when k is large enough. A direct method for solving such equations is described in [30]. In many cases it is advantageous to employ an iterative method. However, standard iterative methods are not applicable to indefinite linear systems. A conjugate gradient type method applicable to such systems is described in [31]. Another iterative method applicable to systems of this type is the multigrid method (see [32] and [33]).

FOOTNOTES AND REFERENCES

*Work Supported By U.S. Department of Energy under contract EY-76-C-02-0016

[1] J. Keller: "Progress and prospects in the theory of linear wave propagation," SIAM Rev. **21**, No. 2, 229-245 (1979)

[2] *Electromagnetic and Acoustic Scattering by Simple Shapes*, ed. by J.J. Bowman, T.B.A. Senior, P.L.E. Uslenghi (North-Holland, Amsterdam, 1969), pp. 349-352

[3] While "direct discretization" refers to finite difference as well as finite element methods, we shall emphasize the finite element approach

[4] B. Engquist, A. Majda: "Absorbing boundary conditions for the numerical simulation of waves," Math. Comp. **31**, No. 139, 629-651 (1977)

[5] B. Engquist, A. Majda: "Radiation boundary conditions for acoustic and elastic wave calculations," to appear

[6] A. Bayliss, E. Turkel: "Radiation conditions for wave-like equations," to appear

[7] C.I. Goldstein: "The finite element method for unbounded domains," to appear

[8] I. Babuska, A.K. Aziz: "Survey lectures on mathematical foundations of the finite element method" in *The Mathematical Foundations of the Finite Element Method with Applications to Partial Differential Equations*, ed. by A.K. Aziz (Academic, New York, 1972)

[9] P.G. Ciarlet: *The Finite Element Method for Elliptic Problems* (North-Holland, Amsterdam, 1978)

[10] G. Strang, G. Fix: *An Analysis of the Finite Element Method* (Prentice Hall, Englewood Cliffs, New Jersey, 1973)

[11] C.I. Goldstein: "Eigenfunction expansions associated with the Laplacian for certain domains with infinite boundaries, I and II," Trans. Am. Math. Soc. **135**, 1-50 (1969)

[12] C.I. Goldstein: "Eigenfunction expansions associated with the Laplacian for certain domains with infinite boundaries, III," Trans. Am. Math. Soc. **143**, 283-301 (1969)

[13] C.I. Goldstein: "Scattering theory in waveguides," in *Scattering Theory in Mathematical Physics,* ed. by J. A. La Vita, J.-P. Marchand (Reidel, Dordrecht, 1974)

[14] J.C. Guillot, C.H. Wilcox: "Steady-state wave propagation in simple and compound waveguides," Math. Z. **160**, 89-102 (1978)

[15] W.C. Lyford: "Spectral analysis of the Laplacian in domains with cylinders," Math. Ann. **218**, 229-251 (1975)
[16] G.J. Fix, S.P. Marin: "Variational methods for underwater acoustic problems," J. Comp. Phys. **28**, 253-270 (1978)
[17] D.S. Jones: *The Theory of Electromagnetism* (Pergamon, Oxford, 1964)
[18] C.I. Goldstein, H. Hahn, W. Bauer: "On the theory of iris-loaded waveguides," Archiv für Elektronik und Übertragungstechnik (Electronics and Communication) AEU **30**, 297-302 (1976)
[19] O.D. Kellogg: *Foundations of Potential Theory* (Ungar, New York, 1929)
[20] A.J. Burton: "Numerical solution of scalar diffraction problems," in *Numerical Solution of Integral Equations*, ed. by L.M. Delves, J. Walsh (Clarendon Press, Oxford, 1974)
[21] R.E. Kleinman, G.F. Roach: "Boundary integral equations for the three-dimensional Helmholtz equation," SIAM Rev. **16**, No. 2, 214-236 (1974)
[22] S.P. Marin: "A finite element method for problems involving the Helmholtz equation in two-dimensional exterior regions," Ph.D. thesis, Carnegie-Mellon University (1978)
[23] D. Greenspan, P. Werner: "A numerical method for the exterior Dirichlet problem for the reduced wave equation," Arch. Rational Mech. Anal. **23**, 288-316 (1966)
[24] A. Bayliss: "On the use of coordinate stretching in the numerical computation of high frequency scattering," J. Sound Vib. **60**(4), 543-553 (1978)
[25] J. Giroire: "Integral equation methods for exterior problems for the Helmholtz equation," École Polytéchnique, Rapport Interne 40 (1978)
[26] O.C. Zienkiewicz, D.W. Kelly, P. Bettess: "The coupling of the finite element method and boundary solution procedure," Int. J. Num. Meth. Eng. **11**, 355-375 (1977)
[27] C.E. Grosch, S.A. Orszag: "Numerical solution of problems in unbounded regions: coordinate transforms," J. Comp. Phys. **25**, 273-296 (1977)
[28] J.B. Keller: "Rays, waves, and asymptotics," Bull Am. Math. Soc. **84**, 727-750 (1978)
[29] C.I. Goldstein, H. Berry: "A numerical study of the weak element method applied to the Helmholtz equation," Brookhaven National Laboratory Report BNL 50746 (1977)
[30] I.S. Duff, N. Munksgaard, H.B. Nielson, J.K. Reid: "Direct solution of sets of linear equations whose matrix is sparse, symmetric, and indefinite," Harwell Report CSS 44 (1977)
[31] C.C. Paige, M.A. Saunders: "Solutions of sparse indefinite systems of linear equations," SIAM J. Num. Anal. **12**, 617-629 (1975)
[32] A. Brandt: "Multi-level adaptive solutions to boundary value problems," Math. Comp. **31**, 333-390 (1977)
[33] R.A. Nicolaides: "On the l^2 convergence of an algorithm for solving finite element equations," Math. Comp. **31**, 892-906 (1977)

THE ROLE OF SURFACE WAVES IN SCATTERING PROCESSES OF NUCLEAR PHYSICS AND ACOUSTICS

H. Überall

Department of Physics
Catholic University,
Washington, District of Columbia 20064

ABSTRACT

What classical and quantum physics have in common is that they deal with material objects, be they elastic bodies imbedded in a medium, atomic nuclei, or even elementary particles. Each of these target objects, when hit by an incident signal (a propagating wave, or a beam of particles), will be excited into eigenvibrations, and in its vibrating state will reemit waves or particles which will carry along with them information about the vibration properties of the target. We shall discuss here the mechanism by which the eigenvibrations are caused, demonstrating that the incident signal produces attenuated surface waves on the object which circumnavigate the latter repeatedly. If their wavelength is such that the surface waves match phases after each circumnavigation, then a standing surface wave is set up which represents the eigenvibration, and which causes peaks of finite width in the scattering amplitude when plotted as a function of frequency or energy, respectively.

INTRODUCTION

The resonance theory which will be employed in the following is patterned after the BREIT-WIGNER theory of nuclear scattering [1,2]. Its first application to a classical scattering problem was made by us, following a suggestion of L. FLAX, for the case of acoustic-wave scattering from an infinite elastic cylinder [3]. This investigation immediately clarified the very irregular structure of the acoustic backscattering cross section when obtained as a function of frequency, either by calculation [4] or by experiment [4]-[6], demonstrating that the scattering amplitude consists of a nonresonant "geometrical" background corresponding to scattering from the target object as if the latter were impenetrable, plus a series of superimposed, and interfering, resonance terms in each normal-mode (or "partial wave") contribution. In this way, the resonances could still be identified in the total backscattering cross section [3,7].

Following this first application, numerous other examples of classical resonances scattering have been treated. These include acoustic-wave scattering from elastic cylindrical [8,9] and spherical shells [10], from gas bubbles [11-13] and from fluid and elastic layers [14,15], as well as elastic-wave scattering from cylindrical [16] and spherical fluid-filled cavities, [17-23] and from solid spherical inclusions [24]. In some of these studies, we proceeded to explain the appearance of the resonances in terms of circumferential waves which are generated in the scattering process and repeatedly circumnavigate the target object. At the resonance frequencies, their wavelength is such that phase matching, and hence resonant reinforcement, occurs during the circumnavigation.

This picture was also used in order to describe surface waves on a finite fluid cylinder [25] and, finally, a problem of nuclear physics has been treated in the same way. The recently discovered higher-multipole giant resonances of nuclei [26], together with the well-known giant dipole resonance [27], were shown [28] to be interpretable as the manifestation of giant nuclear surface waves which, when the energy of the incident projectile coincides with a nuclear resonances energy, leads to a resonant reinforcement due to phase matching, in the same fashion as in classical scattering problems. This example illustrates the fact that there is a certain unity that pervades various disciplines of physics and that permits the utilization of methods developed in one subfield to be applied to other subfields, with fruitful results. It is fortunate that the present conference, with its interdisciplinary viewpoint, permits the discussion of both classical and nuclear-physics applications of the resonance and surface-wave theory in one and the same paper. In closing this section, we remark that resonance and surface-wave pictures also find a place in heavy-ion scattering [29] and in elementary-particle structure problems [30].

1. CLASSICAL RESONANCE THEORY

Resonance theory in classical scattering will here be illustrated by the example of acoustic-wave scattering from an infinite elastic cylinder [3]. If a plane acoustic wave exp $i(kx-\omega t)$, with propagation constant $k = \omega/c$, is incident on a solid elastic cylinder of radius a with its axis parallel to the z-axis, the total pressure field at the position (r,ϕ) is given by the normal-mode (or partial-wave) expansion:

$$p = \sum_{n=0}^{\infty} (2-\delta_{no}) i^n \left[J_n(kr) + \frac{b_n}{D_n} H_n^{(1)}(kr) \right] \cos n\phi, \tag{1}$$

the second term in brackets representing the scattered wave. The 3×3 determinants b_n and D_n are specified by the elastic boundary conditions, and they contain cylinder functions with arguments $x \equiv ka = \omega a/c$, $x_L \equiv k_L a = \omega a/c_L$ and $x_T \equiv k_T a = \omega a/c_T$, where c_L and c_T are the velocities of compressional and shear waves in the cylinder (of density ρ_c), respectively. We introduce the scattering function

$$S_n = 1 + 2b_n/D_n, \tag{2}$$

in terms of which the scattered pressure amplitude becomes

$$p_{sc} = \frac{1}{2} \sum_{n=0}^{\infty} (2-\delta_{mo}) i^n (S_n - 1) H_n^{(1)}(kr) \cos n\phi. \tag{3}$$

In the far field $(kr >> n)$, one has

$$p_{sc} \sim (a/2r)^{1/2} e^{ikr} f_\infty(\phi), \tag{4}$$

with the "form function"

$$f_\infty(\phi) = \sum_{n=0}^{\infty} f_n(\phi), \tag{5a}$$

$$f_n(\phi) = (i\pi ka)^{-1/2} (2-\delta_{no})(S_n - 1) \cos n\phi. \tag{5b}$$

If $|f_\infty(\pi)|$ is evaluated numerically for an evacuated cylindrical aluminum shell in water [4] [for which (1)-(5) apply except that b_n and D_n are now 6×6 determinants; a is the outer radius and b the inner radius of the shell], one obtains the graph of Fig. 1, plotted vs. the dimensionless frequency parameter $x \equiv ka$. Resonances are clearly visible, corresponding to the excitation of the eigenfrequencies of the target.

In principle, the eigenfrequency spectrum, if known, will give us complete information about the geometry and composition of the target. It can be determined from the scattered wave, whose amplitude carries all the resonance information with it; this may be analyzed as follows.

Expanding the determinants b_n and D_n leads to a representation of the S-function

$$S_n = S_n^{(0)} \frac{F_n^{-1} - z_2^{-1}}{F_n^{-1} - z_1^{-1}}, \tag{6a}$$

where

$$S_n^{(0)} = -\frac{H_n^{(2)\prime}(x)}{H_n^{(1)\prime}(x)} \equiv e^{2i\xi_n} \tag{6b}$$

represents the S-function for scattering from a rigid cylinder. Here,

$$z_i = xH_n^{(i)\prime}(x)/H_n^{(i)}(x), \quad i = 1,2, \tag{7a}$$

and

$$F_n(x) = \frac{\rho_w}{\rho_c} x_T^2 \frac{D_n^{(1)}}{D_n^{(2)}}, \tag{7b}$$

where ρ_w is the density of the ambient fluid, and $D_n^{(i)}$ are appropriate 2×2 subdeterminants of D_n. For a lossless cylinder, F_n is real and the resonances of

$$S_n \equiv S_n^{(0)} \frac{F_n^{-1} - \mathrm{Re}\, z_1^{-1} + i\mathrm{Im}\, z_1^{-1}}{F_n^{-1} - \mathrm{Re}\, z_1^{-1} - i\mathrm{Im}\, z_1^{-1}} \tag{8}$$

are determined by the real eigenvalue equation

$$F_n^{-1} = \mathrm{Re}\, z_1^{-1}, \tag{9}$$

whose solutions are the resonance frequencies $x_n^{(r)}$, labeled by $r = 1, 2, 3, \ldots$. Note that in the customary approach, the equation $F_n^{-1} = z_1^{-1}$ leads to complex eigenfrequencies. In the situation where resonance theory applies, however, $\mathrm{Im}\, z_1^{-1}$ is small and the resonances are high and narrow, as in Fig. 1, so that (9) is the appropriate approximation.

A linear expansion of the denominator in S_n about $x_n^{(r)}$ leads to the partial-wave amplitudes

$$f_n(\phi) = 2i(2 - \delta_{no})(i\pi ka)^{-1/2} e^{2i\xi_n} \tag{10}$$
$$\times \left[\sum_r \frac{\frac{1}{2}\Gamma_n^{(r)}}{x_n^{(r)} - x - \frac{1}{2} i\Gamma_n^{(r)}} + e^{-i\xi_n} \sin \xi_n \right] \cos n\phi,$$

where the resonance widths are

$$\Gamma_n^{(r)} = -2(Im\; z_1^{-1}) \Big/ \frac{d}{dx}(F_n^{-1} - \mathrm{Re}\, z_1^{-1}). \tag{11}$$

This demonstrates that the scattering amplitude consists of a nonresonant background term ($\propto e^{-i\xi_n} \sin \xi_n$) upon which a series of pole terms representing resonances ($\sum_r$) are superimposed, and with which they interfere. The background can be seen to correspond to the scattering amplitude from a rigid cylinder ($\rho_w/\rho_c \rightarrow 0$); in nuclear physics it is known as "potential scattering."

A similar theory can be developed for elastic-wave scattering [17], where, e.g., for an incident compressional (p) wave, one may have a scattered p wave, or also a scattered shear (s) wave ("mode conversion"). Figure 2a shows, for a water-filled spherical cavity of radius a imbedded in an aluminum matrix, the $n = 0$ and $n = 1$ partial-wave amplitude moduli for $p \rightarrow p$ scattering (left side). The right-hand side presents the same amplitude for an evacuated cavity, demonstrating that the resonance features on the left are due to the eigenvibrations of the water filling the cavity. If now the evacuated-cavity background is subtracted from each partial-wave amplitude before taking the modulus, then the pure resonances are obtained (left portion of Fig. 2b). The right-hand side shows the resonances in the $p \rightarrow s$ (mode conversion) amplitude, demonstrating that the resonance frequencies are the same in both cases (since they are due to the vibrations of the same cavity filter), but their heights are different, because of different excitation mechanisms. The curves are plotted vs. the dimensionless frequency variable $x \equiv k_d a$ where $k_d = \omega / c_d$, c_d being the speed of compressional waves in the matrix.

The total backscattering amplitude moduli summed over 16 modes are shown in Fig. 3 for $p \rightarrow s$ scattering (top) and for $p \rightarrow p$ scattering (bottom). In spite of the interference of all these modes and resonances, the latter can be discerned in Fig. 3, and are labeled corresponding to the resonance peaks such as shown in Fig. 2b.

The scattering amplitudes may, however, be considered as functions of the two variables x and the mode number n (which is now considered a continuous variable). Figure 4 shows three-dimensional plots of the $p \rightarrow s$ (top) and the $p \rightarrow p$ (bottom) scattering amplitudes, after subtraction of the continuous background, plotted vs. x (increasing towards the lower right) and n (increasing towards the upper rear). In these plots, the resonances manifest themselves as parallel ridges that are inclined to either axis. (The jagged nature of the ridges is due to the unitary property of the S-matrix, which causes zeros in addition to the resonance poles). If these surfaces are sliced at constant n, the frequency resonances of Fig. 2 are obtained. If the surfaces are sliced at constant frequency x, the same ridges give rise to resonances in the mode-number variable n. Mathematically, this can be shown as follows.

In the denominator of the resonance term in (10), the resonance frequency $x_n^{(r)}$ is a function of the variable n, in which it may be expanded around the value $n_o^{(r)}$:

$$x_n^{(r)} = x_{n_o}^{(r)} + (n - n_o^{(r)})\, x_{n_o}^{(r)\prime}; \tag{12}$$

$n_o^{(r)}$ will be determined from the condition $x_{n_o}^{(r)} = x$. This leads to

$$\left[x_n^{(r)} - x - \frac{1}{2} i\Gamma_n^{(r)} \right]^{-1} = (x_{n_o}^{(r)\prime})^{-1} \left[n - n_o^{(r)} - \frac{1}{2} i\hat{\Gamma}_{n_o}^{(r)} \right]^{-1}, \tag{13a}$$

which is a resonance denominator in the n variable, with a width

$$\hat{\Gamma}_{n_o}^{(r)} = \Gamma_{n_o}^{(r)} / x_{n_o}^{(r)'}. \tag{13b}$$

It corresponds to a pole in the complex n plane, located at

$$\hat{n}_o^{(r)} = n_o^{(r)} + \frac{1}{2} i \hat{\Gamma}_{n_o}^{(r)}, \tag{13c}$$

which in the nuclear literature is known as a "Regge pole." A residue evaluation of amplitudes such as (3) shows that two attenuated surface waves

$$\exp(\pm i\hat{n}_o^{(r)}\phi) \equiv \exp(\pm [in_o^{(r)}\phi - (1/2)\hat{\Gamma}_{n_o}^{(r)}]) \tag{14}$$

are present for each pole (labeled by r), which propagate in opposite directions. Their wavelength is given by

$$\lambda_r = 2\pi a / n_o^{(r)}, \tag{15a}$$

their decay angle by

$$\phi_r = 2/\hat{\Gamma}_{n_o}^{(r)}, \tag{15b}$$

and their phase velocity by

$$c_r = \omega a / n_o^{(r)}. \tag{15c}$$

This is valid for a cylindrical target, and (15a) shows that at resonance, where $n_o^{(r)} = n$ (see (13a)), one has $\lambda_r = 2\pi a/n$, so that n wavelengths of the surface wave fit over the circumference, leading to phase matching and, hence, to a resonant reinforcement in the course of the repeated circumnavigations. For a spherical target, one finds again (15a) and (15c), but with $n_o^{(r)}$ replaced by $n_o^{(r)} + 1/2$ in the denominators. Simultaneously, however, a phase jump of a quarter-wavelength can be shown to take place [10,12,19,28] at the north and south pole of the sphere where the surface waves converge, so that phase matching again takes place after each circumnavigation.

Equation (15c) represents an equation for the dispersion curves of the surface waves, which are plotted in Fig. 5 for a water-filled spherical cavity in an aluminum matrix.

2. NUCLEAR GIANT RESONANCES

It appears remarkable that exactly the same procedures which in the preceding were shown to be so fruitful in classical scattering theory, may be used to describe nuclear reactions. What we have in mind is an analysis of the nuclear giant multipole resonances, of which those of electric-dipole type have been known experimentally for a long time [27], while the higher multipole resonances, although predicted [31] in 1966 or even earlier, have been discovered experimentally only much later [26].

These giant resonances, which may be described by the collective model of GOLDHABER and TELLER [27] and its generalization to the nuclear four-fluid model [32], can be excited in many different nuclear reactions, since they are a property of the target only. We shall illustrate their excitation by the inelastic electron-scattering process, e.g., for an ^{16}O target:

$$e + {}^{16}O \rightarrow {}^{16}O^* + e', \tag{16}$$

and show in Fig. 6 a theoretical [31] (top) and experimental [33] spectrum of the final electrons emerging at a fixed scattering angle, which directly reflects the spectrum in the nuclear excitation energy $\hbar\omega_L$. (Henceforth, we shall set $\hbar = 1$.)

Collective multipole vibrations of a spherical nucleus, of multipolarity L, take the form of a radial oscillation (breathing mode or monopole, $L = 0$), a rigid back-and-forth motion (dipole, $L = 1$), ellipsoidal elongation or oblation (quadrupole, $L = 2$), pear-shaped deformations (octupole, $L = 3$), etc. The four-fluid model is illustrated by the picture of quadrupole oscillations in Fig. 7, where any two combinations of the proton fluids with spin up or down (short arrows), and of the neutron fluids with spin up or down, can oscillate together against the other two combinations, leading to the modes i (isospin, I), si (spin-isospin, II) and s (spin wave, III); in addition, a vibration in unison of all four fluids can take place, leading to the c (compressional, IV) mode. Note that in the si and s modes a total spin $S = 1$ is involved which combines with the matter-vibration multipolarity to a total nuclear spin $J = L + 1$, L or $L - 1$. In the i and c cases, one has $J = L$.

The transition in which the incident electron lifts the nucleus into an excited state is described [34] by Coulomb and transverse matrix elements. The Coulomb matrix element, e.g., depends on the transition charge density $\rho(\mathbf{r})$. A multipole expansion of the latter gives [31] (using a reference radius R):

$$\rho(\mathbf{r}) = \sum_{L,M} \alpha_{LM} r \frac{d\rho_0(r)}{dr} \left(\frac{r}{R}\right)^{L-2} Y_{LM}{}^*(\hat{r}), \tag{17}$$

where $\rho_0(r)$ is the ground-state charge density (assumed to be spherically symmetric), $r = |\mathbf{r}|$, $\hat{r} = r^{-1}\mathbf{r}$, and * denotes complex conjugation. The collective nuclear motion will now be described by an oscillator Hamiltonian

$$H = \sum_{L,M} [(1/2\mu_L)\beta_{LM}^\dagger \beta_{LM} + (1/2)\mu_L \omega_L^2 \alpha_{LM}^\dagger \alpha_{LM}], \tag{18}$$

where α_{LM} is the generalized coordinate and β_{LM} the conjugate momentum; the mass parameter μ_L is found as [31]

$$\mu_L = \frac{2L+1}{L} \frac{Am}{4\pi} \frac{<r^{2L-2}>_0}{R^{2L-4}}, \tag{19}$$

where $<\cdot>_0$ is an average with respect to the ground state, A is the nuclear mass number, and m the nucleon mass. We decompose

$$\alpha_{LM} = (1/2\mu_L\omega_L)^{1/2}[a_{LM}^\dagger + (-1)^m a_{L-M}] \tag{20}$$

to obtain $a_{LM}^\dagger$, a_{LM} as the creation and annihilation operators of a giant-resonance phonon. With this development, the Coulomb matrix element becomes

$$M_L^C = \frac{1}{2}\left(\frac{2\pi LA}{<r^{2L-2}> m\omega_L}\right)^{1/2} q \int_0^\infty r^{L+1} \rho_0 j_{L-1}(qr)\, dr \; \frac{(\Gamma_L/2\pi)^{1/2}}{E - \omega_L + \frac{1}{2}i\Gamma_L}$$

$$\equiv \overline{M}_L^C(q) \frac{(\Gamma_L/2\pi)^{1/2}}{E - \omega_L + \frac{1}{2}i\Gamma_L}, \tag{21}$$

where q is the magnitude of the momentum transfer vector, E the continuum excitation energy, and ω_L and Γ_L the resonance energy and width, respectively, which may be taken from experiment (see Fig. 6). The resonance energy was found as a function of the multipolarity L by SATCHLER [35]:

$$\omega_L = C[L(L+3)]^{1/2} A^{-1/3}, \tag{22}$$

C being a constant independent of L and A.

From (21) and the analogous transverse matrix elements given by the collective generalized Goldhaber-Teller model [31], we may calculate the electroexcitation cross section as a function of excitation energy E for given values of the nuclear vibration multipolarity L. This is done [28] in Fig. 8 for an ^{16}O target, for $L = 1$, 2, and 3, at an electron scattering angle of 150° and incident energies of (a) 60 MeV and (b) 100 MeV. The labels $L_\pm$ in Fig. 8 denote $L = J \pm 1$, and L_0 denotes $L = J$, for the case of the si mode. (The excitation of the s mode is negligible here.)

As in the classical case, the scattering amplitude may be considered as a surface, to be plotted vs. E and L. This surface is sliced at constant E and the result plotted vs. the variable L in Fig. 9, for constant excitation energies $E = 20$, 25, and 30 MeV. The mathematics is as before, with the resulting Coulomb matrix element

$$\hat{M}_J^C(q) = \overline{M}_J^C(q) \frac{(\hat{\Gamma}_J/2\pi)^{1/2}}{L - L_E - \frac{1}{2}i\hat{\Gamma}_J}, \tag{23a}$$

with a resonance value L_E chosen so that $\omega_J(L_E) = E$, and a width

$$\hat{\Gamma}_J = \Gamma_J/\omega_J'(L_E). \tag{23b}$$

The expression of (23a) has a pole in the upper half of the complex L-plane, located at

$$\hat{L}_E = L_E + \frac{1}{2}i\hat{\Gamma}_J. \tag{23c}$$

Its location is a function of E, and the locus of $\hat{L}_E$ in the complex L-plane as E is varied (known as a Regge trajectory) is plotted in Fig. 10 for the various modes of the four-fluid model, using (22) and taking empirical widths from Fig. 6 (bottom), which are assumed to increase linearly with E. Moving along its trajectory with increasing excitation energy, each pole will first pass close to the integer $L = 1$ and generate the dipole resonance. Subsequently, it will pass close to $L = 2$, where the quadrupole resonance is generated, and so forth. In this way, the existence of the giant resonances of all multipolarities is seen to be a consequence of the motion of a few Regge poles through the complex L-plane.

As in the classical case, each Regge pole is seen [28] to represent a giant surface wave of wavelength

$$\lambda = 2\pi R/(L_E + 1/2), \tag{24a}$$

with decay angle

$$\theta_e = 2/\hat{\Gamma}_J, \tag{24b}$$

and phase velocity

$$c(E) = RE/(L_E + 1/2). \tag{24c}$$

The nucleus being spherical, the surface waves converge at the two poles of the sphere where they undergo a $\pi/2$-phase jump each time as discussed earlier. Therefore, at a resonance where $L = L_E$, there will be $L + 1/2$ wavelengths of the corresponding surface wave fitting over the nuclear circumference, and the resulting phase-matching causes the resonance.

The phase velocities of the surface waves are given by (24c). In Fig. 11, we plot the corresponding dispersion curves for the various Regge poles of the hydrodynamical model.

The phase-matching condition may be viewed somewhat differently, namely, as a coincidence condition for phase velocities. The vibrational modes depicted, e.g., in Fig. 7, being standing circumferential waves around the nuclear circumference, may each be decomposed into a pair of waves propagating in opposite directions with phase velocity c_L, as shown in Fig. 12 for the quadrupole mode ($L = 2$). The nuclear reaction generates a series of surface waves with speeds $c(E)$ around the nucleus given by (24c), while the modal speeds are given by

$$c_L = RE/(L + 1/2). \tag{24d}$$

From (24c), it is seen that a resonance occurs at that energy ω_L at which the speed of one of the surface waves launched in the course of the reaction coincides with the phase velocity of the modal wave corresponding to the Lth natural multipole vibrational mode of the collective model. Again, this coincidence condition for phase velocities may be viewed as an eigenvalue condition for the resonance energy ω_L.

SUMMARY

Similar methods were shown to be applicable for a description of resonance phenomena in macroscopic elastic obstacles, excited by incident acoustic or elastic waves, and in atomic nuclei, excited by incident particle beams. We first sketched an application of the Breit-Wigner nuclear resonance theory to classical scattering processes. The resulting resonance expressions in the frequency domain (with corresponding nuclear resonance expressions in the excitation energy domain having been known already) were then transformed into Regge-pole expressions, i.e., poles in the complex mode number (or angular momentum) domain. Their motion along trajectories in this plane was shown to successively produce physical resonances of various multipolarities. This method was also applied to nuclear scattering processes.

The corresponding scattering amplitudes could be interpreted as surface waves circumnavigating the target, and dispersion curves for the surface waves have been obtained. Finally, it was demonstrated that a phase matching of the surface waves during their repeated circumnavigation generated the resonances by constructive interference or resonant reinforcement.

ACKNOWLEDGEMENTS

The classical physics portion of this work has been supported by Code 421 of the Office of Naval Research and the nuclear physics portion by the Theoretical Physics Section of the National Science Foundation.

REFERENCES

[1] G. Breit, E.P. Wigner: Phys. Rev. **49**, 519 (1936)
[2] H. Feshbach, D.C. Peaslee, V.F. Weisskopf: Phys. Rev. **71**, 145 (1947)
[3] L. Flax, L.R. Dragonette, H. Überall: J. Acoust. Soc. Am. **63**, 723 (1978)
[4] K.J. Diercks, R. Hickling: J. Acoust. Soc. Am. **41**, 380 (1967)
[5] W.G. Neubauer, R.H. Vogt, L.R. Dragonette: J. Acoust. Soc. Am. **55**, 1123 (1974)
[6] L.R. Dragonette: Thesis, Catholic University, (1978) and Naval Research Laboratory Report 8216, (September, 1978); J. Acoust. Am. **65**, 1570 (1979)
[7] H. Überall, L.R. Dragonette, L. Flax: J. Acoust. Soc. Am. **61**, 711 (1977)
[8] J.D. Murphy, E.D. Breitenbach, H. Überall: J. Acoust. Soc. Am. **64**, 677 (1978)
[9] J.D. Murphy, J. George, H. Überall: J. Wave Motion **1**, 141 (1979)
[10] J.D. Murphy, J. George, A. Nagl, H. Überall: J. Acoust. Soc. Am. **65**, 368 (1979)
[11] K.A. Sage, J. George, H. Überall: J. Acoust. Soc. Am. **65**, 1413 (1979)
[12] H. Überall, J. George, A.R. Farhan, G. Mezzorani, A. Nagl, K.A. Sage: J. Acoust. Soc. Am., in press
[13] G. Gaunaurd, K.P. Scharnhorst, H. Überall: J. Acoust. Soc. Am. **65**, 573 (1979)
[14] R. Fiorito, H. Überall: J. Acoust. Soc. Am. **65**, 9 (1979)
[15] R. Fiorito, W. Madigosky, H. Überall: J. Acoust. Soc. Am., submitted (1979)
[16] A.J. Haug, S.G. Solomon, H. Überall: J. Sound Vib. **57**, 51 (1978)
[17] G.C. Gaunaurd, H. Überall: J. Acoust. Soc. Am. **63**, 1699 (1978)
[18] G.C. Gaunaurd, K.P. Scharnhorst, H. Überall: J. Acoust. Soc. Am. **64**, 1211 (1978)
[19] G.C. Gaunaurd, H. Überall: J. Appl. Phys., in press
[20] D. Brill, G.C. Gaunaurd, H. Überall: J. Acoust. Soc. Am., submitted (1979)
[21] G.C. Gaunaurd, H. Überall: Science, in press
[22] G.C. Gaunaurd, H. Überall: J. Appl. Mech., to be published
[23] G.C. Gaunaurd, H. Überall: Appl. Phys., submitted (1979)
[24] L. Flax, H. Überall: J. Acoust. Soc. Am., submitted (1979)
[25] E. Tanglis, G. Gaunaurd, H. Uberall: J. Acoust. Soc. Am., submitted (1979)
[26] F.E. Bertrand: Annu. Rev. Nucl. Sci. **26**, 457 (1976)
[27] M. Goldhaber, E. Teller: Phys. Rev. **74**, 1046 (1948)
[28] A.R. Farhan, J. George, and H. Überall: Nucl. Phys. **A305**, 189 (1978)
[29] K.W. McVoy: Phys. Rev. C **3**, 1104 (1971)
[30] V.A. Matveev: "Hydrodynamic model of collective resonances in hadronic matter," Fermi National Accelerator Laboratory preprint FERMILAB-pub-76/52-THY (June, 1976); B. Schrempp, F. Schrempp: "Strong interactions-a tunneling phenomenon," CERN preprint Ref. TH 2573-CERN (September, 1978)
[31] R. Raphael, H. Überall, C. Werntz: Phys. Rev. **152**, 899 (1966)
[32] H. Überall: Phys. Rev. **B137**, 502 (1965); Nuovo Cimento **41**, 25 (1966)
[33] A. Goldmann, M. Stroetzel: Phys. Lett. **31B**, 287 (1970)
[34] H. Überall: *Electron Scattering from Complex Nuclei* (Academic, New York, 1971)
[35] R.G. Satchler: Nucl. Phys. **A195**, 1 (1972)

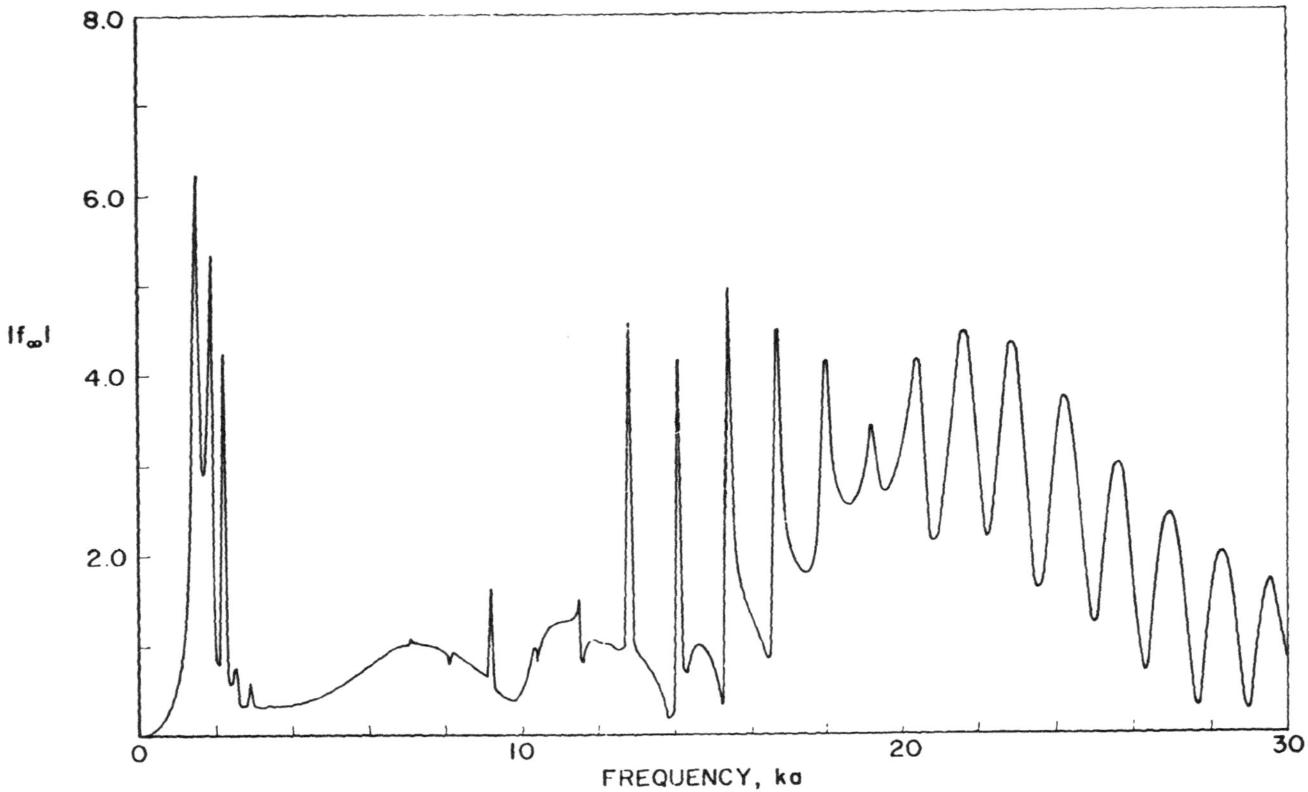

Fig. 1. Resonances in the form-function modulus $|f_\infty(\pi)|$ plotted vs. ka, for an evacuated aluminum shell with $b/a=0.95$ (from [4]).

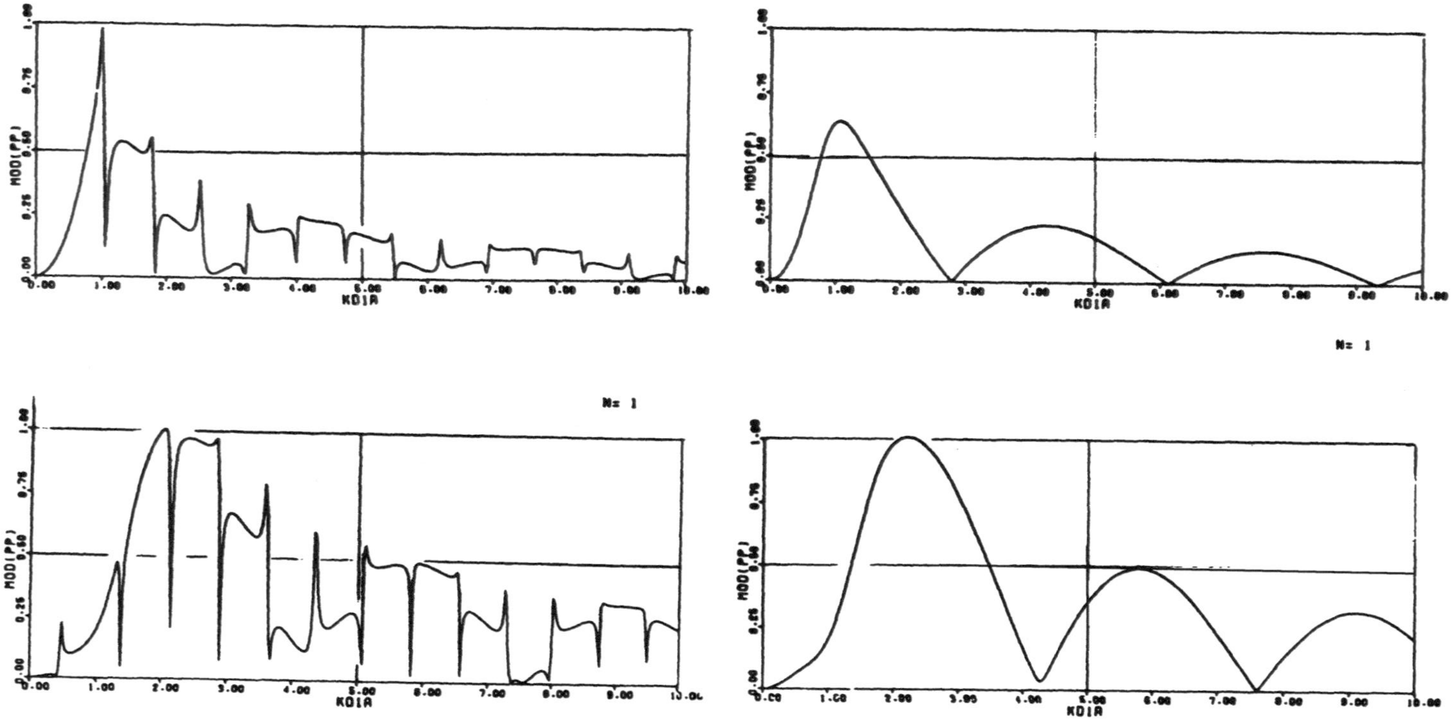

Fig. 2(a). Partial-wave amplitude moduli ($n = 0$ and 1) of compressional wave scattering from a water-filled spherical cavity in aluminum (left), and from an evacuated cavity (right) (from Ref. [17]).

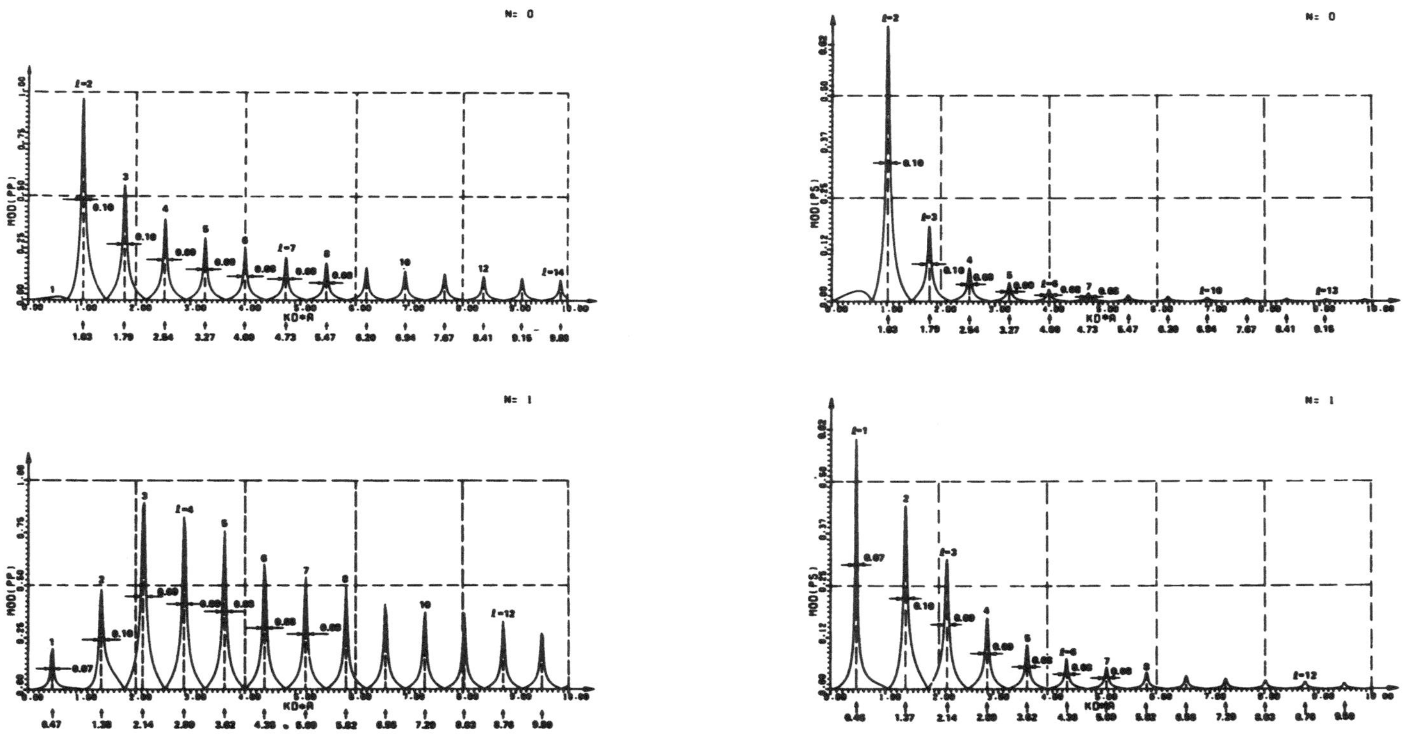

Fig. 2(b). Moduli of the partial-wave amplitudes after subtraction of the empty-cavity background, exhibiting the pure resonances for $p \rightarrow p$ (left) and for $p \rightarrow s$ scattering (right) (from [19]).

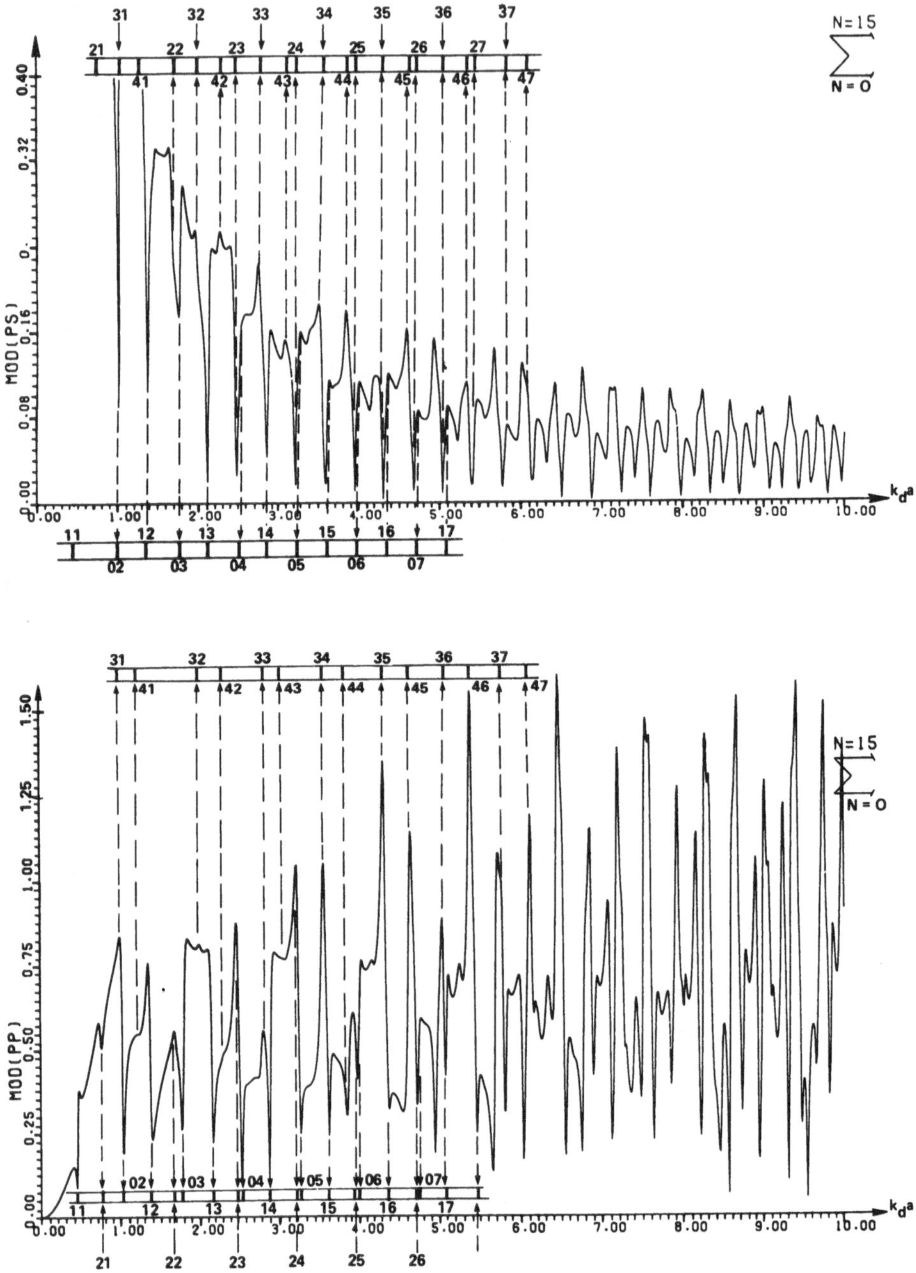

Fig. 3. Modulus of backscattering amplitudes for p → s (top) and p → p scattering (bottom), summed over the first 16 modes, corresponding to the situation of Fig. 2 (from [17]).

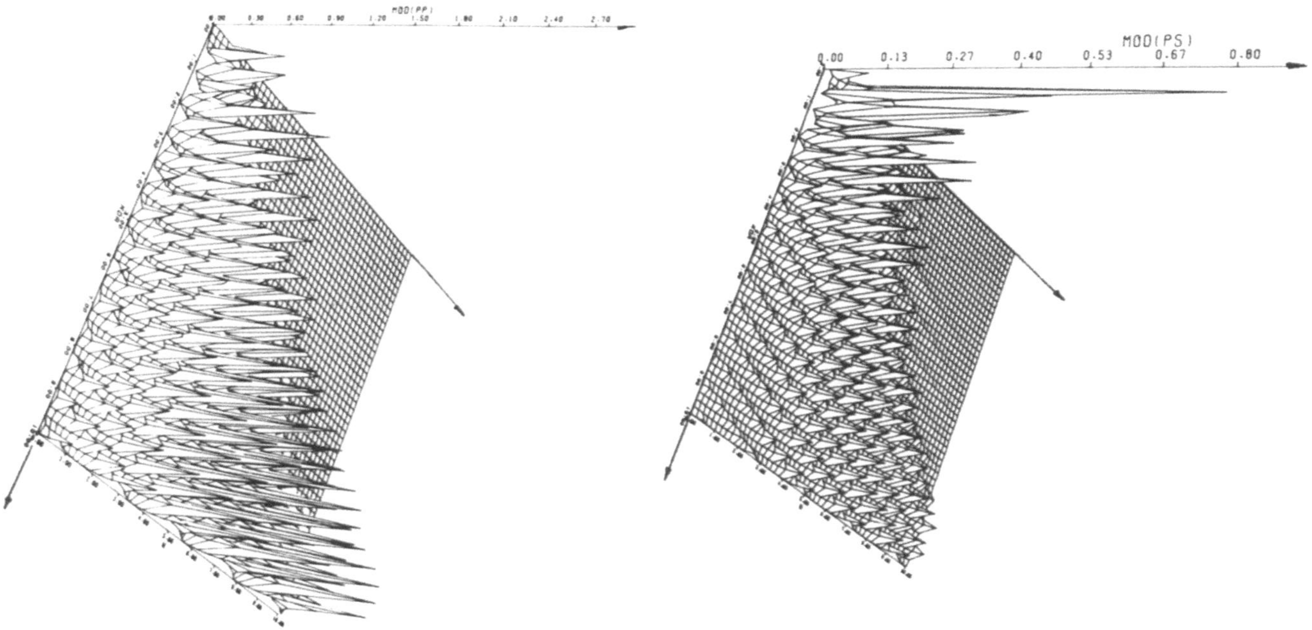

Fig. 4. Three-dimensional plots of p → s (top) and p → $\dot{p}$ (bottom) scattering amplitude moduli (after subtraction of empty-cavity background) for a water-filled cavity in an aluminum matrix plotted vs. frequency x and mode number n (from [17]).

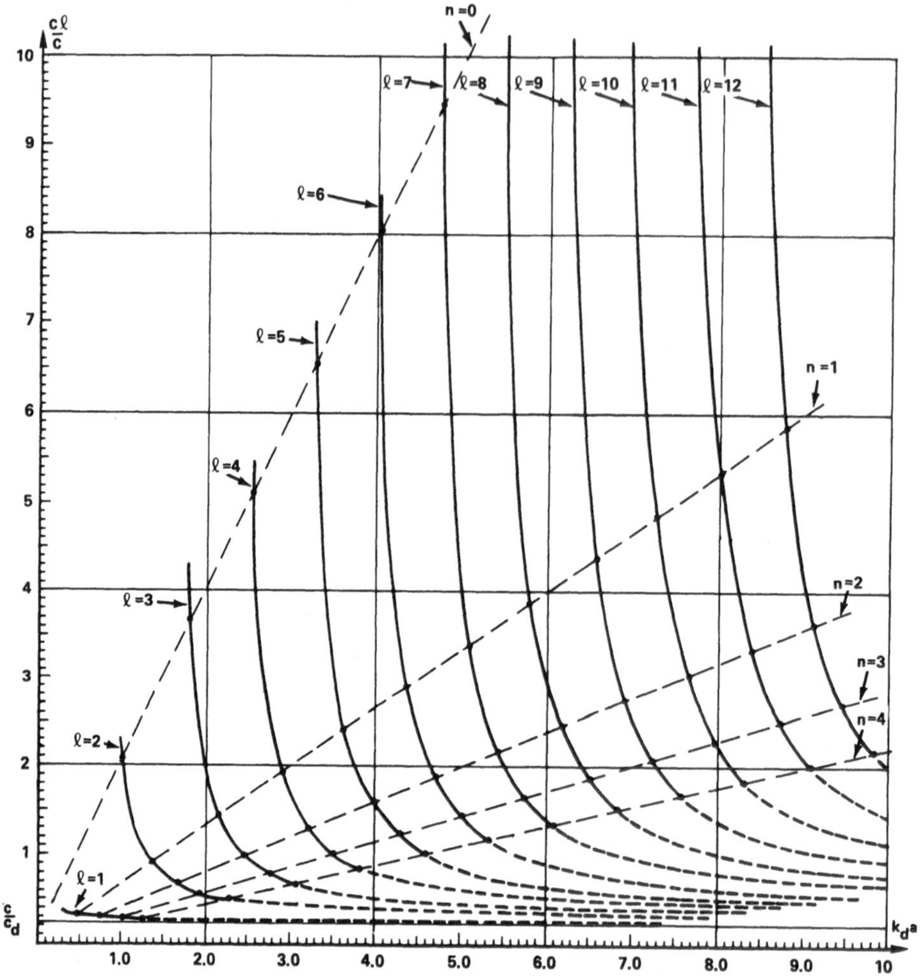

Fig. 5. Dispersion curves of surface waves (labeled by l) on a water-filled spherical cavity in an aluminum matrix (from [17]).

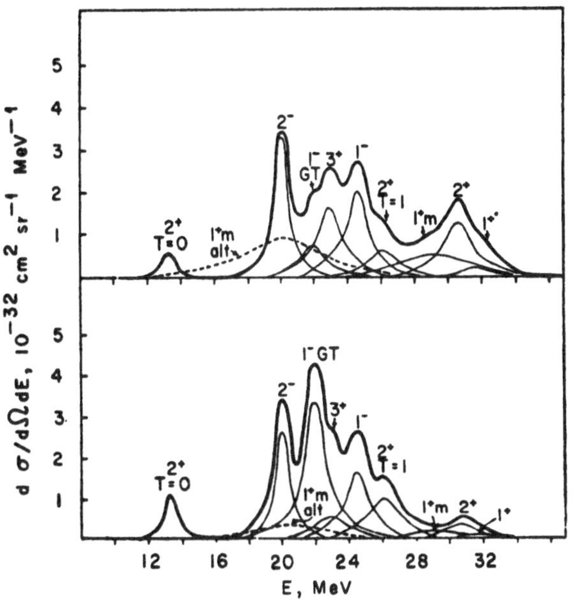

(a)

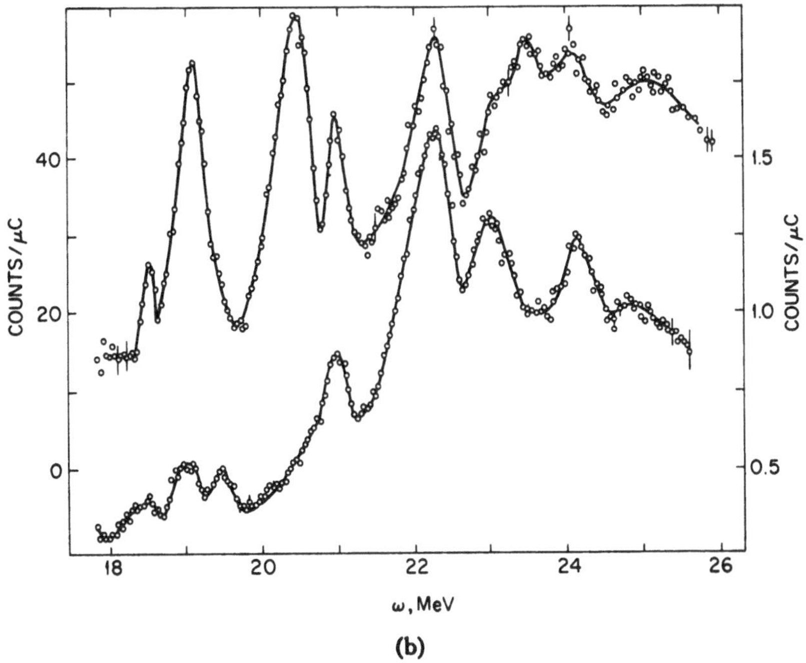

(b)

Fig. 6. Top: Giant resonance in electron scattering from ^{16}O at a scattering angle $\theta = 180°$ containing dipole and quadrupole states of the generalized Goldhaber-Teller model, at $E = 85$ MeV (upper portion) and 60 MeV (lower portion) of incident electron energy (Ref. 31). Bottom: Experimental ^{16}O giant resonance at $\theta = 165°$, $E = 54.3$ MeV (upper portion), and 81°, 45.6 MeV (lower portion) (from [33]).

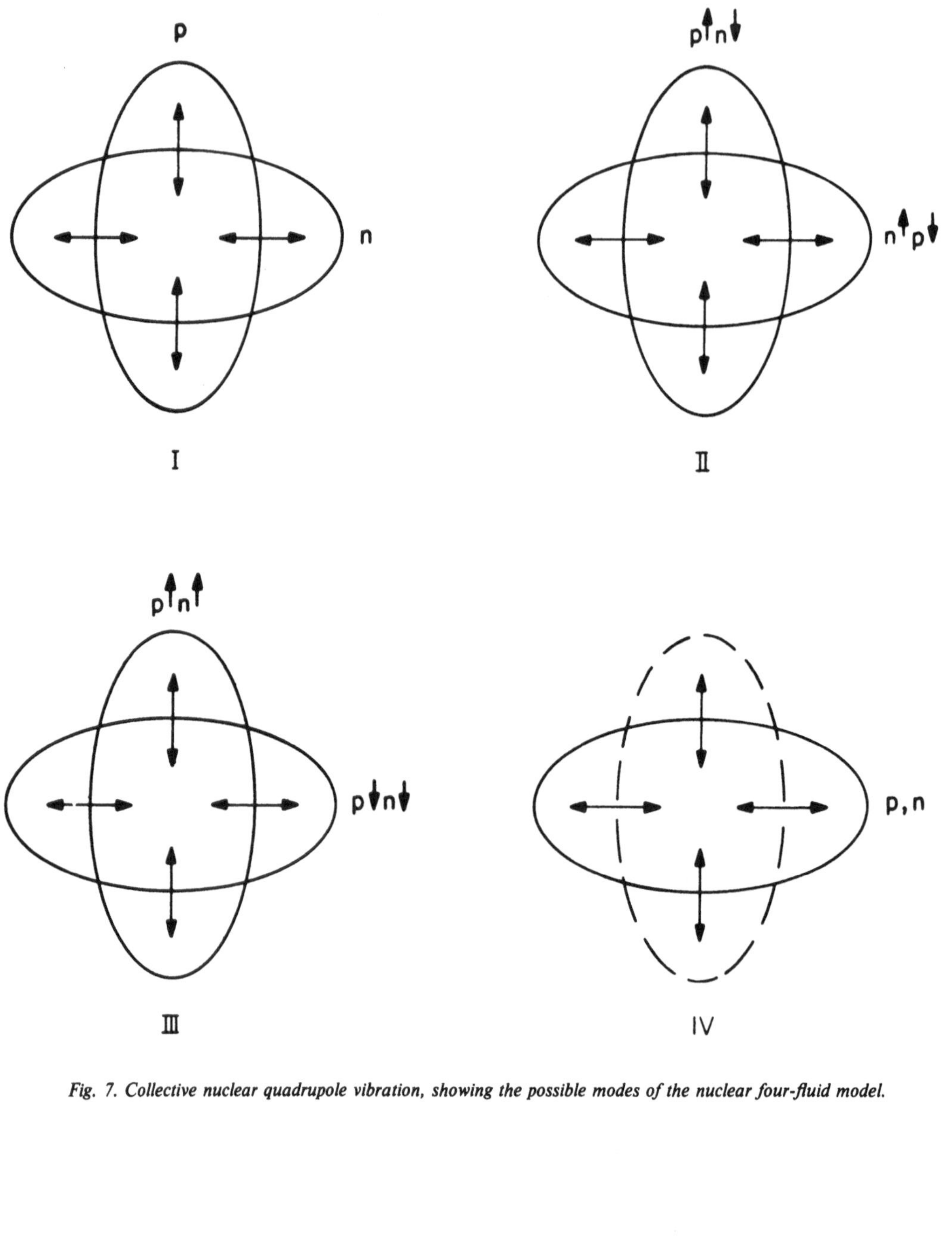

Fig. 7. Collective nuclear quadrupole vibration, showing the possible modes of the nuclear four-fluid model.

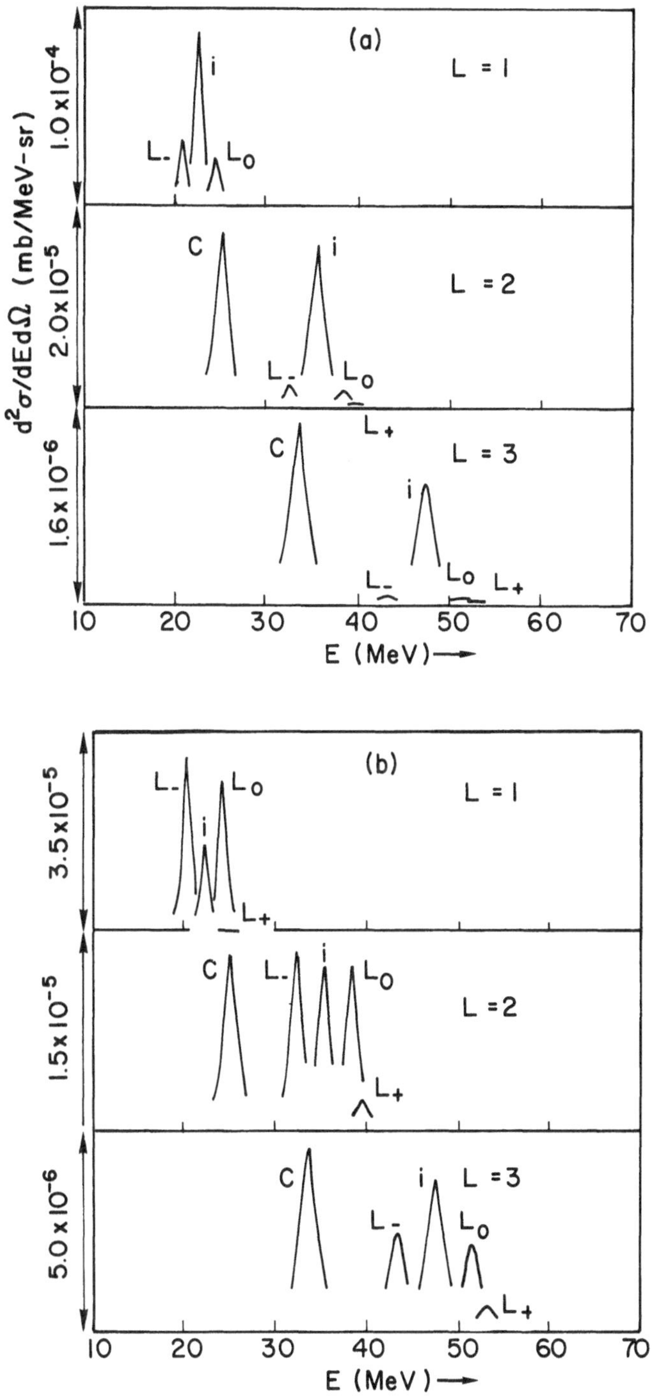

Fig. 8. Giant resonances of several multipolarities L in the electron scattering cross section of ^{16}O, *for a scattering angle* $\theta = 150°$ *and incident energy (a) 60 MeV and (b) 100 MeV, showing several modes of the collective model (from [28]).*

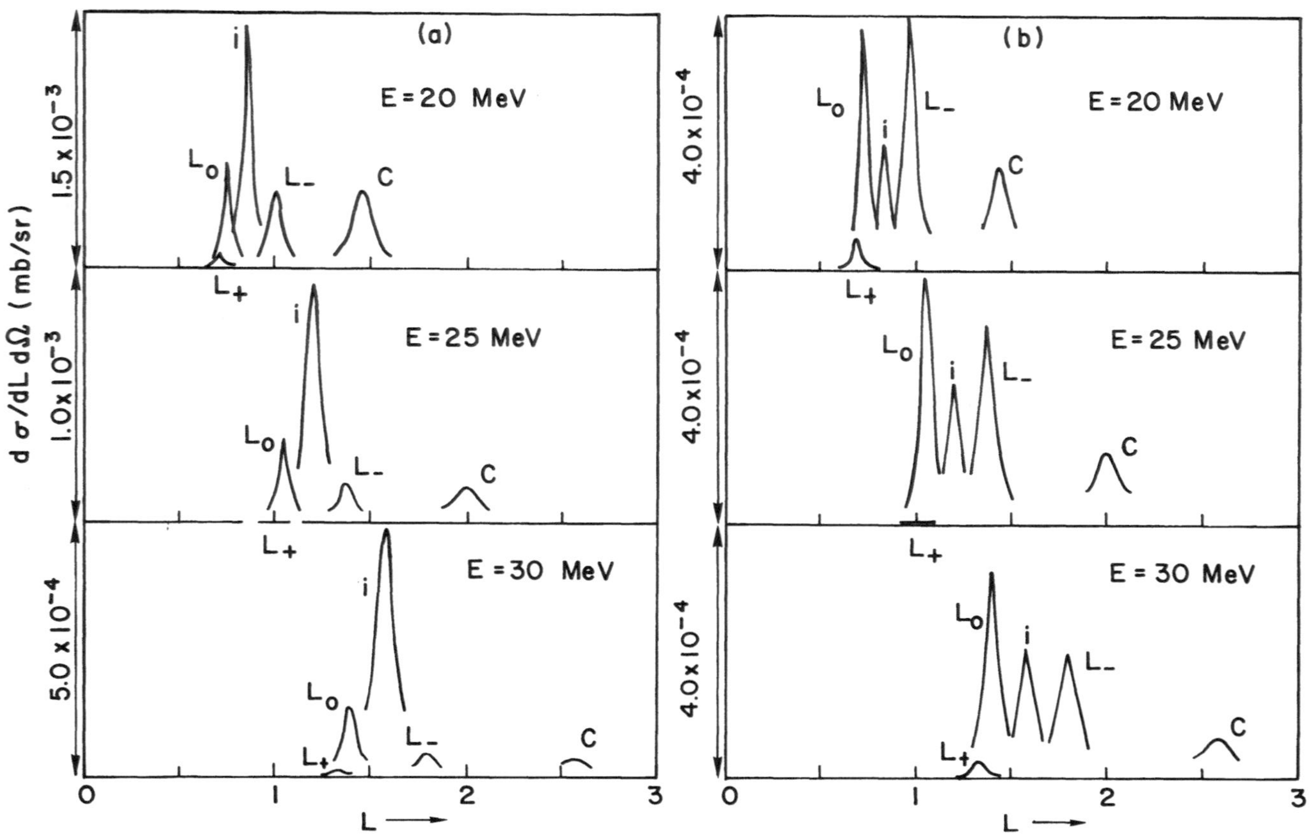

Fig. 9. The same cross sections as in Fig. 7, but plotted vs. the continuous variable L for given excitation energies E = 20, 25, and 30 MeV (from [28]).

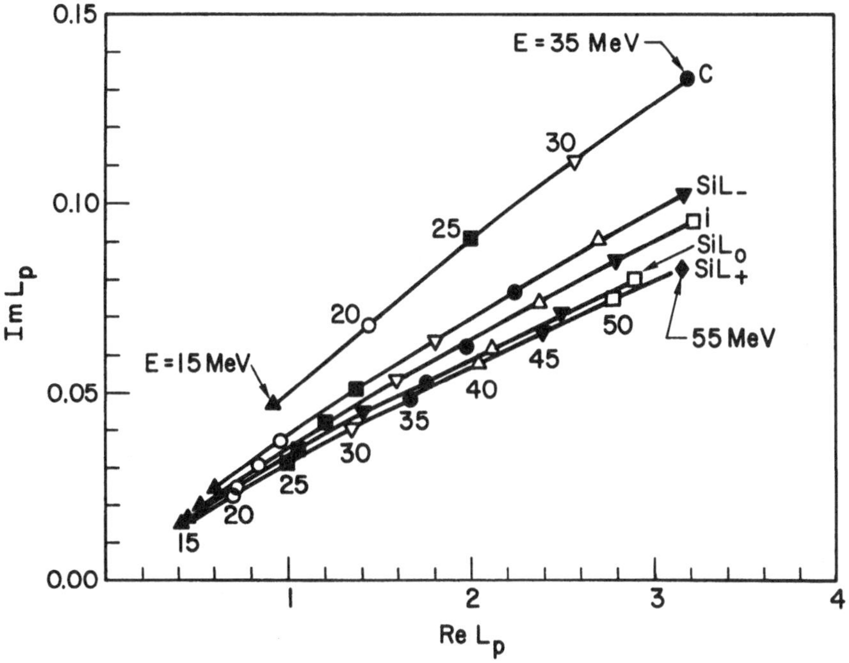

Fig. 10. Trajectories in the complex L plane of five Regge poles that generate the giant multipole resonances of the generalized Goldhaber-Teller model for ^{16}O *(from [28]).*

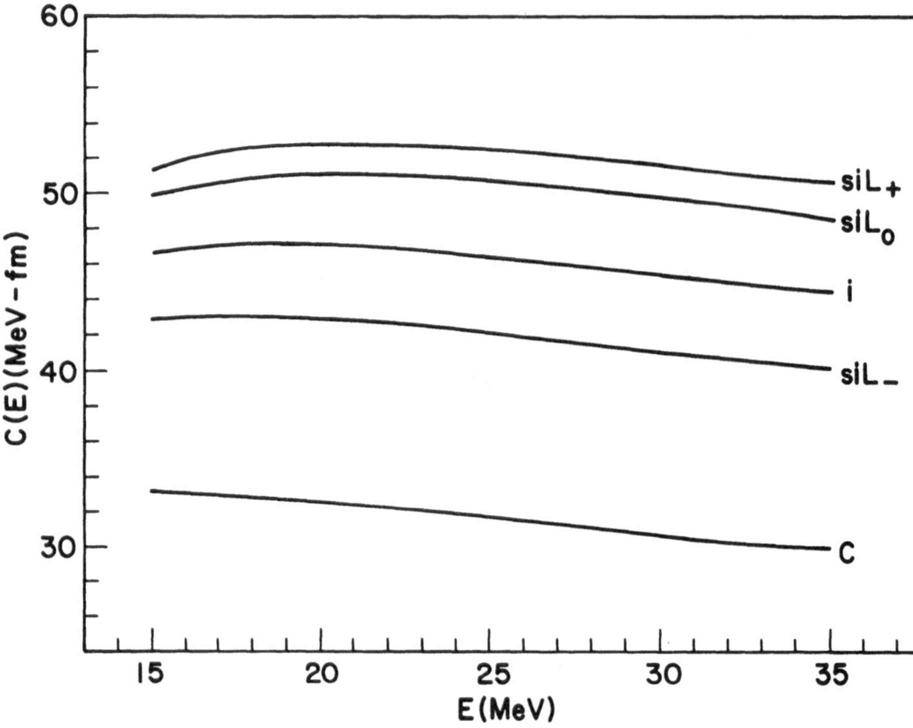

Fig. 11. Dispersion curves of the phase velocities of surface waves, for the five Regge poles of the generalized Goldhaber-Teller model for ^{16}O (from [28]). We used $R = 1.25\ A^{1/3}$ fm.

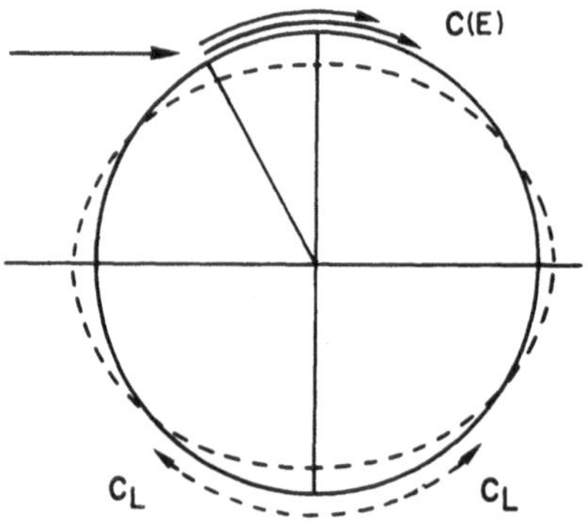

Fig. 12. Phase velocities c_L of standing modal waves, and $c(E)$ of surface waves.

TOPICS IN RADAR SCATTERING THEORY

W.S. Ament

Naval Research Laboratory
Washington, District of Columbia 20375

ABSTRACT

A number of problems, conjectures, and results in radar scattering theory and related fields are reported. Some experiments useful for suggesting theoretical problems or confirming theoretical results are also sketched.

INTRODUCTION

We report here a number of problems, conjectures, and results from investigations during the epoch 1943-63 into microwave propagation, radar backscatter, and tangentially related fields. At the start of that period, the major advances in theory were being made by quantum theorists; during it, tropospheric scatter propagation was discovered and became an identifiable field for symposia, Ph.D. theses, and expensive field measurements, with limited practical application in long-range communications. For that practical function, the communication satellites introduced during the 1960's are vastly superior, so that support for scatter propagation research evaporated, leaving problems unsolved and theoretical ideas undeveloped.

At the close of that epoch, I was trying to assemble a Monte Carlo calculation for long-range tropospheric scatter fields, and had intended to sketch this attempt at the present conference, because the mathematical problem is essentially that of a quantum particle in a radioactive nucleus subjected to time-varying potentials. During the conference I heard, but have not assimilated, several new ideas applicable to the Monte Carlo calculation. Rather than writing up this still nebulous material, I shall report on two or three of its subproblems and then turn to tangentially related questions and conjectures. The paper is, thus, to be mined for suggestions and problems, and not to be treated as a formal scientific paper.

1. FORWARD SCATTERING

Before considering scattering irregularities in the atmosphere, one has the problem of solving the reduced wave equation system

$$\phi_{xx} + \phi_{yy} + k^2[1 + M(y)]\phi = \delta(x,y - y_0) \equiv S,$$
$$\phi(x,0) = 0, \quad \phi(x,y) \text{ upgoing for } y > y_0, \tag{1.1}$$

where $M(y)$ is the "modified refractive index," behaving as $M(y) \approx y$ for large heights y above the smooth surface $y = 0$, x is horizontal distance, and k is the propagation constant, with k^2 equivalent to the quantum particle's mass in the quantum analogue. All important rays emanated from the source S travel mainly horizontally, so the forward-scatter or parabolic approximation is invoked: Let $\phi = \exp(ikx)\psi$, substitute the form into (1.1), and throw away the ψ_{xx} term as "small," obtaining

$$\psi_{yy} + 2ik\psi_x + k^2M\psi = S; \; \psi(x,0) = 0, \; \psi(x,y) \text{ upgoing for } y > y_0. \tag{1.2}$$

This "parabolic approximation" is 1-1 with a one-dimensional time-dependent Schrödinger equation, but in a sense it is not an approximation at all, because from ψ one can retrieve ϕ via an integral:

$$\phi(x,y) = \int_a^b \{\psi(u,y)u^{-1/2} \exp\,[ik(x^2 + u^2)/2u]\}\,du, \text{ where } \{\ldots\}\Big|_a^b = 0. \tag{1.3}$$

This result has easy generalizations and several close relatives [1].

For Monte Carlo purposes, one uses a ray solution of (1.3) valid in the large-k limit: $y(x)$ being the ray trajectory and $s = s(x) = dy/dx$ the ray slope, we have $dy/dx = s$, $ds/dx = dM/dy$, to be forward-integrated to give the trajectory. $M(y) = a_0y^2$ gives a "parabolic atmosphere" or (when a_0 is negative) the quantum-mechanical "parabolic potential well". In that case, (1.2) has a closed-form solution [2], the Wigner distribution

[3,4] is positive definite, translating into plausible joint height-slope statistics for y,s; and the rays from a point source may form foci but do not form caustics. Problems: What is the relation between caustics and the Wigner distribution? Is the quadratic $M(y)$ the most general for which there are no caustics?

The proposed Monte Carlo calculation [5] leads to a joint height-slope distribution $H(y,s)$ known only through a collection of samples (y_i,s_i). From these one needs to estimate the integral $\int_{-\infty}^{\infty}\int_0^{\infty} H(y,s)\,H(y,-s)\,dyds$, for which a Wigner-style distribution, appropriate for finite k, may eventually be useful. Short of that, the simpler, more abstract problem is the following: Given a probability distribution $P(x)$ known only through N random samplings $(x_1, x_2, \ldots, x_N)$, estimate "optimally" the convolution integral $C(y) = \int_{-\infty}^{\infty} P(x)P(x+y)\,dx$.

2. INVERSE PROBLEMS

For time dependence $\exp(-i\omega t)$, the reduced Maxwell's equations in a general stationary medium are

$$\nabla \times \mathbf{E} - i\omega\mu(x,y,z)\mathbf{H} = 0,$$
$$\nabla \times \mathbf{H} + i\omega\epsilon(x,y,z)\mathbf{E} = 0. \tag{2.1}$$

If no quantity varies with y or z, these equations become transmission line equations and one may change variables to $X = X(x) = \int_0^x \sqrt{\mu(x')\epsilon(x')}\,dx'$, where X measures "electrical length" down the transmission line. With the time dependence $\exp(i\omega t)$, the equations take the form

$$\frac{dE}{dX} - i\omega e^{u(X)}H = 0,$$
$$\frac{dH}{dX} + i\omega e^{-u(X)}E = 0, \tag{2.2}$$

where $Z = e^{u(X)}$ is the impedance. Now let

$$u(X) = u_0 > 0 \text{ in } X \geqslant 1, \quad u(X) = -u_0 \text{ in } X \leqslant 1, \tag{2.3}$$

and be a prescribed or unknown $u(X)$ in $-1 < X < 1$, where it forms an "impedance matching section" between two uniform transmission lines of different impedances. So the solutions of (2.2) have the form $E = \exp(i\omega X) + R(w)\exp(-i\omega X)$ in $X \leqslant -1$, $E = T(\omega)\exp(i\omega X)$ in $X \geqslant 1$, and something more abstruse in $-1 < X < 1$.

Problem: Given $R(\omega)$ for a positive ω, find $u(X)$. A more realistic version of this inverse-scattering problem is: Given noise-contaminated observations $O(\omega)$ of $R(\omega)$ over some incomplete set of ω-values, use $O(\omega)$, together with estimated noise statistics and the *prior* information that $u(X)$ varies only in $-1 < X < 1$, to form some best estimate of $u(X)$. For this, one starts with some answer to: What restrictions on the complex function $R(\omega)$ are imposed by the information (2.3)? (Use of prior information is, or should be, of major importance in many kinds of mathematical inference: Wiener and Gabor filters are analytically restricted by having outputs depending on the previous histories of their inputs. Satisfactory X-ray diffraction analyses incorporate in the inversion mathematics the fact that the electron densities being sought are nonnegative. In deconvolving radio telescope data, it is useful to know that radio temperature is a nonnegative function of postion in the sky, etc.).

To me, a more haunting problem is: Given a spectrum $B(\omega) \geqslant 0$ to be used for communications along the transmission line (2.2), (2.3), what is the "optimum matching section" $u(X)$, of the prescribed electrical length 2, for which $\int_0^{\infty} B|R|^2\,d\omega$ is minimum, or (equivalently?) for which $\int_0^{\infty} B|T|^2 d\omega$ is maximum? Conjecture: the optimum $u(X)$ will be found to be an odd function of X.

3. RANDOM MEDIA

A random medium M may have "small" fluctuations of both permeability and dielectric constant:

$$\mu = \mu_0(1 + s'(x,y,z)), \; \epsilon = \epsilon_0(1 + s(x,y,z)). \tag{3.1}$$

The 'unperturbed medium' M_0 is that for which s,s' vanish everywhere. With

$$k_0^2 = \omega^2\mu_0\epsilon_0, \;\; \mathbf{p},\mathbf{q} \text{ position vectors,} \tag{3.2}$$

and with

$$G = G(\mathbf{p},\mathbf{q}) = G(\mathbf{q},\mathbf{p}) = \frac{e^{ik_0|\mathbf{p}-\mathbf{q}|}}{4\pi|\mathbf{p}-\mathbf{q}|}, \tag{3.3}$$

the scalar Green's function for M_0, we write

$$\mathbf{E}(\mathbf{p}) = \mathbf{E}_0(\mathbf{p}) + \int_V s(\mathbf{q})\mathbf{E}(\mathbf{q})[k^2G + \nabla_\mathbf{q}\cdot\nabla_\mathbf{q}G]d\mathbf{q} + i\omega\mu_0\int_V s'(\mathbf{q})\mathbf{H}(\mathbf{q})\times\nabla_\mathbf{q}G\,d\mathbf{q},$$
$$\mathbf{H}(\mathbf{p}) = \mathbf{H}_0(\mathbf{p}) + \int_V s'(\mathbf{q})\mathbf{H}(\mathbf{q})[k^2G + \nabla_\mathbf{q}\cdot\nabla_\mathbf{q}G]d\mathbf{q} - i\omega\epsilon_0\int_V s(\mathbf{q})\mathbf{E}(\mathbf{q})\times\nabla_\mathbf{q}G\,d\mathbf{q}. \tag{3.4}$$

Here, $\mathbf{E}_0$, $\mathbf{H}_0$ satisfy source conditions and Maxwell's equations (2.1) for M_0. Equations (3.4) are the integral equivalents [6] to (2.1) for the prescription (3.1) in that, if one substitutes the right-hand sides for the left-hand sides in the Maxwell's equations for M_0 and uses the delta-function properties of G, Maxwell's equations for M result on finally replacing the left-hand sides of (3.4) by the right-hand sides.

In our opinion, these integral equivalents are to be used in theories of coherence, scattering, and depolarization of electromagnetic waves in random media. We see no alternative for treating the (academic?) problem of propagation in a medium of constant refractive index N having small fluctuations in characteristic impedance Z only. Such a medium is closely approximated by setting $s' \equiv -s$, so that the relative fluctuations in Z, N are respectively $O(s)$ and $O(s^2)$.

4. ALLEN'S EXPERIMENT

Good experiments lead to good theoretical questions. In tribute to the late Philip J. Allen and to the late John W. Wright, we discuss theoretical problems arising from their ingenious experiments. Allen patented a microwave-frequency translator allowing the continuous-wave (CW) microwave frequency F from a conventional oscillator to be shifted, or "translated," to a frequency $F \pm 2R$, where R is the rotation rate of a matched' dipole, transverse to the axis of a single-mode circular waveguide, about the guide's axis [7]. Allen's ingoing single mode is a circularly polarized (CP) transverse-electric mode; the ingoing power is partly reflected by the dipole as a linearly polarized (LP) mode with E-vector parallel to the dipole, at least at the dipole. (See MOON and SPENCER [8] p. 204, Fig. 9.25 for the sense of "parallel" here). An LP mode with E-vector transverse to the dipole passes by to be reflected by a plate or wall situated at distance $\lambda_g/4$ behind the dipole (λ_g = guide wavelength). The LP mode reflected by the plate combines with that reflected by the dipole to give a net CP reflected mode in which the sense or rotation is opposite that of the incoming CP mode. Rotating the symmetric dipole 180° about the guide axis reproduces the initial geometry, but during the rotation the phase of the outoging CP has been advanced or retarded by 360°, according as the rotation is against or with that of the incoming CP. In Allen's frequency translator, the dipole is rotated steadily by an adjustable-speed motor, and standard microwave hardware separates the frequency-shifted reflected wave from the incident.

Allen also measured the torque absorbed by the dipole as consequence of the angular momentum fluxes of ingoing and outgoing waves [7].

In a plane CP light wave of angular frequency ω, each photon carries energy $\hbar\omega$ and angular momentum $\hbar$ or $-\hbar$ depending on convention as to handedness (and perhaps convention as to axes). Thus, Allen suspected that the angular momentum flux (AF) of his single CW, CP mode at microwave angular frequency ω would be related to its constant power flux P through $AF = P/\omega$. Allen suspended his dipole ingeniously in an oil drop, so that the torque of the CP mode would rotate the dipole at some rate R, observable in the frequency of a reflected wave. When the reflecting plate of his translator is replaced with a distant matched load, the dipole passes and reflects only LP modes, each having zero AF. The AF = AF_0 of the incident wave is presumably absorbed as torque on the dipole to produce a rotation rate R_0. With the quarter-wave plate reinserted, the incident AF_0 is converted to flux $-AF_0$ in the oppositely rotating reflected mode, so that a doubled torque applied to the dipole is anticipated to produce rotation rate $2R_0$. Allen's measured rate was $2.8R_0$, discrepant well beyond any inaccuracies. (Allen left this discrepancy buried in the text of his paper, as he had hoped to explain it himself [9].)

To explain the discrepancy, we sought some mechanism by which torque could also be applied to the waveguide, with or without the quarter-wave plate. In either case, the sought torque would in effect attempt to rotate a perfectly conducting figure of revolution about its axis. That this must be possible is seen in considering AF removed from a plane incident CP wave by a perfectly conducting large sphere, or by a right-angle cone pointing into the incident wave.

At least one text [10] develops Maxwell's equations from a four-vector potential, combining the **E**, **B** fields into a bivector and getting AF in a bivector form. Not knowing the four-vector relativistic prescription for a circular waveguide, we write down Maxwell's equations again in Table I, starting from conventional potentials **A**, ϕ, related through the Lorentz gauge condition of line 2.

TABLE I
Maxwell's equations

$$\mathbf{A},\ \phi:$$
$$\nabla\cdot\mathbf{A} + \mu\epsilon\phi_t = 0$$
$$\mathbf{E} \stackrel{d}{=} -\mathbf{A}_t - \nabla_\phi,\quad \mathbf{B} \stackrel{d}{=} \nabla\times\mathbf{A}$$
$$\nabla\times\mathbf{E} + \mathbf{B}_t \equiv 0,\quad \nabla\cdot\mathbf{B} \equiv 0$$
$$\mathbf{H} \stackrel{d}{=} \frac{\mathbf{B}}{\mu},\quad \mathbf{D} \stackrel{d}{=} \epsilon\mathbf{E}$$
$$\mathbf{J} \stackrel{d}{=} \nabla\times\mathbf{H} - \mathbf{D}_t,\quad q \stackrel{d}{=} \nabla\cdot\mathbf{D}$$
$$\nabla\cdot\mathbf{J} + q_t \equiv 0$$
$$\nabla^2\mathbf{A} - \mu\epsilon\mathbf{A}_{tt} \equiv -\mu\mathbf{J}$$
$$\nabla^2\phi - \mu\epsilon\phi_{tt} \equiv \frac{-q}{\epsilon}$$

One is asked to believe that this diaphanous array of definitions and identities has some connection with the "real" physics of ponderable bodies. In fact, the only introduction of a metric or background reference-space is in the constitutive relations, line 5 of Table I [11]. The physical connection is said to come through further identities, conservation principles, quadratic in the field quantities. Table II lists all such identities that we were able to find or derive. On the left are integrals over general volumes V having smooth surfaces S with outward-pointing unit normal $\hat{\mathbf{n}}$; the integrals on the right are surface integrals over S.

TABLE II
Electromagnetic Vector Identities

$$\int \mathbf{E}\cdot\mathbf{J} + \frac{\partial}{\partial t}\int \frac{\epsilon E^2 + \mu H^2}{2} \equiv \oint \hat{\mathbf{n}}\cdot\mathbf{E}\times\mathbf{H}$$
$$\int \{\mathbf{E}q + \mathbf{J}\times\mathbf{B}\} + \frac{\partial}{\partial t}\int \{\mathbf{D}\times\mathbf{B}\} \equiv \oint \left\{\mathbf{E}(\mathbf{D}\cdot\hat{\mathbf{n}}) + \mathbf{H}(\mathbf{B}\cdot\hat{\mathbf{n}}) - \frac{\hat{\mathbf{n}}}{2}(\mathbf{D}\cdot\mathbf{E} + \mathbf{H}\cdot\mathbf{B})\right\}$$
$$\int \mathbf{R}\times\{\ \} + \frac{\partial}{\partial t}\int \mathbf{R}\times\{\ \} \equiv \oint \mathbf{R}\times\{\ \}$$
$$\int \mathbf{A}\times\mathbf{J} + \epsilon\frac{\partial}{\partial t}\int (\mathbf{A}\times\mathbf{E} + \mathbf{B}\phi) \equiv \oint \{\hat{\mathbf{n}}(\mathbf{A}\cdot\mathbf{H}) - \mathbf{A}(\hat{\mathbf{n}}\cdot\mathbf{H}) - \mathbf{H}(\hat{\mathbf{n}}\cdot\mathbf{A}) + \phi\mathbf{D}\times\hat{\mathbf{n}}\}$$
$$\int (\mathbf{A}\cdot\mathbf{J} - \phi q) + \int (\mathbf{E}\cdot\mathbf{D} - \mathbf{H}\cdot\mathbf{B}) + \frac{\partial}{\partial t}\int \mathbf{A}\cdot\mathbf{D} \equiv \oint (\mathbf{H}\times\mathbf{A} - \phi\mathbf{D})\cdot\hat{\mathbf{n}}$$
$$\int \left(\phi\mu\mathbf{J} - \mathbf{A}\frac{q}{\epsilon}\right) + \frac{\partial}{\partial t}\int (\phi\mu\mathbf{D} - \mathbf{A}\times\mathbf{B}) \equiv \oint [\phi\hat{\mathbf{n}}\times\mathbf{B} + \hat{\mathbf{n}}(\mathbf{A}\cdot\mathbf{E}) - \mathbf{E}(\hat{\mathbf{n}}\cdot\mathbf{A}) - \mathbf{A}(\hat{\mathbf{n}}\cdot E)]$$

In Table II, the first line gives Poynting's theorem; the second, the Lorentz force theorem. Insertion of position-vector **R** under the integral signs of the latter gives the third-line identity between "orbital" torque on the left and net "orbital" AF on the right. Line 4 gives the same identity between "spin" torque and "spin" AF. Taken together, lines 3 and 4 amount to a six-component bivector conservation-of-angular-momentum statement. (The foregoing is clearer for CW where the explicit time-derivatives vanish.) The time-integral of line 5 has the dimension of action, and at least the first integral then becomes a classical version of the action principle for a Dirac electron of zero mass. Line 6, if true, has no interpretation known here.

A plane CW, CP wave falls normally on a half-space filled with finitely conducting material, and is eventually absorbed therein. At first glance, **J** and **A** are parallel in the medium, so there should be no $\mathbf{A}\times\mathbf{J}$ spin torque absorbed in the medium. But **A** and **J** are vectors rotating about the plane wave's propagation vector, one lagging behind the other when the conductivity is finite [12]; the "spin" AF lost from the CP wave is indeed transferred as **A**×**J** torque to the lossy medium. When the CP wave falls on the perfectly conducting sphere or cone, one sees no term in the "orbital" form, line 3, by which tangential traction is applied to rotate the sphere; the third term on the right of the "spin" expression, line 4, applies the requisite traction. For a more elementary calculation: the plane CP wave falls obliquely on a conductive half-space; the surface integrals are now over the plane interface, and should always agree with the volume-integrated "real" **A**×**J** torque transfer. This is also the case in the perfectly-conducting limit.

Using the identities of Table II, we consider angular momentum problems raised by Allen's experimental results. With z measuring distance along a general circular wave guide, r measuring distance from its axis, and ϕ measuring azimuth with respect to some fixed reference direction, the natural coordinates are the polar coordinate system (r,ϕ,z). Any propagating CP, CW mode in the guide has an $\exp(in\phi - i\omega t)$ dependence. The AF of any mode with this dependence through any cross section $z = C$ of the guide is related to the power flux P through C through $AF = nP/\omega$, where the AF calculation is given through the "orbital" flux of the surface integral, over C, of line 2, Table II. No two propagating modes combine to given any net torque on the guide walls. Perhaps no two evanescent modes do, either. Allen's dipole, with his CP propagating mode incident, sets up a system of evanescent modes. In the presence of an evanescent plus a propagating mode, the $\mathbf{H}(\hat{\mathbf{n}}\cdot\mathbf{A})$ third term of the right side of line 4, table II, will produce, on z,ϕ integration over the guide walls, some net z-component of torque applied to those walls. Allen's quarter-wave reflecting plate, behind the dipole at the small distance $\lambda_g/4$, reflects evanescent waves to the transverse plane of the dipole, and the interaction of these waves with the propagating modes, leading to (unmeasured) torque applied to the walls, accounts qualitatively for the 2.8 vs. 2 discrepancy.

By moving the plate back to $(2n + 1)\lambda_g/4$, where n is a positive integer, the reflected evanescent modes can be made arbitrarily weak at the dipole, whereas the output's opposite-sense CP is unaltered. Prediction: for a sufficiently large n, the reflected evanescent modes become negligible in the region of the dipole, the net torque applied to the guide walls becomes negligible by the symmetry of the evanescent waves about the dipole, and the anticipated $2R_0$ rotation-rate will be observed. Problem: Allen's dipole is "matched" in the sense that it acts for the propagating guide mode in the manner of a Faraday screen for a plane CP wave, reflecting a parallel, purely LP mode and allowing a transverse, purely LP mode to pass by. Conjecture: there is no passive structure insertable into a circular wave-guide that has this Faraday-screen behavior without at the same time creating evanescent waves.

Other results of pondering Allen's experiment: In an $\mathbf{R} = (R,\theta,\phi)$ spherical coordinate system consider a sphere of radius R_0 having charge $q = q_0 \sin\theta \sin\phi$ pasted on. Let this sphere be set into rotation so that the charge seen at time t at fixed azimuth ϕ is $q = q_0 \sin\theta \sin(\phi - \omega t)$. The resulting fields interior and exterior to the sphere are describable through the spherical wave functions of STRATTON [13] and Morse. At the spherical surface $R = R_0$, the fields may be matched to the rotating source, q, to satisfy all source and radiation conditions. The difficulty is that this matching holds whether or not the equatorial velocity of the sphere exceeds the velocity of light, c. (Our calculation aimed at finding the power $\omega\mathbf{R}\times(\mathbf{E}q + \mathbf{J}\times\mathbf{B})$ needed to keep the sphere in rotation at constant ω, and I do not recall looking at the growth of standing wave energy interior to the sphere as ωr approached c — the calculation should be redone.) SCHWINGER [14] calculated the classical synchrotron radiation of a *point* charge moving in circular orbit, finding that the radiated power increases indefinitely as orbital velocity v approaches c. Our reading of Schwinger's impressive Fourier transform calculations is that there is no such $v = c$ barrier for an extended charge, such as the pasted-on q of the foregoing sphere. Arbitrarily high space frequencies in the Fourier transform of the *point* charge participate in the power calculation, whereas there is a high-frequency tailing-off in the Fourier transform of the *extended* charge. Perhaps the classical notion of "extended charge" should be abandoned as leading to unphysical behavior; or perhaps general-relativistic methods [15] can be invoked to save the notion.

By-product: the radial dependence of CW spherical waves of a given $P_n^m(\theta)\exp(im\phi - i\omega t)$ type is found through recursions [13, p. 406, eq. 35] starting from $\exp(ikR)/R$ for outgoing waves and $\sin(kR)/R$ for standing waves finite at $R = 0$. For a general time dependence, the same recursions will work for the (R,t)-dependence starting from $f(ct - R)/R$ for outgoing waves and $[g(ct + R) - g(ct - R)]/R$ for standing waves, where f and g, if adequately differentiable, are arbitrary, and c is the velocity of light.

5. WRIGHT'S EXPERIMENT

Jack Wright used Allen's frequency translator to provide the reference CW for accurate measurements of Doppler-shifted radar echoes from surface waves moving on a liquid [16]. In his initial experiments, a wave generator oscillating at frequency F created waves on the x,y plane of the liquid's surface having the approximate form wave height = $h(x,t) = h_0 \sin(Kx - \Omega t)$. In first approximation, the CW radar illuminated the surface with a plane wave $\exp[ikx\cos g - iky\sin g - i\omega t]$. Echoes from successive wave crests arrive in phase at the radar when the Bragg condition 2cos $g\cdot$ (surface wavelength L) = (radar wavelength λ), or $2k\cos g = K$, is satisfied. The geometry is reproduced cyclically at frequency $F = \Omega/2\pi$ so that to produce a reference CW for homodyne detection, Allen's dipole rotates at rate $F/2$, against the input mode's rotation if the surface waves advance toward the radar. The homodyne detection permited Wright to filter out echoes from stationary objects around the laboratory in favor of the wave echo. Appropriate optics monitored wave shape and height h_0. For adjustably small values of h_0, the echo powers were observed to be proportional to h_0^2.

Wright's LP radar measured wave echoes as function of grazing angle g for horizontal (H) and vertical (V) polarizations. (For V, the E-vector is in the x,z plane of incidence; for H, it is normal to this plane and parallel to wave crests.) Water has microwave dielectric constant $\epsilon \simeq 80$, the Brewster's angle g_b being about 7°, too near grazing incidence for good measurements. The jet fuel JP5 (a form of kerosene) has dielectric constant 2.1, g_b being about 36°. While Wright's absolute echo power measurements at fixed angle on either V or H were somewhat uncertain, the ratio of power at V to that at H cancels much of the uncertainty and the result is relatively accurate. This V-to-H power ratio, plotted against g, shows [16, Fig. 7] what we claim to be a definite cusp-like dip at g_b. The available small-h_0 theory, for which the writer has some responsibility, showed no such cusp [16,17]. (Wright claimed his experiment was not good enough to rule out the theory on this evidence, and, furthermore, he had no wish to rerun the measurements with improved equipment because JP5 was unpleasant and unsafe to work with. By that time his interest lay in questions of wind-wave interaction, a field where his experiments and theory made first-rank contributions.)

The theory is further suspect in its prediction for the grazing limit $g \to 0$. If one lets the dielectric constant ϵ increase to infinity, the liquid becomes in effect a perfect conductor, and g_b goes to zero. For the double limit $\epsilon \to 0$, then $g \to 0$, the theoretical answer differs from that for the reverse order $g \to 0$, then $\epsilon \to 0$; the theory says (incorrectly) that it matters whether one arrives at the double limit with $g > g_b$ or $g < g_b$. In any event, the finite limit for $g > g_b$ violates what is claimed to be a requirement of the second law of thermodynamics [18].

We point out Wright's Brewster's angle cusp and the foregoing grazing-limit anomalies as items to be accounted for satisfactorily in any acceptable theory.

6. OTHER ROUGH-SURFACE PROBLEMS

D.D. CROMBIE [19] introduced the Bragg condition to the radar world, and radar to the oceanographers as a tool for studying surface wave spectra, after he had analyzed the Doppler-shifted backscatter recorded in echo data from an 13.65-MHz vertically polarized radar observing the sea at what was, geometrically, grazing angle $\simeq 0$. There is no violation of any second-law precept here as the radar wave propagated not as a free plane wave, but as a ground wave, or lateral wave, attached to the ocean sea surface.

I do not know of any theory for the effect of the waves on the lateral wave's propagation. The coherent-incident illumination would increase with height on the wave, a phenomenon called "shadowing" when the radar is at microwave frequencies. I.W. FULLER [20] measured ocean wave echoes with a short-pulse 3-cm wavelength radar at extreme grazing incidences, where shadowing definitely enters. Curiously, horizontal polarization resolved wave crests considerably more sharply than did vertical, though the net echoed power per wave was roughly the same.

T.B.A. SENIOR [21] derived an effective boundary impedance Z for waves propagating over a rough surface, though the roughness scales for which his Z appears valid are definitely too small compared with radio wavelength to apply to Crombie's experiment. Senior's Z applies to coherent scattering by the "average" rough surface. HIATT, SENIOR, and WESTON (HSW) [22] made antenna-range measurements of backscatter from rotating metal spheres (of diameter 10 inches) intended to confirm predictions based on Senior's Z. HSW, however, did not use the inteferometric measurements appropriate for coherent waves, but measured total backscattered power as function of azimuth. Naturally enough, for the three wavelengths (10.5, 3.1, and 1.3 cm) used, there were more and deeper wiggles, or fadings of the echoed power, per 360° rotation at the shorter wavelengths. For lack of a good theoretical explanation for the reproducible wiggles, one is, as usual, led to suggest improvements in the experiment. Let the rough sphere rotate about its polar axis. Then (extrapolating from Fuller's observations) there should be differences in the fluctuation characteristics when an incident radar LP is polarized parallel to the axis or lies in the equatorial plane, the rates being higher in the latter case as arising from points farther around the equator. The fluctuations come from moving echoers, and their locations can be sorted out to fair extent by their Doppler shifts using Allen's frequency translator and Wright's homodyne detection — the general principle was used to locate echoers on the Moon with long-pulse radars.

The radar echo from a smooth large sphere is composed of a direct convex-mirror reflection from the front surface plus contributions from waves, excited by rays tangential to the sphere, creeping 180° around the back of the sphere, and relaunching tangential rays in the direction of the radar. These creeping waves are the curved-surface version of Crombie's ground wave; intuitively, they are affected by the HSW sphere's roughness in two conflicting ways: (1) the roughness provides a wave-slowing average structure that enhances the propagation, while (2) introducing scatterers that transfer power in the creeping wave into rays scattered in random directions. In our opinion, way (1) is the winner for small-scale roughness. Continuously changing the shorter HSW wavelengths over a 10% range should change the phase of a creeping wave echo, relative to that of the front-face echo, over a full 360°, enabling their relative magnitudes to be established at CW — though interferometric measurement may be needed to isolate these coherent echoes from the random scatter.

Notable recent advances in theoretical methods for attacking rough surface scattering problems are described in the papers of J.A. DeSanto and R.H. Andreo in these proceedings. I trust that the behavior of rough surface scattering conjectured here and elsewhere [18] will soon be within reach of quantitative theory, and that it has been appropriate to suggest confirming experiments.

FOOTNOTES AND REFERENCES

[1] G.N. Watson: *Bessel Functions* (Cambridge, 1966), Sec. 6.2; J.A. DeSanto: "Relation between the solutions of the Helmholtz and parabolic equations for sound propagation," J. Acoust. Soc. Am. **62**, 295-297 (1977)
[2] K. Husimi: "Miscellanea in elementary quantum mechanics," Progr. Theoret. Phys. **9**, 238, 281; **10**, 173 (1953)
[3] J.E. Moyal: "Quantum mechanics as a statistical theory," Proc. Camb. Phil. Soc. **45**, 99 (1949)
[4] C. Morette: "On the definition and approximation of Feynman's path integrals," Phys. Rev. **81**, 848 (1951)
[5] W.S. Ament: "A statistical normal mode theory of transhorizon scatter propagation," *Proceedings of the 1964 World Conference on Radio Meteorology*, pp. 422-425
[6] C. Muller: "Zur mathematischen Theorie elektromagnetischer Schwingungen," Abh. Deutches Akad. Wiss. Berlin 1945/46, No. 3, 5-56 (1950)
[7] P.J. Allen: "A radiation torque experiment," Am. J. Phys. **34**, 1185-1192 (1966)
[8] P. Moon, D.E. Spencer: *Foundations of Electrodynamics* (Van Nostrand, New York, 1960)
[9] P.J. Allen: private conversation
[10] F. Rohrlich: *Classical Charged Particles* (Addison-Wesley, Reading, Massachusetts, 1965)
[11] C. Truesdell, R. Toupin: *The Classical Field Theories*, Handbunch der Physik (Springer, Berlin, 1960), Vol. III/1, p. 677
[12] N. Bloembergen: private correspondence
[13] J.A. Stratton: *Electromagnetic Theory*, (McGraw-Hill, New York, 1941). See the recursions on p. 406, (35).
[14] J. Schwinger: "On the classical radiation of accelerated electrons," Phys. Rev. **75**, 1912-1925 (1949)
[15] J. Van Bladel: "Electromagnetic fields in the presence of rotating bodies," Proc. IEEE **76**, 301-318 (1976)
[16] J.W. Wright: "Backscattering from capillary waves with application to sea clutter," Trans. IEEE **AP-14**, 749-754 (1966)
[17] F.G. Bass, I.M. Fuks, A.I. Kalmykov, I.E. Ostrowsky, A.D. Rosenberg: "Very high frequency radiowave scattering by a disturbed sea surface", Trans. IEEE **AP-16**, 554-559 (1968). Their equation (20) gives a different VV backscatter without any Brewster effect.
[18] W.S. Ament: "Forward- and back-scattering from certain rough surfaces," *Proceedings of the Electromagnetic Wave Theory Symposium*, Trans. IEEE **AP-4**, 369-373 (1956) (Appendix A)
[19] D.D. Crombie: "Doppler spectrum of sea echo at 13.65 Mc/s," Nature **175**, 4459 (1955)
[20] I.W. Fuller: "Radar data films," private communication (1953)
[21] T.B.A. Senior: "Impedance boundary conditions for statistically rough surfaces," Appl. Sci. Res. **8B**, 437-462 (1960)
[22] R.E. Hiatt, T.B.A. Senior, V.H. Weston: "A study of surface roughness and its effect on the backscattering cross section of spheres," Proc. IRE **48**, 2008-2016 (1960)

COHERENT SCATTERING FROM ROUGH SURFACES

J. A. DeSanto

Naval Research Laboratory
Washington, D.C. 20375

INTRODUCTION

Several topics pertaining to rough-surface scattering are discussed. Among them are the region of validity of the Rayleigh hypothesis, which treats the field expansion in a concavity in terms of only outgoing waves, and a solvable model for plane wave scattering from a thin-comb grating. Both involve periodic, deterministic surfaces. In addition, for an arbitrary deterministic surface, the T-matrix is shown to satisfy a three-dimensional Lippmann-Schwinger integral equation for a noncentral and complex "potential." Specializing this to a homogeneous Gaussian distributed random surface yields a one-dimensional integral equation for the ensemble average of the T-matrix (Dyson equation) which can be solved. The solution is compared to experimental measurements and illustrates the importance of multiple scattering in explaining surface scattering phenomena.

1. THE RAYLEIGH HYPOTHESIS

For this discussion it is helpful to keep in mind the sinusoidal surface illustrated in Fig. 1a, although any analytic periodic surface will do. The velocity potential (or acoustic pressure field) ψ is decomposed into an incident plane wave (ψ_i) plus scattered field (ψ_{sc}) as

$$\psi(x,z) = \psi_i(x,z) + \psi_{sc}(x,z) \quad , \tag{1}$$

where ψ_{sc} satisfies the two-dimensional Helmholtz equation

$$(\partial_x^2 + \partial_z^2 + k^2)\psi_{sc}(x,z) = 0 \quad , \tag{2}$$

with k the wavenumber, and the boundary condition

$$\psi(x,h(x)) = 0 \quad . \tag{3}$$

Define the following set of functions satisfying (2)

$$\phi_m(x,z) = \exp[ik(\alpha_m x + \beta_m z)] \quad , \tag{4}$$

with $m = 0, \pm 1, \ldots,$ and $\alpha_m^2 + \beta_m^2 = 1$ ($\alpha_m = \sin\theta_m$ and $\beta_m = \cos\theta_m$ where θ_m is the scattered angle). Periodicity implies the grating equation, $\alpha_m = \alpha_o + n\Lambda$, where $\Lambda = \lambda/L$, λ is the incident wavelength and L the surface period. In addition, we assume Re $\beta_m \geqslant 0$ or Im $\beta_m \geqslant 0$ so that, with a (suppressed) time dependence $\exp(-i\omega t)$, the set $\{\phi_m\}$ consists either of waves which propagate in the positive z direction or exponentials which decay in that direction. RAYLEIGH [1] in 1903 assumed the set $\{\phi_m\}$ to be complete for $z \geqslant h(x)$. That is, any function, in particular ψ_{sc}, can be expanded in the $\{\phi_m\}$ alone everywhere in the wells of and on the surface, and the expansion coefficients determined from the boundary value problem. This has come to be known as the Rayleigh Hypothesis (RH).

LIPPMANN [2] and URETSKY [3] assumed that it was also necessary to use a set of downgoing waves in the wells, but did not investigate any specific region of validity of the RH. MILLAR [4] stated that: a necessary and sufficient condition for the RH to hold is that

$$\psi_{sc}(x,z) = \sum_{n=-\infty}^{\infty} A_n\phi_n(x,z), \tag{5}$$

is an analytic function of x and z in $V_1 + V_2$ (see Fig. 1a). That is, since (5) is a function of a real variable, ψ_{sc} can be expanded in a uniformly convergent power series. He as well as PETIT and CADILHAC [5] and VAN DEN BERG and FOKKEMA [6] showed a specific region of validity of the RH in terms of a bound on the surface slopes. We now study the convergence of (5) following the latter reference.

Assume first that h and ψ_i are both analytic. If either is not, the RH fails. Note that as $n \to \infty$, $\alpha_n \sim n\Lambda$ and $\beta_n \sim in\Lambda$. We only study the $n \geqslant 0$ part of (5). Conclusions for the remainder of the sum follow in a similar way. Equation (5) must converge for $z = -d$. This follows if

$$\liminf_{n\to\infty} |A_n|^{-1/n} \exp(-\Lambda kd) > 1 \ . \tag{6}$$

In addition, if s is the arc length on the boundary, and is assumed to be a complex variable, so that $x = f(s)$ and $z = g(s)$ on the boundary, then

$$\lim_{n\to\infty} |\phi_n(s)| = |w| = |\exp\{k\Lambda[i\,f(s) - g(s)]\}| \ , \tag{7}$$

where $\phi_n(s) = \exp\{ik[\alpha_n f(s) + \beta_n g(s)]\}$. With the boundary condition (3), (5) becomes

$$-\psi_i(s) = \sum_{n=-\infty}^{\infty} A_n\, \phi_n(s) \ . \tag{8}$$

Now, the convergence of the bounding series $v(w) = \sum_{n=-\infty}^{\infty} A_n w^n$ fails if the mapping $w \to v$ does not exist, i.e., if $\partial v/\partial w = \partial v/\partial s\, (\partial w/\partial s)^{-1}$ does not exist. The latter occurs if $\partial w/\partial s = 0$ and from (7) this implies

$$d/ds\, [i\,f(s) - g(s)] = 0 \ , \tag{9}$$

which is satisfied for some $s = s_p$. Hence, the radius of convergence of $v(w)$ is $w(s_p)$ so that

$$\liminf_{n\to\infty} |A_n|^{-1/n} = |w(s_p)| = |\exp\{k\Lambda[if(s_p) - g(s_p)]\}| \ . \tag{10}$$

Combining (6) and (10) implies convergence (i.e., RH holds) if

$$\mathrm{Re}[i\,f(s_p) - g(s_p) - d] > 0 \ . \tag{11}$$

If we return to representing the surface profile as a function of x, $z = h(x) = -d\zeta(x)$, then (9) and (11) imply

$$h' = i; \text{ and } \mathrm{Re}[ix_p + d\zeta(x_p) - d] > 0 \ . \tag{12}$$

The RH fails then for some $d = \bar{d}$ such that

$$\mathrm{Re}[ix_p + \bar{d}\zeta(x_p) - \bar{d}] = 0 \ , \tag{13}$$

where x_p follows from $h' = i$. If $\bar{d} > d$, the RH holds everywhere in the wells and on the surface, and the A_n coefficients can be found from (8). If $\bar{d} < d$, the RH expansion can be continued only up to $\bar{d}$, where it has a singularity. Note that this is not a singularity in the physics of the problem. It simply signifies that the mathematical expansion we have chosen is inadequate, and that, physically, standing waves occur in the wells.

As an example, consider the sinusoidal surface in Fig. 1a, $h(x) = -d/2\ [1 + \cos(2\pi x/L)] = -d\zeta(x)$. For this surface, $h' = i$ in (12) yields the transcendental equation $(\pi d/L) \sinh(2\pi \bar{x}_p/L) = 1$, where $x_p = i\,\bar{x}_p$. This is solved for $\bar{x}_p$. Then (13) yields $\bar{d} = \bar{x}_p[\zeta(i\bar{x}_p) - 1]^{-1}$. Simple algebra then shows that the RH fails if $\pi d/L > 0.448$ where $\pi d/L$ is the maximum slope of the surface. Note that this criterion is independent of wavenumber and incidence angle. It only depends on the geometric properties of the surface.

2. THIN COMB GRATING

Here we present a solvable example of scattering from a thin comb grating (see Fig. 1b) where the non-analyticity of the surface requires different field expansion methods in the surface wells. The velocity potential ψ satisfies (2) with the additional conditions: (a) $|\nabla\psi| = O\,(r^{-1/2})$ as an edge of the surface is approached radially. This edge condition [7] guarantees that the edges do not serve as additional sources of energy; (b) ψ and $\nabla\psi$ are continuous at $z = 0$; (c) $\psi = 0$ on the surface [8]; and (d) ψ_{sc} satisfies the radiation condition. In region A ($z \geqslant 0$), ψ_{sc} is rigorously expanded in Rayleigh eigenfunctions so that with plane wave incidence

$$\psi_A(x,z) = \exp[ik(\alpha_o x - \beta_o z)] + \sum_{n=-\infty}^{\infty} A_n\, \phi_n(x,z). \tag{14}$$

The solution in region B ($-d \leqslant z \leqslant 0$) is expanded in standing wave eigenfunctions which include both up- and down-going waves in the wells and satisfy condition (c),

$$\psi_B(x,z) = \sum_{j=1}^{\infty} B_j \sin(p_j kx) \sin[q_j k(z + d)] \ , \tag{15}$$

with $p_j = j\pi/kL$ and $p_j^2 + q_j^2 = 1$. Substituting these expansions into the continuity conditions (b) yields the sets of linear equations

$$\sum_{n=-\infty}^{\infty} A_n \left\{ \frac{1}{\beta_n - q_j} \pm \frac{\exp(2ikdq_j)}{\beta_n + q_j} \right\} - \left\{ \frac{1}{\beta_o + q_j} \pm \frac{\exp(2ikdq_j)}{\beta_o - q_j} \right\} = \begin{cases} \sim B_j , \\ 0 \; , \end{cases} \tag{16}$$

where the equations with the upper sign are proportional to B_j on the rhs, and the equations with the lower sign vanish.

These equations can be solved using the residue calculus method [9]. Consider closed contour (C) integrals of a meromorphic function $f(w)$ of the form

$$\int_C \frac{f(w)}{w - q_j}\, dw \pm \exp(2ikdq_j) \int_C \frac{f(w)}{w + q_j}\, dw \tag{17}$$

If $f(w)$ has the following properties:

(a') simple poles at β_n $(n = 0, \pm 1, \ldots)$ and $-\beta_0$,

(b') simple zeroes at $\bar{q}_j = q_j + \epsilon_j$ where the ϵ_j are determined numerically via the symmetry relation $f(q_i) = f(-q_j)\exp(2ikdq_j)$, and

(c') $f(w)$ vanishes algebraically as $|w| \to \infty$ such that the edge condition holds,

then, as the contour $C \to \infty$, (17) are zero. The resulting residue series reproduces (16) provided $A_n = RES[f(\beta_n)]$ and $B_j \sim f(q_j)$ where $RES[f]$ is the residue of the function f. The problem then becomes one of constructing $f(w)$, in particular its zeroes, using an iterative algorithm which follows from condition (b'). This can be done [8] and, in addition, we can prove the energy conservation relation (unitarity)

$$\sum_n R_n = \sum_n (\beta_n/\beta_o)|A_n|^2 = 1 \; , \tag{18}$$

where the summation is over those n values for which β_n is real (i.e. over upgoing waves which carry energy away from the surface), and the R_n are scattered power coefficients.

An example with three propagating orders and the Brewster angle effect (where $R_o = 0$) is illustrated in Fig. 2a. It is also possible to design gratings for which $R_o = 0$ for both TE and TH polarizations [10]. The smooth behavior of the R_n are resonance effects. Rayleigh or threshold effects are illustrated in Fig. 2b. This is the cusp-like behavior (observable in all R_n by (18) but most prominent in R_0) when, due to a parameter change, an evanescent wave propagating along the surface and decaying for $z > 0$ becomes a propagating order for $z > 0$ (and hence is illustrated in the figure) or vice versa (i.e., a zero of some β_n). Note that (18) is satisfied at any abscissa value in both figures.

Other examples of solvable problems for scattering from rectangularly corrugated surfaces are available [11], as well as applications to atomic beam scattering from crystals [12] radiation losses in an electron ring accelerator [13] and surface plasmons [14].

3. DETERMINISTIC ROUGH SURFACES

We next treat the scattering from the deterministic rough surface (in general, nonperiodic) illustrated in Fig. 1c. It separates two semi-infinite media of different density [15]. The Green's functions G_j in regions V_j $(j = 1, 2)$ satisfy

$$(\partial_m \partial_m + k_j^2) G_j(\mathbf{x}, \mathbf{x}'') = -\delta(\mathbf{x} - \mathbf{x}'') \; , \tag{19}$$

for $\mathbf{x} \in V_j$ where $\partial_m = (\partial/\partial x, \partial/\partial y, \partial/\partial z)$, and continuity conditions at the interface

$$G_1(\mathbf{x}_s, \mathbf{x}'') = \rho\, G_2(\mathbf{x}_s, \mathbf{x}'') \; , \tag{20}$$

$$N_1(\mathbf{x}_s, \mathbf{x}'') = N_2(\mathbf{x}_s, \mathbf{x}'') \; , \tag{21}$$

where

$$\rho = \rho_2/\rho_1 \, , \quad \mathbf{x}_s = (x, y, h(x, y)) = (x_\perp, h(x_\perp)) \; ,$$

$$N_j(\mathbf{x}_s, \mathbf{x}'') = n_m(x_\perp)\, \partial_m\, G_j(\mathbf{x}_s, \mathbf{x}'') \; , \tag{22}$$

and $n_m(x_\perp) = \delta_{m3} - \partial_{m\perp} h(x_\perp)$ is a vector in the direction of the surface normal. Equations (20) and (21) correspond to the continuity of pressure and normal velocity components, respectively.

Using the free-space Green's function

$$G^0(\mathbf{x}',\mathbf{x}'') = \exp[ik_1|\mathbf{x}'-\mathbf{x}''|]/4\pi|\mathbf{x}'-\mathbf{x}''| \tag{23}$$

and Green's theorem in V_1 (assume that the source point is also in V_1) yields

$$G_1(\mathbf{x}',\mathbf{x}'') = G^0(\mathbf{x}',\mathbf{x}'') - \int G^0(\mathbf{x}',\mathbf{x}_s)N_1(\mathbf{x}_s,\mathbf{x}'')dx_\perp - \int \partial'_m G^0(\mathbf{x}',\mathbf{x}_s)n_m(x_\perp)\, G_1(\mathbf{x}_s,\mathbf{x}'')dx_\perp \ , \tag{24}$$

a standard integral relation between the field and surface values of G_1 and its normal derivative on the surface. Green's theorem in V_2, with receiver point $\mathbf{x}' \in V_1$, yields

$$0 = \int G^0(\mathbf{x}',\mathbf{x}_s)N_2(\mathbf{x}_s,\mathbf{x}'')dx_\perp + \int \partial'_m G^0(\mathbf{x}',\mathbf{x}_s)n_m(x_\perp)\, G_2\,(\mathbf{x}_s,\mathbf{x}'')dx_\perp \ , \tag{25}$$

a nonlocal impedance-type boundary condition, also called an extended boundary condition [16] or an extinction coefficient [17]. Using the continuity conditions to combine (24) and (25), and taking the surface limit $\mathbf{x}' \rightarrow \mathbf{x}'_s$ yields

$$G^s(\mathbf{x}'_s,\mathbf{x}'') = G^0(\mathbf{x}'_s,\mathbf{x}'') - R\int P(\mathbf{x}'_s,\mathbf{x}_s)G^s(\mathbf{x}_s,\mathbf{x}'')dx_\perp \ , \tag{26}$$

where $G_1 = (2\rho/\rho + 1)G^s$, $R = (\rho - 1)/(\rho + 1)$ and the function P represents the principal value of the normal derivative of G^0.

Introducing the Fourier transform

$$G^s(\mathbf{x}',\mathbf{x}'') = (2\pi)^{-6}\iint G(\mathbf{k}',\mathbf{k}'')\exp[i\mathbf{k}'.\mathbf{x}' - i\mathbf{k}''.\mathbf{x}'']d\mathbf{k}'d\mathbf{k}''$$

and those of G^0 and P in (26), setting the integrand equal to zero (only a sufficient condition for this surface-restricted Fourier transform), and defining the singularity-free scattered function Γ via

$$G(\mathbf{k}',\mathbf{k}'') = (2\pi)^3 G^0(k')\{\delta(\mathbf{k}'-\mathbf{k}'') + \Gamma(\mathbf{k}',\mathbf{k}'')G^0(k'')\} \ , \tag{27}$$

yields a Lippmann-Schwinger integral equation [18] with a noncentral and complex "potential" term for Γ,

$$\Gamma(\mathbf{k}',\mathbf{k}'') = V(\mathbf{k}',\mathbf{k}'')A(\mathbf{k}'-\mathbf{k}'') + \int V(\mathbf{k}',\mathbf{k})A(\mathbf{k}'-\mathbf{k})G^0(k)\Gamma(\mathbf{k},\mathbf{k}'')d\mathbf{k} \ , \tag{28}$$

where the "potential" term VA contains a (kinematical) vertex function V

$$V(\mathbf{k},\mathbf{k}'') = \frac{-2i}{(2\pi)^3}R\left\{k_{m\perp} + \delta_{m3}P\left[\frac{k_1^2 - k_\perp^2}{k_z}\right]\right\}\frac{k_m - k''_m}{k_z - k''_z} \ , \tag{29}$$

with P again representing the principal value, and the phase-modulation amplitude spectrum A,

$$A(\mathbf{k}) = \int \exp[-ik_\perp \cdot x_\perp - ik_z h(x_\perp)]dx_\perp \ , \tag{30}$$

expressing the dynamical properties of the surface interaction. Most importantly, $\Gamma(\mathbf{k}',\mathbf{k}'')$ for $k'_z = \sqrt{k_1^2 - k_\perp'^2} = -k''_z$ is proportional to the T-matrix for the scattering problem.

There are several limiting cases of the above formalism. For $\rho = \infty$, the hard-boundary-value problem follows [19]. Corresponding elastic [20] (torsion-free boundary) and electromagnetic [21] (TH polarization) problems have been developed. For $\rho = 0$, the soft-surface results follow. For $\rho = 1$, $R = 0$ and the free-field result follows. For $h = 0$, the integral term in (28) vanishes and the usual flat-surface results ensue [22]. Also, in one dimension, if $h(x)$ is sinusoidal, A is proportional to a Bessel function, and this is just the Rayleigh amplitude for scattering from a sinusoid [23].

We now simplify this three-dimensional integral equation by assuming $h(x_\perp)$ to be a random variable.

4. RANDOM SURFACE

Now let $h(x_\perp)$ be a Gaussian-distributed random variable and, from (28), find the integral equation satisfied by the coherent part of Γ, $<\Gamma>$. The bracket signifies the ensemble average [24]. We briefly outline the method. Attach to each term in (28) a Feynman-diagram notation illustrated in Fig. 3. The ensemble average of Γ considered as a Born expansion of the integral equation (see Fig. 3f) involves the calculation of ensemble averages of products of A-functions. For a centered Gaussian probability distribution function (PDF) of surface heights $h(x_\perp)$, and homogeneous statistics, the correlation function is

$$\gamma(x_{i\perp} - x_{j\perp}) = \langle h(x_{i\perp})\, h(x_{j\perp})\rangle \ , \tag{31}$$

which is translationally invariant. For Gaussian statistics, all moments can then be expressed in terms of this two-point moment. The ensemble average of products of A's requires a cluster decomposition [25], and is best illustrated with examples. We have that

$$<A(\mathbf{k}_1)> = A_1(\mathbf{k}_1) = (2\pi)^2\delta(k_{1\perp})E(k_{1z}) \ , \tag{32}$$

where

$$E(k_z) = \exp[-1/2\,\gamma(0)\,k_z^2] \ , \tag{33}$$

and $\gamma(0) = \sigma^2$, where σ is the rms height of the surface. For two A functions we have that

$$<A(\mathbf{k}_1)A(\mathbf{k}_2)> = A_1(\mathbf{k}_1)A_1(\mathbf{k}_2) + A_2(\mathbf{k}_1, \mathbf{k}_2) \ , \tag{34}$$

$$A_2(\mathbf{k}_1, \mathbf{k}_2) = (2\pi)^2\delta(k_{1\perp} + k_{2\perp})E(k_{1z})E(k_{2z})R_2(\mathbf{k}_1, \mathbf{k}_2) \ , \tag{35}$$

and

$$R_2(\mathbf{k}_1, \mathbf{k}_2) = \int \exp(-ik_{1\perp}\cdot\rho_\perp)\cdot\{\exp[-\gamma(\rho_\perp)k_{1z}k_{2z}] - 1\}\,d\rho_\perp \ . \tag{36}$$

Note that if we expand the bracketed exponential in (36) in powers of the surface height, R_2 is proportional to the Fourier transform of the correlation function, i.e. to the spectral function of the surface, an important experimental measure of surface randomness. The general term A_n can also be calculated [19] (also see Fig. 4a). The result of the ensemble average of Γ is then resummed (partial summation method) to yield the Dyson equation

$$<\Gamma(\mathbf{k}', \mathbf{k}'')> = M(\mathbf{k}', \mathbf{k}'') + \int M(\mathbf{k}', \mathbf{k})G^0(k)<\Gamma(\mathbf{k}, \mathbf{k}'')>d\mathbf{k} \ , \tag{37}$$

where

$$M(\mathbf{k}', \mathbf{k}'') = \sum_{j=1}^{\infty} M_j(\mathbf{k}', \mathbf{k}'') \ , \tag{38}$$

is called the mass operator [24]. Its first two terms are illustrated in Fig. 4b. All its terms correspond to connected diagrams, i.e., those which cannot be decomposed into products of simpler diagrams.

The δ-function factorization in (32) and (35) is a general result so we can write

$$M_j(\mathbf{k}', \mathbf{k}'') = \delta(k'_\perp - k''_\perp)\ u_j\ (k'_z, k''_z) \tag{39}$$

and

$$<\Gamma(\mathbf{k}', \mathbf{k}'')> = \delta(k'_\perp - k''_\perp)\ T\ (k'_z, k''_z), \tag{40}$$

where the $k_\perp'$ dependence in u_j and T is suppressed. Scaling the resulting T-matrix [26] and adding subscripts to T enables us to write the class of one-dimensional singular integral equations

$$T_{mn}(\xi', \xi'') = V_m(\xi', \xi'') + \int V_n(\xi', \xi)K(\xi)T_{mn}(\xi, \xi'')d\xi \tag{41}$$

where

$$V_m(\xi', \xi') = \sum_{j=1}^{m} u_j(\xi', \xi'') \ , \tag{42}$$

and

$$K(\xi) = \lim_{\epsilon\to 0^+} (\pi i)^{-1}R(\xi^2 - 1 - i\epsilon)^{-1}P(1/\xi) \ . \tag{43}$$

Here, m and n refer to the number of terms in the Born or kernel parts of (41) respectively and are not summed over ($n = 0$ means no integral term at all). For example, for plane wave incidence at angle θ

$$u_1(\xi', \xi'') = \exp[-1/2\ \Sigma^2(\xi' - \xi'')^2] \ , \tag{44}$$

where $\Sigma = k_1\sigma \cos\theta_i$ is the Rayleigh roughness parameter. This corresponds to T_{10} in (41), and is just the result found by Ament [27]. Higher order terms can also be computed.

The coherent specular intensity is then given by

$$I_c(k_\perp) = R^2|T_{mn}(1, -1)|^2\delta(k_\perp - k_\perp^{in}) \ , \tag{45}$$

for plane wave incidence. To compare some of the above results with experiment, we first note that experimental PDF's are nearly but not exactly Gaussian [28,29]. In Fig. 5, the experimental results (x) from CLAY, MEDWIN, and WRIGHT [28] are compared with several theoretical models: (a), $|T|^2 = |T_{10}(1, -1)|^2 = \exp(-4\Sigma^2)$, the Ament result which is single scatter from a surface distributed with a Gaussian PDF (essentially the result is a Fourier transform of a Gaussian); (b), T is proportional to the Fourier transform of the experimentally measured PDF of the surface; (c), T is proportional to the Fourier transform of the experimental PDF modified using shadowing theory [28], and contains the free parameter $\gamma''(0)$ which is proportional to the correlation function of surface slopes at zero separation. It cannot be measured directly. It must be extrapolated from separated measurements. Its value such the T fits the data is about two orders of magnitude different from the extrapolated value; (d), $|T|^2 = |T_{11}(1,-1)|^2$, our [19,26] multiple scatter solution for a Gaussian PDF which essentially assumes an infinite correlation distance ($u_2 = 0$) and can thus be related to scattering from a Gaussian distribution of flat surfaces, i.e., to a phase screen model [30]. It has no free parameters; (e), $|T|^2 = I_0^2(2\Sigma^2)\exp(-4\Sigma^2)$ where I_0 is the modified Bessel function. This is the single scatter eikonal approximation for a Gaussian PDF derived elsewhere [31].

Some conclusions are possible. First, single scatter is very sensitive to the form of the PDF. This is seen by comparing (a) and (b) (in Fig. 5) which differ by only the Gaussian (theoretical) and slightly non-Gaussian (experimental) PDF's. Second, the shadowed result (c) looks good but is really a curve fit. Third, the eikonal result (e) consistently predicts more coherent return than is measured, and, fourth, our multiple scattering result (d) works well for large roughness whereas the single scatter theories fail for one reason or another.

5. SUMMARY

We have, thus, demonstrated the region of validity of the Rayleigh hypothesis for analytic periodic surfaces, and given an example of scattering from a surface which violates it. Numerical results for the latter illustrate some of the many grating effects which occur, and their similarity to quantum scattering results corresponding to resonance and threshold effects. The T-matrix for scattering from an arbitrary surface was shown to satisfy a Lippmann-Schwinger equation where the kinematical and dynamical properties of the surface interaction formed the non-central and complex "potential". The restriction to Gaussian distributed surfaces provided a multitude of mathematical techniques as well as interesting conclusions when compared with experimental results on the sensitivity of single scattering and the necessity of multiple scattering for coherent rough surface scattering.

ACKNOWLEDGMENTS

This work was supported by the Naval Research Laboratory and the Office of Naval Research.

FOOTNOTES AND REFERENCES

[1] J.W.S. Rayleigh: *The Theory of Sound* (Dover, New York, 1945), Vol. 2, pp. 89-96

[2] B.A. Lippman: "Note on the theory of gratings," J. Opt. Soc. Am. **43**, 408 (1953)

[3] J.L. Uretsky: "The scattering of plane waves from periodic surfaces," Ann. Phys. (N.Y.) **33**, 400-427 (1965)

[4] R.F. Millar: "On the Rayleigh assumption in scattering by a periodic surface," Proc. Camb. Phil. Soc. **65**, 773-791 (1969); "On the Rayleigh assumption in scattering by a periodic surface II," Proc. Camb. Phil. Soc. **69**, 217-225 (1971); "The Rayleigh hypothesis and a related least-squares solution to scattering problems for periodic surfaces and other scatterers," Radio Sci. **8**, 785-796 (1973)

[5] R. Petit, M. Cadilhac: "Sur la diffraction d'une onde plane par un réseau infiniment conducteur," C.R. Acad. Sci. Paris **262B**, 468-471 (1966)

[6] P.M. van den Berg, J.T. Fokkema: "The Rayleigh hypothesis in the theory of reflection by a grating," J. Opt. Soc. Am. **69**, 27-31 (1979)

[7] C.J. Bouwkamp: "A note on singularities occurring at sharp edges in electromagnetic diffraction theory," Physica **12**, 467-474 (1946)

[8] J.A. DeSanto: "Scattering from a periodic corrugated structure: thin comb with soft boundaries," J. Math. Phys. **12**, 1913-1923 (1971) for the soft (Dirichlet, TE polarization) case in the text. The hard (Neumann, TH) case is in "Scattering from a periodic corrugated structure II: thin comb with hard boundaries," ibid. **13**, 336-341 (1972)

[9] R. Mittra, S.W. Lee: *Analytical Techniques in the Theory of Guided Waves* (Macmillan, New York, 1971)

[10] E.V. Jull, J.W. Heath, G.R. Ebbeson: "Gratings that diffract all incident energy," J. Opt. Soc. Am. **67**, 557-560 (1977)

[11] J.A. DeSanto: "Scattering from a periodic corrugated surface: semi-infinite alternately filled plates," J. Acoust. Soc. Am. **53**, 719-734 (1973) and "Scattering from a periodic corrugated surface: finite-depth alternately filled plates," ibid. **56**, 1336-1341 (1974)

[12] N. García, N. Cabrera: "New method for solving the scattering of waves from a periodic hard surface: Solutions and numerical comparisons with the various formalisms," Phys. Rev. B **18**, 576-589 (1978); R.I.

Masel, R.P. Merrill, W. H. Miller: "Atomic scattering from a sinusoidal hardwall: Comparison of approximate methods with exact quantum results," Lawrence Berkeley Laboratory Technical Report LBL-4969 (April, 1976); G. Boato, P. Cantini, V. Garibaldi, A.C. Levi, L. Mattera, R. Spadacini, G.E. Tommei: "Diffraction and rainbow in atom-surface scattering," J. Phys. C: Solid State Phys. **6** L 394-398 (1973)

[13] R.D. Hazeltine, M.N. Rosenbluth, A.M. Sessler: "Diffraction radiation by a line charge moving past a comb: a model of radiation losses in an electron ring accelerator," J. Math. Phys. **12**, 502-514 (1971)

[14] J.J. Cowan, E.T. Arakawa: "Diffraction of surface plasmons in dielectric-metal coatings on concave diffraction gratings," Z. Phys. **235**, 97-109 (1970)

[15] K.M. Mitzner: "Acoustic scattering from an Interface between media of greatly different density," J. Math. Phys. **7**, 2053-2060 (1966)

[16] P.C. Waterman: "Scattering by periodic surfaces," J. Acoust. Soc. Am. **57**, 791-802 (1975)

[17] D.N. Pattanayak, E. Wolf: "Resonance states as solutions of the Schroedinger equation with a nonlocal boundary condition," Phys. Rev. D **13**, 2287-2290 (1976)

[18] R.G. Newton: *Scattering Theory of Waves and Particles* (McGraw-Hill, New York, 1966)

[19] G.G. Zipfel, J.A. DeSanto: "Scattering of a scalar wave from a random rough surface: a diagrammatic approach," J. Math. Phys. **13**, 1903-1911 (1972)

[20] J.A. DeSanto: "Scattering from a random rough surface: diagram methods for elastic media," J. Math. Phys. **14**, 1566-1573 (1973)

[21] J.A. DeSanto: "Green's function for electromagnetic scattering from a random rough surface," J. Math. Phys. **15**, 283-288 (1974)

[22] L.M. Brekhovskikh: *Waves in Layered Media* (Academic, New York, 1960)

[23] J.A. DeSanto: "Scattering from a sinusoid: derivation of linear equations for the field amplitudes," J. Acoust. Soc. Am. **57**, 1195-1197 (1975)

[24] Analogous techniques are used in random-volume scattering theory. See V. Frisch: "Wave propagation in random media," in *Probabilistic Methods in Applied Mathematics*, ed. by A.T. Bharucha-Reid (Academic, New York, 1968), Vol. 1

[25] K. Huang: *Statistical Mechanics* (Wiley, New York, 1963)

[26] J.A. DeSanto, O. Shisha: "Numerical solution of a singular integral equation in random rough surface scattering theory," J. Comp. Phys. **15**, 286-292 (1974)

[27] W.S. Ament: "Forward- and back-scattering from certain rough surfaces," Trans. IRE AP-4, 369-373 (1956); P. Beckmann, A. Spizzichino: *The Scattering of Electromagnetic Waves from Rough Surfaces* (Pergamon, New York, 1963)

[28] C.S. Clay, H. Medwin, W.M. Wright: "Specularly scattered sound and the probability density function of a rough surface," J. Acoust. Soc. Am. **53**, 1677-1682 (1973)

[29] J.G. Zornig: "Physical modeling of underwater acoustics," in *Ocean Acoustics*, Topics in Current Physics, vol. 8, ed. by J.A. DeSanto (Springer, New York, 1979)

[30] E. Jakeman, P.N. Pusey: "Non-Gaussian fluctuations in electromagnetic radiation scattered by a random phase screen I. Theory," J. Phys. A: Math. Gen. **8**, 369-381 (1973)

[31] R.M. Brown, A.R. Miller: "Geometric optics theory for coherent scattering of microwaves from the ocean"; Naval Research Laboratory Report 7705 (1974)

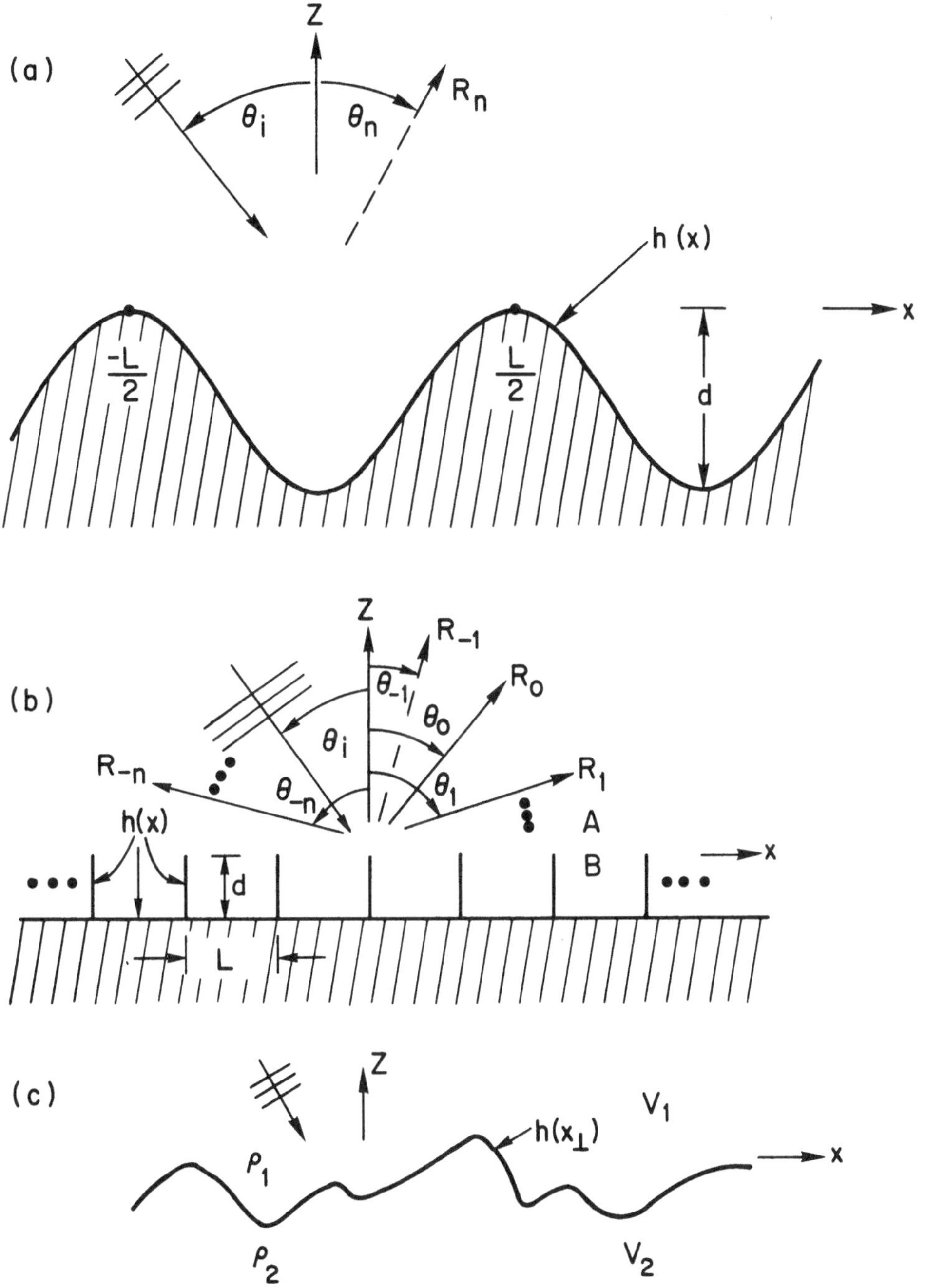

Fig. 1. Plane wave scattering from periodic surfaces (a) (sinusoid) and (b) (thin comb grating), and from an arbitrary rough interface (c). Surfaces (a) and (b) have period L, depth d, and are one-dimensional ($z = h(x)$). The θ_n are the direction angles of the Bragg orders, θ_i the incidence angle, and $R_n = (\beta_n/\beta_o)|A_n|^2$ where $\beta_n = \cos\theta_n$ and A_n is the scattering amplitude in the θ_n direction. The surface in (c) is a one-dimensional slice of a two-dimensional surface ($z = h(x_\perp)$, $x_\perp = (x, y)$) and separates two media V_j of different density ρ_j ($j = 1, 2$).

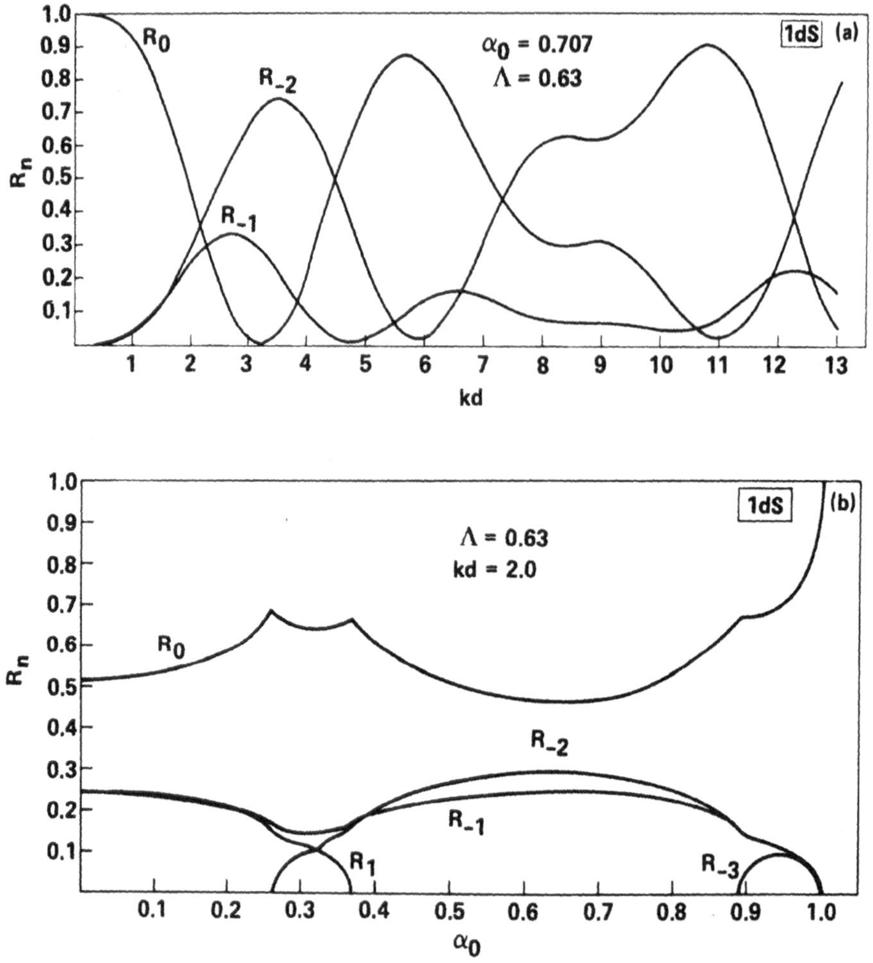

Fig. 2. Examples of plane wave scattering from the surface illustrated in Fig. 1(b). Part (a) illustrates three propagating orders ($\beta_n > 0$, $n = 0, -1, -2$) for 45° incidence ($\alpha_0 = \sin \theta_i$) and a wavelength shorter than the surface period ($\Lambda = \lambda / L$). The zero of the specular coefficient R_0 illustrates the Brewster angle effect. Part (b) illustrates Rayleigh or threshold effects as the incidence angle varies. These effects occur when some β_n passes through zero so that its associated scattering order changes from one that propagates away from the surface to a surface waver or vice versa.

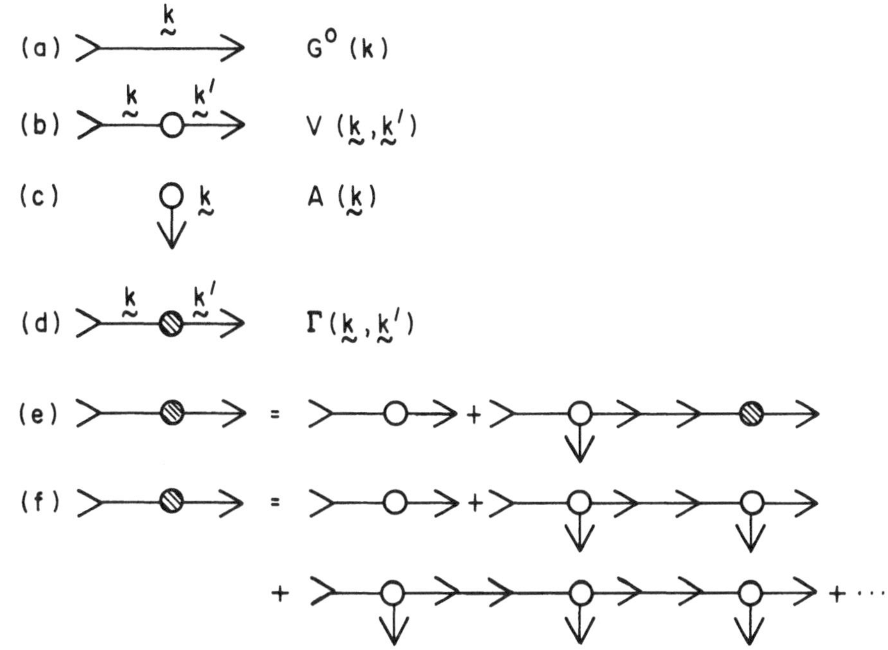

Fig. 3 — Diagram notation associated with scattering from an arbitrary surface. Part (a) is the propagator, (b) the vertex, (c) the interaction term, (d) the full scattering amplitude, (e) the representation of the integral equation (28), and (f) the representation of the Born-series expansion of (28).

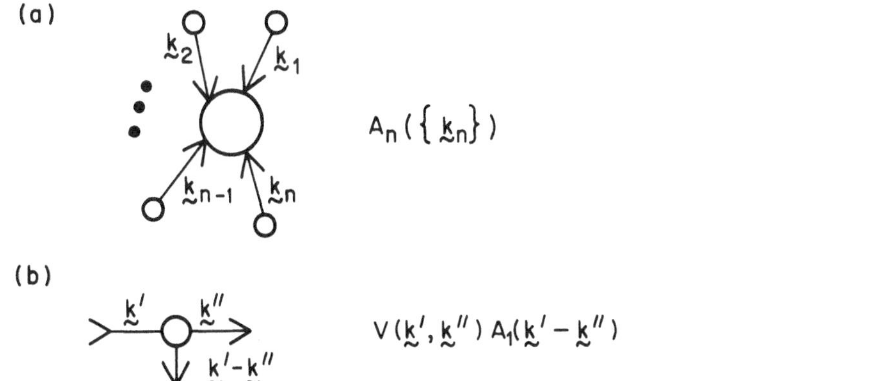

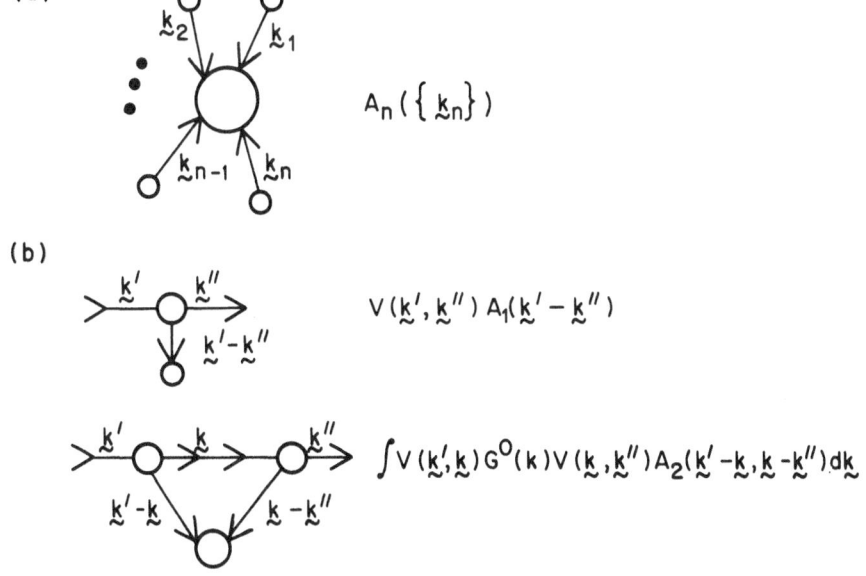

Fig. 4 — Statistical diagrams associated with scattering from a Gaussian-distributed random surface. Part (a) illustrates the connected multipoint interaction term A_n resulting from a cluster decomposition, with $\mathbf{k}_n$ the momentum lines. Part (b) illustrates the two lowest order connected diagrams in the Born expansion and their functional correspondence.

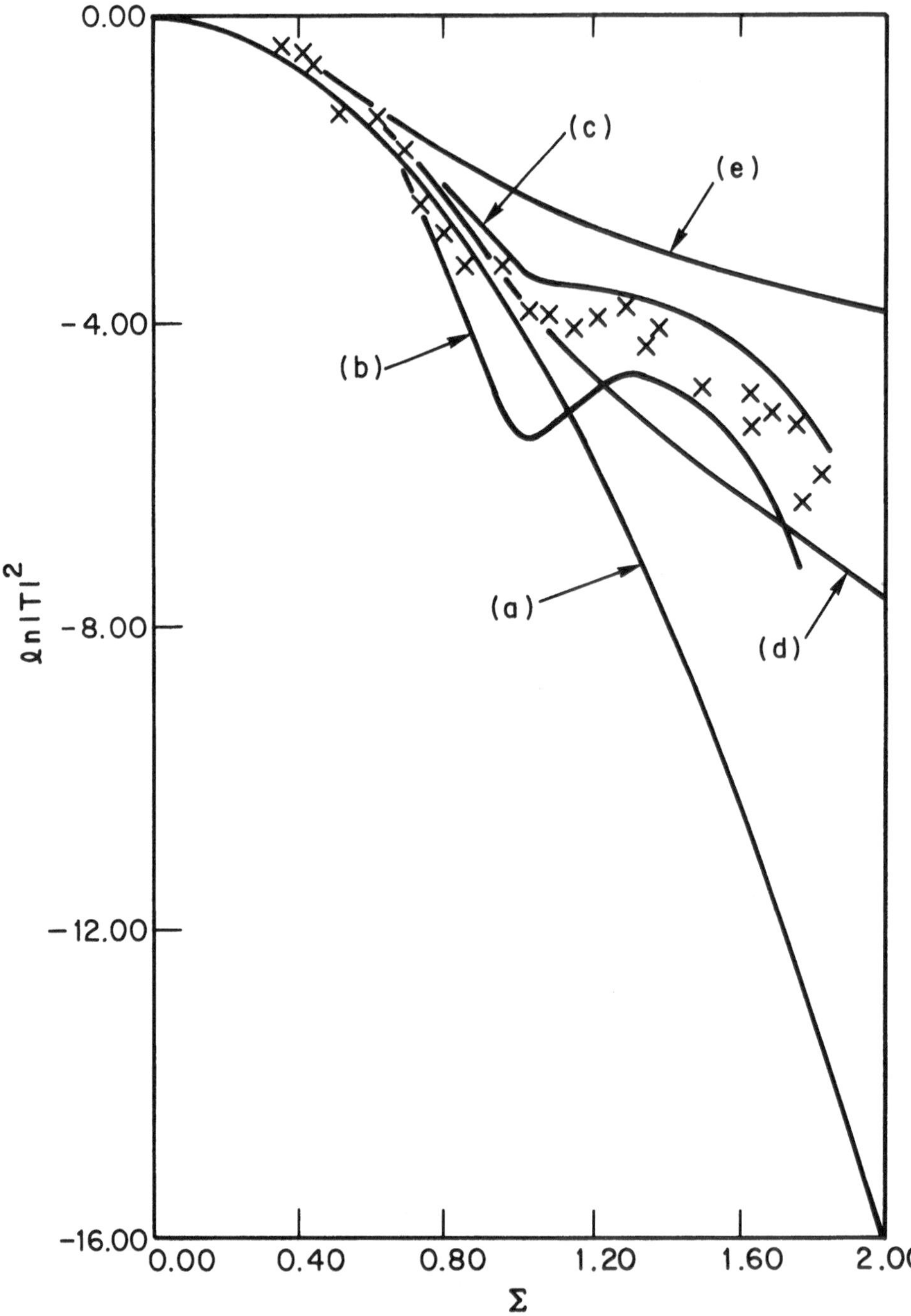

Fig. 5. Coherent specular intensity versus the Rayleigh roughness parameter Σ for scattering from a (theoretical) Gaussian distributed random surface. The (×) are data from CLAY, MEDWIN, and WRIGHT [28], (a) is the Ament single scatter model, (b) and (c) are single scatter Fourier transforms of the experimental surface probability distribution function with no shadowing (b) and shadowing (c) (see the text). Part (d) is our multiple scattering result, and (e) is the eikonal approximation.

APPLICATION OF A TECHNIQUE OF FRANKLIN AND FRIEDMAN TO SOME PROBLEMS IN ACOUSTICS

D. C. Stickler

Courant Institute of Mathematical Sciences
New York University, New York, New York 10012

INTRODUCTION

In their 1957 paper, FRANKLIN and FRIEDMAN [1] introduced a method for obtaining an improved asymptotic expansion for Laplace transforms, which as they showed is sometimes also convergent. In their introduction, they listed three potential disadvantages of the usual Watson result [2]. First, the resulting series is sometimes divergent; second, it is always singular as the transform variable approaches zero; and third, the first term alone is often virtually useless for numerical calculations, except for very large values of the transform variable. They then introduced an improved asymptotic expansion which is sometimes convergent and whose first term is more accurate than the first term in the Watson expansion. More precisely, the remainder after one term of the Franklin-Friedman expansion is one order of magnitude smaller than the first term in the Watson expansion. However, as a penalty, the first term of the Franklin-Friedman expansion is more complicated. The purpose of this paper to motivate the ansatz chosen by Franklin and Friedman, to describe some extensions to other integrals, and to examine some examples. These examples are chosen to illustrate the improved accuracy and sometimes uniform results this method yields.

Recall that the first term plus remainder in the Watson expansion of the Laplace integral,

$$I = \int_0^\infty f(t) t^{\lambda-1} e^{-xt} dt, \lambda > 0, \; x \to \infty, \tag{1}$$

with f sufficiently smooth, is given by

$$I = f(0) \frac{\Gamma(\lambda)}{x^\lambda} + O(x^{-\lambda-1}). \tag{2}$$

FRANKLIN and FRIEDMAN [1] introduced without motivation the first term,

$$f(\lambda/x) \frac{\Gamma(\lambda)}{x^\lambda}, \tag{3}$$

and showed that the remainder is one order of magnitude smaller, namely,

$$I = f\left(\frac{\lambda}{x}\right) \frac{\Gamma(\lambda)}{x^\lambda} + O(x^{-\lambda-2}). \tag{4}$$

There are two rather straightforward ways to justify the choice in (3) and, once this is done, the extension of their ideas to other kernels can be obtained.

The first is to expand $f(t)$ as follows:

$$f(t) = f(\alpha) + h(t)(t-\alpha), \tag{5}$$

where $h(t) = \dfrac{f(t) - f(\alpha)}{t - \alpha}$ and α is an undetermined parameter. Note that because $f(t)$ is smooth so is $h(t)$. The parameter α is then chosen so that the remainder

$$R = \int_0^\infty t^{\lambda-1} h(t)(t-\alpha) e^{-xt} dt \tag{6}$$

is smaller than the Watson remainder. This is the case if

$$\int_0^\infty (t-\alpha) t^{\lambda-1} e^{-xt} dt = 0 \tag{7}$$

or

$$\alpha = \lambda/x. \tag{8}$$

A second way to motivate their *Ansatz* is to write out the first two terms in the Watson expansion,

$$I = f(0)\frac{\Gamma(\lambda)}{x^{\lambda}} + f'(0)\frac{\Gamma(\lambda+1)}{x^{\lambda+1}} + O(x^{-\lambda-2}), \tag{9}$$

which to $O(x^{-\lambda-2})$ can be rewritten

$$I = \frac{\Gamma(\lambda)}{x^{\lambda}}\left[f(0) + \frac{\lambda}{x} f'(0)\right] + O(x^{-\lambda-2}) = \frac{\Gamma(\lambda)}{x^{\lambda}} f(\lambda/x) + O(x^{-\lambda-2}). \tag{10}$$

It is important to point out that $f(\lambda/x)$ contains terms of all orders in $(1/x)$ and that while it cannot be guaranteed that the approximation is not better than the remainder would suggest, this is often the case in practice.

With this insight, it is possible to extend their ideas to other kernels and some of these results are considered below.

1. FOURIER INTEGRALS

The first extension is to integrals of the form

$$I = \int_{\alpha}^{\beta} f(t)(t-\alpha)^{\lambda-1}(\beta-t)^{\mu-1}e^{ixt}dt, \quad \lambda,\mu > 0, \tag{11}$$

where $f(t)$ is sufficiently smooth in some complex neighborhood of both α and β. Erdélyi has derived an asymptotic expansion for I, and the first term plus reminder is given by [3]

$$I = f(\alpha)\Gamma(\lambda)e^{i\lambda\pi/2}(\beta-\alpha)^{\mu-1}\frac{e^{ix\alpha}}{x^{\lambda}} + f(\beta)\Gamma(\mu)e^{-i\mu\pi/2}(\beta-\alpha)^{\lambda-1}\frac{e^{ix\beta}}{x^{\mu}} + O(x^{-\lambda-1} + x^{-\mu-1}). \tag{12}$$

Application of the ideas presented in the last section yield the following expansion, which was given by STICKLER [4],

$$I = f(\alpha + i\lambda/x)\Gamma(\lambda)e^{i\lambda\pi/2}(\beta - \alpha - i\lambda/x)^{\mu-1}\frac{e^{ix\beta}}{x^{\lambda}}$$

$$+ f(\beta + i\mu/x)\Gamma(\mu)e^{-i\mu\pi/2}(\beta - \alpha + i\mu/x)^{\lambda-1}\frac{e^{ix\beta}}{x^{\mu}} + O(x^{-\lambda-2}) + O(x^{-\mu-2}). \tag{13}$$

Note that the coefficients are more complicated and that the remainder is one order of magnitude smaller than the Erdélyi expansion in (12).

In order to obtain a feeling for the computational advantage this method provides, this result was applied to the Poisson integral representation for $J_\nu(x)$ and compared for $\nu = 9$ and $x \geqslant 9$ with both the exact and usual expansions. The Poisson integral representation is given by [5]

$$J_\nu(x) = \frac{(x/2)^\nu}{\sqrt{\pi}\Gamma(\nu+1/2)}\int_{-1}^{+1}(t+1)^{\nu-1/2}(1-t)^{\nu-1/2}e^{ixt}dt. \tag{14}$$

Application of (13) yields

$$J_\nu(x) = \sqrt{\frac{2}{\pi x}}A(x)\cos(x - (\nu+1/2)\frac{\pi}{2} + \Phi(x)) + O(x^{-\nu-1/2}), \tag{15}$$

where

$$A(x) = [1 + ((\nu+1/2)/2x)^2]^{(\nu-1/2)/2}, \tag{16}$$

$$\Phi(x) = (\nu - 1/2)\tan^{-1}((\nu+1/2)/2x). \tag{17}$$

Note that if $A(x)$ were replaced by unity and $\Phi(x)$ by zero, (15) would be the "usual" first term large-argument approximation to $J_\nu(x)$, but with the remainder estimate replaced by $O(x^{-3/2})$. Figure 1 compares the exact, the "usual", and the improved expansion of (15) for $9 \leqslant x \leqslant 20$. Note that the "usual" expansion is essentially "out of phase" with the exact solution in this range while (15) is significantly better. A calculation of the first two terms in the usual expansion offers little improvement. Using the expansion in [5], p. 85, Eq. 3, for $\nu = 9$ and $x = 50$, the difference between the "usual" expansion and that of (15) is still large. Equation (15) yields $J_9(50) = -0.02825$, and the usual expansion yields $J_9(50) = -0.097927$.

Return now to the Erdélyi expansion of (11), given in (12), and note that it is not in general valid as $\beta \to \alpha$. This is because the derivation of (12) assumed that α and β were isolated. If this assumption is not made and $f(t)$ is expanded in a Lagrange expansion about α and β,

$$f(t) = f(\alpha) + \frac{f(\beta) - f(\alpha)}{\beta - \alpha}(t - \alpha) + R, \tag{18}$$

then the first term of a uniform expansion is obtained. Note that the remainder R vanishes at both $t = \alpha$ and $t = \beta$. The first term is given by

$$I = e^{ix\alpha}(\beta - \alpha)^{\lambda+\mu-1}\frac{\Gamma(\mu)\Gamma(\lambda)}{\Gamma(\lambda+\mu)}[f(\alpha)\Phi(\lambda;\lambda + \mu;ix(\beta - \alpha)) + (f(\beta) - f(\alpha))\Phi'(\lambda;\lambda + \mu;ix(\beta - \alpha))] + O(x^{-1}), \tag{19}$$

where Φ is the confluent hypergeometric function [6] and Φ' indicates the derivative with respect to the third argument. This result is believed to be new.

For $\beta \neq \alpha$, and as $x \to \infty$, the uniform expansion in (19) reduces to (12), the Erdélyi result. This uniform expansion has the proper behaviour as $\beta \to \alpha$, if $x \to \infty$ sufficiently rapidly. If the idea introduced by Franklin and Friedman is used on this uniform expansion, then the first term of the expansion becomes

$$I = e^{ix\alpha}(\beta - \alpha)^{\lambda+\mu-1}\frac{\Gamma(\mu)\Gamma(\lambda)}{\Gamma(\lambda+\mu)}[f(\alpha_1)\Phi(\lambda,\lambda + \mu,ix(\beta - \alpha)) + (f(\beta_1) - f(\alpha_1))\Phi'(\lambda,\lambda + \mu,ix(\beta - \alpha))] + O(x^{-2}), \tag{20}$$

where

$$\alpha_1 = \alpha + \frac{i\lambda}{x}\left[1 - \frac{f(\beta) - f(\alpha)}{(\beta - \alpha)f'(\alpha)}\right], \quad f'(\alpha) \neq 0, \tag{21}$$

$$\beta_1 = \beta - \frac{i\mu}{x}\left[1 - \frac{f(\beta) - f(\alpha)}{(\beta - \alpha)f'(\beta)}\right], \quad f'(\beta) \neq 0. \tag{22}$$

Note that in this expansion the argument of $f(\cdot)$ is not only complex but depends on $f'(x)$ at α and β. However, as before, the remainder is one order of magnitude smaller than in (19).

2. OTHER KERNELS

It is apparent that these ideas can be extended to other transformation, such as the sine and cosine transforms, and stationary phase integrals. In addition, it has been applied to the finite Hankel transform defined by

$$I = \int_0^\beta (\beta - t)^{\mu-1} z^\lambda f(t) H_0^{(1)}(xt)\,dt, \quad \lambda,\mu > 0, \tag{23}$$

and the K_0 transform,

$$I = \int_0^\infty z^{\lambda-1} f(t) K_0(xt)\,dt. \tag{24}$$

In fact, it appears that this method can be used whenever $f(t)$ is sufficiently smooth and the moments exist.

3. SOMMERFELD MODEL

In the remainder of this paper, two examples from acoustics are considered, both of which are classical problems. The first is called the Sommerfeld problem and is described as follows. An upper-half space ($z > 0$) is filled with an isospeed, constant density medium $[c_1,\rho_1]$ and the lower half-space ($z < 0$) is filled with a second isospeed, constant density medium $[c_2,\rho_2]$. The problem is to find the response to a point harmonic source.

Let a point harmonic source be placed in the upper medium ($z > 0$) and assume that the speed in the upper medium is slower than that in the lower medium $[c_1 < c_2]$. It is well known that in this situation a lateral or head wave is present and takes the form

$$p_{lat} = \frac{i}{4\pi}\frac{2n}{k_1 m(1 - n^2)}\frac{e^{i\Phi}}{\sqrt{r}L_2^{3/2}}, \quad \theta > \theta_c, \tag{25}$$

where $n = c_1/c_2$, $m = \rho_2/\rho_1$, $k_j = \omega/c_j$, where ω is the angular source frequency, r is the radial distance between source and receiver, θ is measured from the z axis with vertex at $-z_0$, and Φ and L_2 are described below. The phase Φ is given by

$$\Phi = k_1L_1 + k_2L_2 + k_1L_3, \tag{26}$$

where the paths L_1, L_2 and L_3 are shown in Fig. 2, i.e., energy leaves the source z_0 at the critical angle $\theta_c = \sin^{-1}(c_1/c_2)$, travels along the interface at speed c_2, and leaves the interface at the critical angle to reach the observation point z. As the range r — the horizontal distance between the source and receiver — decreases, the distance L decreases and as the observation point approaches the critical ray — shown as a dashed line in Fig. 2 — the lateral wave becomes singular. In addition, as $n \rightarrow 1$, i.e., as the sound speed in the two media become equal, the lateral wave also becomes unbounded. It is these two difficulties that are discussed below.

For the first part of this example, the response is determined for both the source and observation point on the interface $z = 0$. The response in this case is given by two integrals [4]:

$$p(r) = \frac{ik_1}{4\pi}\int_0^\infty \frac{2m^2q^2}{(m^2-1)q^2+(1-n^2)} H_0^{(1)}(k_1r\sqrt{1-q^2})\,dq$$
$$-\frac{ik_2}{4\pi}\int_0^\infty \frac{2mq^2}{(m^2-1)q^2+m^2(1-n^2)/n^2} H_0^{(1)}(k_2r\sqrt{1-q^2})\,dq, \tag{27}$$

where $H_0^{(1)}(x)$ is the Hankel function of the first kind.

The asymptotic expansion of the integrals in (27) can be obtained as a special case of the asymptotic expansion of

$$I = \int_0^\infty f(q)\cos(zq)H_0^{(1)}(\rho\sqrt{1-q^2})\,dq, \tag{28}$$

where $f(q)$ is smooth and even in q.

The expansion is obtained by writing

$$f(q) = f(\alpha) + \frac{f(q)-f(\alpha)}{q^2-\alpha^2}(q^2-\alpha^2) \tag{29}$$

and asking for an α which makes the remainder small. Such an α has been found [4]:

$$\alpha^2 = \cos^2\phi - \frac{i}{R}(1+\frac{i}{R})(1-3\cos^2\phi), \tag{30}$$

$$\cos\phi = z/R, \quad R = \sqrt{z^2+\rho^2}, \tag{31}$$

and the expansion corresponding to this α is

$$I = -if(\alpha)\frac{e^{iR}}{R} + O\left\{\frac{\sin 2}{R^2} + \frac{1}{R^3}\right\}. \tag{32}$$

Note that for R large, α is approximately at the contributing saddle point, if the exponential form of the cosine and the asymptotic form of the Hankel function are substituted in (28). Application of (28)-(32) to (27) yields

$$p(r) = \frac{-2im^2(1+i/k_1r)}{(k_1r)(1-n^2)-i(m^2-1)(1+i/k_1r)}\frac{e^{ik_1r}}{4\pi r}$$
$$+\frac{2imn^2(1+i/k_2r)}{(k_2r)m^2(1-n^2)-in^2(m^2-1)(1+i/k_2r)}\frac{e^{ik_2r}}{4\pi r} + O((k_1r)^{-3}+(k_2r)^{-3}). \tag{33}$$

First, it should be noted that it is the second term on the rhs of (33) which corresponds to the lateral or head wave and that, for r sufficiently large, it reduces to (25). Second, as either $n \rightarrow 1$ or $m \rightarrow 1$, this representation reduces to the exact result. In other words, it is uniform in both m and n, whereas (25) was not. Thus, while the coefficients in (33) contain terms in r^{-3} and higher, they do so in a particularly useful manner which illustrates the advantage of this technique.

When the source and observation point are again in the slower medium, then it is necessary to obtain a uniform expansion of integrals with an algebraic singularity near a saddle point,

$$I = \int_C t^r g(t) e^{-i\kappa(t^2/2+\gamma t)}\,dt, \; r > -1, \tag{34}$$

and an appropriate integration contour C ending in valleys.

Such an expansion has been given by BLEISTEIN [7], and the application of this expansion to the Sommerfeld problem yields a scattered field which is bounded in the neighborhood of the critical ray and reduces asymptotically to the expected ray-theoretic result away from the critical ray. Thus, one of the difficulties in (25) is removed. However, the result is not uniform in the index n. This was remedied by STICKLER [4], using both the Bleistein result and a modification of the Franklin-Friedman idea. Recall that the Franklin-Friedman type modification to the uniform Fourier integral (20) required knowledge of $f'(x)$ at the end-points. The modification used by STICKLER [4] does not, but as a penalty does not yield a remainder which is always smaller than the original expansion.

The expansion is, however, uniform in n. The Franklin-Friedman type modification to the Bleistein result will now be presented. It has not been applied to the Sommerfeld problem. It is obtained, as expected, by the proper combination of the two leading terms of the Bleistein expansion and is given by

$$I = g(\eta_1)\,U_r(u) + \frac{g(\eta_2) - g(\eta_1)}{u}\,U_r'(u) + O(x^{-2}), \tag{35}$$

where

$$U_r(u) = \int_C t^r e^{-ix(t^2/2 + \gamma t)}\,dt, \quad u = \sqrt{x}\,\gamma e^{i\pi/4}, \tag{36}$$

and is simply related to the parabolic cylinder function $D_r(u)$, and where

$$\eta_1 = \frac{1}{ix}\frac{r+1}{\gamma}\left[1 + \frac{g(-\gamma) - g(0)}{\gamma g'(0)}\right], \; g'(0) \neq 0, \tag{37}$$

and

$$\eta_2 = -\gamma - \frac{1}{ix}\frac{1}{g'(-\gamma)}\left\{\frac{r}{\gamma}\left[g'(-\gamma) + \frac{g(-\gamma) - g(0)}{\gamma}\right] - \frac{1}{2}g''(-\gamma)\right\}, \; g'(-\gamma) \neq 0. \tag{38}$$

Note that η_1 is near the branch point at $t = 0$ and that η_2 is near the saddle point at $t = -\gamma$. Furthermore, as the saddle point $\gamma \to 0$, i.e., as the observation point approaches the critical ray,

$$\eta_2 = \eta_1 = \frac{r+1}{ix}\frac{g''(0)}{2g'(0)}. \tag{39}$$

Hence, the coefficient of $U_r'(u)$ is bounded as $\gamma \to 0$.

4. PEKERIS MODEL

The last section of this paper is concerned with the uniform evaluation of an integral which has near its path many first order poles and it has been applied to the Pekeris model. It has, of course, other applications.

The Pekeris model [8] is described as follows: an isospeed, constant density layer $[c_1, \rho_1]$ occupies the space $0 \leqslant z < L$, and the pressure satisfies a pressure-release condition at $z = 0$,

$$p(x,y,0) = 0. \tag{40}$$

The space $z > L$ is a second, isospeed, constant-density $[c_2, \rho_2]$ half-space. The pressure and normal component of acoustic velocity are required to be continuous at $z = L$ and the pressure satisfies a radiation condition at infinity. For some applications of interest, it is necessary to determine the pressure field in the layer $0 \leqslant z < L$ due to a point harmonic source also in the layer. For this problem, the pressure field can be represented by the sum of, at most, a finite number of proper or square integrable modes plus the contribution of the continuous spectrum. For this problem, the continuous spectrum can be expressed in terms of a branch cut integral. This integral is characterized by the fact that near the path of integration there can be many simple pole singularities, whose presence introduces significant structure in the branch integral. That is, these poles are critical points of the integrand.

The pole singularities are of two types. Those of the first type are the finite number of proper or square integrable modes and those of the second the infinite number of improper (non-square integrable) modes. These improper modes are sometimes called leaky modes or, in quantum physics, Regge poles [9], [10], [11]. As the frequency increases, some of these leaky-wave poles move from the "improper" branch and become proper poles on the "proper" branch and in so doing pass through the branch point, i.e., the endpoint of the branch integral. The frequency at which this transition occurs is called the cutoff frequency. It is this behaviour which it is desired to describe uniformly.

Using the techniques described in this paper, STICKLER and AMMICHT [12] have obtained two uniform asymptotic expansions of this integral. To obtain these expansions, the integrand is represented by a Mittag-Leffler expansion and the resulting integrals are transformed to integrals of the form

$$I = \int_z^\infty f(t) e^{ixt^2} dt, \tag{41}$$

where $f(t)$ is even, x real, and z complex.

For Re $z > 0$, the main contribution to this integral is an endpoint contribution. For Re $z < 0$ the main contribution is from the saddle point at the origin. The cutoff condition occurs when the singularity is at the saddle point. What is needed, then, is an expansion which is uniform in frequency. Two methods will be used to obtain expansions of this type and the results compared with a numerical integration procedure [13].

An application of the Franklin-Friedman idea yields

$$I = f(\alpha)\operatorname{erfc}(\sqrt{x}z) + O\left\{\frac{1}{x^{5/2}}\operatorname{erfc}(\sqrt{x}z)\right\}, \quad |\arg z| < 3\pi/4, \tag{42}$$

where

$$\alpha^2 = \frac{1}{2x} + \frac{z}{\sqrt{\pi x}} \frac{1}{e^{xz^2}\operatorname{erfc}(\sqrt{x}z)}. \tag{43}$$

For large z and Re $z > 0$, α tends to be near the endpoint, while for Re $z < 0$, α is near the saddle point at the origin.

An alternate uniform expansion is obtained by expanding $f(t)$ about the two critical points in a Lagrange expansion

$$f(t) = f(0) + \frac{f(z) - f(0)}{z^2 - 0} t^2 + R, \tag{44}$$

Substitution of (44) into (41) yields

$$I = f(0)\left\{\frac{1}{2}\sqrt{\frac{\pi}{x}}\operatorname{erfc}(\sqrt{x}z)\right\} + \frac{f(z^2) - f(0)}{z^2} \frac{d}{dx}\left\{\frac{1}{2}\sqrt{\frac{\pi}{x}}\operatorname{erfc}(\sqrt{x}z)\right\} + O\left\{\frac{1}{x^{3/2}}\operatorname{erfc}(\sqrt{x}z)\right\}. \tag{45}$$

For the examples considered, the two uniform methods have yielded results identical to four significant figures, but the expansion based on the Franklin-Friedman idea is easier to implement in this case.

In Fig. 3, a plot is shown of the modulus — converted to db — of the pressure field as a function of ranges with frequency as a parameter. For frequencies less than 54.27 Hz, there are no proper modes and at the frequency the first proper mode emerges from the "improper" sheet. This plot shows the transmission loss at two frequencies well above and below cutoff, repectively, and at two frequencies just a fractional part of a Hertz above and below cutoff, respectively. It is clearly seen that the transition through the cutoff frequency is smooth.

ACKNOWLEDGMENT

This research was supported by the Office of Naval Research.

REFERENCES

[1] J. Franklin, B. Friedman: "A convergent asymptotic representation for Laplace integrals," Proc. Camb. Phil. Soc. **53**, 612 (1957)

[2] A. Erdélyi: *Asymptotic Expansions* (Dover, New York, 1956), pp. 29-34

[3] See [2], pp. 49-50

[4] D.C. Stickler: "Reflected and lateral waves for the Sommerfeld model," J. Acoust. Soc. Am. **60**, 1061 (1976)

0[5] A. Erdélyi *et al.*: *Higher Transcendental Functions* (McGraw-Hill, New York, 1953), Vol. 2, p. 81
0[6] A. Erdélyi *et al.*: *Higher Transcendental Functions* (McGraw-Hill, New York, 1953), Vol. 1, p. 248
0[7] N. Bleistein, R.A. Handelsman: *Asymptotic Expansions of Integrals* (Holt, Rinehart, and Winston, New York, 1975), pp. 380-385
0[8] C.L. Pekeris: in *Propagation of Sound in the Ocean*, Mem. Geo. Soc. Am. **27** , 1 (1948)
0[9] L.B. Felsen, N. Marcuvitz: *Radiation and Scattering of Waves* (Prentice-Hall, Englewood Cliffs, New Jersey, 1973)
[10] R.G. Newton: *Scattering Theory of Waves and Particles* (McGraw-Hill, New York, 1966), pp. 401-415
[11] N. Marcuvitz: "On field representations in terms of leaky modes or eigenmodes," IEEE Trans. **AP-4**, 192 (1956)
[12] D.C. Stickler, E. Ammicht: "A uniform asymptotic evaluation of the continuous spectrum contribution for the Pekeris model," submitted to the J. Acoust. Soc. Am.
[13] D.C. Stickler: "Normal mode program with both the discrete and branch line contributions," J. Acoust. Soc. Am. **57**, 856 (1975)

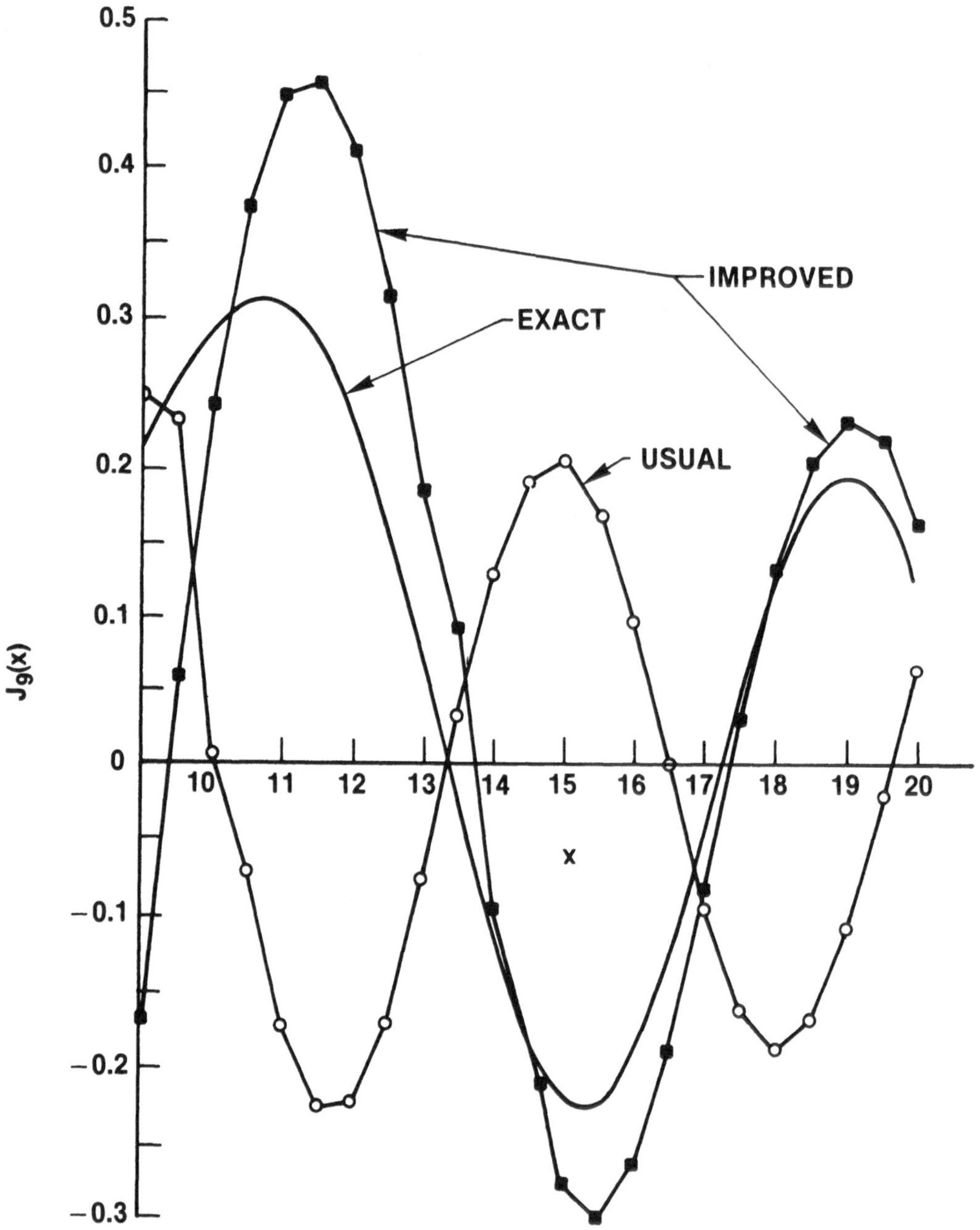

Fig. 1. A comparison of $J_9(x)$ with the first term of the "usual" asymptotic expansion and the expansion given in (15).

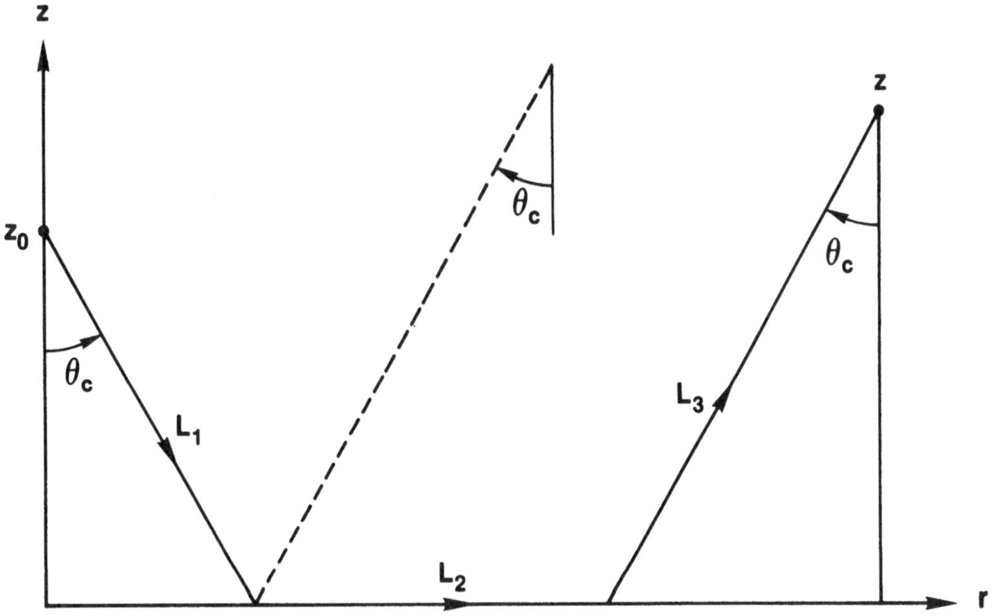

Fig. 2. The lateral wave path is shown by the segments L_1, L_2, and L_3.

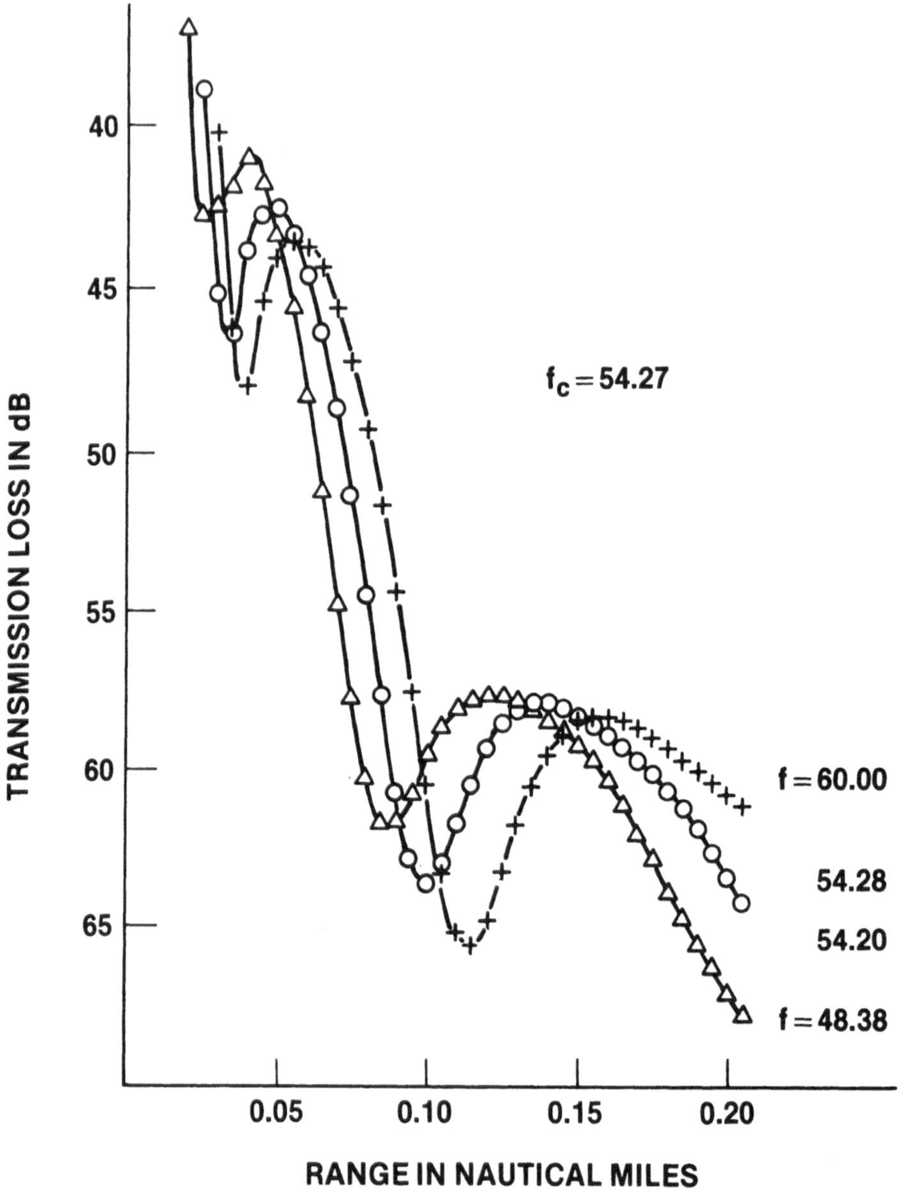

Fig. 3. For this example, the layer depth is 150 feet and $c_1 = 5000$ ft/sec, $c_2 = 5060$ ft/sec, and $c_2/c_1 = 1.4$. The source depth is 20 ft and the observation depth is 40 ft.

HIGH-FREQUENCY SIGNAL PROPAGATION AND SCATTERING IN GUIDING CHANNELS

L. B. Felsen

Department of Electrical Engineering
Polytechnic Institute of New York
Farmingdale, New York 11735

ABSTRACT

The earth's environment contains propagation channels wherein waves can be guided because of the presence of transverse boundaries or transverse refractive index gradients. By a new approach, high-frequency guiding by a single concave surface or by the boundary of an inhomogeneous surface duct has recently been analyzed in terms of a judiciously chosen combination of rays and modes. In essence, the modes account for the cumulative effect of those rays that have experienced a great many reflections on the boundary. This hybrid formulation is appealing in that it requires far fewer rays and far fewer modes than if the field representation involves rays only or modes only, as has been customary. By an extension of the theory, it has now been shown that the hybrid method can also be applied to channels wherein guiding occurs between multiple transverse boundaries. Here, modes near cutoff can represent efficiently all those rays that have made many excursions between the channel walls. As an illustration, results are presented for a parallel plane waveguide excited by a line source.

INTRODUCTION

Terrain probing along the earth's surface or in subsurface layers may involve propagation channels that can support electromagnetic or acoustic waves guided along a single surface or between multiple surfaces. In the high-frequency range, the propagation phenomena may be described either in terms of guided modes and continuous spectra (when required), or in terms of multiply reflected ray-optical fields. The calculations may require the summation over many modes or many rays, and they become even more complicated when profile perturbations introduce scattering centers or nonuniformities along the propagation path.

In a recently developed theory, a judiciously chosen mixture of modal and ray fields has been shown to provide an effective method for calculating signal strengths and providing new insights into the physical mechanism of propagation and scattering. The ray-mode mixture involves far fewer rays and modes than when only rays or only modes are employed to characterize the field. The theory was first constructed for source and observation points on concave terrain contours where the relevant modal fields are whispering-gallery modes guided along a single boundary described by a surface impedance [1]. Subsequent extensions have dealt with ducts having an inhomogeneous refractive index profile that causes field trapping near a planar surface [2]. In both cases, the ray fields that undergo many reflections between the source point Q and the observation point P cannot be calculated by geometrical optics because the local plane wave assumption underlying the geometric optical model cannot be satisfied for such fields. In qualitative terms, those rays lying within the duct of the most closely bound whispering gallery mode (Fig. 1) must be excluded from a ray-optical calculation and accounted for in some other way. Various schemes involving canonical integrals (the analogues of Fock integrals for shadow boundary effects on convex surfaces), guided modes and the above-noted appropriately chosen mixture of rays and modes have been explored in this context [1,2]. Among these formulations, the hybrid ray-mode model is simplest and physically most appealing. For details of the analysis and for numerical comparisons of the accuracy of the various alternative formulations, which confirm the validity of the hybrid model, see [1,2].

When guiding is produced by two boundaries, there exist rays that experience many reflections between these boundaries as they progress from Q to P. While the ray-optical description of the field along these rays is legitimate, their summation poses numerical problems. Here, it is found that the cumulative effect of such rays can be accounted for in terms of an appropriately chosen numbers of waveguide modes near cutoff.

1. HYBRID FORMULATION FOR PARALLEL PLANE WAVEGUIDES

For illustration, consider a homogeneously filled parallel plane waveguide with perfectly conducting boundaries, excited by a magnetic line source on one of the boundaries. The exact Green's function can here be expressed as a modal sum involving the propagating and evanescent waveguide modes. The Green's function can also be expressed as a Fourier transform with respect to the axial coordinate z. In the resulting integral representation, there exists a resonant denominator whose zeros provide pole singularities that correspond to the waveguide modes. If the integration path is deformed around those singularities and the residue theorem is invoked, one obtains the above-mentioned modal sum, each mode being identified by the characteristic angle θ_m (Fig. 2). On the other hand, if the resonant denominator is expanded in a geometric series, thereby removing the poles, one obtains a series of integrals. When these are evaluated asymptotically by the saddle point method, each furnishes a contribution that can be identified exactly with a geometric-optical ray field characterized by the ray angle θ_n (Fig. 2).

If one does not wish to include all of the multiply reflected rays in the formulation, one may expand the resonant denominator into a partial sum up to θ_N plus a remainder term. By contour deformation in the resulting remainder integral such that the contour coincides with the steepest descent path corresponding to the last ray with θ_N, and by subsequent asymptotic (saddle point) evaluation, one may show that the remainder integral is well approximated $(-1/2)$ of the contribution from the last ray provided that $\theta_N \not\approx \theta_M$ or θ_{M-1}, where θ_M and θ_{M-1} are the characteristic angles of those waveguide modes that lie on either side of θ_N (i.e., $\theta_M > \theta_N > \theta_{M-1}$) (see Fig. 1). These steps are directly analogous to those performed in dealing with the surface-guided fields [1,2]. Thus, the exact solution yields for the field G on the lower boundary of Fig. 2, (see [3])

$$G = \sum_{m=0}^{\overline{M}} (\text{propagating waveguide modes}) + \sum_{m=\overline{M}+1}^{\infty} (\text{evanescent waveguide modes}). \tag{1}$$

Unless the lowest evanescent mode with $m=\overline{M}+1$ is very near the propagating regime, the evanescent mode sum can be neglected for large enough ranges between P and Q in Fig. 2. The hybrid ray-mode formulation yields

$$G \sim \sum_{m=M}^{\overline{M}} (\text{propagating waveguide modes}) + \sum_{m=M+1}^{\infty} (\text{evanescent waveguide modes})$$
$$+ \sum_{n=0}^{N} (\text{ray fields}) - \frac{1}{2}\,(N\text{th ray field}). \tag{2}$$

A negligible remainder term, which is also found to appear [1,2] has been omitted in (2). The criterion for choosing M and N is schematized in Fig. 2: if θ_N is the departure angle of the last included ray (the one with the highest number of reflections), then the modes needed are those with $\theta_M > \theta_N$. It is evident that if N is kept constant but the range z between the source point and observation point is increased, the departure angle θ_N decreases and will eventually approach the characteristic angle θ_{M-1} of the first excluded mode. When z is such that $\theta_N < \theta_{M-1}$, this latter mode must be added to the mode sum. Analogous considerations prevail when z decreases, thereby leading eventually to the removal of the mode with θ_M. Alternatively, if the number of modes $(\overline{M} - M)$ is kept constant, the number of included rays N must be adjusted with varying range z such that all possible rays with $\theta_N < \theta_M$ are included. When $\theta_N \approx \theta_M$ or θ_{M-1}, the saddle point for the last ray in the contour integral representation approaches the pole for the nearest waveguide mode. In this transition region, the simple ray-mode model fails and must be patched up by a transition function (a Fresnel integral; see [4]) that accounts for the proximity of a pole and a saddle point. Note that at least one of the propagating modes (the one closest to cutoff) must be included, whence $\theta_N < \theta_M$. The conclusions summarized above may also be confirmed by applying partial Poisson summation to the ray series in (1), thereby converting it into the hybrid formulation in (2) (see [3]). The procedure is directly analogous to that reported in [2].

Typical numerical results [3] obtained from the exact and hybrid formulations are depicted in Fig. 3, where $|G|$ is plotted over a normalized range kz from 250-290, k being the wavenumber in the medium. The normalized waveguide height is $ka = 50$ and admits $\overline{M} = 16$ propagating modes. The exact field is calculated from (1); for this case, the evanescent modes can be omitted, since they contribute negligibly over the specified range of observation points. Also plotted in Fig. 3 is the hybrid ray-mode solution obtained from (2) when $N=2$ as in Fig. 2 (i.e., the direct ray and the rays with one and two reflections at the upper boundary have been retained). From the criterion $\theta_m \geqslant \theta_N = \theta_2$, it is found that $M = 13$ for $250 < kz < 261$, whereas $M = 12$ for $261 < kz < 290$, in view of the fact that the characteristic mode angle θ_{13} for $m=13$ coincides with the ray departure angle θ_2 when $kz = 261$. The hybrid ray-mode formulation in (2) is discontinuous at $\theta_M = \theta_N$, but the jump is not very large and can be eliminated approximately by drawing a smooth curve through the average value at $\theta_M = \theta_N$. By a more rigorous procedure, one may employ the Fresnel integral transition function that provides continuous field values through the transition region surrounding $kz = 261$. Both the approximate and rigorous methods are seen to furnish excellent agreement between the exact and hybrid solutions. Note that whereas the exact solution requires 16 propagating modes, the hybrid solution requires 3 rays and three or four modes, respectively, for $250 < kz < 261$ and $261 < kz < 290$. When $N=3$ so that rays with three reflections at the upper boundary are included as well, one finds $\theta_3 = \theta_{14}$ at $kz = 282$. Consequently, two propagating modes (with $m = 14,15$) are required for $250 < kz < 282$ while three modes (with $m = 13\text{-}15$) are required for $282 < kz < 290$. The numerical accuracy is comparable to that depicted in Fig. 3. Similar results are obtained for other values of N and other range intervals [3]. When the first non-propagating mode is very close to cutoff, its contribution must be included in (1) and (2). Numerical comparisons again confirm the validity of the hybrid calculation in that event [3].

2. A NEW CLASS OF DIFFRACTION PROBLEMS

The hybrid ray-mode formulation of propagation leads to a new class of diffraction problems when a scatterer is located inside the duct. In previous studies [5], a ray-optical method was devised to account for the presence of obstacles or strong scattering centers within the guiding structure. The field in each of the incident modes is represented by its congruence of rays, and the scattering of these rays by the obstacle is calculated from the geometrical theory of diffraction. The diffracted rays, which experience multiple reflection at the waveguide boundaries or continuous refraction within the duct, are then converted into guided modes to establish the scattering matrix elements that describe the effect of the obstacle on reflection and transmission of the incident modes as well as coupling to other modes. It is to be emphasized that *all* of the rays are converted into the propagating guided modes by this scheme.

In the hybrid formulation, each of the retained incident modes and each of the incident ray species will give rise to a ray-optical scattering process as described above. The multiply reflected ray fields excited by the scattering center can be decomposed into another hybrid mixture that may be preferable for calculation at various ranges of the observation point with respect to the scatterer. The hybrid selection could be made so as to minimize the complexity of the scattering calculation. The flexibility introduced thereby is illustrated in the example of Fig. 4. The principal difference between the mode theory of diffraction and the hybrid theory is the dependence of the ray-mode mixture on range. For the incident field, the relevant range is from the source to the obstacle, whereas, for the scattered field, it is from the obstacle to the observer.

3. CONCLUSIONS

The hybrid ray-mode formulation affords a new approach to source-excited high-frequency propagation in guiding regions formed by transverse refractive index inhomogeneities and (or) bounding surfaces. Conventional methods have expressed the field either in terms of rays or in terms of (discrete and continuous) modes. The hybrid ray-mode mixture improves upon these methods since it requires far fewer rays and modes than when only modes or only rays are considered. This facilitates numerical treatment of the problem. It also grants new physical insights since the formulation implies that propagation processes characterized by rays with many reflections can be treated collectively in terms of a few modes, while processes characterized by many modes can be expressed succinctly in terms of a few rays. In effect, the hybrid formulation *quantifies* the truncation error of a mode series in terms of rays, or equivalently, the truncation error of a ray series in terms of modes. Moreover, since the number of modes in a modal expansion can now be suitably restricted, the eigenvalue problem in a complicated ducting environment may often be reduced to a simpler form for the retained cluster of modes. This feature economizes on computer time and required computer capacity. The hybrid formulation also appears to be well suited to treatment of lateral inhomogeneities along the duct provided that these occur gradually over a length interval equal to the local wavelength of the signal spectrum. Finally, the ray-mode field representation provides a new approach to scattering from strong inhomogeneities since physical insights derived from scattering of either a modal field or a ray field can be exploited to advantage.

REFERENCES

[1] T. Ishihara, L. B. Felsen: "High frequency fields excited by a line source located on a concave cylindrical impedance surface," IEEE Trans. **AP-27**, 172-179 (1979)

[2] L. B. Felsen, T. Ishihara: "Hybrid ray-mode formulation of ducted propagation", J. Acoust. Soc. Am. **65**, 595-607 (1979)

[3] A. Kamel, L. B. Felsen: "Hybrid ray-mode representation of the high-frequency parallel plane waveguide Green's function," in preparation

[4] L. B. Felsen, N. Marcuvitz: *Radiation and Scattering of Waves* (Prentice Hall, Englewood Cliffs, New Jersey, 1973), Chap. 4

[5] D. V. Batorsky, L. B. Felsen: "Ray optical calculation of modes excited by sources and scatterers in weakly inhomogeneous ducts," Radio Science **6**, 911-923 (1971)

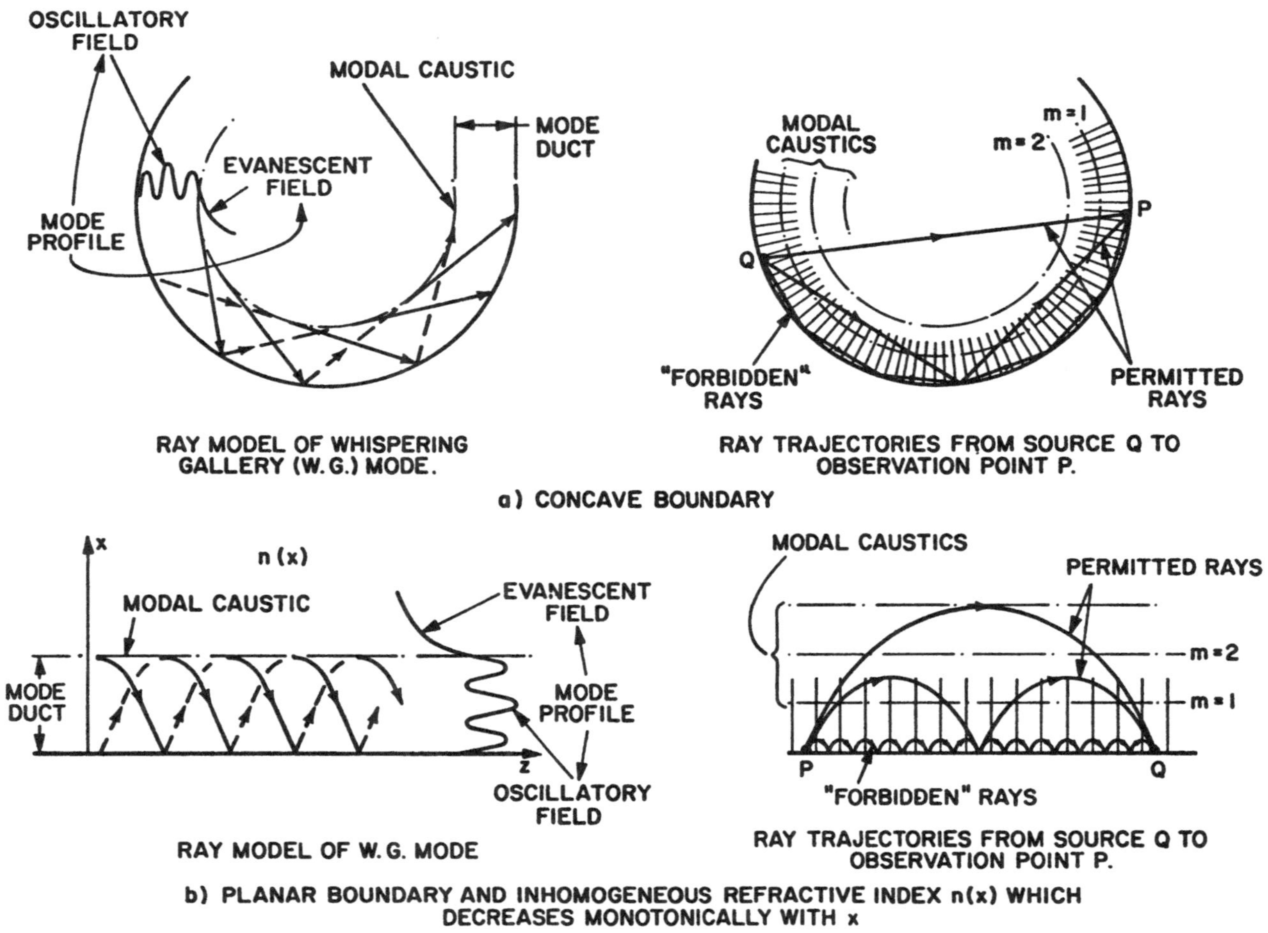

Fig. 1. Whispering gallery modes and rays. → -→ modal ray congruences. Rays proceeding entirely within the shaded region that includes the duct width of the most tightly bound ($m = 1$) W.G. mode must be excluded from the ray-optical field representation.

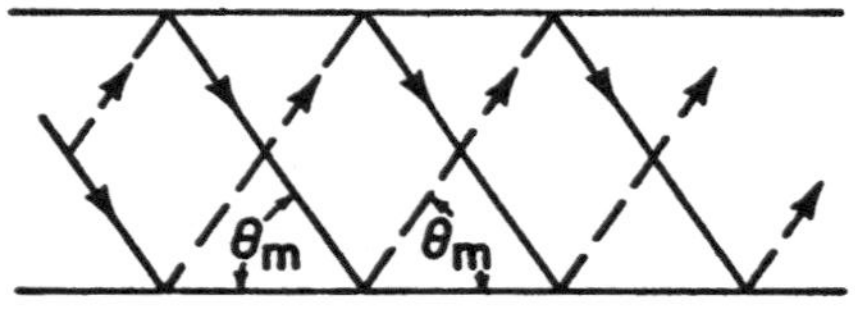

Ray model of waveguide mode. - - -> ——> modal ray congruences. θ_m is the characteristic (modal) angle.

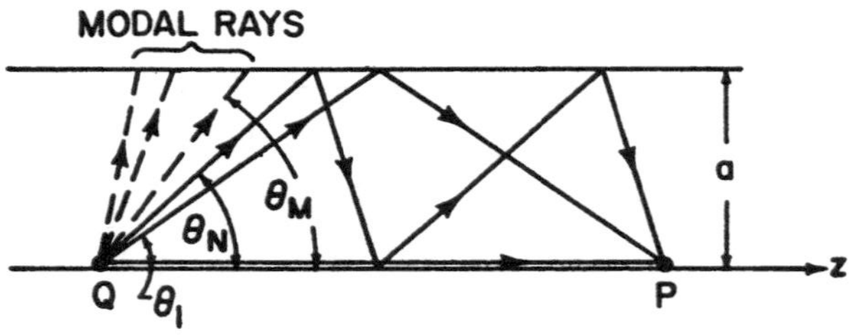

Ray trajectories from Q to P, including direct and multiply reflected rays. The last included ray is described by the angle θ_N. Rays with more reflections ($\theta_n > \theta_N$) are accounted for by all those waveguide modes with $\theta_m \geq \theta_M$, where θ_M is the mode whose characteristic angle is closest to, but larger than, θ_N. The figure shows truncation of the ray series after the reflected ray with $\theta_N = \theta_2$.

Fig. 2. Hybrid ray-mode model for parallel plane waveguide

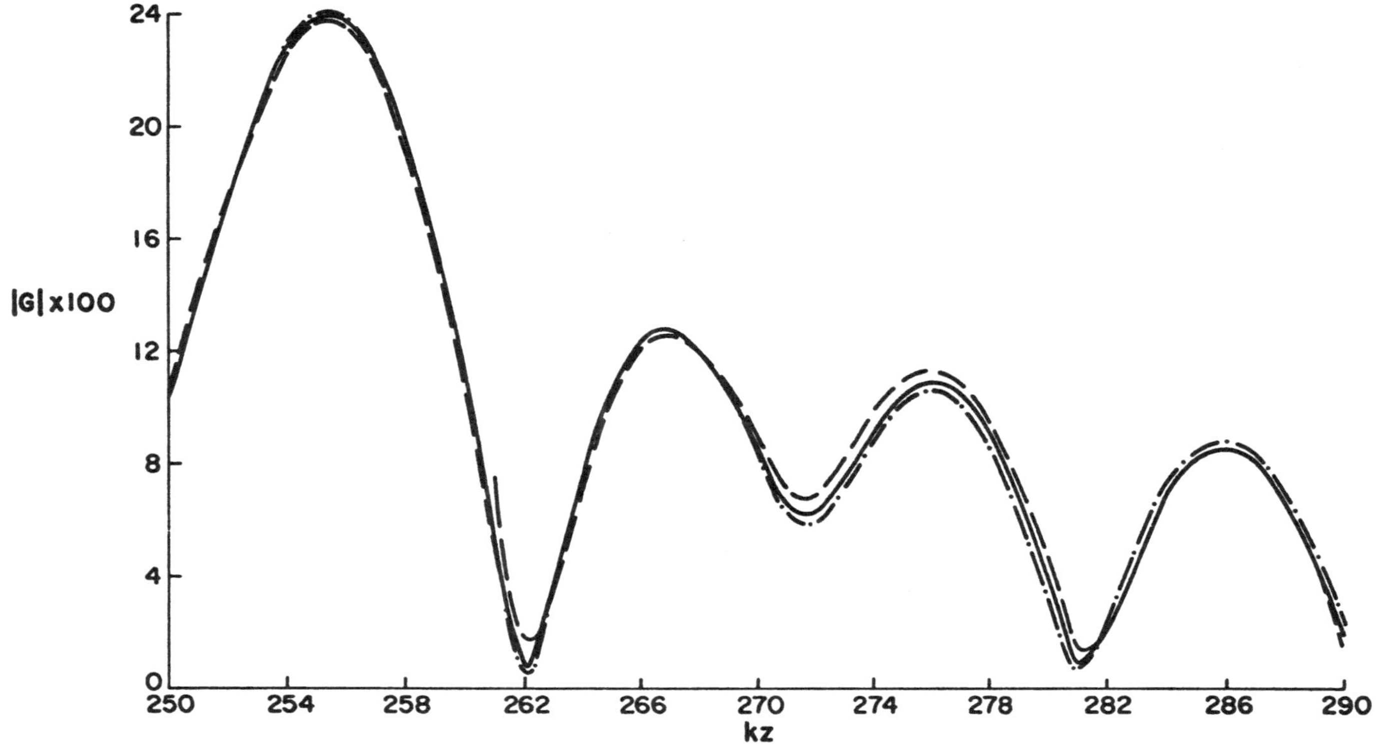

Fig. 3. Field magnitude in the range interval $250 < kz < 290$, when $ka = 50$. ——— *Exact results calculated from (1), with $\overline{M} = 16$.* ——— ——— *Hybrid ray-mode result calculated rom (2), with $N = 2$.* —— · —— · —— *Hybrid ray-mode result calculated from (2), with $N = 2$, when Fresnel transition function is used (when curve is ot shown, it coincides with the exact result).*

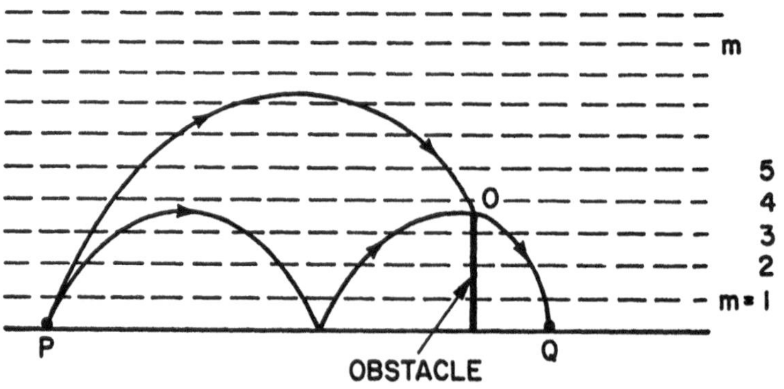

Fig. 4. Hybrid ray-mode formulation of scattering by a vertical strip obstacle in a surface duct. The modes retained (M = 3) are those whose caustics lie below the edge since the reflection of these modes is as from an infinite plane and therefore evaluated trivially. The modes whose field strength are appreciable at the edge are accounted for by the two incident rays. The scattered fields are evaluated as from an equivalent non-isotropic line source located at the edge O. Only a single ray proceeds from there to the observation point Q, and the excited modal fields (m = 1,2,3) are negligible.

SCATTERING BY MOVING BODIES: THE QUASI-STATIONARY APPROXIMATION

Jeffery Cooper*

Dept. of Mathematics
University of Maryland
College Park, Maryland 20742

INTRODUCTION

Scattering of electromagnetic radiation by moving bodies occurs in many situations of engineering importance. The article of VAN BLADEL [1] treats rotating bodies and has a lengthy list of references of other articles in the physics and engineering literature. The customary method of approximating the field scattered by a moving body is to calculate the stationary field scattered by the body for each time t, as if the body were "frozen" at that time. This procedure yields a time dependent sequence of stationary fields usually referred to as a quasi stationary field. In the engineering literature, it is assumed that the quasi stationary field provides a good approximation to the exact scattered field. Recently the mathematical theory of scalar scattering of LAX and PHILLIPS [2] has been extended to the case of moving bodies [1,3]. Within this framework, it is now possible to consider the validity of the quasi stationary approximation.

In this paper, we shall discuss the representative case of a rotating, convex body in the presence of a time-harmonic plane wave. Let $\mathbb{R}^3$ denote three-dimensional Euclidean space with coordinates $x = (x_1,x_2,x_3)$ and $r = |x| = \sqrt{x_1^2 + x_2^2 + x_3^2}$. Let $O(t)$ denote the set of points occupied by a rotating convex body O_o at time t. We assume that $O(0) = O_o$ and that the body is rotating with constant angular velocity Ω about the x_3-axis. Furthermore, O_o contains the origin and $O_o \subset \{x: r \leqslant \rho\}$ for some constant $\rho > 0$. To obtain a well-posed problem, we shall assume that $\Omega\rho < 1$. Let $E(t)$ denote the exterior region at time t,

$$\hat{E} = \bigcup_{-\infty < t < \infty} E(t)x\{t\}; \ \Sigma = \partial E$$

and $\Sigma = \partial\hat{E}$ the space-time boundary of the obstacle.

Now let a time-harmonic plane wave $e^{i\sigma(t - x\cdot\eta)}$ be incident on the body, where σ is the frequency and η is a unit vector. The speed of propagation c is taken to be unity. We consider the problem of approximating the scattered field given by the solution of

$$\begin{cases} \psi_{tt} - \Delta\psi = 0 \text{ in } \hat{E} , \\ \psi = -e^{i\sigma(t-x\cdot\eta)} \text{ on } \Sigma . \end{cases} \tag{1}$$

Before we discuss the quasi stationary approximation to ψ, we must review some properties of the exact solution to (1).

1. THE EXACT SOLUTION

Because of the moving boundary, it is not possible to separate variables in the usual manner to solve (1). In [4], a notion of *outgoing* was defined for solutions of (1) which generalizes the Sommerfeld radiation conditions. It was shown that outgoing solutions of (1) exist and are unique, provided the local energy of finite-energy solutions of the Cauchy problem decays exponentially. For a rotating convex body, this can be proved for Ω sufficiently small (see [3]). In this case, we have the representation

$$\psi(x,t) = \int_{-\infty}^{t} w(x,t,s)e^{i\sigma s}ds - \chi(x)e^{i\sigma(t-x\cdot\eta)} , \tag{2}$$

where w is the finite energy solution of

$$\begin{cases} w_{tt} - \Delta w = 0 \text{ in } \hat{E} , \\ w = 0 \text{ on } \Sigma , \\ w(x,s,s) = o, \ w_t(x,s,s) = q(x,\sigma) \text{ in } E(s) . \end{cases}$$

$\chi(x)$ is a smooth function such that $\chi = 1$ for $|x| \leqslant \rho$ and $\chi = 0$ for $|x| \geqslant \rho+1$. Here, $q(x,\sigma) = e^{-i\sigma x\cdot\eta} \times(2i\sigma\eta\cdot\nabla\chi - \Delta\chi)$. The integral in (2) converges in the local-energy norm for each t. Since the motion of the body is periodic, we may deduce from (2) that ψ has the form

$$\psi(x,t) = e^{i\sigma t}p(x,t) \ ,$$

where for each x, $t \rightarrow p(x,t)$ has period $T = 2\pi/\Omega$. A *far field* for the solution ψ can be defined as follows. Let θ be a unit vector in $\mathbb{R}^3$. The points (x,t) where $x = (t-\tau)\theta$ constitute a space time ray of unit speed which leaves the origin in $\mathbb{R}^3$ at time $t = \tau$. We define the far field of ψ to be

$$\Psi(\tau,\theta) = \lim_{t\rightarrow\infty} t\psi((t-\tau)\theta,t) \ .$$

It can be shown (see [6]) that $\Psi(\tau,\theta) = e^{i\sigma\tau}P(\theta,\tau)$ where $\tau \rightarrow P(\theta,\tau)$ is a function of period T taking values in $L^2(S_2)$, $S_2 = \{\theta \in \mathbb{R}^3 : |\theta| = 1\}$. P depends only on the values of ψ near the obstacle.

2. THE QUASI-STATIONARY APPROXIMATION

The quasi-stationary approximation for the scattered field given by the solution of (1) is the time dependent sequence of stationary fields given by the function

$$\phi(x,t) = e^{i\sigma t}f(x,t),$$

where for each t, $x \rightarrow f(x,t)$ is the outgoing solution of

$$\begin{cases} \Delta f + \sigma^2 f = 0 \text{ in } E(t) \ , \\ f = e^{-i\sigma x\cdot\eta} \text{ on } \partial E(t) \ , \text{ the boundary of } E(t) \end{cases} \tag{3}$$

As in Sec. 1, we may write f as an integral of solutions of the Cauchy problem, this time in $E(t)$. We can compare the two integral representations and exploit the finite speed of propagation to prove

Theorem 1: *There exists* $\Omega_0 > 0$, depending on the shape of the body, such that in the local-energy norm

$$||\psi(t) - \phi(t)||_R \leqslant C\sigma^2(\Omega\rho)^\beta$$

where for a function $u(x,t)$, $||u(t)||_R^2 = \frac{1}{2}\int_{|x|\leqslant R}(u_t^2 + |\nabla u|^2)dx$. *The inequality holds for all* $\beta < 1$ *whenever* $\Omega < \Omega_0$ *and* $|\sigma| > 0$. *The constant* C *depends on* R, *the shape of the body*, σ, ρ, *and* β. *The estimate is uniform in t.* (The proof is given in [6]).

Next, we wish to define a far field for ϕ. The limit along space-time rays $x = (t-\tau)\theta$ as $t \rightarrow \infty$ does not exist for the quasi-stationary approximation. Instead, we define

$$\Phi(\tau,\theta) = e^{i\sigma\tau}\, a(\theta,\tau) \ ,$$

where $a(\theta,\tau)$ is the usual far field for the solution f of the stationary problem (3):

$$a(\theta,\tau) = \lim_{r\rightarrow+\infty} re^{i\sigma r}f(r\theta,\tau) \ .$$

The local energy estimate of Theorem 1 then leads to a comparison of the far fields.

Theorem 2: *For* $\Omega < \Omega_0$, we have

$$||\Psi(\tau,\cdot) - \Phi(\tau,\cdot)||_{L^2(S_2)} \leqslant M\sigma^2(\rho\Omega)^\beta$$

for all $\beta < 1$ *where* M *is a constant which depends on* ρ, β, σ *and the shape of the body. The estimate is uniform in* τ, and in *the incident direction* η.

FOOTNOTES AND REFERENCES

*Research supported in part by the National Science Foundation under Grant MCS 76-06759-A01

[1] J. Van Bladel: "Electromagnetic fields in the presence of rotating bodies," Proc. IEEE **64**, 301-318 (1976)
[2] P. Lax, R.S. Phillips: *Scattering Theory* (Academic, New York, 1967)
[3] J. Cooper, W. Strauss: "Energy boundedness and decay of waves reflecting off a moving obstacle," Indiana Univ. Math. J. **25**, 671-690 (1976)
[4] J. Cooper: "Scattering of plane waves by a moving obstacle," Arch. Rational Mech. Anal. **71**, 113-141 (1979)

[5] J. Cooper, W. Strauss: "Representations of the scattering operator for moving obstacles," to appear in Indiana Univ. Math. J. (1979)

[6] J. Cooper: "Scattering by moving bodies: the quasi-stationary approximation," to appear in Math. Meth. Appl. Sci.

VARIATIONAL METHODS FOR WAVE SCATTERING FROM RANDOM SYSTEMS

R. H. Andreo*

The Johns Hopkins University
Applied Physics Laboratory
Laurel, Maryland 20810

ABSTRACT

The scattering of waves by random surfaces and media has long been of considerable interest from both theoretical and experimental viewpoints. This paper briefly reviews the work of our group toward developing variational principles which are applicable to the scattering of scalar and vector waves from stochastic systems. These principles have the general form $4\pi <T> = <N_1><N_2>/<D>$ for arbitrary scattering statistics. In this expression, T is the far-field scattering amplitude, N_1 is the usual noninvariant integral representation of T, and the ratio of integrals N_2/D is the variational correction factor. Application to a simple model of a random rough surface has shown this stochastic variational approach to account in large measure for multiple scattering. The potential tractability of stochastic variational principles should allow broader application of variational techniques to random scattering problems.

INTRODUCTION

The theoretical description of wave scattering by stochastic systems is of interest in many diverse areas [1]. In order to investigate effects such as interference and multiple scattering, researchers at the Johns Hopkins University Applied Physics Laboratory have developed and tested variational methods for wave scattering from random systems. These stochastic variational principles allow tractable variational evaluation of the statistics of the scattering amplitude and of the differential scattering cross section. The following examples will illustrate their formulation.

1. SCALAR STOCHASTIC VARIATIONAL PRINCIPLE FOR SCATTERING FROM RANDOM ROUGH SURFACES

Consider the well-known variational formulation of the scattering of a plane wave $\psi_{inc}(\vec{x}) = Ae^{i\vec{k}_i \cdot \vec{x}}$ from a closed surface S_0 on which the wavefunction ψ satisfies Neumann boundary conditions, $\partial\psi(\vec{x})/\partial n|_{\vec{x}\in S_0} = 0$. The scattering amplitude $T = iN_1/4\pi A$ can be written in the familiar stationary form [2]

$$4\pi T(\vec{k}_s,\vec{k}_i) = N_1 N_2/D \tag{1}$$

by expressing the amplitude A of the incident plane wave in terms of the "adjoint" wave function $\tilde{\psi}$, where

$$N_1 = \oint dS'\hat{n}' \cdot \vec{k}_s \psi(\vec{x}')e^{-i\vec{k}_s \cdot \vec{x}'}, \; N_2 = -\oint dS\, \hat{n} \cdot \vec{k}_i \tilde{\psi}(\vec{x})e^{i\vec{k}_i \cdot \vec{x}},$$

and

$$D = \oint dS \oint dS' \tilde{\psi}(\vec{x}) \left[\frac{\partial^2}{\partial n \partial n'} G_0(\vec{x},\vec{x}') \right] \psi(\vec{x}'),$$

with $G_0(\vec{x},\vec{x}')$ the usual free-space Helmholtz Green function. As is well-known [2-4], the importance of (1) for calculations is that it is insensitive to errors made in approximating ψ and $\tilde{\psi}$ on S_0.

In the application of scattering theory to random systems, the measureable quantity of interest is generally a statistical moment, e.g., the ensemble average $<T>$. The direct application of (1) to rough surfaces requires the prohibitive average

$$4\pi <T> = \left\langle \frac{N_1 N_2}{D} \right\rangle. \tag{2}$$

However, HART and FARRELL [3] have used the nonstochastic nature of the amplitude $A = iD/N_2$ of the incident plane wave to derive the *exact* result

$$4\pi < T > = < N_1/(-iA) > = < N_1 > /(-iA) = < N_1 > < N_2 > / < D > . \tag{3}$$

Furthermore, by virtue of the nonstochastic nature of the amplitude $\tilde{A} = iD/N_1$ of the plane wave for the adjoint problem, they were able to demonstrate that variations of (3) cancel to first order, thereby establishing (3) as a valid stochastic variational expression for scattering from rough surfaces. Similar results can be obtained for the higher statistical moments of both the scattering amplitude and of the differential cross section $d\sigma/d\Omega \equiv |T|^2$. See [5].

2. TEST CASE – A RANDOM SURFACE MODEL

In order to gain some insights into the new stochastic variational approach, GRAY, HART, and FARRELL [6] considered electromagnetic-wave scattering from a model random rough surface [7] consisting of a large number N of nonoverlapping parallel hemicylindrical bosses of equal radii a, randomly distributed over a length L of an infinite conducting plane. They compared the first-order perturbation approximation of the differential cross section for scattering of the normally incident *TM* mode to the variational improvement of this appproximation using the same plane wave trial function. In the Rayleigh limit and to terms linear in the area fraction $\nu = 2aN/L$ occupied by the hemicylinders, they found the average differential scattering cross section to have the form

$$< |T|^2 > = C(1 - \nu\alpha),$$

where C is a function of the incident and scattering angles, and where the coefficient α has the value $\alpha = 2$ for the perturbational approximation and $\alpha = 4.08$ for the variational improvement.

In order to investigate this difference, KRILL and FARRELL [8] have computed the exact value, in the Rayleigh limit and to first order in ν, for the scattering from a special case of this model with only two hemicylinders present, and have compared it with the perturbation and variational results. They found for this surface, with $N = 2$, the same expression for the differential scattering cross section, with now the value $\alpha = 1.89$ for the exact solution, $\alpha = 1$ for the perturbational approximation, and $\alpha = 2.04$ for the variational improvement. [Note that the large N limit does not apply for this ($N = 2$) surface.] An examination of the exact solution revealed that the closed agreement between the variational and the exact results is due to the fact that the variational approximation accounts, in large measure, for multiple scattering effects, whereas the perturbational approximation which it improves does not.

3. VECTOR-WAVE SCATTERING

In order to account explicitly for the vector nature of the electromagnetic field, e.g., to investigate polarization effects, KRILL and ANDREO [5] have developed vector stochastic variational principles for the scattering of a plane electromagnetic wave $\vec{E}_{inc} = A\hat{e}_i e^{i\vec{k}_i \cdot \vec{x}}$ from an inhomogeneous, anisotropic, conducting dielectric. The scattering amplitude for radiation polarized along a specfied direction $\hat{e}_s$ is

$$T(\vec{k}_s,\vec{k}_i) = \frac{1}{4\pi A}\int d^3x\left\{\hat{e}_s \cdot \overline{\overline{I}}_{\hat{k}_s} e^{-i\vec{k}_s \cdot \vec{x}'}\right\} \cdot \overline{U}(\vec{x}') \cdot \vec{E}(\vec{x}') \equiv \frac{1}{4\pi A} N_1,$$

where the dyadic operator $\overline{U}$ depends of the tensor conductivity and permitivity. As for the scalar case, this scattering amplitude can be written in an invariant form by expressing A in terms of the adjoint field $\tilde{\vec{E}}$, and one finds $4\pi T = N_1 N_2/D$, with

$$N_2 = \int d^3x \tilde{\vec{E}}(\vec{x}) \cdot \overline{U}(\vec{x}) \cdot \left(\overline{\overline{I}}_{\hat{k}_i} \cdot \hat{e}_i e^{i\vec{k}_i \cdot \vec{x}}\right),$$

and

$$D = \int d^3x \tilde{\vec{E}}(\vec{x}) \cdot \overline{U}(\vec{x}) \cdot \vec{E}(\vec{x}) - \int d^3x \int d^3x' \tilde{\vec{E}}(\vec{x}) \cdot \overline{U}(\vec{x}) \cdot \overline{\overline{G}}_0(\vec{x},\vec{x}') \cdot \overline{U}(\vec{x}') \cdot \vec{E}(\vec{x}').$$

Because the amplitude A is nonstochastic, the exact stochastic expression (3) can be derived for vector waves just as in the scalar case, and, furthermore, is readily demonstrated to be stationary about the exact fields $\vec{E}$ and $\tilde{\vec{E}}$.

These results can be extended in a straightforward manner to vector-wave scattering from rough perfect conductors.

4. SUMMARY AND CONCLUSIONS

Stochastic variational principles have been developed for plane-wave scattering from conducting dielectric surfaces and volumes. These formulations have the general form $4\pi<T> = <N_{1}><N_2>/<D>$, and express the statistical moments of the scattering amplitude and the differential cross-section as quotients of other, more readily evaluated statistical moments. Their potential tractability promises to allow broader application of variational methods to scattering from random systems.

FOOTNOTES AND REFERENCES

*This work was supported by the Department of the Navy, Naval Sea Systems Command, under Contract N00024-78-C-5384

[1] A. Ishimaru: *Wave Propagation and Scattering in Random Media* (Academic, New York, 1978), Vols. 1, 2
[2] P. M. Morse, H. Feshbach: *Methods of Theoretical Physics* (McGraw-Hill, New York, 1953), Vol. II, Chap. 9
[3] R. W. Hart, R. A. Farrell: "A variational principle for scattering from rough surfaces," IEEE Trans. **AP-25**, 708-710 (1977)
[4] H. Levine, J. Schwinger: "On the theory of electromagnetic wave diffraction by an aperture in an infinite plane conducting screen," Comm. Pure Appl. Math **3**, 355-391 (1950)
[5] J. A. Krill, R. H. Andreo: "Vector stochastic variational principles for electromagnetic wave scattering," to be submitted to IEEE Trans. Ant. Prop. (1979)
[6] E. P. Gray, R. W. Hart, R. A. Farrell: "An application of a variational principle for scattering by random rough surfaces," Radio Science **13**, 333-348 (1978)
[7] V. Twersky: "On scattering and reflection of electromagnetic waves by rough surfaces," IRE Trans. **AP-5**, 81-90 (1957)
[8] J. A. Krill, R. A. Farrell: "Comparisons between variational, perturbational, and exact solutions for scattering from a random rough surface model," J. Opt. Soc. Am. **68**, 768-774 (1978)

RESONANCE THEORY AND APPLICATION

Louis R. Dragonette and Lawrence Flax

Naval Research Laboratory
Washington, District of Columbia 20375

Normal-mode series solutions for the scattering of a plane acoustic wave by submerged elastic spheres and cylinders have existed since 1951 [1-4], and curves of backscattered pressure vs frequency have been computed up to ka values of 1000 for materials with negligible acoustic absorption [5]. Here ka is $2\pi a/\lambda$ where a is the target radius and λ is the wavelength of the incident sound. The normal-mode series expression for the scattering of an incident plane wave $p_o e^{ikx}$ by an infinite elastic cylinder in the geometry described in Fig. 1 is given by [4]

$$p_s(\theta) = -p_o \sum_{n=0}^{\infty} \epsilon_n (i)^n \left[\frac{J_n(Z)L_n - ZJ_n'(Z)}{H_n(Z)L_n - ZH_n'(Z)} \right] H_n(kr) \cos n\theta. \tag{1}$$

In (1), $Z \equiv ka$, r is the range, ϵ_n is the Neumann factor, the J_n are Bessel functions, the H_n are Hankel functions of the first kind, and the primes denote derivatives with respect to the argument. The L_n are defined by:

$$L_n = \frac{\rho}{\rho_s} \frac{\begin{vmatrix} a_{11} a_{13} \\ a_{21} a_{23} \end{vmatrix}}{\begin{vmatrix} a_{11} a_{13} \\ a_{31} a_{33} \end{vmatrix}}, \tag{2}$$

where ρ and ρ_s are the densities of water and the target material respectively, and the a_{ij} are given in [4].

For the problem of backscattering in the far field, the expression in Fig. 1 can be simplified to [4]

$$p_s(\pi) = -p_o e^{ikr} \left[\frac{2}{\pi kr} \right]^{1/2} e^{i\pi/4} \sum_{n=0}^{\infty} \epsilon_n (-1)^n G_n(Z). \tag{3}$$

Here, $G_n(Z)$ is the expression in the brackets in (1). The backscattered form function $f_\infty(\pi)$ is the dimensionless pressure variable defined for an infinite cylinder as

$$f_\infty(\theta) = \left(\frac{2r}{a} \right)^{1/2} \frac{p_s(\theta)}{p_o}, \tag{4}$$

and from (3) and (4) the individual partial waves $f_n(\pi)$ are given by

$$f_n(\pi) = \frac{-2}{(i\pi Z)^{1/2}} (\epsilon_n (-1)^n G_n(Z)), \tag{5a}$$

$$f_\infty(\pi) \equiv \sum_{n=0}^{\infty} f_n(\pi). \tag{5b}$$

For Dirichlet and Neumann boundary conditions, (5) reduces to a simple form, as seen in Fig. 2. The steady state form function in Fig. 2 is made up of the interference of specular reflection and a FRANZ [6]-type circumferential wave [7]. This diffractive contribution is much stronger in the case of a rigid cylinder (Neumann boundary conditions) and can be isolated by experiments performed in air where rigid-boundary conditions can be satisfied [8,9].

For bodies immersed in water, the form function curves are more complicated than those seen in Fig. 2, as demonstrated by the form function for a submerged aluminum cylinder given in Fig. 3. It was noted by previous researchers that the major features of the elastic form-function curves are related to the free modes of vibration of the target [1,10,11]. These features are identified in Fig. 3 by the labels (n, l), where n is the normal mode number $n = 1, 2, \ldots$ and l is the particular eigenfrequency, $l = 1, 2, \ldots$.

The form functions for various materials are generally greatly different [2,3], and can be measured to within a high degree of accuracy [9,12,13] for both submerged spheres [9,12] and cylinders [13,14]. Despite the disparity in appearence between form functions for various materials, it can be demonstrated that in general all solid targets made of materials whose density and shear speed are greater than the density and sound speed of water can be described in similar fashion. For such targets, the scattering solution can be interpreted in terms of a resonance behavior superimposed on a rigid background [15]. Solid elastic bodies can be regarded as rigid bodies except in the *ka* intervals over which resonances occur. This is demonstrated in Fig. 4, which shows a plot resulting from the subtraction of the $n = 2$ partial wave in the normal mode series solution for a ridid cylinder from the $n = 2$ partial wave term for an aluminum cylinder. The amplitude is nearly zero except in the *ka* region in which the (2,1), (2,2), and (2,3) resonances are excited.

The resonance theory was mathematically formalized in terms analogous to the existing resonance formalism of nuclear scattering theory and this formalism can be found in [8,15]. The result derived in these latter references shows that an individual partial waves can be written from (5) as

$$f_n(\theta) = \frac{2i\xi_n}{(i\pi ka)^{1/2}} e^{2i\xi_n} \left[\frac{1/2\Gamma_n}{Z_n - Z - 1/2i\Gamma_n} + e^{i\xi_n} \sin \xi_n \right] \cos n\theta, \tag{6}$$

where the first term represents the resonance contribution and the second term is the result obtained for the Neumann boundary condition. In (6), ξ_n is the scattering phase shift for a rigid cylinder, the Z_n are the *ka* values at which resonances occur, and the Γ_n are the resonance widths. The results exemplified by Fig. 4 are, thus, formalized mathematically by (6), and apply to solid submerged objects whose densities and sound speeds satisfy the conditions described earlier. Circumferential wave description of the scattering by simple shapes were given by ÜBERALL and collaborators [16-17] As discussed above, the rigid body form function results from the interference of specular reflection and a purely geometric circumferential wave. Elastic-body [16,17] scattering in water can be described in terms of R-type or Rayleigh type circumferential waves and a unification of the resonance formalism [8,15] and the circumferential wave results [16] has been accomplished [8,11].

A particular elastic circumferential wave, labeled $R_1, R_2 \cdots R_l$ by DOOLITTLE [16], is related to all resonances having the same eigenfrequency label l, i.e., all the $(n, 1)$ resonances are related to the R_1 circumferential wave, all the $(n, 2)$ resonances are related to the R_2 circumferential wave, etc. The resonances occur when the circumference of the cylinder or sphere is exactly n wavelengths in length. In resonance terms the n is related to the number of circumferential nodes and the l to the number of radial nodes. Of particular interest is the R_1 circumferential wave, since this is related to the leaky Rayleigh surface wave, as demonstrated in [18]. Observation of Fig. 3 shows that backscattered form-function features related to $(n, 1)$ resonances are significant only at low *ka*. Experimental isolation of the leaky Rayliegh wave was attempted based on a consideration of the $(n, 1)$ resonances in Fig. 3, and the wave was observed experimentally [19], as seen in Fig. 5, which shows the backscattered specular reflection and R_1 circumferential wave contributions isolated at $k_o a = 13.5$ for an aluminun cylinder. Here, k_o is $2\pi f_o / c$ where f_o is the center frequency of the incident pulse.

As seen in Fig. 3, the excitation of the (2,1) resonance marks the end of the purely rigid-like behavior of the form-function curve. By this is meant of that for *ka* values below the excitation of the (2,1) resonance the form-function curve for an aluminum cylinder in water is similar to that of the rigid cylinder (Fig. 2), and in general the excitation of the (2,1) resonance for any material having the properties discussed earlier marks the end of the purely rigid behavior. If aluminum were taken as a standard, the *ka* position of the (2,1) resonance $(ka)_{2,1}$ would be given closely by [8]

$$(ka)_{2,1} \text{ (material)} = (ka)_{2,1} \text{ (Al)} \cdot \frac{C_{\text{shear}} \text{ (material)}}{C_{\text{shear}} \text{ (Al)}}.$$

The $(n, 1)$ resonances are the lowest-frequency resonances strongly excited by a plane wave incident in water. They are related to a known mechanism, the Rayleigh surface wave, and shifts in postition of these resonances could give clues to the presence and position of flaws in materials.

A simple, though not practical calculation was made to demonstrate the above possibility. Figure 6 compares the form function vs *ka* curves for an infinte solid iron cylinder and an iron cylinder with a center hole whose diameter is 0.1 of the diameter of the cylinder. Note the frequency shift in the (2,1) resonance whereas the (3,1) resonance is changed in amplitude but not shifted in frequency. This is due to the fact that at $ka \approx 4.78$ [the (2,1) resonance position] the radius of the cylinder is approximately 1/2 the Rayleigh wavelength, so that the Rayleigh surface wave is interacts strongly with the hole whereas at $ka \approx 7.5$ [the position of the (3,1) resonance] the cylinder radius is a larger fraction of a wavelength and the interaction is smaller. As frequency is further increased, the Rayleigh surface wave has no interaction with the hole, i.e., the (4,1) and higher order $(n,1)$ resonances are not affected. If the hole were off center, the interactions should be different and the effect on the (3,1) and (4,1) resonances would become more substantial. It has been shown previously [9,10,12,13] that the *ka* position of the $(n,1)$ resonance nulls can be obtained with great accuracy and in a real-time framework [13]. The *ka* position of resonances with higher order eigenfrequencies l are also affected by the presence of the hole, but these are not considered here.

The resonance formalism can also be developed and applied to problems in which the scatterer is contained in a solid matrix. A review of elastic wave scattering of both P and S waves from cavities and inclusions in a solid matrix is given in [20]. Of interest here are the results obtained from a resonance analysis of the scattering of a P wave by inclusions in a metal matrix. As in the case of the solid cylinder in water, the individual partial waves can be separated into a background term and a resonant term, and the *ka* position of the resonances is determined by solving an eigenvalue equation, which for the case of nonabsorbing media is a real equation with real solutions. Resonance solutions for the resulting P-wave scattering amplitude and the S-wave wave scattering amplitude were obtained for an iron inclusion in an aluminum matrix [21]. In the case of the P-wave scattering amplitude the results are analagous to those obtained for the submerged body cases considered above, i.e., the background term is a rigid body solution. The derivation of the resonance expression for the spherical inclusion problem is given in [22]. The results are given for $n = 1$ and $n = 2$ in Fig. 7, and show the isolation of the resonance behavior. The relationship between resonances and circumferential waves in the solid matrix case is discussed in [22].

SUMMARY

The scattering by submerged solid elastic bodies whose density and shear speeds are larger than the density and speed of sound in water can be described in terms of a resonance behavior. The bodies act as rigid scatterers, except in the *ka* regions where free-body resonances are excited. This resonance behavior has been mathematically formalized in terms analogous to nuclear reaction theory. The relationship between resonance behavior and circumferential waves has been established and used to predict the presence of backscattering on aluminum cylinders due to the Rayleigh surface wave at *ka* values below 15. The ease of measurement of the $(n,1)$ resonances and shifts in their position due to the presence of flaws give a possible method of flaw detection.

The resonance solution to the scattering of a P wave by iron inclusions in an aluminum matrix bears some similarity to the description of bodies in a fluid in that for scattered P waves the background behavior is found to be rigid.

REFERENCES

[1] J.J. Faran, Jr.: "Sound scattering by solid cylinders and spheres," J. Acoust. Soc. Am. **23**, 405 (1951).
[2] R. Hickling: "Analysis of echoes from a solid elastic sphere in water," J. Acoust. Soc. Am. **34**, 1582 (1962).
[3] R. Hickling, "Analysis of echoes from a hollow metallic sphere in water." J. Acoust. Soc. Am. **36**, 1124 (1964).
[4] L. Flax, W. G. Neubauer: "Acoustic reflection from layered elastic absorptive cylinders," J. Acoust. Soc. Am. **61**, 307 (1977).
[5] L. Flax: "High *ka* scattering of elastic cylinders and spheres," J. Acoust. Soc. Am. **62**, 1502 (1977).
[6] W. Franz, "Über die Greenschen Funktionen des Zylinders und der Kugel," Z. für Naturforsch **9a**, 705 (1954).
[7] H. Überall, R.D. Doolittle, J.V. McNicholas: "Use of sound pulses for a study of circumferential waves," J. Acoust. Soc. Am. **39**, 564 (1966).
[8] L. R. Dragonette: "Evaluation of the relative importance of circumferential or creeping waves in the acoustic scattering from rigid and elastic solid cylinders and from cylindrical shells," Naval Research Laboratory Report 8216 (1978).
[9] L. R. Dragonette: R. H. Vogt, L. Flax, W. G. Neubauer, "Acoustic reflection from elastic spheres and rigid spheres and spheroids II. Transient analysis," J. Acoust. Soc. Am. **55**, 1130 (1974).
[10] R. H. Vogt, W. G. Neubauer: "Relationship between acoustic reflection and vibrational modes of elastic spheres," J. Acoust. Soc. Am. **60**, 15 (1976).
[11] H. Überall, L. R. Dragonette, L. Flax: "Relation between creeping waves and normal modes of vibration of a curved body," J. Acoust. Soc. Am. **61**, 711 (1977).

[12] W. G. Neubauer, R. H. Vogt, L. R. Dragonette: "Acoustic reflection from elastic spheres. I. Steady-state signals," J. Acoust. Soc. Am. **55**, 1123 (1974).
[13] H. D. Dardy, J. A. Bucaro, L. S. Schuetz, L. R. Dragonette: "Dynamic wide-bandwidth acoustic form-function determination," J. Acoust. Soc. Am. **62,** 1373 (1977).
[14] L. S. Schuetz, W. G. Neubauer: "Acoustic reflection from cylinders—nonabsorbing and absorbing," J. Acoust. Soc. Am. **63,** 513 (1977).
[15] L. Flax, L. R. Dragonette, H. Überall: "Theory of elastic resonance excitation by sound scattering," J. Acoust. Soc. Am. **63,** 723 (1978).
[16] R. D. Doolittle, H. Überall, P. Ugincius: "Sound scattering by elastic cylinders," J. Acoust. Soc. Am. 43, 1 (1968).
[17] P. Ugincius, H. Überall: "Creeping-Wave analysis of acoustic scattering by elastic cylindrical shells," J. Acoust. Soc. Am. **43**, 1025 (1968).
[18] G. V. Frisk, H. Überall: "Creeping waves and internal waves in acoustic scattering by layer elastic cylinders," J. Acoust. Soc. Am. **59**, 46 (1976).
[19] L. R. Dragonette: "The influence of the Rayleigh surface wave on the backscattering by submerged aluminum cylinders," J. Acoust. Soc. Am. **65**, 1570 (1979).
[20] G.C. Gaunaurd, H. Überall: "Numerical evaluation of modal resonances in the echoes of compressional waves scattered from fluid-filled spherical cavities in solids," J. Appl. Phys., in press.
[21] L. Flax, J. George, H. Überall: "Resonating fields inside elastic scattering objects," J. Acoust. Soc. Am. **65**, Suppl. 1, S128 (1979).
[22] L. Flax, H. Überall: "Resonant scattering of elastic waves from spherical solid inclusions," submitted to J. Acoust. Soc. Am.

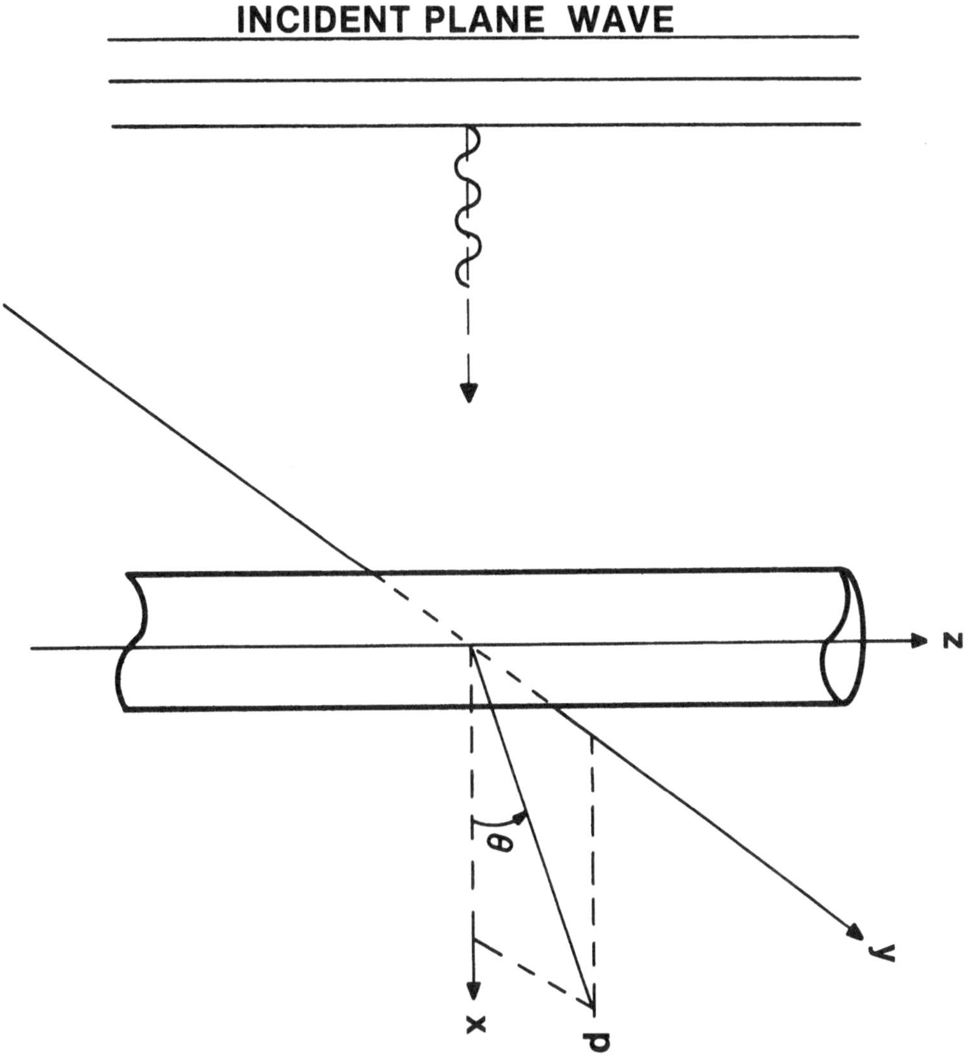

Fig. 1. The geometry of the cylindrical problem

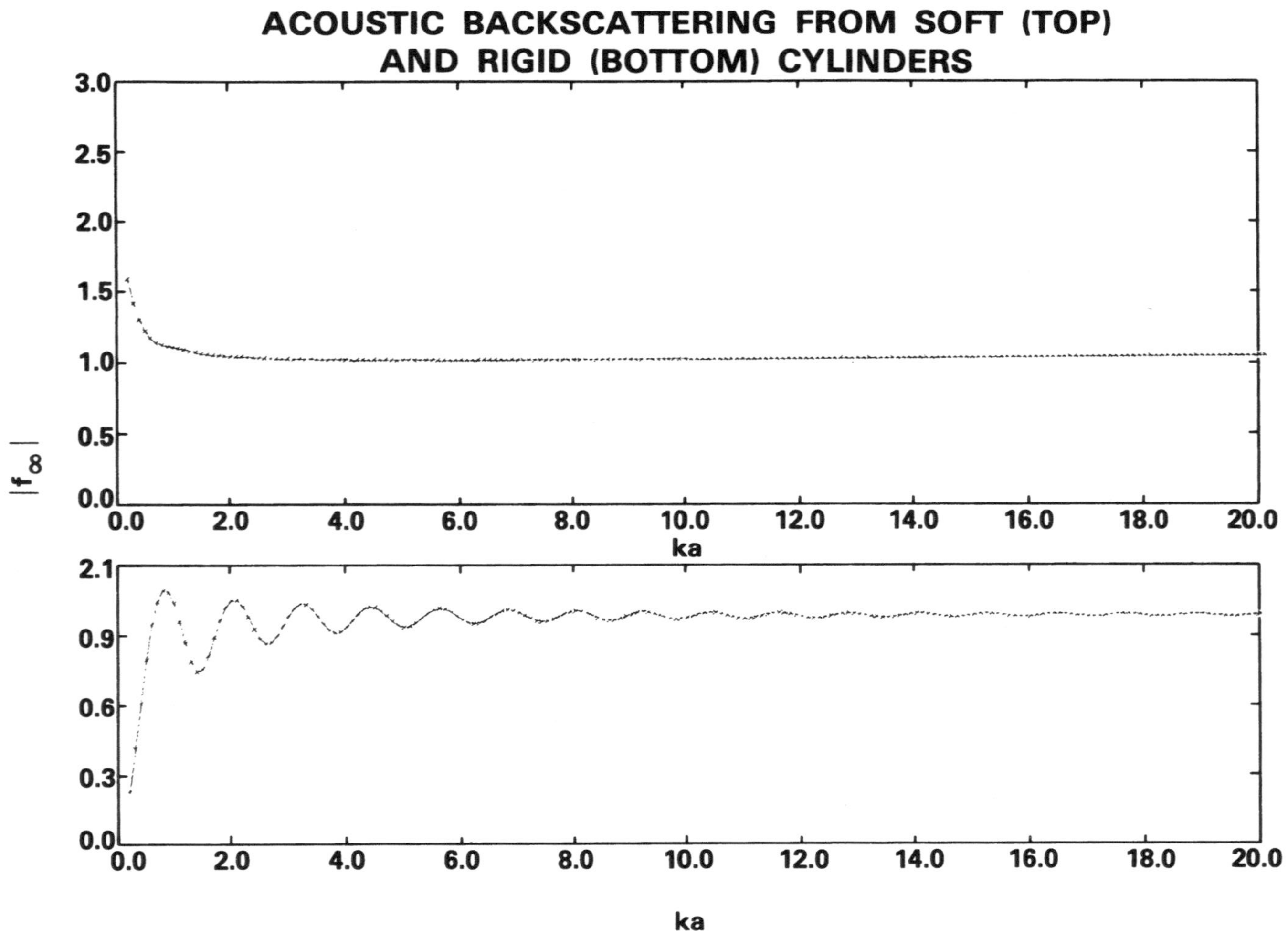

Fig. 2. The form function for acoustically soft (upper) and rigid (lower) cylinders.

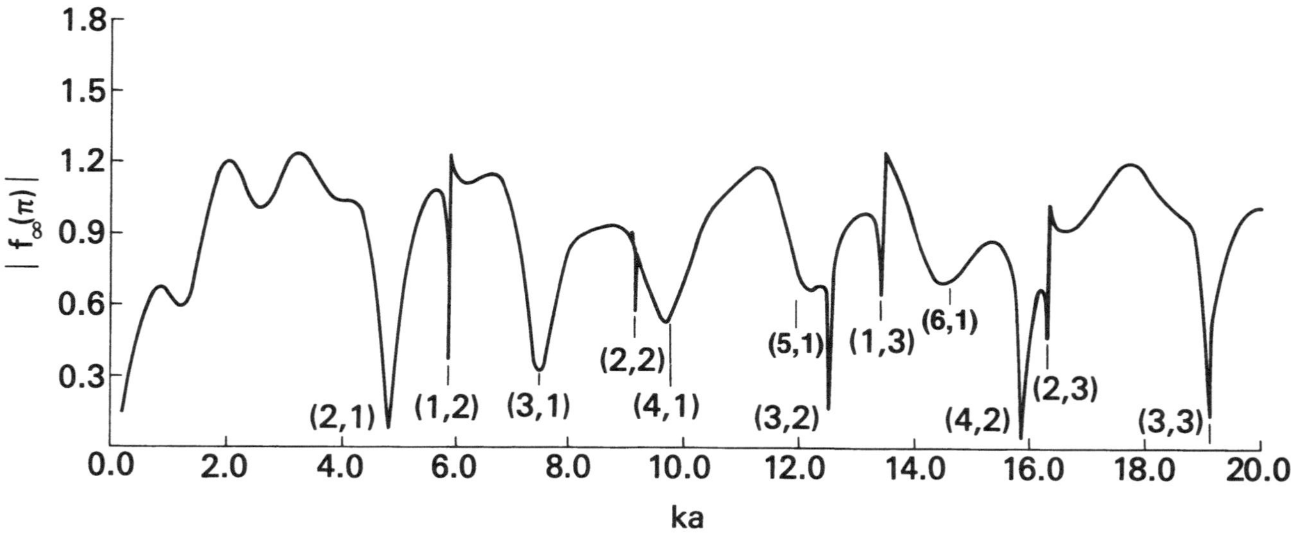

Fig. 3. The form function for an aluminum cylinder in water.

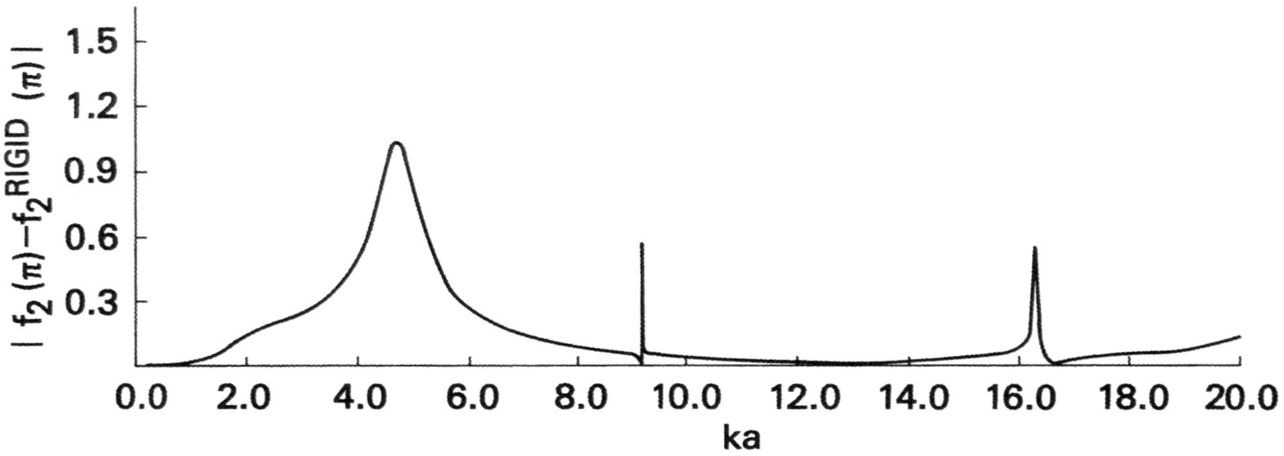

Fig. 4. Isolation of the (2,l) resonances obtained by subtracting the f_2 partial wave amplitudes for an aluminum cylinder from the f_2 term for a rigid cylinder.

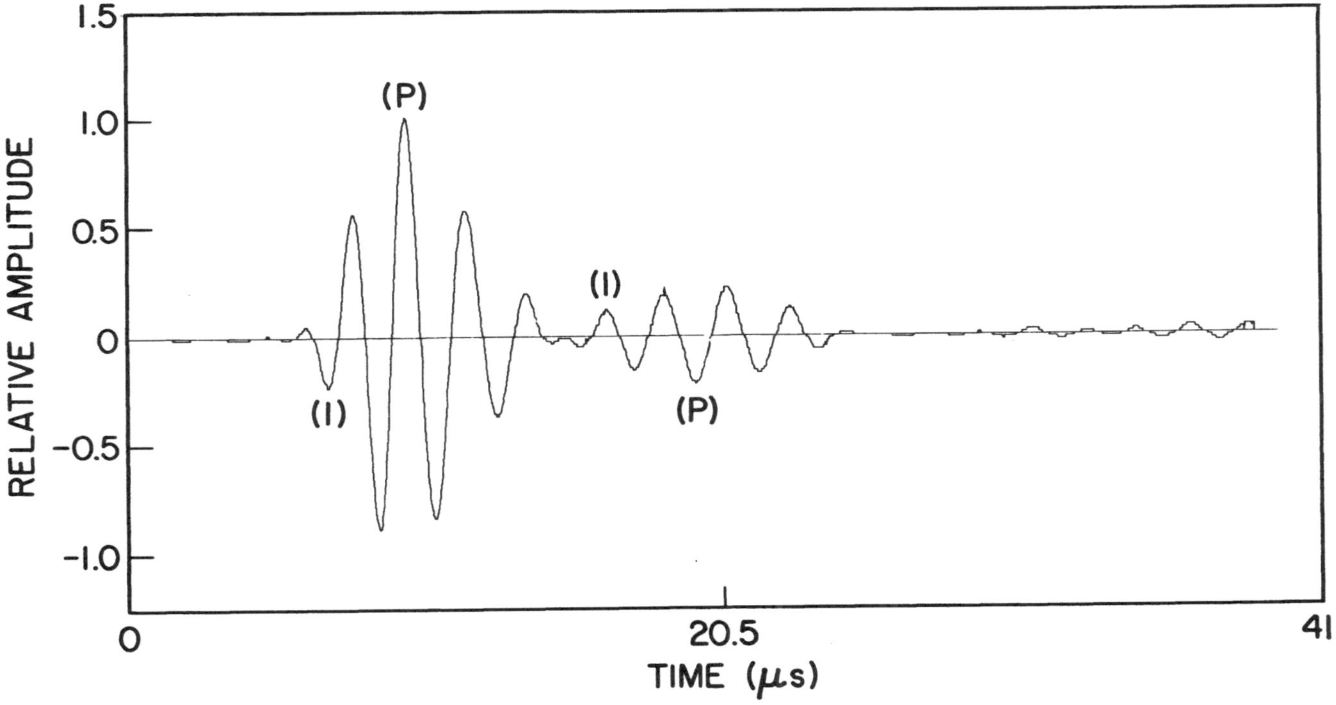

Fig. 5. Experimental isolation of the circumferential wave associated with the Rayleigh surface wave at ka = 13.5.

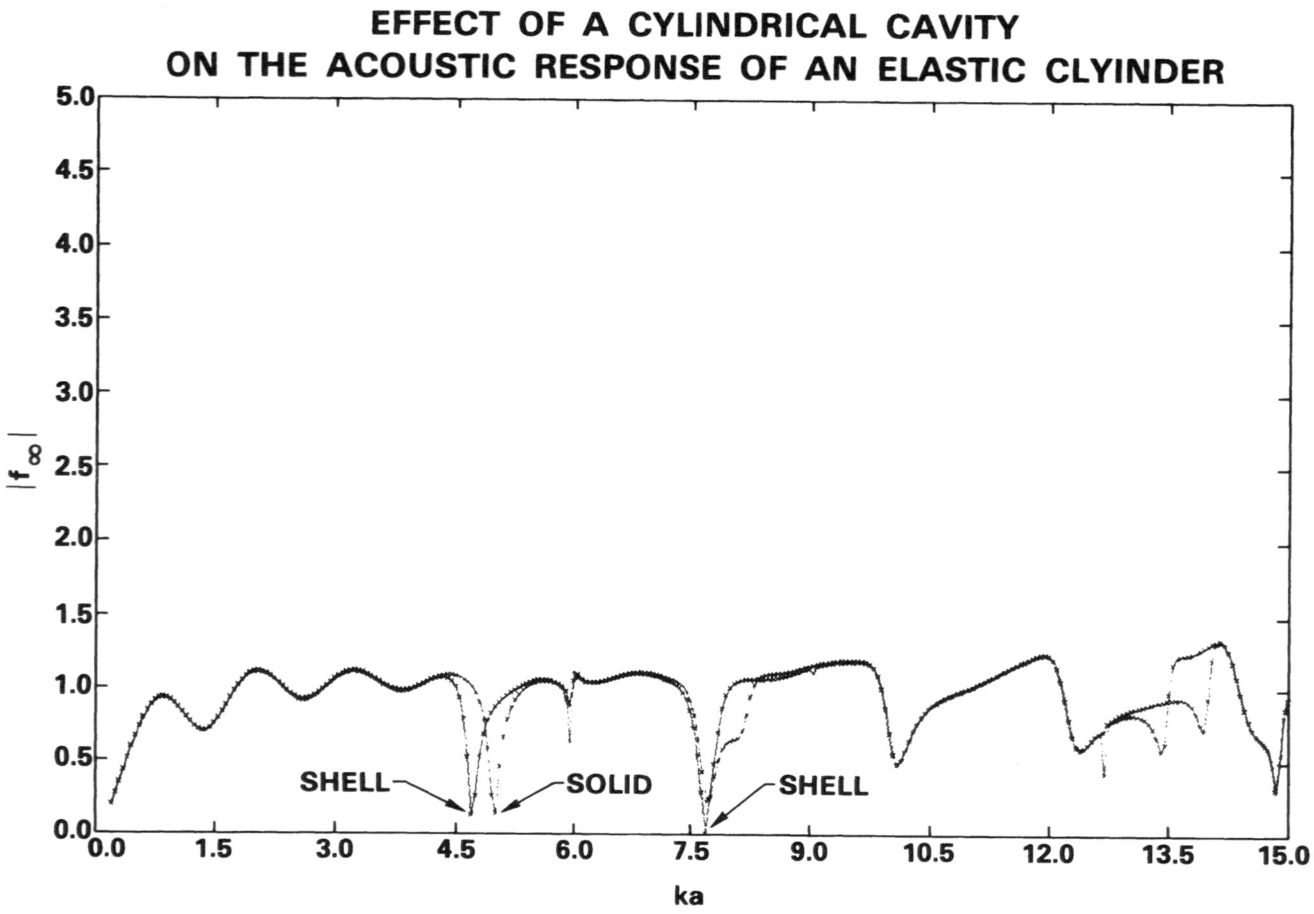

Fig. 6. The effect of a centrally located hole on the resonances associated with the Rayleigh surface wave on an aluminum cylinder.

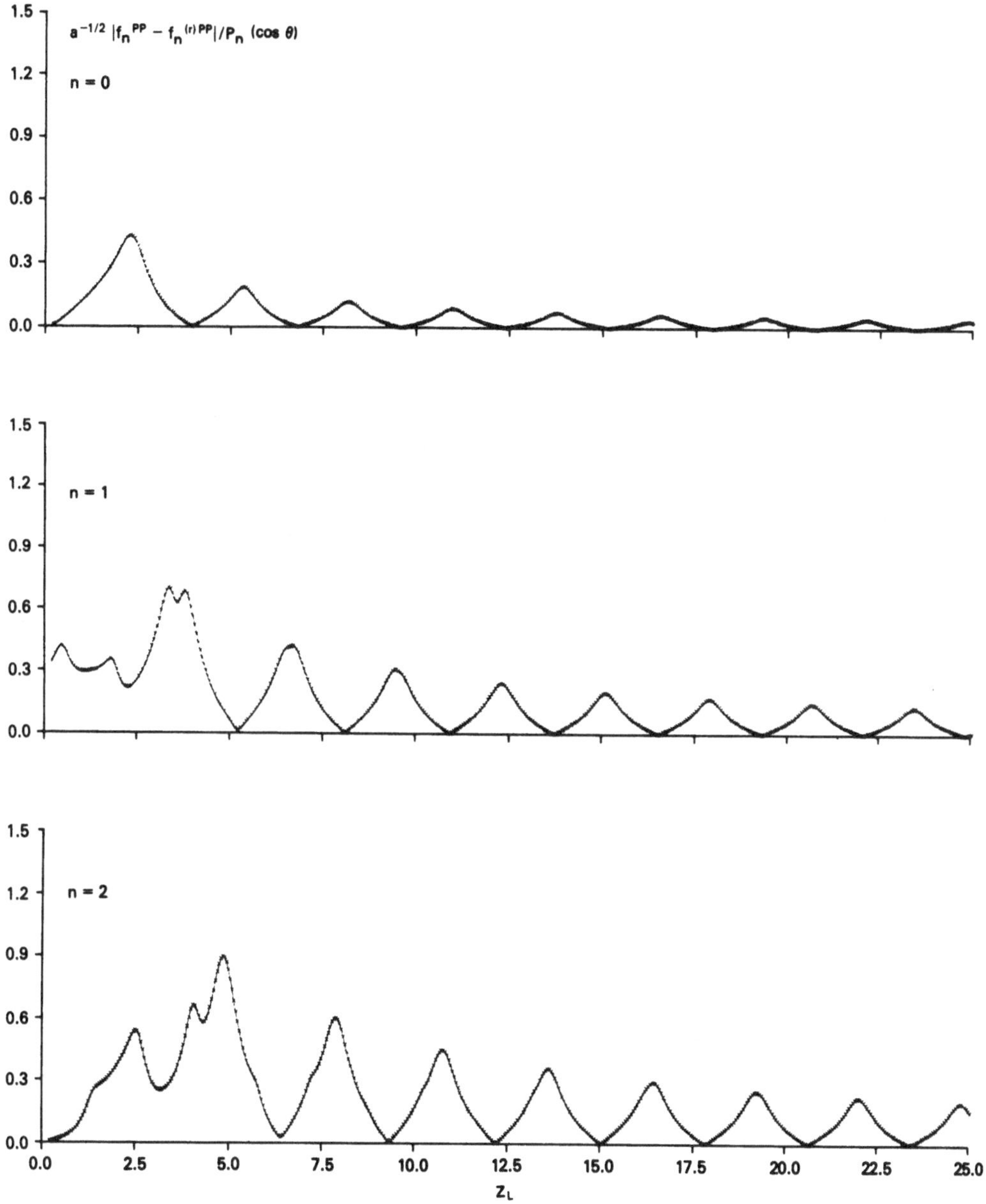

Fig. 7. Isolation of the n = 1 and n = 2 resonances for a spherical ion inclusion in an aluminum matrix.

APPLICATION OF ELASTODYNAMIC RAY THEORY TO DIFFRACTION BY CRACKS: THEORY AND EXPERIMENT

J.D. Achenbach

The Technological Institute
Northwestern University
Evanston, Illinois 60201

L. Adler

Department of Physics
University of Tennessee
Knoxville, Tennessee 37916

INTRODUCTION

In the high-frequency domain the diffraction of elastic waves by cracks can be analyzed conveniently on the basis of elastodynamic ray theory. For time-harmonic wave motion, ray theory provides a method to trace the amplitude of a disturbance as it propagates along a ray. In a homogeneous, isotropic, linearly elastic solid the rays are straight lines, which are normal to the wavefronts. An unbounded solid can support rays of longitudinal and transverse wave motion. These rays are denoted as L-rays and T-rays, respectively.

In analogy with geometrical optics, the simplest theory for diffraction of elastic waves by cracks may be called geometrical elastodynamics (GE). In GE a crack acts as a screen which creates a shadow zone of no motion, and zones of reflected waves. The geometrical theory of diffraction (GTD) provides a first correction to GE.

For *plane* longitudinal and transverse waves, which are under arbitrary angles of incidence with a traction-free *semi-infinite* crack, the fields on the diffracted rays can be obtained by asymptotic considerations, as shown by ACHENBACH *et al.* [1,2]. The results can be expressed in terms of diffraction coefficients which relate the diffracted fields to the incident fields. Geometrical diffraction theory provides modifications to the semi-infinite crack results, to account for curvature of incident wave-fronts and curvature of crack edges, and finite dimensions of the crack. In the usual terminology the results for diffraction of plane waves by a semi-infinite crack are the canonical solutions. For incident waves with curved wavefronts and for curved diffracting edges, the cones of diffracted rays have envelopes, at which the rays coalesce and the fields become singular. The envelopes are called caustics, and GTD breaks down at caustics.

Results obtained on the basis of GTD have been presented in [3-5]. In [3] results obtained by elastodynamic ray theory have been compared with results obtained by numerical solution of a governing singular integral equation.

1. EXPERIMENT

Experimental results in the high-frequency range that are suitable for comparison with theoretical results have been reported by ADLER *et al.* (see, e.g., [6,7]). The sample was a circular disk (2.5×10 cm) of titanium alloy which contained a penny-shaped crack of radius 2500μ parallel to the flat faces, and located at the center of the disk. The disk was immersed in water. A transmitter launched a longitudinal wave to the water-titanium interface under normal incidence. This wave was transmitted into the solid, diffracted by the crack, and the diffracted waves were transmitted back into the fluid, where they were received by a second transducer. The experimental setup and the processing of the data are discussed in some detail in [7].

In the experimental work, the nature of the diffracted signals is determined by their arrival times. Since the first arriving signals are related to longitudinal waves in the solid, it is possible to gate out and separate the purely longitudinal diffracted signals from subsequent signals. By appropriate processing of the experimental data, as described in [7], the amplitude-spectrum is obtained for the longitudinal diffracted waves only. Thus, for the present comparison of analytical and experimental results we need to consider only the primary diffracted body-wave rays in our analytical work.

2. THEORETICAL RESULTS FOR A PENNY-SHAPED CRACK

The interference patterns for the first arriving longitudinal waves in the fluid are generated by phase differences and amplitude differences on the direct rays from the two crack tips (see Fig. 1). Adding the primary diffracted longitudinal fields from the points O_1 and O_2 we obtain at the point B in the far field:

$$\mathbf{u}_L \sim F(\theta,\theta_o)\exp[i\omega(S/c_L + \bar{S}/c_F) + i\pi/4]\,U_o\mathbf{i}_F, \tag{1}$$

$$F(\theta,\theta_o) = H_1\exp[-i(\omega a/c_L)(\cos\theta - \sin\theta_o)] + H_2\exp[i(\omega a/c_L)(\cos\theta - \sin\theta_o)], \tag{2}$$

$$H_j = \frac{\mathrm{sgn}(\cos\theta_j)\,T(\theta_L^j)\,|D_L^L(\theta_j;\theta_f^j)|}{(\omega S_j/c_L)^{1/2}(1+S_j/C)^{1/2}(1+\bar{S}_j/\bar{E})^{1/2}(1+\bar{S}_j/\bar{C})^{1/2}},\ j = 1,2. \tag{3}$$

In (1), $\mathbf{u}_L$ is the diffracted longitudinal field and $\mathbf{i}_F$ is defined in Fig. 1. Moreover, ω is the circular frequency, $\bar{S} = \overline{AB}$, U_o represents the incident wave at point 0, c_L and c_F are the velocities of longitudinal waves in the solid and fluid, respectively, and a is the crack radius. The geometrical quantities are indicated in Fig. 1. In (3), $T(\theta_L^j)$ is the transmission coefficient at the solid-fluid interface, and $D_L^L(\theta_j;\theta_f^j)$ is the diffraction coefficient. For details of the derivation of (1) - (3), and the definition of C, $\bar{E}$, and $\bar{C}$, we refer to [7]. It should be noted that one of the terms H_i is imaginary, since the ray has crossed a caustic. Of particular interest is the absolute magnitude of F,

$$|F| = \{|H_1|^2 + |H_2|^2 + 2|H_1||H_2|\sin[2(\omega a/c_L)(\cos\theta - \sin\theta_o)]\}^{1/2}. \tag{4}$$

Here we have taken into account that either H_1 or H_2 is imaginary.

3. COMPARISONS WITH EXPERIMENTAL DATA

Theoretical results obtained from (4) have been plotted together with experimental data in Fig. 2. The frequency varies from 2 MHz to about 14 MHz. The angles in the solid are $\theta'(= \pi/2 - \theta) = 35°, 45°, 55°$ and $60°$, respectively. The amplitudes of the first few cycles agree well. At higher frequencies (above 6 MHz) the experimental results are lower than predicted by theory. One possible explanation is the effect of attenuation which is not accounted for in the theory. In all cases the positions of maxima and minima of the spectra agree well. The locations of the maxima are significant for the inversion process.

INVERSE PROBLEM

The discussion of the previous sections has been concerned with the direct problem, that is the computation of the scattered field when the size, shape and orientation of the crack are known. We will conclude with a few comments on the inverse problem for plane waves incident on penny-shaped cracks, for the special case that the incident wave is known to be in a plane of symmetry of the crack. The geometry in the plane of symmetry is then as shown in Fig. 1, where the incident wave is under an oblique angle with the plane of the crack. For a given point of observation, say the point B in Fig. 1, the unknowns then are θ_o, a, and b.

The theoretical expression for the amplitude spectrum given by (4) implies that the amplitude of the primary diffracted field is modulated with respect to ω/c_L, with period

$$P = \pi/[a|\cos\theta - \sin\theta_o|]. \tag{5}$$

It is of interest to apply (5) to the experimental measurements. Since we know that at $\theta_o = 0$ each amplitude spectrum will give a number for a from the periodicity of the modulation. We have

$$a = \frac{c_L}{2\sin(\theta')\Delta f_{ave}}, \tag{6}$$

where $\theta' = \pi/2 - \theta$ and Δf_{ave} is the average frequency spacing between two consecutive maxima.

The results of the size determination are given in Table I. The agreement between actual crack radius ($a = 2500\mu$) and the predicted values is excellent.

TABLE I
Crack radius a computed from Eq. (6) for a penny-shaped crack in titanium ($c_L = 6330$ m/s)

$\theta' = \pi/2 - \theta$	Δf_{ave}	computed a in μ
35°	2.18	2530
40	1.87	2630
45	1.83	2450
50	1.68	2460
55	1.60	2410
60	1.47	2500
65	1.39	2510

ACKNOWLEDGMENTS

This paper was written in the course of research sponsored by the Center for Advanced NDE operated by the Science Center, Rockwell International, for the Advanced Research Projects Agency and the Air Force Materials Laboratory under Contract F33615-74-C-5180.

FOOTNOTES AND REFERENCES

[1] J.D. Achenbach, A.K. Gautesen: "Geometrical theory of diffraction for three-d elastodynamics," J. Acoust. Soc. Am. **61**, 413-421 (1977)

[2] A.K. Gautesen, J.D. Achenbach, H. McMaken: "Surface wave rays in elastodynamic diffraction by cracks," J. Acoust. Soc. Am. **63**, 1824-1831 (1978)

[3] J.D. Achenbach, A.K. Gautesen, H. McMaken: "Application of elastodynamic ray theory to diffraction by cracks," in *Modern Problems in Elastic Wave Propagation*, ed. by J. Miklowitz and J.D. Achenbach (Wiley-Interscience, New York, 1978), pp. 219-238

[4] J.D. Achenbach, A.K. Gautesen, H. McMaken: "Diffraction of point-source signals by a circular crack," Bull. Seism. Soc. Am. **68**, 889-905 (1978)

[5] J.D. Achenbach, A.K. Gautesen, H. McMaken: "Diffraction of elastic waves by cracks—analytical results," in *Elastic Waves and Non-Destructive Testing of Materials*, ed. by Y.H. Pao, AMD **29** (Am. Soc. of Mech. Eng., New York, 1978), pp. 33-52

[6] L. Adler, H.L. Whaley: "Interference effect in a multifrequency ultrasonic pulse echo and its application to flaw characterization," J. Acoust. Soc. Am. **51**, 881-887 (1972)

[7] J.D. Achenbach, L. Adler, D. Kent Lewis, H. McMaken: "Diffraction of ultrasonic waves by penny-shaped cracks in metals: theory and experiment," J. Acoust. Soc. Am., in press

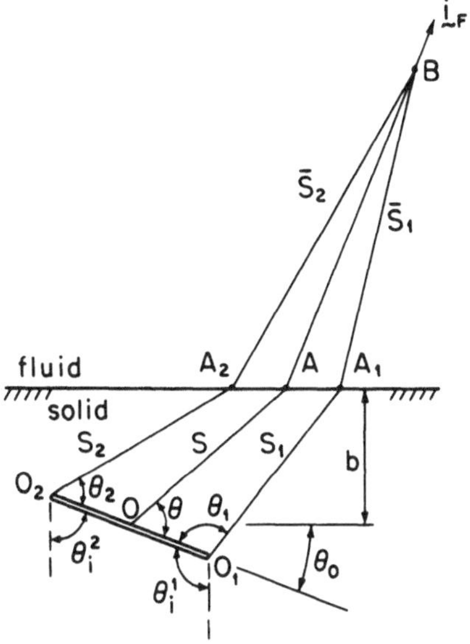

Fig. 1. Penny-shaped crack and associated ray diagram .

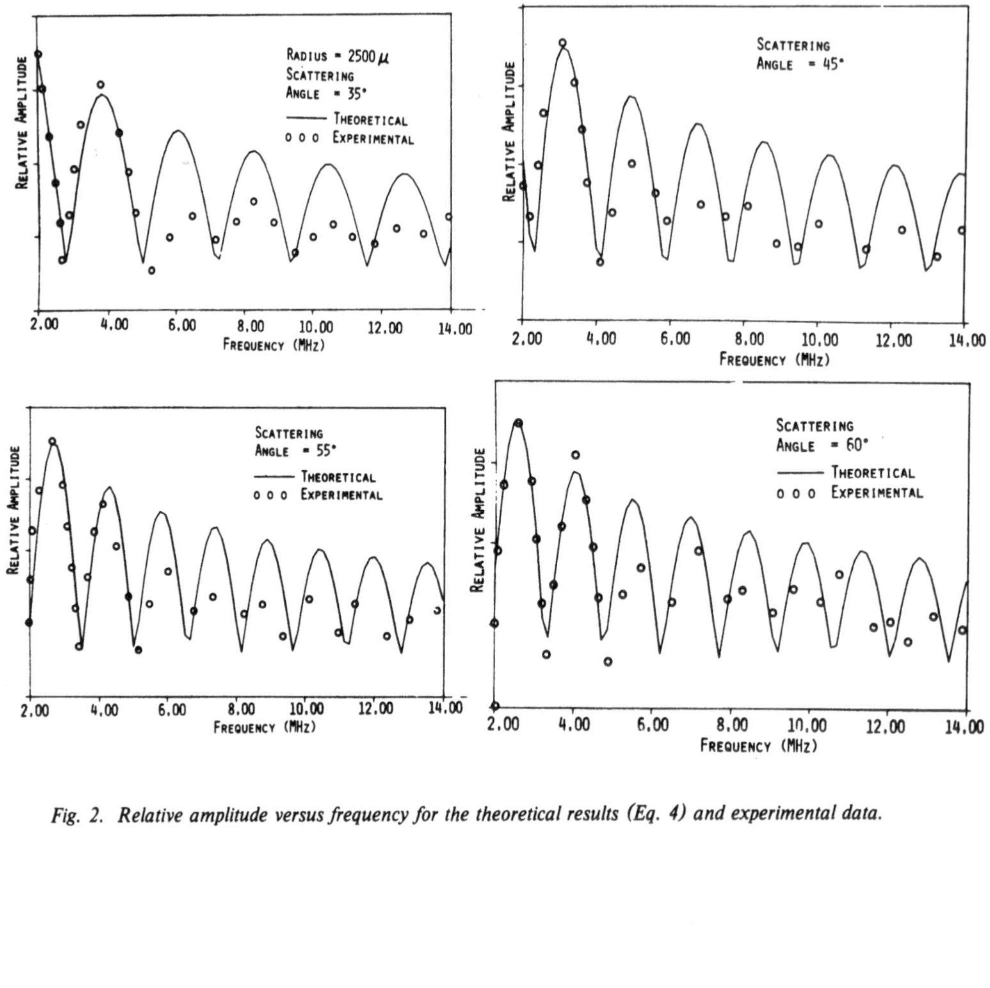

Fig. 2. Relative amplitude versus frequency for the theoretical results (Eq. 4) and experimental data.

SCATTERING OF ACOUSTIC WAVES BY ELASTIC AND VISCOELASTIC OBSTACLES OF ARBITRARY SHAPE IMMERSED IN WATER

Bo. Å. Peterson, Vasundara V. Varadan, and Vijay K. Varadan

Department of Engineering Mechanics
Ohio State University
Columbus, Ohio 43210

INTRODUCTION

In this article, we present an analysis of scattering of acoustic waves by elastic and viscoelastic obstacles of arbitrary shape immersed in a fluid. This problem is difficult because the wave equation governing the regions inside and outside the scatterer are different and admit different types of wave solutions, and these solutions must be coupled at the boundary by the continuity conditions for the pressure and particle velocity. To the authors' knowledge, no detailed discussion with numerical results for both elastic and viscoelastic obstacles is available in the literature, except for circular and spherical obstacles.

Here we apply the *T*-matrix or null field method to study our problem. The *T*-matrix method has been successful and computationally very efficient for studying the scattering of acoustic, electromagnetic, and elastic waves by single obstacles, finite numbers of obstacles, layered obstacles, and statistical distributions of obstacles [1]. In deriving the *T*-matrix formalism starting with the Helmholtz formulas in the fluid space, one arrives at a *Q*-matrix which is not square and, hence, cannot be inverted. Additional representations of the scattered and refracted field must be considered to overcome this difficulty. The analysis is explained in detail in [2,3].

1. FORMULATION OF THE PROBLEM

We consider an elastic or viscoelastic scatterer of arbitrary surface S (with a continuously-turning unit normal $\hat{n}$) immersed in an invicid fluid. The elastic properties of the scatterer are given by the Lamé constants λ and μ, and the mass density ρ, while the properties of the fluid are given by the compressibility λ_f and the mass density ρ_f. If the scatterer is viscoelastic, λ and μ are complex and frequency-dependent.

A plane acoustic wave of unit amplitude, frequency ω and wave number k_f is incident obliquely to the scatterer. We denote the incident-wave and scattered-wave displacements by $\vec{u}^0$ and $\vec{u}^s$, respectively. The starting point of our *T*-matrix formalism is to express the interior and exterior Helmholtz integral representations for the fluid and solid as follows [2,3]:

$$\vec{u}^0(\vec{r}) + \int_S \{\vec{u}_{f'} \cdot [\hat{n}' \cdot \vec{\vec{\Sigma}}_f(\vec{r} \cdot \vec{r}')] - \hat{n}' \cdot \vec{\vec{\tau}}_{f'} \cdot \vec{\vec{G}}_f(\vec{r}, \vec{r}')\} \, ds' = \begin{cases} \vec{u}_f(\vec{r}), \ \vec{r} \text{ outside } S, \\ 0, \ \vec{r} \text{ inside } S. \end{cases} \tag{1}$$

For the solid,

$$-\int_S \{\vec{u}' \cdot \left[\hat{n}' \cdot \vec{\vec{\Sigma}}(\vec{r}, \vec{r}')\right] - \hat{n} \cdot \vec{\vec{\tau}}' \cdot \vec{\vec{G}}(\vec{r}, \vec{r}') \, dS'\} = \begin{Bmatrix} \vec{u}(\vec{r}), \ \vec{r} \text{ inside } S \\ 0, \ \vec{r} \text{ outside } S \end{Bmatrix}, \tag{2}$$

where $\vec{u}$ is the displacement vector, $\vec{\vec{\tau}}$ is the stress tensor and $\vec{\vec{\Sigma}}$ and $\vec{\vec{G}}$ are the Green's stress and displacement tensors. The terms with subscripts f refer to the corresponding quantities in the fluid.

The philosophy of the *T*-matrix approach is to expand all the terms appearing in the integral representations in terms of spherical vector basis functions, $\vec{\psi}_{1\sigma mn}$, $\vec{\psi}_{2\sigma mn}$, $\vec{\psi}_{3\sigma mn}$, $\vec{\psi}_{\tau n}$, $\tau = 1, 2, 3$. The subscript $\tau = 1$ refers to the compressional wave functions, while $\tau = 2, 3$ refer to shear wave functions. Since the fluid supports only compressional waves, we need only the $\tau = 1$ component of the basis functions denoted by $\vec{\psi}_{fn}$.

We expand the incident and scattered wave fields, the field inside the scatterer, and the Green's tensors in terms of the basis functions:

$$\vec{u}^0(\vec{r}) = \sum_n A_n \mathrm{Re}\vec{\psi}_{fn}(\vec{r}), \tag{3}$$

$$\vec{u}^S(\vec{r}) = \sum_n f_n \mathrm{Ou}\vec{\psi}_{fn}(\vec{r}), \tag{4}$$

$$\vec{u}(\vec{r}\,') = \sum_n \alpha_{\tau n} \mathrm{Re}\vec{\psi}_{\tau n}(\vec{r}\,'), \tag{5}$$

and

$$\vec{\vec{G}}_f(\vec{r}, \vec{r}\,') = \frac{ik_f}{\rho_f \omega^2} \sum_n \mathrm{Ou}\vec{\psi}_{fn}(r>) \mathrm{Re}\vec{\psi}_{fn}(r<). \tag{6}$$

In the above equations, the symbols Ou and Re will represent the outgoing and regular functions, respectively. For example, Ou $\vec{\psi}_q = \vec{\psi}_q$, but, in contrast, Re$\vec{\psi}_q$ means that, instead of using h_n, we use the regular function j_n (at the origin). Substituting these expansions (3)-(6) in (1), and using the continuity and boundary conditions at the surface of the scatterer in the integral equation for the fluid, we obtain

$$-A_n = \sum iQ_{n,\tau n'}(\mathrm{Ou},\mathrm{Re})\alpha_{\tau n'}, \tag{7}$$

$$f_n = \sum iQ_{n,\tau n'}(\mathrm{Re},\mathrm{Re})\alpha_{\tau n'}, \tag{8}$$

where the matrix Q is given by

$$Q_{n,\tau n}\begin{pmatrix}\mathrm{Ou} & \\ & \mathrm{Re} \\ \mathrm{Re} & \end{pmatrix} = \frac{k_f}{\rho_f \omega^2}\int_S \left\{ \lambda_f \nabla \cdot \begin{pmatrix}\mathrm{Ou} & \\ & \vec{\psi}_{fn} \\ \mathrm{Re} & \end{pmatrix} \hat{n} \cdot \mathrm{Re}\vec{\psi}_{\tau n'} - \hat{n} \cdot \begin{pmatrix}\mathrm{Ou} & \\ & \vec{\psi}_{fn} \\ \mathrm{Re} & \end{pmatrix} \hat{n} \cdot \vec{\vec{\tau}}(\mathrm{Re}\vec{\psi}_{\tau n'}) \cdot \hat{n} \right\} dS \tag{9}$$

The Q-matrix has a 1×3 substructure and, hence, cannot be inverted. In order to obtain the desired T-matrix connecting A_n and f_n, we must use (2) until we arrive at a set of matrix equations that are invertible. To this end, we follow the work outlined in [2,3] to obtain:

$$\sum P_{\tau n, n'} d_{n'} + \sum R_{\tau n, \tau' n'} \alpha_{\tau' n'} = 0, \tag{10}$$

where the matrices P and R are given by

$$P_{\tau n, n'} = \frac{k_s}{\rho \omega^2} \int_S [\{\hat{n} \cdot \mathrm{Re}\vec{\psi}_{fn'}\}\{\hat{n} \cdot \vec{\vec{\tau}}(\mathrm{Re}\vec{\psi}_{\tau n}) \cdot \hat{n}\}] dS, \tag{11}$$

and

$$R_{\tau n, \tau' n'} = \frac{k_s}{\rho \omega^2} \int_S \{ (\mathrm{Re}\vec{\psi}_{\tau' n'})_{tang} \cdot \hat{n} \cdot \vec{\vec{\tau}}(\mathrm{Re}\vec{\psi}_{\tau n}) - (\hat{n} \cdot \vec{\vec{\tau}}(\mathrm{Re}\vec{\psi}_{\tau' n'}) \cdot \hat{n}) \hat{n} \cdot \mathrm{Re}\vec{\psi}_{\tau n} \} dS. \tag{12}$$

From (7), (8), and (10), we then obtain the following relationship between the incident and scattered field coefficients:

$$f = TA, \tag{13}$$

where

$$T = -Q(\mathrm{Re},\mathrm{Re})R^{-1}P[Q(\mathrm{Ou},\mathrm{Re})R^{-1}P]^{-1}. \tag{14}$$

The T-matrix defined by (2) is applicable to both elastic and viscoelastic obstacles of arbtirary shape immersed in a fluid. The wavenumbers for a viscoelastic solid are complex and frequency dependent. The T-matrix is symmetric for both elastic and viscoelastic scatterers, but the scattering matrix $S = 1-2T$ is no longer unitary if the material has a complex elastic modulus.

Once the scattered field coefficients are known from (13), the quantities of interest, such as backscattering and bistatic, absorption, and extinction cross sections, can be computed as a function of the frequency of the incident wave. One could also extend this formalism to layered obstacles immersed in a fluid [4]. Extensive numerical results are presented in [3,4].

ACKNOWLEDGMENTS

B.A. Peterson was supported by a post doctoral fellowship from the Graduate School and the Department of Engineering Mechanics of Ohio State University. Use of the Instructional and Research Computer Center at

OSU are gratefullly acknowledged. V. V. Varadan and V. K. Varadan were supported in part by the Office of Naval Research under Contract No. N00014-78-C-0559. Helpful discussions with Prof. Staffan Ström and Mr. Anders Boström are gratefully acknowledged.

FOOTNOTES AND REFERENCES

[1] *Recent Developments in Classical Wave Scattering: Focus on the T-matrix Approach*, ed. by V.V. Varadan, V.K. Varadan (Pergamon, New York, 1979)

[2] A. Boström: "Scattering of acoustic waves by elastic obstacles in water," Institute of Theoretical Physics Report 78-43, Chalmers, Göteborg, Sweden (1978)

[3] B. Peterson, V.V. Varadan, V.K. Varadan: "Scattering of acoustic waves by elastic and viscoelastic obstacles immersed in water," to appear in J. Wave Motion

[4] B. Peterson, V.V. Varadan, V.K. Varadan: "On the multiple scattering of waves from obstacles with solid-fluid interfaces," in [1]

MULTIPOLE RESONANCES IN ELASTIC WAVE-SCATTERING FROM CAVITIES AND IN ACOUSTIC WAVE-SCATTERING FROM BUBBLES AND DROPLETS

G. Gaunaurd

Naval Surface Weapons Center
White Oak, Silver Spring, Maryland 20910

ABSTRACT

By means of the Resonance Theory of viscoelastic wave-scattering from cavities in solids, we examine several multipole contributions to the sonar cross sections of fluid-filled cavities in solid rubbers. The analysis is done first ignoring, and then accounting for mode-conversion in the solid rubber matrix. The results are *analytically* particularized to the simpler cases of: a) the nth multipole of a gas-bubble in water, b) the nth multipole of a liquid droplet in a gas, c) the zeroth order multipole (i.e., monopole) contribution for a gas-filled cavity in solid rubber. Results for other multipoles are then *numerically* examined and displayed, up to the quadrupole case. The program we have developed can generate similar results for any higher-order multipole and any combination of substances, in any of the above situations.

INTRODUCTION

Compressional plane waves traveling through elastic media can be scattered by a fluid obstacle contained within the medium. We have studied the case of spherically shaped obstacles [1], accounting for and ignoring the presence of mode-conversion in the solid matrix. The resonance theory for this situation is well documented elsewhere [1,2] and will not be repeated here. We have also studied the monopole mode of vibration (i.e., $n=0$) in some detail [3] and found that the splitting of each modal contribution into "backgrounds and resonances" that characterizes this theory, was performed for the *monopole* case in an exact fashion without any linearization. In this contribution we extend those findings to other multipoles and to the particular situations mentioned in the Abstract.

1. THEORY

The normalized bistatic (i.e., differential) scattering cross-section of an obstacle in a nonviscous elastic material was found to be [2]

$$\frac{1}{a^2}\frac{d\sigma}{d\theta} = \left|\frac{1}{a}f^{pp}(\theta)\right|^2 + \frac{\kappa_s}{\kappa_d}\left|\frac{d}{d\theta}\left[\frac{1}{a}f^{ps}(\theta)\right]\right|^2, \tag{1}$$

where the quantity a is the obstacle radius, κ_d and κ_s are, respectively, the dilatational and shear wavenumbers of the material exterior to the obstacle, and the quantities $f^{pp}(\theta)$, $f^{ps}(\theta)$ are the scattering amplitudes (or normalized form-functions) of the returned waves in the absence or presence of mode-conversion, respectively. These latter quantities were defined in (19) of [1] in terms of coefficients A_n, B_n, as follows:

$$\frac{1}{a}f^{pp}(\theta) = \frac{1}{i\kappa_d a}\sum_{n=0}^{\infty}(2n+1)A_nP_n(\cos\theta),$$

$$\frac{1}{a}f^{ps}(\theta) = \frac{1}{i\kappa_s a}\sum_{n=0}^{\infty}(2n+1)B_nP_n(\cos\theta). \tag{1a}$$

Coefficients A_n,B_n were determined [1] by direct evaluation into the boundary conditions at the obstacle's surface, as the following ratios of 3 × 3 determinants:

$$A_n = -\frac{1}{D_n}\begin{vmatrix} \mathrm{Re}\, d_{11} & d_{12} & d_{13} \\ \mathrm{Re}\, d_{21} & d_{22} & d_{23} \\ \mathrm{Re}\, d_{31} & d_{32} & 0 \end{vmatrix}, \quad B_n = -\frac{1}{D_n}\begin{vmatrix} d_{11} & \mathrm{Re}\, d_{12} & d_{13} \\ d_{21} & \mathrm{Re}\, d_{22} & d_{23} \\ d_{31} & \mathrm{Re}\, d_{32} & 0 \end{vmatrix}, \tag{2}$$

where

$$D_n = \begin{vmatrix} d_{11} & d_{12} & d_{13} \\ d_{21} & d_{22} & d_{23} \\ d_{31} & d_{32} & 0 \end{vmatrix} \tag{3}$$

and the elements d_{ij} are listed in the Appendix.

The interior field inside the cavity is controlled (see Eq. (11) of [1]) by a coefficient C_n of the form

$$C_n = -\frac{1}{D_n}\begin{vmatrix} d_{11} & d_{12} & \mathrm{Re}\, d_{11} \\ d_{21} & d_{22} & \mathrm{Re}\, d_{21} \\ d_{31} & d_{32} & \mathrm{Re}\, d_{31} \end{vmatrix}. \tag{4}$$

Using Eqs. (1a) and the definition of backscattering (i.e., sonar) cross section, it can be verified that each partial-wave contribution contained within (1) reduces to

$$\left[\frac{\sigma}{4\pi a^2}\right]_n = \left|\frac{1}{a} f_n^{pp}(\pi)\right|^2 = \left|\frac{A_n}{ix}\right|^2, \tag{5}$$

where $x = \kappa_d a$ is a real quantity in the absence of absorption in the outer medium. Clearly, the second term in (1) has no contribution in the backscattering direction $\theta = \pi$ (because $dP_n(-1)/d\theta = 0$), or in the case where the host elastic medium can support no shear waves (i.e., an inviscid liquid of vanishing shear speed $c_s = 0$ or $\kappa_s = \infty$). This formulation already contains all the above mentioned particular cases, as we will see next.

2. EXAMPLES

We have shown earlier [3] that the coefficient A_n given in (2) can be written in the convenient form

$$A_n = \frac{1}{2}\left\{-\frac{h_n^{(2)}(x)}{h_n^{(1)}(x)}\left[\frac{L_n^{(2)} - M_n^{(2)}F_n}{L_n^{(1)} - M_n^{(1)}F_n}\right] - 1\right\}, \tag{6}$$

where

$$L_n^{(i)} = \hat{z}_n^{(i)} + \frac{n(n+1)(1-\hat{z}_n^{(1)})}{n(n+1) - 1 - (x_s^2/2) - \tilde{z}_n^{(1)}}$$

$$M_n^{(i)} = 1 + \frac{4\hat{z}_n^{(i)}}{x_s^2} - \frac{2n(n+1)}{x_s^2}\left[1 - \frac{(1-\hat{z}_n^{(i)})(1-\hat{z}_n^{(1)})}{n(n+1) - 1 - (x_s^2/2) - \tilde{z}_n^{(1)}}\right]$$

and

$$F_n = \frac{\rho}{\rho_f}\beta x\frac{j_n'(\beta x)}{j_n(\beta x)}, \quad \hat{z}_n^{(i)} = \frac{x h_n^{(i)\prime}(x)}{h_n^{(i)}(x)}, \quad \tilde{z}_n^{(i)} = \frac{x_s h_n^{(i)\prime}(x_s)}{h_n^{(1)}(x_s)},$$

and where we defined $x = \kappa_d a$, $x_s = \kappa_s a$, $\beta = k_f/\kappa_d$, and $i = 1$ or 2. The wavenumber in the fluid interior to the obstacle is $k_f = \omega/c_f$, and c_f is the sound speed in it, and c_d is the dilatational speed in the matrix. An analogous expression can be found for B_n.

Case 1

The *monopole* case for a fluid-filled cavity in an elastic solid. In this case n=0 and we can show that

$$L_o^{(i)} = \frac{x h_o^{(i)\prime}(x)}{h_o^{(i)}(x)}, \quad M_o^{(i)} = 1 + \frac{4}{x_s^2}\left(\frac{x h_o^{(i)\prime}(x)}{h_o^{(i)}(x)}\right), \quad F_o = \frac{\rho}{\rho_f}\beta x\frac{j_o'(\beta x)}{j_o(\beta x)}, \tag{7}$$

where for an elastic matrix we have $\beta = c_d/c_f$. Coefficient A_o then takes the simpler form

$$A_o = -\frac{\rho_f x j_o'(x) j_o(\beta x) - \rho\beta x j_o'(\beta x)\{j_o(x) + (4x/x_s^2) j_o'(x)\}}{\rho_f x h_o^{(1)\prime}(x) j_o(\beta x) - \rho\beta x j_o'(\beta x)\{h_o^{(1)}(x) + (4x/x_s^2) h_o^{(1)\prime}(x)\}}, \tag{8}$$

in agreement with [3]. Coefficient C_o can be shown to take the form

$$C_o = -\frac{(\rho/ix)}{\rho_f x j_o(x_f) h_o^{(1)\prime}(x) - \rho x_f j_o'(x_f)\{h_o^{(1)}(x) + (4x/x_s^2) h_o^{(1)\prime}(x)\}}, \tag{9}$$

where $x_f = \beta x = k_f a$. Equations (8) and (9) have the same denominators which show that the same resonances of the filler are communicated to the cavity wall.

Case 2

The general multipole case when the outer medium is another inviscid *fluid.* Here, $x_s \to \infty$ (or $c_s = 0$) and we can show that

$$L_n^{(i)} = \frac{x h_n^{(i)\prime}(x)}{h_n^{(i)}(x)}, \quad M_n = 1, \quad F_n = \frac{\rho}{\rho_f} \beta x \left\{ \frac{j_n'(\beta x)}{j_n(\beta x)} \right\}. \tag{10}$$

Coefficient A_n then takes the form

$$A_n = -\frac{\rho_f x j_n'(x) j_n(\beta x) - \rho \beta x j_n'(\beta x) j_n(x)}{\rho_f x h_n^{(1)\prime}(x) j_n(\beta x) - \rho \beta x j_n'(\beta x) h_n^{(1)}(x)}. \tag{11}$$

This is the coefficient controlling the sonar cross section of gas bubbles in liquids or of liquid droplets in gases. It already accounts for the mobility of the scatterers due to radiation pressure, and the monopole subcase is merely obtained by setting n=0 in (11). This equation also contains the "rigid" and "soft" sphere results which are, respectively,

$$\rho_f = \infty, \; A_n = -\frac{j_n'(x)}{h_n^{(1)\prime}(x)}, \quad \text{and } \rho_f = 0, \; A_n = -\frac{j_n(x)}{h_n^{(1)}(x)}. \tag{12}$$

Coefficient C_n can be shown to take the form

$$C_n = \frac{\begin{vmatrix} d_{11} & \mathrm{Re}\, d_{11} \\ d_{21} & \mathrm{Re}\, d_{21} \end{vmatrix}}{\begin{vmatrix} d_{11} & d_{13} \\ d_{21} & d_{23} \end{vmatrix}} = -\frac{(\rho/ix)}{\rho_f x j_n(x_f) h_n^{(1)\prime}(x) - \rho x_f j_n'(x_f) h_n^{(1)}(x)}, \tag{13}$$

as $x_s = \infty$. This coefficient controls the behavior of the fluid inside the bubble or the droplet during the scattering process. The monopole case is recovered setting n=0 in (13). As we discussed in Case 1, (11) and (13) have the same denominators.

3. NUMERICAL CALCULATIONS

The numerical calculations are performed using the program we developed for the general forms of the coefficient as given in (2)-(4). Equations (8), (9), (11) and (13) give the explicit forms of the coefficients which are computed whenever either n=0, or $c_s = 0$, but these forms need not be explicitly programmed to generate the graphs displayed here, or any similar ones.

In Fig. 1 we plot the modulus of the normalized partial-wave backscattering amplitudes as in Eqs. (20) of [1]. These quantities are:

$$\mathrm{MOD(PP)} \equiv \left| \frac{1}{a} f_n^{pp}(\pi) \right| = \left| \frac{1}{\kappa_d a} (2n+1) A_n \right|, \tag{14a}$$

and

$$\mathrm{MOD(PS)} \equiv \left| \frac{1}{a} f_n^{ps}(\pi) \right| = \left| \frac{1}{\kappa_s a} (2n+1) B_n \right|. \tag{14b}$$

They are plotted versus $\kappa_d a$ for an air-filled cavity in solid rubber (assumed lossless). The material parameters for air, rubber, and later for water, were all given earlier [1,3]. Columns two and three respectively of Fig. 1 display the plots of MOD(PP) and MOD(PS). The first row of Fig. 1 corresponds to the monopole (n=0) case. The second and third rows, respectively, correspond to the dipole and quadrupole cases (i.e., n=1 and n = 2). The graphs all have logarithmic ordinates (i.e., dB) and since for air-in-rubber the resonances are so narrow, the plots have been produced in the narrow region around the first resonance of each multipole contribution, so that their width becomes visible in the graphs. For example, the dipole (n=1) fundamental resonance shown in the central graph of Fig. 1 is seen to occur for $0.508 \leqslant \kappa_d a \leqslant 0.516$. The first column of Fig. 1 is analogous to the second, but now for an air bubble in *water*. The monopole plot in the top row of the first column is the well known result [4] for the "giant monopole resonance" of air bubbles in water. Its peak occurs at $\kappa_d a \cong 0.013$ and has value MOD(PP) $\cong$ 71. All the plots in Fig. 1 are for the fundamental resonance (l = 1) of the three multipole contributions (n=0,1,2) shown. Similar plots can also be generated for the overtone (l = 2,3, ...) resonances in each multipole. We have displayed some of these in our earlier work [3]. The results in Fig. 1 and in [3] have shown the dominance of the monopole contribution over that of all higher multipoles. The resonances of the higher order multipoles manifest themselves, in summed cross-section plots, as small spikes superimposed on the graph for the monopole contribution to the cross-section. Analogous results for water drops in air (not displayed here) are obtained from (11) in identical fashion as for air bubbles in water, but reversing the labels for air and water. The overall appearance of these plots is similar to that of the plots in the first column of Fig. 1.

The monopole situation (n=0) of gas-filled cavities in nonabsorbing rubber in the second and third columns of the first row of Fig. 1 deserves further attention. The first row of Fig. 2 shows the non-mode-converted results (i.e., MOD(PP)), while the second displays the mode-converted ones (i.e., MOD(PS)), all for the same air-filled cavity in rubber plotted versus $\kappa_d a$ in the range $0 \leqslant \kappa_d a \leqslant 3.0$. The first column shows the composite modal contribution and the third column exhibits the smooth "background" of an evacuated cavity. The central column, obtained as the modulus of the difference between the quantities whose moduli are displayed in the first and third columns, shows the isolated resonances (fundamental and overtones) of the filler substance (i.e., air), in the form now typical of a problem analyzed by the Resonance Theory of Scattering. All resonances of the MOD(PP) and MOD(PS) plots (viz. four are visible in these plots), occur at the same locations, as explained before [3].

4. CONCLUSIONS

The computerized code we have developed to analyze the scattering of p (and s) waves from fluid-filled cavities in elastic or viscoelastic solids can handle all the particular situations we have described here. These include all the multipole contributions to the sonar cross section of fluid spheres contained in different fluids. We can study any possible combination of two fluids, or of a fluid within a solid. There is no difficulty in analyzing each and every resonance (fundamental or overtone) of the composite multipole contributions, by splitting them into their background and resonance parts. For bubbles or cavities the backgrounds are usually "soft", and for liquid droplets they are "rigid". All these features are possible by merely adjusting some of the parameters of our general Program, and we have displayed many plots to illustrate these points.

ACKNOWLEDGMENTS

I wish to thank our Center Independent Research Board for support and encouragement and John Barlow of the Dahlgren Computational Group for his programing assistance in the generation of the figures.

APPENDIX

The elements d_{ij} used in (2)-(4) and found in [1] are

$$d_{11} = \kappa_d a h_n^{(1)\prime}(\kappa_d a),$$

$$d_{12} = n(n+1) h_n^{(1)}(\kappa_s a),$$

$$d_{13} = -k_f a j'_n(k_f a),$$

$$d_{21} = [2n(n+1) - \kappa_s^2 a^2] h_n^{(1)}(\kappa_d a) - 4\kappa_d a h_n^{(1)\prime}(\kappa_d a),$$

$$d_{22} = 2n(n+1)[\kappa_s a h_n^{(1)\prime}(\kappa_s a) - h_n^{(1)}(\kappa_s a)],$$

$$d_{23} = (\rho_f/\rho)\kappa_s^2 a^2 j_n(k_f a),$$

$$d_{31} = \kappa_d a h_n^{(1)\prime}(\kappa_d a) - h_n^{(1)}(\kappa_d a),$$

$$d_{32} = \left[n(n+1) - 1 - \frac{1}{2}\kappa_s^2 a^2\right] h_n^{(1)}(\kappa_s a) - \kappa_s a h_n^{(1)\prime}(\kappa_s a),$$

$$d_{33} = 0.$$

REFERENCES

[1] G. Gaunaurd, H. Überall: "Theory of resonant scattering from spherical cavities in elastic and viscoelastic media," J. Acoust. Soc. Amer. **63**, 1699-1712 (1978)

[2] G. Gaunaurd, H. Überall: "Numerical evaluation of modal resonances in the echoes of compressional waves scattered from fluid-filled spherical cavities in solids", J. Applied Phys. **52**, in press

[3] G. Gaunaurd, K. P. Scharnhorst, H. Überall: "Giant monopole resonances in the scattering of waves from gas-filled spherical cavities and bubbles," J. Acoust. Soc. Amer. **65**, 573-594 (1979)

[4] *Physics of Sound in the Sea*, Pt. IV, ed. by R. Wildt, Summary Technical Report Div. 6, Vol. 8 (AD-200624), Chap. 28, Fig. 1 (1946)

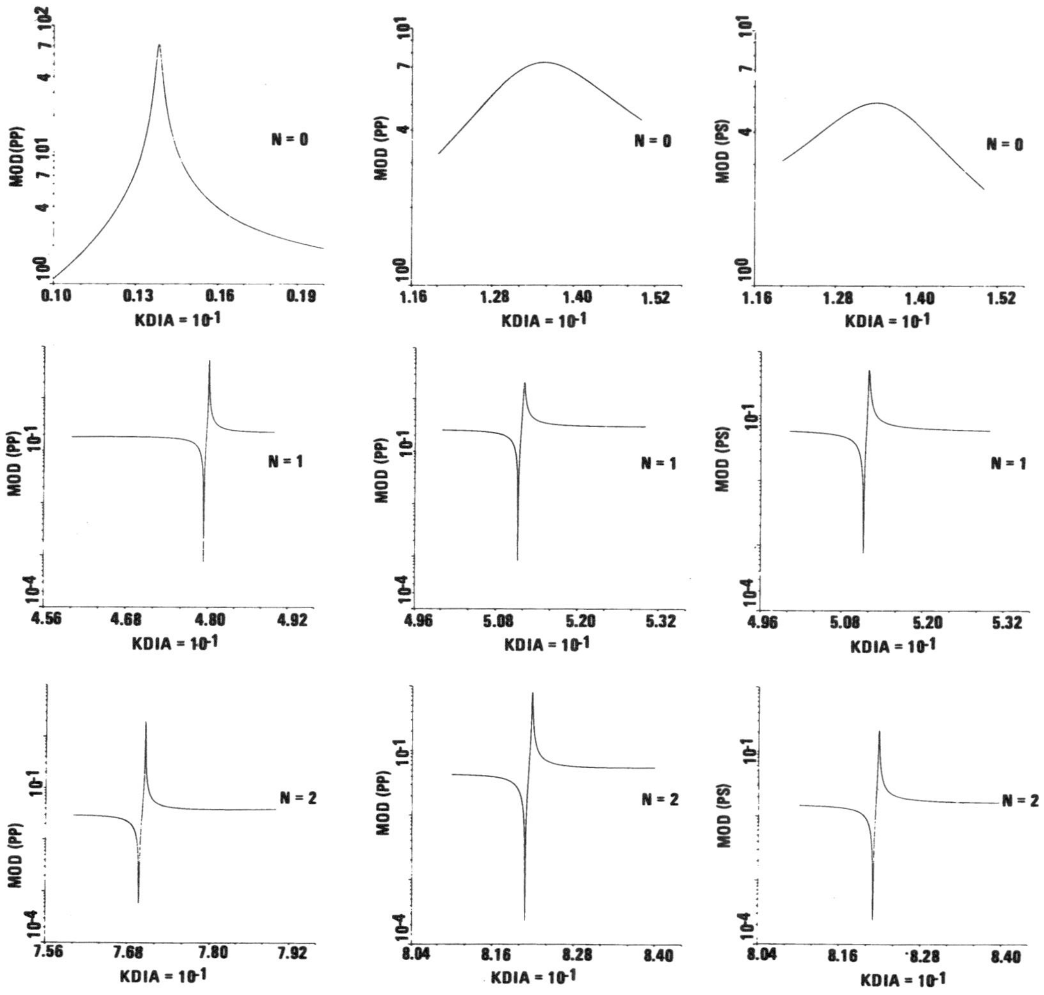

Fig. 1. Modulus of the normalized partial-wave backscattering amplitudes (in dB) versus $\kappa_d a$ *for monopole (N = 0), dipole (N = 1), and quadrupole (N = 2) cases.*

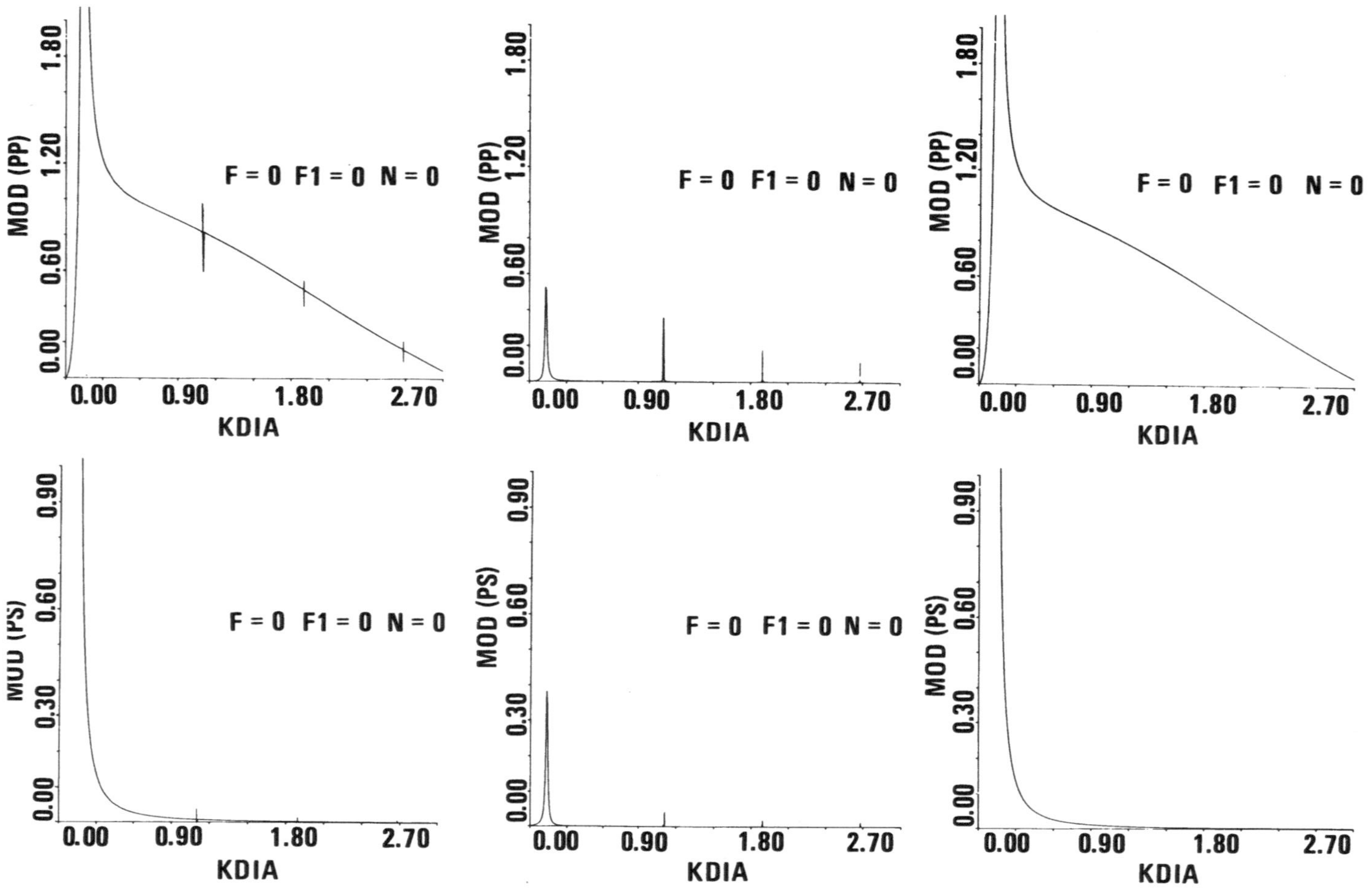

Fig. 2. Same as Fig. 1 for the monopole situation of gas-filled cavities in nonabsorbing rubber.

QUANTUM-MECHANICAL SCATTERING THEORY

SCATTERING THEORY IN THE MIXED REPRESENTATION

Roger G. Newton*

Institute for Advanced Study
Princeton, New Jersey 08540

and

Physics Department, Princeton University
Princeton, New Jersey 08540

ABSTRACT

The recently proposed mixed representation in quantum mechanics is discussed and applied to the scattering of a particle by a potential in three dimensions. Such scattering becomes equivalent to a one-dimensional reflection problem with a nonlocal potential.

The mixed representation was recently introduced [1] into quantum mechanics with the idea that it may be useful in scattering theory. It is situated partly in momentum space, specifying the direction of the $3n$-dimensional momentum vector of n particles and thereby the directions of their n momenta ($2n$ variables) as well as their energy fractions ($n-1$ variables), and partly in configuration space, specifying a position along the "direction of motion." Thus, the representation fixes exactly those momentum variables that are independently specified as the initial and final ones of an n-particle scattering amplitude, and it leaves one variable in configuration space for the description of the motion of a wave packet. Such a representation is possible because the unit vector in the direction of the momentum commutes with the projection of the position on the momentum:

$$[\mathbf{q}\cdot\mathbf{p}/|\mathbf{p}|, \mathbf{p}/|\mathbf{p}|] = 0 .$$

For an odd number of particles, the mixed representation is closely related to the Radon transform; for an even number of particles, to a generalized Euler transform. Contrary to well-known restrictions [2] on the range of the Radon transform, the transformation from the coordinate representation to the mixed representation has been proved [1] to be a one-to-one unitary mapping of $L^2(\mathbb{R}^m)$ onto $L^2(S^{m-l}\times \mathbb{R}_+)$, S^{m-l} being the unit sphere in m dimensions and $\mathbb{R}_+$ the positive reals.

Let $\mathbf{q}_\alpha$, $\alpha = 1, \ldots, n$, be the position vectors of n particles of masses m_α, and let $\mathbf{p}_\alpha$ be their momentum vectors. It is convenient to use a configuration space $\mathbb{R}^{3n}$ made up of the vectors $\mathbf{z}_\alpha = m_\alpha^{1/2}\mathbf{q}_\alpha$, $\alpha = 1, \ldots, n$, and a momentum space made up of $\mathbf{k}_\alpha = m_\alpha^{-1/2}\mathbf{p}_\alpha$. We then have

$$\mathbf{k}\cdot\mathbf{z} = \sum_{\alpha=1}^{n} \mathbf{p}_\alpha\cdot\mathbf{q}_\alpha ,$$

where the left-hand side denotes the inner product in $\mathbb{R}^{3n}$ and the right-hand side in $\mathbb{R}^3$. Then,

$$\frac{1}{2}k^2 \equiv \frac{1}{2}|\mathbf{k}|^2 = \sum_{l}^{n} \frac{|\mathbf{p}_\alpha|^2}{2m_\alpha} = \sum_{l}^{n} E_\alpha = E ,$$

where E is the total kinetic energy and [3]

$$\boldsymbol{\theta}_\alpha = (E_\alpha/E)^{1/2}\, \hat{p}_\alpha = \mathbf{k}_\alpha/|\mathbf{k}| ,$$

with $\hat{p}_\alpha = \mathbf{p}_\alpha/|\mathbf{p}_\alpha|$ a vector in $\mathbb{R}^3$ whose direction is that of $\mathbf{p}_\alpha$ and whose squared magnitude is the energy fraction of particle α. Thus, the vector $\hat{\theta}$ in $\mathbb{R}^{3n}$ made up of the components $\boldsymbol{\theta}_\alpha$, $\alpha = 1, \ldots, n$, in $\mathbb{R}^3$, is of unit magnitude and denotes a point in S^{3n-1}. It specifies the directions of the momenta of all the particles as well as their energy fractions. Let

$$\tilde{f}(\mathbf{k}) = \left(\prod_\alpha m_\alpha\right)^{3/4} \langle \mathbf{p}_1, \ldots, \mathbf{p}_n | \rangle$$

be a momentum-space wave function (where the right-hand side denotes a normalized wave function in Dirac notation). We then form

$$\mathring{f}(\hat{k},k) = k^{\frac{3n-1}{2}}\,\tilde{f}(\mathbf{k})$$

for $k \geqslant 0$ and extend it to $k < 0$ by the symmetry

$$\mathring{f}(\hat{k},k) = (-1)^{\frac{3n-1}{2}}\,\mathring{f}(-\hat{k},-k)$$

for odd n, and by

$$\mathring{f}(\hat{k},k) = (-1)^{\frac{3n}{2}}\,\mathring{f}(-\hat{k},-k)$$

for even n.

The wave function in configuration space is then given by

$$<\mathbf{q}_1,\ldots,\mathbf{q}_n|> = \Big(\prod_\alpha m_\alpha\Big)^{3/4} f(z),$$

where

$$f(\mathbf{z}) = (2\pi)^{\frac{-3n}{2}} \int d^{3n}\mathbf{k}\; e^{i\mathbf{k}\cdot\mathbf{z}}\,\tilde{f}(\mathbf{k})\;,$$

and the wave function in the mixed representation by

$$\bar{f}(\hat{\theta},x) = (2\pi)^{-1/2}\int_{-\infty}^{\infty} dk\; e^{ikx}\,\mathring{f}(\hat{\theta},k)\;.$$

If $f \in L^2(\mathbf{R}^{3n})$, then $\bar{f} \in L^2(S^{3n-1}\times\mathbf{R}_+)$ and

$$\int_0^\infty dx \int d\hat{\theta}\,|\bar{f}(\hat{\theta},x)|^2 = \int d^{3n}\mathbf{z}\,|f(\mathbf{z})|^2\;.$$

For negative values of x we have

$$\bar{f}(\hat{\theta},x) = \begin{cases} (-1)^{\frac{3n-1}{2}}\;\bar{f}(-\hat{\theta},-x) & \text{if } n \text{ is odd,}\\ (-1)^{\frac{3n}{2}}\;\bar{f}(-\hat{\theta},-x) & \text{if } n \text{ is even.}\end{cases}$$

The direct relation between $\bar{f}$ and f may be written formally [1] for odd n

$$\bar{f}(\hat{\theta},x) = (2\pi)^{\frac{1-3n}{2}}\left(\frac{1}{i}\,\frac{\partial}{\partial x}\right)^{\frac{3n-1}{2}} \int d^{3n}\mathbf{z}\; f(\mathbf{z})\;\delta(x-\hat{\theta}\cdot z)\;,$$

and for even n

$$\bar{f}(\hat{\theta},x) = (2\pi)^{\frac{-3n}{2}}\left(\frac{1}{i}\,\frac{\partial}{\partial x}\right)^{\frac{3n}{2}} \int d^{3n}\mathbf{z}\;\frac{f(\mathbf{z})}{|x-\hat{\theta}\cdot\mathbf{z}|^{1/2}}\;.$$

These formulas show the relation of the mixed representation to the Radon transform or Euler transform, respectively. The transformation function from one representation to the other may then be written

$$<\hat{\theta},x|\mathbf{z}> = (2\pi)^{\frac{1-3n}{2}}\Big(\prod_\alpha m_\alpha\Big)^{3/4}\left(\frac{1}{i}\,\frac{\partial}{\partial x}\right)^{\frac{3n-1}{2}}\delta(x-\hat{\theta}\cdot z)$$

for odd n, and

$$<\hat{\theta},x|\mathbf{z}> = (2\pi)^{-\frac{3n}{2}}\Big(\prod_\alpha m_\alpha\Big)^{3/4}\left(\frac{1}{i}\,\frac{\partial}{\partial x}\right)^{\frac{3n}{2}}|x-\hat{\theta}\cdot\mathbf{z}|^{-1/2}$$

for even n.

The principal virtue of the mixed representation is that in it the kinetic energy operator becomes simply

$$H_0 = -\,\frac{1}{2}\,\frac{d^2}{dx^2}\;.$$

The price for this, however, is that the potential becomes complicated and generally nonlocal in x.

Let us discuss a one-particle scattering problem with an external potential V. The transformation function gives us immediately that in the mixed representation the potential is represented by the integral kernel

$$<\hat{\theta},x|V|\hat{\theta}',x'> \equiv V(\hat{\theta},\hat{\theta}';x,x') = (2\pi)^{-2}\frac{\partial^2}{\partial x\partial x'}\overline{V}(\hat{\theta},\hat{\theta}';x,x')\ ,$$

where

$$\overline{V}(\hat{\theta},\hat{\theta}';x,x') \equiv \int d^3\mathbf{r}\ V(\mathbf{r})\ \delta(x-\hat{\theta}\cdot\mathbf{r})\ \delta(x'-\hat{\theta}'\cdot\mathbf{r})\ ,$$

in units such that $m = 1$. For orientation purposes, notice that if V vanishes outside a sphere of radius R then $V(\hat{\theta},\hat{\theta}',x,x')$ vanishes when $|x-x'| > 2R$. Also, notice that the local (i.e., multiplicative) nature of the potential in configuration space manifests itself in the mixed representation by the fact that for $\hat{\theta} = \hat{\theta}'$ V becomes a distribution concentrated at the point $x = x'$, and for $\hat{\theta} = -\hat{\theta}'$ it is concentrated at $x = -x'$. If x and x' are allowed to take on negative values, then V has the obvious symmetries

$$V(\hat{\theta},\hat{\theta}',x,x') = -V(-\hat{\theta},\hat{\theta}',-x,x') = V(-\hat{\theta},-\hat{\theta}',-x,-x').$$

It is also symmetric in its variables,

$$V(\hat{\theta},\hat{\theta}',x,x') = V(\hat{\theta}',\hat{\theta},x',x)\ ,$$

and real; thus, hermitian.

The function $\overline{V}$ is defined as an integral over the intersection of the two planes $\hat{\theta}\cdot\mathbf{r} = x$ and $\hat{\theta}'\cdot\mathbf{r} = x'$, whose normals are $\hat{\theta}$ and $\hat{\theta}'$, and whose normal distances from the origin are x and x', respectively. The integral is, therefore, a one-dimensional one, over the line of intersection of the two planes. It can be written as follows:

$$\overline{V}(\hat{\theta},\hat{\theta}',y,y') = \frac{1}{\gamma}\ \hat{V}\left[\frac{(\hat{\theta}'y-\hat{\theta}y')\times\hat{\theta}''}{\gamma},\hat{\theta}''\right]\ ,$$

where

$$\hat{\theta}'' = (\hat{\theta}\times\hat{\theta}')/\gamma,\ \gamma = |\hat{\theta}\times\hat{\theta}\ |\ ,$$

and

$$\hat{V}(\mathbf{z},\hat{\theta}) = \int_{-\infty}^{\infty} dt\ V(\mathbf{z}+\hat{\theta}t)\ ,$$

which is known as the x-ray transform [4], because it is used extensively in radiology.

The transformation function also immediately leads to the inversion formula

$$(2\pi)^{-2}\int d\hat{\theta}d\hat{\theta}'\ V_{xx'}\ (\hat{\theta},\hat{\theta}';\hat{\theta}\cdot\mathbf{r},\hat{\theta}'\cdot\mathbf{r}) = V(\mathbf{r})\ \delta^3(\mathbf{r}-\mathbf{r}')\ ,$$

in which the subscripts x and x' indicate differentiations with respect to the variables x and x', x being subsequently set equal to $\hat{\theta}\cdot r$, and x' to $\hat{\theta}'\cdot\mathbf{r}$. The fact that this integral over two unit spheres comes out to be a distribution concentrated at $\mathbf{r} = \mathbf{r}'$ is attributable solely to the distributional nature of $V(\hat{\theta},\hat{\theta}';x,x')$ when $\hat{\theta} = \pm\hat{\theta}'$.

The Schrödinger equation in the mixed representation reads

$$-\psi''(\hat{\theta},x) + 2\int d\hat{\theta}'\int_0^{\infty} dx' V(\hat{\theta},\hat{\theta}';x,x')\psi(\hat{\theta}',x') = k^2\psi(\hat{\theta},x)\ ,$$

on the half-line $0 < x < \infty$. We may extend it to the whole real line by demanding that

$$\psi(\hat{\theta},x) = -\psi(-\hat{\theta},-x)\ .$$

Because of the symmetry of V, it then follows that

$$-\psi''(\hat{\theta},x) + \int d\hat{\theta}'\int_{-\infty}^{\infty} dx'\ V(\hat{\theta},\hat{\theta}',x,x')\ \psi(\hat{\theta}',x') = k^2\psi(\hat{\theta},x)\ .$$

At large distances, ψ is supposed to become equal to the transform of $e^{i\mathbf{k}\cdot\mathbf{r}}$ plus outgoing waves. The Lippmann-Schwinger equation for the physical wave function in the mixed representation becomes, therefore,

$$\psi(\hat{\theta}',\hat{\theta},x) = \psi_0(\hat{\theta}',\hat{\theta},x) - \frac{i}{k}\int_{-\infty}^{\infty} dx'\int_0^{\infty} dx''\ e^{ik|x-x'|}\int d\hat{\theta}''\ V(\hat{\theta}',\hat{\theta}'';x',z'')\ \psi(\hat{\theta}'',\hat{\theta},x'')\ ,$$

where

$$\psi_0(\hat{\theta}',\hat{\theta},x) = \frac{1}{2}\left[\delta(\hat{\theta}',\hat{\theta})\ e^{ikx} - \delta(\hat{\theta}',-\hat{\theta})e^{-ikx}\right]\ ,$$

which corresponds to $\psi_0(\mathbf{k},\mathbf{r}) = (k/4\pi)\, e^{i\mathbf{k}\cdot\mathbf{r}}$. The antisymmetry of the potential implies that a solution of this LS equation for $x > 0$, defined for $x < 0$ by the same equation, automatically satisfies the correct antisymmetry condition. Thus, the integral over x'' may be replaced by one half of the integral from $-\infty$ to ∞, and the equation may be considered as an equation for ψ for $-\infty < x < \infty$.

The scattering problem now becomes completely equivalent to a one-dimensional reflection problem with a nonlocal potential. If we define a solution f_1 by the equation

$$f_1(\hat{\theta}',\hat{\theta};x) = \delta(\hat{\theta}',\hat{\theta})\, e^{ikx} - \frac{i}{2k}\int_{-\infty}^{\infty} dx' \int_{-\infty}^{\infty} dx'' \int d\hat{\theta}'' e^{ik|x-x'|}\, V(\theta',\theta'';x',x'')\, f_1(\hat{\theta}'',\hat{\theta}'x''),$$

then the antisymmetric part of f_1 just equals ψ,

$$\psi(\hat{\theta}',\hat{\theta},x) = \frac{1}{2}[f_1(\hat{\theta}',\hat{\theta},x) - f_1(-\hat{\theta}',\hat{\theta},-x)],$$

while its symmetric part equals the symmetric part of the inhomogeneity.

Let us suppress the dependence on $\hat{\theta}$, analogous to a matrix notation. Then the integral equation for f_1 reads

$$f_1(x) = \mathbf{1}\, e^{ikx} - \frac{i}{2k}\int_{-\infty}^{\infty} dx' \int_{-\infty}^{\infty} dx''\, e^{ik|x-x'|}\, V(x',x'') f_1(x'').$$

This function describes a transmission-reflection problem for incidence from the left, because for $x \to \infty$ it contains only waves travelling to the right. The transmission and reflection amplitudes are given by

$$T \equiv T_l = \mathbf{1} - \frac{i}{2k}\int_{-\infty}^{\infty} dx' \int_{-\infty}^{\infty} dx''\, e^{-ikx'}\, V(x',x'')\, f_1(x''),$$

$$R \equiv R_l = -\frac{i}{2k}\int_{-\infty}^{\infty} dx' \int_{-\infty}^{\infty} dx''\, e^{ikx'}\, V(x',x'') f_1(x'').$$

We may similarly define

$$f_2(x) = \mathbf{1}\, e^{-ikx} - \frac{i}{2k}\int_{-\infty}^{\infty} dx' \int_{-\infty}^{\infty} dx''\, e^{ik|x-x'|}\, V(x',x'') f_2(x''),$$

which leads to the transmission and reflection amplitudes from the right,

$$T_r = \mathbf{1} - \frac{i}{2k}\int_{-\infty}^{\infty} dx' \int_{-\infty}^{\infty} dx''\, e^{ikx'}\, V(x',x'') f_2(x''),$$

$$R_r = -\frac{i}{2k}\int_{-\infty}^{\infty} dx' \int_{-\infty}^{\infty} dx''\, e^{-ikx'}\, V(x',x'') f_2(x'').$$

Now, it follows from the symmetry of V that

$$f_2(\hat{\theta}',\hat{\theta},x) = f_1(-\hat{\theta}',-\hat{\theta},-x).$$

Introducing the "matrix" I whose explicit representation is $\delta(-\hat{\theta}',\hat{\theta})$, we may write

$$f_2(x) = I f_1(x) I.$$

It follows that

$$T_r = ITI, \qquad R_r = IRI.$$

Furthermore,

$$T = \mathbf{1} - IR,$$

and use of the integral equation in R shows that R is symmetric,

$$R = \tilde{R}.$$

Expressing f_2^* as a linear combination of f_1 and f_2 leads to the equation

$$T_r R_l^* + R_l T_r^* = 0,$$

(where the asterisk denotes complex conjugation) or equivalently,

$$2RR^* = RI + IR^*.$$

The S-matrix for the one-dimensional problem is defined by

$$S = \begin{pmatrix} T_l & R_r \\ R_l & T_r \end{pmatrix} = \begin{pmatrix} \mathbf{1} - IR & IRI \\ R & \mathbf{1} - RI \end{pmatrix} .$$

One easily finds from the above equations that S is unitary,

$$SS^\dagger = S^\dagger S = \mathbf{1} .$$

Finally, by returning to the customary expression for the scattering amplitude

$$A(\hat{\theta}',\hat{\theta}) = -\frac{1}{2\pi} \int d^3\mathbf{r}\, \phi_0^*(\hat{\theta}',\mathbf{r})\, V(\mathbf{r})\, \phi(\hat{\theta},\mathbf{r})$$

in the present units, one finds that

$$A = (4\pi i/k) IR .$$

The reciprocity theorem [5] and the generalized optical theorem [6] then follow from the listed properties of R.

The outlined formalism demonstrates the equivalence of the three-dimensional scattering problem, in this representation, to the one-dimensional reflection problem with a nonlocal potential. The demonstration of existence and uniqueness of solutions requires L^2-Fredholm methods and iterating the equation once. The principal effect of the nonlocality of the potential is that analyticity properties are impossible to prove unless the potential has compact support. This is, of course, a serious disadvantage and it is not likely to be a temporary lack of success. The reason is that, even though the three-dimensional wave function is the boundary value of an analytic function of k, regular in the upper half plane, its Radon transform

$$h_k(\hat{\theta},x) = \int d^3\mathbf{r}\, \psi_k(\mathbf{r})\, \delta(x - \hat{\theta}\cdot\mathbf{r})$$

need not have such an analytic extension because for complex values of k the integral diverges.

The use of the mixed representation in the three-particle scattering problem is the next step. It has not yet been taken.

ACKNOWLEDGMENT

It is a pleasure to acknowledge the hospitality of the Institute for Advanced Study and of the physics department of Princeton University, where most of this work was done. Partial support by the National Science Foundation, under Grant PHY 77-25337, is also acknowledged.

FOOTNOTES AND REFERENCES

*Permanent address: Physics Department, Indiana University, Bloomington, Indiana 47405

[1] R.G. Newton: Physica **96A**, 271-279 (1979)
[2] See, for example, I.M. Gelfand, M.I. Graev, N. Ya Vilenkin: *Generalized Functions* (Academic, New York, 1966), Vol. 5
[3] We always denote vectors in $\mathbb{R}^3$ or in $\mathbb{R}^{3n}$ of unit magnitude by a letter with a caret. The integral over S^{3n-1} will be written $\int d\hat{\theta}$
[4] See, for example, K. T. Smith, D.C. Solmon, S. L. Wagner: Bull. Am. Math. Soc. **83**, 1227 (1977)
[5] See, for example. R. G. Newton: *Scattering Theory of Waves and Particles* (McGraw-Hill, New York, 1966), p. 271
[6] *Ibid.*, p. 190

GREEN AND LANFORD REVISITED

John D. Dollard and Brian Bourgeois

Mathematics Department
University of Texas
Austin, Texas 78712

INTRODUCTION

This article is concerned with the approach to quantum-mechanical scattering theory in which one tries to deduce existence of the Møller wave operators by studying the stationary-state wave-functions of the Hamiltonian. Various applications of this approach can be found in the literature. For nonradial potentitals, perhaps the most famous work is by IKEBE [6]. For the radial case, with which we shall be almost exclusively concerned here, the paper of GREEN and LANFORD [5] is probably the outstanding reference. In recent years, the approach mentioned above has been somewhat overshadowed by others, in particular the powerful and successful trace-class methods. Our theme here is that the approach via stationary-state wave-functions still has its attractions. In particular, if it is successful, one obtains a concrete expression for the Møller wave operators in terms of the stationary-state wave-functions, providing a connection between the conceptually satisfying time-dependent theory and the experimentally important time-independent theory. Also, if radial potentials are studied, weak asymptotic completeness is usually obtained "free". Finally, and these are the points which we shall emphasize: (i) for radial potentials, the aproach is extremely simple conceptually: modulo a few technicalities, the entire problem boils down to the question "How fast does a solution of the radial Schrödinger equation approach 'free' behavior as $r \to \infty$?"; (ii) the approach has recently been used [1-3] to analyze scattering problems which had not been dealt with by other methods, the premier examples being scattering by radial potentials of the VON NEUMANN-WIGNER-type [12] and other oscillatory potentials of slow decay. These are not "short range" potentials, and are difficult to analyse by methods which focus only on the "size" of the potential. The oscillations of the potentials, nevertheless, mitigate their total impact on the scattering process, and a naturai way to assess this total impact is to study the stationary-state wave-functions, whose behavior reflects not just the value of the potential at a point but its net effect over an interval.

We begin with a heuristic discussion of scattering in three dimensions, and then pass to the case of radial potentials, which effectively reduces the problem to one dimension and the method of GREEN and LANFORD [5].

1. REMARKS ON SCATTERING IN THREE DIMENSIONS

As stated above, this section is heuristic. Under appropriate assumptions, it can be rigorized. Let H_o denote Laplace's operator defind in the usual way on $L^2(\mathbb{R}^3)$ so as to be self-adjoint. Let

$$H = H_o + V, \tag{1}$$

where V is a real multiplicative operator which is well-enough behaved so that H can be defined as a self-adjoint operator on $L^2(\mathbb{R}^3)$. We wish to establish existence of the strong limits

$$W^{\pm} = \operatorname*{s-lim}_{t \to \pm\infty} e^{iHt}\, e^{-iH_o t}. \tag{2}$$

If the limits in (2) exist, there is an expected connection between $W^{\pm}$ and the "stationary-state wave functions" $\psi^{\pm}(\mathbf{k},\mathbf{x})$ for H. The expected connection is this: define $\psi^{\pm}(\mathbf{k},\mathbf{x})$ as the solution of the equation ($k = |\mathbf{k}|$)

$$\psi^{\pm}(\mathbf{k},\mathbf{x}) = e^{i\mathbf{k}\cdot\mathbf{x}} - \frac{1}{4\pi}\int \frac{e^{\mp ik|\mathbf{x}-\mathbf{x}'|}}{|\mathbf{x}-\mathbf{x}'|} V(\mathbf{x}')\psi^{\pm}(\mathbf{k},\mathbf{x}')\,d^3x'. \tag{3}$$

Then, if $g \in L^2(\mathbb{R}^3)$ has Fourier transform $\tilde{g}$, we have (in the sense of limits in the mean)

$$g(\mathbf{x}) = \frac{1}{(2\pi)^{3/2}}\int e^{i\mathbf{k}\cdot\mathbf{x}}\tilde{g}(\mathbf{k})\,d^3k$$

and

$$(W^{\pm}g)\ (\mathbf{x}) = \frac{1}{(2\pi)^{3/2}} \int \psi^{\pm}(\mathbf{k},\mathbf{x})\tilde{g}(\mathbf{k})d^3k. \tag{4}$$

Thus the action of $W^{\pm}$ consists simply in replacing $e^{i\mathbf{k}\cdot\mathbf{x}}$ by $\psi^{\pm}(\mathbf{k},\mathbf{x})$ in the Fourier integral for g.

If one does not know the limits in (2) exist, one can try to prove it using (4). Namely, one *defines* $W^{\pm}$ by (4), and attempts to verify that

$$\lim_{t\to\pm\infty} ||e^{-iHt}W^{\pm}g - e^{-iH_0t}g|| = 0. \tag{5}$$

Now $e^{-iH_0t}g$ is obtained by inserting e^{-ik^2t} in the first integral in (4), and formally, at least, $e^{-iHt}W^{\pm}g$ is obtained by doing the same thing in the second integral in (4). Thus (5) becomes

$$\lim_{t\to\pm\infty} ||\int [\psi^{\pm}(\mathbf{k},\mathbf{x}) - e^{i\mathbf{k}\cdot\mathbf{x}}]e^{-ik^2t}\tilde{g}(\mathbf{k})\ d^3k|| = 0. \tag{6}$$

The truth or falsehood of (6) depends of course on the function

$$R^{\pm}(\mathbf{k},\mathbf{x}) = \psi^{\pm}(\mathbf{k},\mathbf{x}) - e^{i\mathbf{k}\cdot\mathbf{x}}. \tag{7}$$

$R^{\pm}(\mathbf{k},\mathbf{x})$ measures the distance between the stationary state wave function and its asymptotic plane-wave form for large $|\mathbf{x}|$. Intuitively, (6) will hold if $\psi^{\pm}(\mathbf{k},\mathbf{x})$ approaches its asymptotic form "fast enough" as $|\mathbf{x}| \to \infty$. Thus, if the steps leading to (6) can be justified, existence of $W^{\pm}$ is seen to depend on the rate at which $\psi^{\pm}(\mathbf{k},\mathbf{x})$ approaches its asymptotic form. To see how an attempt to prove (6) might go, take the expression for $R^{\pm}(\mathbf{k},\mathbf{x})$ from (3) and insert it into (6). Reverse the integrations over $\mathbf{k}$ and $\mathbf{x}'$, and bring the L^2 norm inside the $\mathbf{x}'$ integral to bound the norm in (6). This results in the following estimate for the norm in (6):

$$h^{\pm}(t) = \frac{1}{4\pi} \int |V(\mathbf{x}')|\ || \frac{1}{|\mathbf{x}-\mathbf{x}'|} \int e^{\mp ik|\mathbf{x}-\mathbf{x}'| - ik^2t}\psi^{\pm}(\mathbf{k},\mathbf{x}')\tilde{g}(\mathbf{k})d^3k\,||d^3x'. \tag{8}$$

The problem is to show that $h^{\pm}(t) \to 0$ as $t \to \pm\infty$. Suppose we assume that $V \in L^1(\mathbb{R}^3)$. Then by Lebesgue's Dominated Convergence Theorem, $h^{\pm}(t)$ will approach zero if we can show that (i) the norm in (8) is bounded by a constant (independent of t and $\mathbf{x}'$) and (ii) for each fixed $\mathbf{x}'$, the norm in (8) approaches zero as $t \to \pm\infty$.

Now changing variables in the $\mathbf{x}$ integral from $\mathbf{x}$ to $\mathbf{y} = \mathbf{x} - \mathbf{x}'$, we obtain the following expression for the square of the norm in (8):

$$4\pi\int_0^{\infty} \left|\int e^{\mp ikr - ik^2t}\psi^{\pm}(\mathbf{k},\mathbf{x}')\tilde{g}(\mathbf{k})d^3k\right|^2 dr. \tag{9}$$

If $\psi^{\pm}(\mathbf{k},\mathbf{x}')$ can be shown to be bounded as a function of $\mathbf{k}$ and $\mathbf{x}'$, there is no difficulty in verifying that the expression in (9) has properties (i) and (ii) above. No crafty analysis is required: (i) follows from Plancherel's theorem and (ii) from standard lore on solutions of the free Schrödinger equation.

Although the argument just given is very easy, it is not, as far as we are aware, a popular one. The reason is that the suppositions made regarding $\psi^{\pm}(\mathbf{k},\mathbf{x})$ are difficult to verify, and this in turn is due to the fact that the integral equation (3) is rather clumsy to work with. If V is a *radial* potential, however, we can trade the integral equation (3) for an ordinary differential equation, and arguments like that sketched above become manageable.

2. RADIAL POTENTIALS: THE METHOD OF GREEN AND LANFORD

In this section we shall proceed rigorously, but at times without complete details when these can be easily found in the literature. We now assume that V is radial and satisfies KATO's condition of being the sum of a square-integrable and a bounded potential [7]. This guarantees that H is self-adjoint with the some domain as H_o. (Weaker assumptions on V can be made [10]. We do not strive for full generality here.) We note that under this condition V satisfies the standard requirement for studying the radial Schrödinger equation: $\int_0^1 r|V(r)|dr < \infty$. By separation of variables, the problem of establishing existence of the Mϕller wave operators of (2) is transformed in a standard way [5] into a succession of problems on $L^2(0,\infty)$: for each $l = 0, 1, 2, \ldots$, let h_{ol} and h_l be the operators

$$h_{ol} = -\frac{d^2}{dr^2} + \frac{l(l+1)}{r^2},$$

$$h_l = h_{ol} + V(r). \tag{10}$$

These operators are defined in the obvious manner on C^∞ functions of compact support in $(0,\infty)$. For $l > 0$ the operators so defined are essentially self-adjoint, and for $l = 0$ they are extended to self-adjoint operators by fixing zero boundary conditions at 0 [5]. Then the existence of the limits in (2) can be proved by showing that for each l the limits

$$w_l^\pm = \underset{t\to\pm\infty}{\text{s-lim}}\; e^{ih_lt}e^{-ih_{ol}t} \tag{11}$$

exist. This can be attempted by the method of GREEN and LANFORD [5], which is just the one-dimensional version of the argument outlined above. Instead of using an integral equation like (3), we seek the stationary-state wave-functions $\psi_l(k,r)$ of h_l as the solutions of the radial Schrödinger equation

$$\left\{-\frac{d^2}{dr^2} + \frac{l(l+1)}{r^2} + V(r)\right\}\psi_l(k,r) = k^2\psi_l(k,r),\; k > 0, \tag{12}$$

which are $0(r^{l+1})$ as $r \to 0$. (If V is not continuous, then (12) may not have a literal solution. In this case, one simply passes to a corresponding integral equation. This is easily done after (12) has been converted to matrix form as in (16) below. This integral equation is not analogous to (3)—it is as easy to manage as (12). For simplicity, we shall write all equations as if V were continuous.) If V has suitable behavior for large r (we shall specify some "suitable" V soon), then any solution of (12) will approach a linear combination of the plane waves $e^{\pm ikr}$ as $r \to 0$. For the solutions $\psi_l(k,r)$ which are $0(r^{l+1})$ as $r \to 0$, this fact is usually written as

$$\psi_l(k,r) = \sin(kr - \frac{l\pi}{2} + \delta_l(k)) + R_l(k,r), \tag{13}$$

where $R_l(k,r) \to 0$ as $r \to \infty$ and $\delta_l(k)$ is, of course, the "phase shift." Now arguments analogous to those given above for the three-dimensional case show that in order to deduce existence of the limits (11) it is enough to show that

$$\lim_{t\to\pm\infty} \left\|\int R_l(k,r)e^{-ik^2t}\hat{g}(k)dk\right\| = 0 \tag{14}$$

for $\hat{g} \in L^2(0,\infty)$, and where the norm intended is, of course, the norm in $L^2(0,\infty)$. (As in the three-dimensional case, some technical points must be settled before concluding that (14) implies existence of the limits in (11). One must show that an integral of the type $\int \psi_l(k,r)\hat{g}(k)dk$ actually belongs to $L^2(0,\infty)$ and that e^{-ih_lt} acting on this integral is realized by multiplying the integrand by e^{-ik^2t}. In the one-dimensional theory these technical points are normally very easy to handle [3], especially since one need actually verify the results only for $\hat{g}$ lying in a dense subspace of "nice" functions, as remarked below. These points are easily dealt with under our hypotheses on V below.) If (14) can be shown to hold, then we have the usual explicit representation for $w_l^\pm$: let $\psi_{ol}(k,r)$ denote the standard solution of (12) with $V = 0$ ($\psi_{ol}(k,r) = krj_l(kr)$). Let $g \in L^2(0,\infty)$ have the expansion

$$g(r) = \left(\frac{2}{\pi}\right)^{1/2}\int_0^\infty \psi_o(k,r)\hat{g}(k)dk.$$

Then

$$(w_l^\pm g)(r) = \left(\frac{2}{\pi}\right)^{1/2}\int_0^\infty e^{\mp i\delta_l(k)}\psi_l(k,r)\hat{g}(k)dk.$$

These formulas are analogous to (4), but they are more revealing in that, if they hold, then it is clear that the ranges of the operators $w_l^\pm$ are identical. If this holds for all l, then the ranges $W^\pm$ are identical, and we have weak asymptotic completeness with no special effort. The fundamental point to be appreciated, however, is that, modulo a few easily handled technical details, the entire analysis boils down to verification of (14), and (14) in turn boils down to the question: "How fast does $\psi_l(k,r)$ approach its asymptotic form?" This is now a problem in ordinary differential equations, and the power of the method of GREEN and LANFORD [5] rests in the fact that ordinary differential equations are so well understood.

We now investigate the validity of (14) under some special hypotheses on V. By a standard argument, in order to obtain existence of the limits in (11), it suffices to verify the existence of the limits on a dense set. Correspondingly, (14) need only be verified for a dense set of functions $\hat{g}$, and we choose $\hat{g}$ to be a C^∞ function of compact support in $(0, \infty)$. A second point is the following: for fixed r, the k-integral in (14) vanishes as $t \to \pm\infty$ by the Riemann-Lebesgue lemma. If $R_l(k,r)$ is bounded for $r \leqslant R > 0$ and k in the support of $\hat{g}$, as will always be the case for the potentials we study, then Lebesgue's dominated convergence theorem shows that the portion of the L^2 norm in (14) referring to $r \leqslant R$ approaches zero as $t \to \pm\infty$. Thus, in (14), we need really only prove convergence to zero of the portion of the L^2 norm referring to $r \geqslant R$, i.e., the norm in $L^2(R,\infty)$. This is a minor point, but is a technical help.

We shall begin by proving (14) under the hypothesis $V \in L^1(1,\infty)$. Existence of the Møller wave operators is, of course, known for this case, but the only proof in the literature appears to be that of KURODA [8], who combined the weaker results originally proved by GREEN and LANFORD [5] with an approximation theorem based on trace-class methods. (Kuroda's method gives additional information not obtained here.) We wish to illustrate that the method of Green and Lanford alone is equal to the task, and that the proof is not hard. (J. Dollard and C. Friedman, submitted.)

We drop the index l and write $w(r) = \dfrac{l(l+1)}{r^2} + V(r)$. Our problem then is to analyze solutions of the radial Schrödinger equation

$$-\psi''(k,r) + w(r)\psi(k,r) = k^2(\psi(k,r). \tag{15}$$

We convert (15) to the equivalent matrix equation: writing $\phi = \begin{pmatrix}\psi\\ \psi'\end{pmatrix}$, (15) becomes

$$\phi'(k,r) = A(k,r)\phi(k,r), \tag{16}$$

with

$$A(k,r) = \begin{pmatrix}0 & 1\\ w(r)-k^2 & 0\end{pmatrix}. \tag{17}$$

Anticipating that $\phi(k,r)$ will asymptotically approach a combination of plane waves, we factor out the plane-wave behavior as follows: setting

$$E(k,r) = \begin{pmatrix}e^{ikr} & e^{-ikr}\\ ike^{ikr} & -ike^{-ikr}\end{pmatrix}, \tag{18}$$

we write

$$\phi(k,r) = E(k,r)\theta(k,r). \tag{19}$$

$\theta(k,r)$ is then found to satisfy the equation

$$\theta'(k,r) = B(k,r)\theta(k,r), \tag{20}$$

with

$$B(k,r) = \frac{iw(r)}{2k}\begin{pmatrix}-1 & -e^{-2ikr}\\ e^{2ikr} & 1\end{pmatrix}. \tag{21}$$

Since $V \in L^1(1,\infty)$, clearly we have $\|B(k,r)\| \in L^1(1,\infty)$ and an elementary theorem in ordinary differential equations states that in this case $\theta(k,r)$ has a limit $\theta_+(k)$ as $r \to \infty$. Further, for any compact set $S \subset (0,\infty)$, $\sup_{k\in S}\|B(k,r)\|$ is, by inspection, $L^1(1,\infty)$, and this allows us to obtain a bound of the type [4]:

$$\|\theta(k,r)\| \leqslant Mk \in S, \quad r \in (1,\infty). \tag{22}$$

We now rewrite (19) as

$$\begin{aligned}\phi(k,r) &= E(k,r)\theta_+(k) + E(k,r)(\theta(k,r) - \theta_+(k))\\ &\equiv E(k,r)\theta_+(k) + R_1(k,r).\end{aligned} \tag{23}$$

This equation displays the fact that $\phi(k,r)$ asymptotically approaches a linear combination of plane waves. We also have the useful formula for the remainder:

$$\begin{aligned}R_1(k,r) &= E(k,r)(\theta(k,r) - \theta_+(k))\\ &= -E(k,r)\int_r^\infty B(k,s)\theta(k,s)ds = -E(k,r)\int_0^\infty \chi_r(s)B(k,s)\theta(k,s)ds,\end{aligned} \tag{24}$$

where $\chi_r(s) = \begin{cases}1 & s \geqslant r\\ 0 & s < r\end{cases}$. The remainder $R(k,r)$ for $\psi(k,r)$ is, of course, just the top entry of $R_1(k,r)$, which we denote $P_+R_1(k,r)$. This function, as given by (24), can now be inserted into (14), where in (14) the norm is the norm in $L^2(1,\infty)$. As in the previous analysis in three dimensions, we reverse the integrations over s and k and bring the L^2 norm inside the s-integral to get the following estimate for the norm in (14):

$$\int_1^\infty |w(s)| \left\| \chi_r(s) \int_0^\infty P_+ \left\{ \frac{E(k,r)}{2k} \begin{pmatrix} -1 & -e^{-2iks} \\ e^{2iks} & 1 \end{pmatrix} e^{-ik^2 t}\theta(k,s) \right\} \hat{g}(k)\,dk \right\| ds \tag{25}$$

Suppose we can show that the norm in (25) is bounded as a function of s and approaches zero as $t \to \pm\infty$ for each fixed s. Then, since $w \in L^1(1,\infty)$, the expression (25) will approach zero as $t \to \pm\infty$ by Lebesgue's dominated convergence theorem, and we will be done. Now, by the form of $E(k,r)$, the integral over k in (25) is a sum of two terms $h_\pm(r,s,t)$ having the form

$$h_\pm(r,s,t) = \int_S e^{\pm ikr - ik^2 t} j_\pm(k,s)\,dk, \tag{26}$$

where S is the compact support of $\hat{g}$ in $(0,\infty)$, and $j_\pm(k,s)$ is $\frac{\hat{g}(k)}{k}$ times a linear combination of products of entries of $\theta(k,s)$ with $e^{\pm 2iks}$, 1, and k. We note that, by Plancherel's theorem on $L^2(\mathbb{R})$,

$$\|h_\pm(r,s,t)\|_{L^2((0,\infty),dr)} \leqslant \|h_\pm(r,s,t)\|_{L^2(\mathbb{R},dr)}$$
$$= 2\pi \|j_\pm(k,s)\|_{L^2(S,dk)} \leqslant \text{const},$$

where the last inequality follows from the form of $j_\pm(k,s)$ and (22). Also, for each fixed s, $h_\pm(r,s,t)$ is a solution of the free Schrödinger equation on $L^2(\mathbb{R})$ and, as is well-known the integral of the square of such a solution over a bounded set approaches zero as $t \to \pm\infty$. Thus,

$$\|\chi_r(s) h_\pm(r,s,t)\|^2_{L^2((0,\infty),dr)} = \int_0^s |h_\pm(r,s,t)|^2 dr \underset{t\to\pm\infty}{\longrightarrow} 0, \tag{28}$$

and we are done. The reader will, of course, recognize that this proof is identical in conception to the earilier proof suggested in three dimensions, the point being that in the present case the "technicalities" are trivial, whereas before they were formidable.

Now we take up a case in which V does not belong to L^1, but has an oscillatory behavior. For concreteness, we assume that V is the VON NEUMANN-WIGNER potential [12], whose specific form will not be important in our argument. The only information we need is that this potential is bounded on $(0,\infty)$ and has the following asymptotic form for large r:

$$V(r) = -\frac{8\sin 2r}{r} + O\left(\frac{1}{r^2}\right), \quad r \to \infty. \tag{29}$$

Again, we wish to verify (14), and we make the same analysis of (15) as given above. We note that now

$$w(r) = \frac{l(l+1)}{r^2} + V(r) = -\frac{8\sin 2r}{r} + O\left(\frac{1}{r^2}\right), \quad r \to \infty. \tag{30}$$

Equation (30) shows that $B(k,r)$ of (21) is no longer in $L^1(1,\infty)$, so that our previous proof does not apply. However, a little elementary calculus shows that if $k \neq 1$ then $B(k,r)$ is *conditionally* integrable over $(1,\infty)$. Further, setting

$$H(k,r) = \int_r^\infty B(k,s)\,ds, \quad k \neq 1, \tag{31}$$

we have $\|H(k,r)\| = 0(1/r)$ for large r, so that

$$\|H(k,r)B(k,r)\| \in L^1(1,\infty), \quad k \neq 1. \tag{32}$$

Condition (32) is enough to show that, for $k \neq 1$, $\theta(k,r)$ has a limit as $r \to \infty$. [4]. Briefly, the reason is this: setting

$$\chi(k,r) = (I + H(k,r))\theta(k,r), \tag{33}$$

we have

$$\chi'(k,r) = (I + H(k,r))'\theta(k,r) + (I + H(k,r))\theta'(k,r)$$
$$= H(k,r)B(k,r)\theta(k,r). \tag{34}$$

Now, as just remarked, $H(k,r) \underset{r\to\infty}{\longrightarrow} 0$, so for large enough r, (say $r \geqslant R$), $I + H(k,r)$ is invertible and (34) can be rewritten

$$\chi'(k,r) = H(k,r)B(k,r)(I + H(k,r))^{-1}\chi(k,r)$$
$$\equiv C(k,r)\chi(k,r). \tag{35}$$

Now for large enough R, $\|C\| \in L^1(R,\infty)$, since $\|C\| \leqslant$ const $\|HB\| \in L^1(R,\infty)$. But then (35) implies that $\chi(k,r)$ has a limit as $r \to \infty$ and, hence, so does $\theta(k,r)$, by (33). In fact, $\theta(k,r)$ has the same limit as $\chi(k,r)$ (see (33)). Setting

$$\theta_+(k) = \lim_{r\to\infty} \theta(k,r), \tag{36}$$

we have

$$\begin{aligned}\theta(k,r) &= \theta_+(k) + (\theta(k,r) - \theta_+(k)) \\ &= \theta_+(k) + (\chi(k,r)\ \ \theta_+(k)) + (\theta(k,r) - \chi(k,r)) \\ &= \theta_+(k) - \int_r^\infty C(k,s)\chi(k,s)ds - H(k,r)\,\theta(k,r)\end{aligned} \tag{37}$$

Multiplying (37) by $E(k,r)$, we find

$$\phi(k,r) = E(k,r)\,\theta(k,r) = E(k,r)\,\theta_+(k) + R_1(k,r), \quad k \neq 1. \tag{38}$$

In (38), $R_1(k,r)$ consists of two terms. The first involves an integral containing the function $C(k,s)$, whose norm is in $L^1(R,\infty)$. This is of the type dealt with in our previous analysis for $V \in L^1(1,\infty)$. In the present case we interpret the norm in (14) to be the norm in $L^2(R,\infty)$ and proceed as before. The only refinement is that one chooses $\hat{g}(k)$ to vanish in a neighborhood of the point $k = 1$, to avoid the difficulty there. The second term in $R_1(k,r)$ is easily shown to be $0(1/r)$ uniformly in k on the support S of $\hat{g}$. This permits the estimate (recall P_+ projects out the top entry)

$$\left| \int_S P_+\{E(k,r)H(k,r)\theta(k,r)\} e^{-ik^2t}\hat{g}(k)dk \right| \leqslant \frac{\text{const}}{r}. \tag{39}$$

Since $\frac{\text{const}}{r}$ is a fixed function in $L^2(R,\infty)$ and the left-hand side of (39) approaches zero for fixed r when $t \to \pm\infty$ by the Riemann-Lebesgue lemma, Lebesgue's dominated convergence theorem shows that the $L^2(R,\infty)$ norm of the left-hand side of (39) approaches zero as $t \to \pm\infty$, finishing the argument.

By successive refinements, this method can be used to prove existence of the Mϕller wave operators and weak asymptotic completeness for a class of potentials of the type $\frac{\lambda \sin \mu x^\alpha}{x^\beta}$, where λ and μ are real, and α and β are positive numbers satisfying $\alpha + \beta > 1, \beta > 1/2$ [1]. (For other conditions on α and β and stronger results on asymptotic completeness, see [2].)

A side benefit of the above analysis is that it is not difficult to show that $\theta_+(k) \neq 0$ unless $\phi(k,r) \equiv 0$, so that a nontrivial plane-wave asymptotic form is established for all solutions of (15), except when k=1. This, in turn, precludes the possibility of a square-integrable solution except for k=1 and, hence, a positive-energy bound state cannot occur except for $E = k^2 = 1$. As is well-known, the von Neumann-Wigner potential does have a positive-energy bound state with E=1.

In summary, the method of Green and Lanford consists, modulo a few tecnicalities, in estimating the rate at which solutions of the radial Schrödinger equation achieve their asymptotic forms as $r \to \infty$. At first glance, the method may seem as bit more "cluttered" than some other methods available, but in certain instances the method is subtler and gives more information.

REFERENCES

[1] B. Bourgeois: "Quantum mechanical scattering theory for long-range oscillatory potentials," Ph.D. thesis, University of Texas at Austin (Spring, 1979). Results to be published in Ann. Phys. (N.Y.)
[2] A. Devinatz, M. Ben-Artzi: J. Math. Phys. **20**, 594 (1979)
[3] J. Dollard, C. Friedman: Ann. Phys. (N.Y.) **111**, 251 (1978)
[4] J. Dollard, C. Friedman: J. Math. Phys. **18**, 1598 (1977)
[5] T. A. Green, O. E. Lanford, III: J. Math. Phys. **1**, 139 (1960)
[6] T. Ikebe: Arch. Rational Mech. Anal. **5**, 1 (1960)
[7] T. Kato: Trans. Am. Math. Soc. **70**, 195 (1951)
[8] S. T. Kuroda: J. Math. Phys. **3**, 933 (1962)
[9] B. Simon: Comm. Pure Appl. Math. **22**, 531 (1969)
[10] B. Simon: *Quantum Mechanics for Hamiltonians Defined as Quadratic Forms* (Princeton University Press, Princeton, New Jersey, 1971)
[11] B. Simon: *Trace Ideals and their Applications*, London Mathematical Society Lecture Note Series, Vol. 35 (Cambridge University Press, Cambridge, 1979)
[12] J. von Neumann, E. P. Wigner: Phys. Z. **30**, 465 (1929)

A TWO-HILBERT-SPACE FORMULATION OF MULTICHANNEL SCATTERING THEORY

Colston Chandler

Department of Physics and Astronomy
University of New Nexico
Albuquerque, New Mexico 87131

A. G. Gibson

Department of Mathematics and Statistics
University of New Mexico
Albuquerque, New Mexico 87131

INTRODUCTION

In this paper, we announce results which appear to us to found a new theoretical approach to nonrelativistic multichannel quantum scattering problems. The approach has the calculational flexibility of the resonating-group method, yet preserves the mathematical rigor of formulations based on integral-equation techniques.

The time-dependent two-Hilbert-space formulation of the scattering theory is first reviewed and the transition to time-independent theory made (Secs. 1-6). Dynamical equations for the transition operator T are then written down (Sec. 7). In these equations the cluster-(or channel-) projection operators are incorporated into the transition operator, a feature characteristic of our two-Hilbert-space formulation. The next step (Sec. 8) is the introduction of a new system of N-body equations, which we call M-operator equations. These equations have a unique solution and are strongly approximation-solvable. The theory of strong approximation-solvability (Sec. 9) with projection type approximations leads to the conclusion that the T- and M-operator equations are also stable. There are other advantages of our equations, when compared (Sec. 10) with, for example, the Faddeev-Yakubovskii equations, that are especially apparent when the number of particles is large but the number of interesting channels is small. Still another advantage is provided by the fact that the solutions to the approximate equations can be related to approximate scattering systems (Sec. 11). This permits insight into the faithfulness with which the approximations represent the physics of the scattering process. The basic development is then brought to a close with a discussion (Secs. 12 and 13) of how to include effects of particle indistinguishability and of Coulomb interactions.

There remains the question of how to implement the basic ideas of the theory of approximation-solvability. Examples are presented (Sec. 14) and interesting features discussed. Among the most interesting aspects of these particular approximation schemes is that the approximate scattering operators are all unitary (Secs. 14 and 15).

Throughout this paper we quote theorems without proofs. These proofs will be published elsewhere.

1. NOTATION

In this paper, we adopt the terminology and notation of [12]. Somewhat different terminologies can be found in [2, 49, 50].

Consider first a scattering process of $N \geqslant 2$ nonrelativistic, spinless, distinguishable particles interacting via short-range pair potentials. Asymptotically, the particles cluster themselves into *fragments* each of which is in a specific quantum mechanical bound state. Denote the different possible incoming and outgoing *clusterings* of fragments of the N particles by A, B, C, ... Let O denote the clustering into N fragments, and let N denote the clustering into one fragment. A *channel* is a specification of a clustering and a bound state for each fragment. Let α, β, γ, ... denote different channels. If channel α has clustering A, we write $\alpha \in A$.

The *total Hamiltonian* of the system (with the center of mass motion removed) is the self-adjoint operator H_N with domain $D(H_N)$ dense in the Hilbert space $\mathscr{H}_N \equiv L^2(\mathbb{R}^{3N-3})$. Each clustering A determines an *external potential* $\bar{V}_A$, which is the sum of interactions between particles in different fragments, and a *cluster Hamiltonian* H_A (see [2, 12, 33, 50]). Then $H_N = H_A + \bar{V}_A$ for every clustering A.

Let $R_N(z) \equiv (z - H_N)^{-1}, R_A(z) \equiv (z - H_A)^{-1}$, and $R_O(z) \equiv (z - H_O)^{-1}$ denote the resolvents of the respective Hamiltonians.

2. CLUSTER SUBSPACES AND P_A OPERATORS

Each channel $\alpha \in A$ determines a *channel subspace* $\mathscr{H}_\alpha \subset \mathscr{H}_N$ which is the closed linear span of vectors of the form

$$\psi_\alpha = \hat{\phi}_\alpha(p_A) \otimes f(q_A). \tag{1}$$

The $\hat{\phi}_\alpha$ are products of the bound state wave functions of the fragments considered as functions of the internal momentum space coordinates p_A. The function $f(q_A)$ belongs to the space $L^2(\mathbb{R}^{3n_A-3})$ of functions of external momentum space coordinates q_A. n_A is the number of fragments in clustering A. Define *channel projection operators* P_α to be the orthogonal projections of $\mathscr{H}_N$ onto $\mathscr{H}_\alpha$.

Define *cluster subspaces* $\mathscr{H}_A \subset \mathscr{H}_N$ to be the closed linear span of all the Hilbert spaces $\mathscr{H}_\alpha$ for which $\alpha \in A$. Define *cluster projection operators* $P_A = \hat{P}_A \otimes I_A^0$ to be the orthogonal projection of $\mathscr{H}_N$ onto $\mathscr{H}_A$ for all A , i.e., for $\phi = \phi(p_A, q_A) \in \mathscr{H}_N$,

$$P_A\phi = \sum_{\alpha \in A} P_\alpha \phi = \sum_{\alpha \in A} \hat{\phi}_\alpha (\hat{\phi}_\alpha, \phi). \tag{2}$$

The integration in the inner product $(\hat{\phi}_\alpha, \phi)$ in (2) is with respect to the internal variables p_A only.

3. ASYMPTOTIC (DIRECT SUM) SPACE AND HAMILTONIAN

In order to keep track of all clusterings simultaneously, define the *asymptotic Hilbert space*

$$\mathscr{H} \equiv \oplus_A \mathscr{H}_A (\mathscr{H}_A = P_A \mathscr{H}_N). \tag{3}$$

Define the *asymptotic* ("free") Hamiltonian H by

$$H\Phi \equiv \oplus_A H_A \phi_A, \tag{4}$$

for $\Phi = \oplus_A \phi_A \in \mathscr{H}$ with $\phi_A \in D(\mathscr{H}_A)$, the domain of H_A . Let $R(z) \equiv (z - H)^{-1}$.

We use the notational convention that all operators with a subscript map (a subspace of) $\mathscr{H}_N$ into $\mathscr{H}_N$. Operators without a subscript have domain and/or range in a direct sum space.

4. INJECTION OPERATORS J AND J*

Communication between the Hilbert spaces $\mathscr{H}$ and $\mathscr{H}_N$ is provided by the injection operator $J: \mathscr{H} \to \mathscr{H}_N$ defined by

$$J\Phi = J\oplus_A \phi_A \equiv \sum_A \phi_A. \tag{5}$$

Then $J^*: \mathscr{H}_N \to \mathscr{H}$ is given by

$$J^*\psi = \oplus_A P_A \psi. \tag{6}$$

Lemma 1 [11,12]. *(i) J maps $D(H)$ into $D(H_N)$. (ii) J^* maps $D(H_N)$ into $D(H)$. (iii) $JJ^* = \sum_A P_A$ has a bounded inverse on $\mathscr{H}_N$.*

5. ASSUMPTIONS

Define *potential operators* $V: D(V) \supset D(H) \to \mathscr{H}_N$ by

$$V \equiv H_N J - JH, \tag{7}$$

and $V^*: D(V^*) \supset D(H_N) \to \mathscr{H}$ by

$$V^* \equiv J^*H_N - HJ^*. \tag{8}$$

We assume that V and V^* satisfy the following conditions.

Assumptions. (A1) For every positive ϵ and ϵ^*, there exist finite constants b and b^* such that

$$||V\Phi|| \leqslant \epsilon||H\Phi|| + b||\Phi|| \text{ (for every } \Phi \in D(H)) \tag{9}$$

and

$$||V^*\psi|| \leqslant \epsilon^*||H_N\psi|| + b^*||\psi|| \text{ (for every } \psi \in D(H_N)). \tag{10}$$

(A2) The two-Hilbert-space *wave operators* $\Omega^{\pm}: \mathscr{H} \rightarrow \mathscr{H}_N$ defined by

$$\Omega^{\pm} \equiv \operatorname*{s-lim}_{t\rightarrow\pm\infty} e^{iH_Nt}Je^{-iHt} \tag{11}$$

exist.

Remark 1. (a) Assumptions (A1) and (A2) are true if, for example, the pair potentials V_{ij} satisfy [12]

$$V_{ij} \in L^2(\mathbb{R}^3) + L^p(\mathbb{R}^3) \text{ for some } 2 \leqslant p < 3. \tag{12}$$

(b) The idea of using two-Hilbert-space wave operators for quantum mechanical scattering was originally due to EKSTEIN [23]. They have been pioneered by COESTER for relativistic scattering [16-18] and by HUNZIKER for nonrelativistic scattering with singular potentials [34]. Two-Hilbert-space wave operators have also been used extensively in acoustical and electromagnetic scattering by KATO [38], BELOPOL'SKII and BIRMAN [3], WILCOX [59], and DEIFT [20], *et al.*

6. TRANSITION TO TIME INDEPENDENT THEORY

A *full system scattering operator* $S : \mathscr{H} \rightarrow \mathscr{H}$ is defined by

$$S \equiv \Omega^{+*}\Omega^{-}. \tag{13}$$

Theorem 1 [11 - 13]. *$S-I$ is given on $\mathscr{H}$ by*

$$S - I = \operatorname*{w-lim}_{\epsilon_1\rightarrow 0+} \operatorname*{s-lim}_{\epsilon_2\rightarrow 0+} (-2\pi i) \int\int dE(\mu)\delta_{\epsilon_1}(\lambda-\mu)T(\lambda+i\epsilon_2)dE(\lambda), \tag{14}$$

where

$$\delta_\epsilon(x) \equiv \frac{\epsilon}{\pi}(x^2 + \epsilon^2)^{-1}, \; x \text{ real}, \tag{15}$$

and where $T(z):D(H) \rightarrow \mathscr{H}$ is defined by

$$T(z) \equiv J^*V + V^*R_N(z)V, \tag{16}$$

for z in $\rho(H_N)$, the resolvent set of H_N.

The *full system transition operator* $T(z)$ given in (16) is the "prior" form. "Post" and "AGS" forms are also possible [12,13]. The restrictions of T which map $\mathscr{H}_A$ into $\mathscr{H}_B$ are called the *cluster matrix elements* of T and are denoted by $T_{BA}(z)$. Then

$$T_{BA}(z) = P_B U_{BA}(z) P_A, \tag{17}$$

where

$$U_{BA}(z) \equiv \overline{V}_A + \overline{V}_B R_N(z) \overline{V}_A \tag{18}$$

is the usual prior form of the transition operator [12].

The next step is the derivation of some solvable equations for the U_{BA} or T_{BA} operators, or for some intermediate operators such as FADDEEV's M^F_{BA} operators (or our M_{BA} operators in Sec. 8 below).

The present N-body equations are of two basic types. The first consists of those equations which iterate through FADDEEV's "trees" [25] and incorporate the exact solution of all N-1 fragment subproblems into the kernel of the N body equations [1, 10, 24-26, 29, 36, 45, 57, 60]. The second consists of those equations which have almost all of the branches of the trees in the kernel (or some iterate of the kernel) [4, 9, 28, 39, 44, 48, 49, 51, 53, 54, 56, 58]. These equations determine unknown operators which act on the full N-body space $\mathscr{H}_N$. The kernel (or some iterate of the kernel) is made connected (and compact) by forming complicated products of operators.

What we do here is very different from these previous methods. The P_A operators are incorporated into the unknowns of the equations and connectedness (and compactness) of the kernel is obtained by modifying the space $\mathscr{H}$, or, equivalently, by modifying the P_A operators. This new option is greatly facilitated by our two-Hilbert-space methods. It allows us to "trim the trees", i.e. we can approximate the problem by omiting some of the branches of the trees, or by limiting the diameter of the branches by restricting the space $\mathscr{H}$. Then, by using better and better approximations, we can watch the trees grow to full size.

7. *T*-OPERATOR EQUATIONS

Theorem 2 [12]. *Define* $L^T = L^T(z) \equiv I - V^*(JJ^*)^{-1}JR(z)$. *If* $z \in \rho(H_N)$ *and* $\Psi \in D(H) \subset \mathscr{H}$, *then* $\Phi = T\Psi$ *is the unique solution in* $\mathscr{H}$ *of the equation*

$$L^T\Phi = J^*V\Psi, \tag{19}$$

i.e., $T_{BA} = T_{BA}(z)$ *are the unique solutions of the system of equations*

$$T_{BA} = P_B \bar{V}_A P_A + P_B \bar{V}_B (JJ^*)^{-1} \sum_C R_C P_C T_{CA}. \tag{20}$$

Because of the P_C operators, the resolvents $R_C(z)$ are simple, i.e., "free" resolvents on the spaces $L^2(\mathbb{R}^{3n_C-3})$.

There are two major objections to (19) and (20). The first is that the operator $(JJ^*)^{-1}$ appears in the kernel. The second is that the kernel is not, in general, compact. The first objection can be removed (Sec. 8) by reformulating the theory in terms of a new operator M. The basic equations for M, which are new N-body equations, do not contain $(JJ^*)^{-1}$. The second objection is overcome (Sec. 9) by using a more general theory of equations that does not always require the kernel to be compact.

8. *M*-OPERATOR EQUATIONS

Define the operator $M(z):D(H) \to \mathscr{H}$ by

$$M(z) \equiv (z-H)J^*(JJ^*)^{-1}R_N(z)V. \tag{21}$$

The cluster matrix elements of $M(z)$ are

$$M_{BA}(z) = P_B(z-H_B)(JJ^*)^{-1}R_N(z)\bar{V}_A P_A. \tag{22}$$

Note that the cluster projection operators are incorporated into the definition of the operators M_{BA}, and, hence, $M_{BA} = P_B M_{BA} P_A$.

Theorem 3 (Existence). *The operator* $M = M(z)$ *is, for* $z \in \rho(H_N)$, *a solution of the equation*

$$(J^*J - J^*VR)M = J^*V. \tag{23}$$

That is, the operators $M_{BA}(z)$ *are solutions of the system of equations*

$$\sum_C P_B(I_N - \bar{V}_C R_C)P_C M_{CA} = P_B \bar{V}_A P_A. \tag{24}$$

Remark 2. In order to prove that (23) has a unique solution, we must restrict the solution space. This does not present a problem, however, because the exact solution is in the restricted space. Moreover, the restriction is automatically enforced when $T(z)$ or $R_N(z)$ is calculated from M.

Now define P to be the orthogonal projection of $\mathscr{H}$ onto $\mathscr{N} \equiv \overline{R(J^*)}$, the closure of the range of J^*. For a fixed $z \in \rho(H_N)$, let $\mathscr{D} = \mathscr{D}(z) \equiv (z-H)PR(z)\mathscr{H}$. Then $\mathscr{D}$ is a linear vector space. For $\Phi, \Psi \in \mathscr{D}$, define the inner product $(\Phi,\Psi)_{\mathscr{M}} \equiv (PR(z)\Phi, PR(z)\Psi)$. Finally, define the Hilbert space $\mathscr{M}$ to be the completion of $\mathscr{D}$ under the inner product $(\cdot,\cdot)_{\mathscr{M}}$.

Theorem 4 (Uniqueness). *Define* $L^M = L^M(z) \equiv J^*J - J^*VR$. *If* $z \in \rho(H_N)$ *and* $\Psi \in D(H) \subset \mathcal{H}$, *then the equation*

$$L^M\Phi = J^*V\Psi \tag{25}$$

has a unique solution $\Phi = M\Psi$ *in* $\mathcal{M}$.

Corollary 4.1. *Suppose that* $M = M(z)$ *is a solution of (23). Let* $X \equiv R^{-1}J^*JRM$ *and* $Y \equiv JR + JRMR$. *Then* $X = T$ *and* $Y = R_NJ$.

Remark 3. (a) The operator $(JJ^*)^{-1}$ does not arise in our M-operator equations.

(b) No mention of the space $\mathcal{M}$ is needed in Corollary 4.1, because the operator JR in JRM annihilates any component of M which is not in $\mathcal{M}$.

(c) The formula $R_NJ = JR + JRMR$, i.e.,

$$R_NP_A = R_AP_A + \sum_B R_BP_BM_{BA}P_AR_A, \tag{26}$$

is of the type encountered in the "limiting-absorption principle" [30-32]. This strongly suggests that M is the proper operator to be studying.

9. APPROXIMATION-SOLVABILITY

In this section the T- and M-operator equations are placed within the context of the theory of A-proper operators and strong approximation-solvability. This theory has been reviewed by PETRYSHYN [47]. It subsumes Fredholm theory, Galerkin methods, monotone operator theory, ball condensing operator theory, and others.

Assumption (Π). The operator $\Pi = \oplus_A \Pi_A : \mathcal{H} \to \mathcal{H}$ has the properties

(a) Π is an orthogonal projection operator,

(b) $[\Pi, H] = 0$,

(c) $(J\Pi J^*)^{-1}$ exists and is bounded on $\overline{\mathcal{R}(J\Pi)}$, the closure of the range of $J\Pi$.

Definition. $\Gamma = \{\mathcal{H}^{(n)}, \mathcal{N}^{(n)}, \Pi^{(n)}, Q\}$ is called a *suitable approximation scheme* for an operator $L: \mathcal{D} \to \mathcal{N}$ if $\{\Pi^{(n)}\}$ is a sequence of operators satisfying Assumption (Π) such that $\Pi^{(n)}\Phi \to \Phi$ for each $\Phi \in \mathcal{H}$, $\mathcal{H}^{(n)} \equiv \Pi^{(n)}\mathcal{H}$, $\mathcal{N}^{(n)} \equiv \overline{\Pi^{(n)}\mathcal{N}}$, and Q is a bounded operator from $\mathcal{H}$ into $\mathcal{D}$.

Remark 4. Our definition of a suitable approximation scheme differs from PETRYSHYN's definition [47] of an admissible approximation scheme in two respects. First, for theoretical purposes we do not always assume that $\mathcal{H}^{(n)}$ and $\mathcal{N}^{(n)}$ are finite dimensional. Second, our Assumptions (Πb) and (Πc) do not play a role in [47].

Definition. For a given $\Psi \in \mathcal{N}$, the equation $L\Phi = \Psi$ is said to be *strongly approximation-solvable* with respect to a suitable approximation scheme Γ if there is an integer $n_0 \geqslant 1$ such that the approximate equation

$$\Pi^{(n)}LQ\Pi^{(n)}\Phi^{(n)} = \Pi^{(n)}\Psi \tag{27}$$

has a solution $\Phi^{(n)} \in \mathcal{H}^{(n)}$ for each $n \geqslant n_0$ such that $Q\Pi^{(n)}\Phi^{(n)} \xrightarrow[\mathcal{M}]{} \Phi$ for some $\Phi \in \mathcal{D}$ with $L\Phi = \Psi$.

Let $P_\Pi^{(n)}$ denote the orthogonal projection onto the closure of the range of $J\Pi^{(n)}J^* = \sum_A \Pi_A^{(n)}$, for every n. We define *approximate total Hamiltonians* $H_N^{(n)}$ on $D(H_N^{(n)}) \subset \mathcal{H}_N^{(n)} \equiv P_\Pi^{(n)}\mathcal{H}_N$ by

$$H_N^{(n)} \equiv P_\Pi^{(n)}H_NP_\Pi^{(n)}, \tag{28}$$

and let $R_N^{(n)}(z) \equiv (z - H_N^{(n)})^{-1}$.

Define $Q^T = Q^T(z) \equiv (z - H)J^*JR(z)$ and $Q^M = Q^M(z) \equiv (z - H)PR(z)$. Then

$$\begin{aligned} L^{(n)} &\equiv \Pi^{(n)}J^*(z - H_N^{(n)})JR\Pi^{(n)} \\ &= \Pi^{(n)}L^T Q^T \Pi^{(n)} = \Pi^{(n)}L^M Q^M \Pi^{(n)}. \end{aligned} \tag{29}$$

Suppose that Assumptions (A1), (A2), and (Π) are satisfied. Then we have the following theorems.

Theorem 5. *The operators $H_N^{(n)}$ are self-adjoint.*

Theorem 6. *The operator H_N is the strong resolvent limit of $H_N^{(n)}$, i.e., for all z with* Im $z \neq 0, R_N^{(n)}(z) \underset{s}{\rightarrow} R_N(z)$ *as* $\Pi^{(n)} \underset{s}{\rightarrow} I$.

Theorem 7. *(i) The T-operator equation* (19) *is strongly approximation-solvable with respect to* $\Gamma^T \equiv \{\mathscr{H}^{(n)}, \mathscr{N}^{(n)}, \Pi^{(n)}, Q^T\}$ *for each* $\Psi \in D(H)$.

(ii) The M-operator equation (25) is strongly approximation-solvable with respect to $\Gamma^M \equiv \{\mathscr{H}^{(n)}, \mathscr{N}^{(n)}, \Pi^{(n)}, Q^M\}$ *for each* $\Psi \in D(H)$.

In either case (i) or (ii), the associated approximate equation is

$$L^{(n)}\Phi^{(n)} = \Pi^{(n)}J^*V\Psi, \tag{30}$$

with $\Psi \in D(H)$ and $\Phi^{(n)} = M^{(n)}\Psi \in \mathscr{H}^{(n)}$. Therefore, the cluster matrix elements $M_{BA}^{(n)} = M_{BA}^{(n)}(z)$: $D(H) \rightarrow \mathscr{H}^{(n)}$ are solutions of the system of approximate M-equations

$$\sum_C \Pi_B^{(n)}(I_N - \bar{V}_C R_C)\Pi_C^{(n)} M_{CA}^{(n)} = \Pi_B^{(n)} \bar{V}_A P_A. \tag{31}$$

Condition (Γ). There exist $c_0 > 0$ and $n_0 \geqslant 1$ such that the suitable approximation scheme Γ satisfies $||J\Pi^{(n)}J^*\psi|| \geqslant c_0||\psi||$ for all $\psi \in \mathscr{H}_N^{(n)}$ and all $n \geqslant n_0$.

Corollary 7.1. *If Condition* (Γ) *is also satisfied, then the operators L^T and L^M are Petryshyn A-proper and one-to-one.*

Remark 5. (a) Condition (Γ) is our analog of the Condition (A) in the Petrov-Galerkin method (see [47]).

(b) The approximate *M*-equation (30) has a unique solution on the subspace $(z-H)\Pi^{(n)}J^*D(H_N)$ of $\mathscr{H}^{(n)}$. Any solution $\Phi^{(n)} \in \mathscr{H}^{(n)}$, however, uniquely determines the approximate transition operator $T^{(n)}(z)$ by $T^{(n)}(z)\Psi \equiv \Pi^{(n)}Q^T(z)\Phi^{(n)}$ and the approximate resolvent operator $R_N^{(n)}(z)$.

(c) Theorem 7 removes the need to obtain equations with $L = I - C$, C a compact operator. The main purpose of having equations with such kernels is to take advantage of the Fredholm alternative. But Theorem 7 together with our uniqueness Theorem 2 or 4, accomplish the same goals.

Theorem 8 (Stability). *Suppose that Assumptions (A1), (A2), and* (Π) *are satisfied, and let Q denote either Q^T or Q^M. Then the projective approximation method given in (30) is stable in the sense that there exist positive constants c_1, c_2 (independent of n and Ψ) such that the perturbed equation*

$$(L^{(n)} + \Pi^{(n)}F^{(n)}Q\Pi^{(n)})\Theta^{(n)} = \Pi^{(n)}J^*V\Psi + \Psi^{(n)} \tag{32}$$

has a solution $\Theta^{(n)} \in \mathscr{H}^{(n)}$ for $n \geqslant n_0$ such that

(i) $Q\Pi^{(n)}\Theta^{(n)} \in \mathscr{D} \subset \mathscr{M}$ is uniquely determined, and

(ii) the inequality

$$||Q\Pi^{(n)}(\Phi^{(n)} - \Theta^{(n)})||_{\mathscr{M}} \leqslant c_1||\Phi^{(n)}|| \, ||F^{(n)}|| + c_1||\Psi^{(n)}|| \tag{33}$$

holds, where $F^{(n)}: \mathscr{M} \rightarrow \mathscr{N}$ is a bounded linear operator perturbation satisfying $||F^{(n)}|| \leqslant c_2$, and $\Psi^{(n)}$ is an arbitrary vector in $\Pi^{(n)}\mathscr{N}$.

0. COMPARISON WITH FADDEEV-YAKUBOVSKII EQUATIONS

We digress to compare our *M*-equations with the FADDEEV-YAKUBOVSKII (FY) equations [26,60]. Similar comparisons with other *N*-body equations may also be made, but, in the interest of brevity, we shall not make them here.

The Faddeev equations for $N = 3$ are

$$M^F_{BA} = \delta_{BA} T_A + T_B R_0 \sum_C \bar{\delta}_{BC} M^F_{CA} \tag{34}$$

where δ_{BA} is the kronecker delta, $\bar{\delta}_{BA} = 1 - \delta_{BA}$, and T_A are the two-fragment transition operators in the three-body space. Because of their complexity, we shall not write down the Yakubovskii equations for $N > 3$.

We put our equation $L^M M = J^* V$ into a form similar to (34) by rewriting it as

$$M = J^* V + (I - L^M) M \tag{35}$$

and taking cluster matrix elements to obtain

$$M_{BA} = P_B \bar{V}_A P_A + \sum_C P_B (\bar{V}_C R_C - \bar{\delta}_{CB}) P_C M_{CA}. \tag{36}$$

We emphasize that (36) is an N-body equation for any $N \geqslant 2$.

Table I contains a comparison of the properties of the equations. We remark that the 3-body Faddeev equations are known to be stable under certain conditions [61], but, to our knowledge, no approximation method such as pole approximation has been proved to be stable.

In Tables II and III, we compare the number of coupled FY equations with the number of equations in the system (36) for N distinguishable particles and for N identical particles. The numbers in the tables are from BENCZE [5]. The rows in Tables II and III labeled $CG < \Sigma_3$ refer to the number of equations in the system (36) if only the two-fragment clusterings are included. This is not expected to be a good approximation. Nevertheless, in calculations below the three-fragment threshold Σ_3, we expect that the number of additional equations required to represent the important virtual breakup states is small. The number of equations needed for a more realistic approximation is, therefore, not much larger than the number given in the tables for $CG < \Sigma_3$.

TABLE I

Comparison of Properties of the Faddeev-Yakubovskii Equation with the Chandler-Gibson Equations.

PROPERTY	FY	CG
Unique solution	Yes	Yes (Theorem 4)
Compact kernel (with additional potential assumptions)	Yes (after iteration)	No for exact equation (but A-proper!) Yes for approximate equations (without iteration)
Simple kernel	Yes (if T_A are known)	Yes
Nonhomogeneous term	T_A	Born approximation
Reduces to Lippmann-Schwinger equation when $N = 2$	Not directly	Yes
Approximation method	Pole approximation (and others)	Projection methods
Unique solution of approximate equations	Not always [40]	Not always (but $T^{(n)}(z)$ are always unique)
Stability of approximation method	?	Yes (Theorem 8)
Unitarity of approximate $S^{(n)}$ Operators	?	Yes (Theorems 15 and 16)
Convergence of approximate $S^{(n)}$ operators to exact S operator	?	Yes (Theorem 10)

TABLE II

Number of Equations for N Distinguishable Particles

Approach \ N	3	4	6	8
FY	3	18	2,700	1,587,600
CG exact	4	14	202	4,139
CG < Σ_3	3	7	31	127

TABLE III

Number of Equations for N Identical Particles

Approach \ N	3	4	6	8
FY	1	2	11	116
CG exact	2	4	10	21
CG < Σ_3	1	2	3	4

Table IV contains a comparison of the dimension of the integrals arising in the FY and CG equations. The last line refers to the approximation method outlined in Sec. 14.

TABLE IV

Dimension of Integration Space

Approach \ N	3	4	6	8	N
FY	$L^2(\mathbb{R}^6)$	$L^2(\mathbb{R}^9)$	$L^2(\mathbb{R}^{15})$	$L^2(\mathbb{R}^{21})$	$L^2(\mathbb{R}^{3N-3})$
CG exact	Varies from $L^2(\mathbb{R}^3)$ to $L^2(\mathbb{R}^{3N-3})$				
CG < Σ_3	$L^2(\mathbb{R}^3)$	$L^2(\mathbb{R}^3)$	$L^2(\mathbb{R}^3)$	$L^2(\mathbb{R}^3)$	$L^2(\mathbb{R}^3)$
CG approx. method	$L^2(\mathbb{R})$	$L^2(\mathbb{R})$	$L^2(\mathbb{R})$	$L^2(\mathbb{R})$	$L^2(\mathbb{R})$

11. INTERPRETATION OF THE APPROXIMATE SOLUTIONS

Any solution $M^{(n)}(z):D(H) \subset \mathscr{H} \to \mathscr{H}^{(n)}$ to the approximate equation

$$L^{(n)}M^{(n)} = \Pi^{(n)}J^*V \tag{37}$$

can be interpreted in terms of an approximate scattering system.

Theorem 9. *Assume that V, V^*, and $\Pi^{(n)}$ are such that Assumptions (A1) and (Π) are satisfied. Suppose further that there is a set $E^{(n)}$, dense in $\mathscr{H}$, such that for each $\Phi \in E^{(n)}$*

$$\int_{-\infty}^{\infty} \| V\Pi^{(n)}e^{-iHs}\Phi \| ds < \infty . \tag{38}$$

Then the wave operators

$$\Omega^{(n)\pm} \equiv \operatorname*{s-lim}_{t\to\pm\infty} e^{iH_N^{(n)}t} J\Pi^{(n)} e^{-iHt} \tag{39}$$

exist, and the scattering operator

$$S^{(n)} \equiv \Omega^{(n)+*}\Omega^{(n)-} \tag{40}$$

is given on $\mathscr{H}^{(n)}$ by

$$S^{(n)} - \Pi^{(n)} = \operatorname*{w-lim}_{\epsilon_1\to 0+} \operatorname*{s-lim}_{\epsilon_2\to 0+} (-2\pi i) \int\int dE(\mu)\delta_{\epsilon_1}(\lambda-\mu) T^{(n)}(\lambda+i\epsilon_2)\Pi^{(n)} dE(\lambda). \tag{41}$$

Here, δ_ϵ is as in (15) and

$$T^{(n)}(z)\Pi^{(n)} = \Pi^{(n)}\{J^*V + V^*R_N^{(n)}P_\Pi^{(n)}V\}\Pi^{(n)}. \tag{42}$$

The operator $T^{(n)}(z)$ is related to $M^{(n)}(z)$ by

$$T^{(n)}(z) = \Pi^{(n)}Q^T(z)M^{(n)}(z). \tag{43}$$

The approximate scattering system corresponding to $M^{(n)}(z)$ thus has the approximate total Hamiltonian $H_N^{(n)}$, the approximate N-particle space $\mathscr{H}_N^{(n)} \equiv P_\Pi^{(n)}\mathscr{H}_N$, and the approximate asymptotic space $\mathscr{H}^{(n)}$. Within this framework, the relation on $\mathscr{H}^{(n)}$ of $M^{(n)}(z)$ to the approximate scattering operator $S^{(n)}$ is formally the same as the relation on $\mathscr{H}$ of the exact $M(z)$ to the exact scattering operator S.

There are two important consequences of this correspondence between $M^{(n)}(z)$ and an approximate scattering system. First, the formal structure of the exact theory is preserved so that certain features of the exact system, such as symmetries and unitarity, can be incorporated into the approximate theory in a manifest way. Second, one is led to expect that $T^{(n)}(z)$ will be a good approximation to the exact $T(z)$ when $\mathscr{H}_N^{(n)}$ contains all the states, both real and virtual, that are physically important for the process being considered. *A priori* physical arguments may thus be used to choose the projection operator $\Pi^{(n)}$ that yields an accurate approximation $T^{(n)}(z)$ to the exact $T(z)$.

It is also true that the sequence of approximate wave operators $\Omega^{(n)\pm}$ converges strongly to the exact wave operator $\Omega^\pm$.

Theorem 10. *Let Γ^T be a suitable approximation scheme for L^T, and let the assumptions of Theorem 9 hold with $E^{(n)}$ being independent of n. Then the wave operators $\Omega^\pm$ exist on $\mathscr{H}$,*

$$\Omega^\pm = \operatorname*{s-lim}_{n\to\infty} \Omega^{(n)\pm}, \tag{44}$$

and

$$S = \operatorname*{w-lim}_{n\to\infty} S^{(n)}. \tag{45}$$

This convergence of the approximate scattering operators to the exact one is, so far as we know, the only such result that has been rigorously proved for general N-body systems.

12. SYSTEMS CONTAINING IDENTICAL PARTICLES

For practical applications, it is important that projectionally suitable approximation schemes are compatible with the symmetry requirements that are placed on the theory when some of the particles are identical.

Let $\mathscr{S}$ be the group of permutations A_s that interchange identical particles. The Young symmetrizer for $\mathscr{S}$ is given by

$$A_N \equiv |\mathscr{S}|^{-1} \sum_{A_s\in\mathscr{S}} f_s A_s, \tag{46}$$

where $|\mathscr{S}|$ is the number of permutations in $\mathscr{S}$, and where f_s is -1 if A_s involves an odd permutation of fermions and is $+1$ otherwise. The space of physically acceptable states is then $A_N\mathscr{H}_N$.

The projection operator A_N induces an equivalent symmetrizing projection operator $Q^{(\mathscr{S})}$ on $\mathscr{H}$.

Theorem 11 [8]. *There exists an orthogonal projection operator* $Q^{(\mathscr{S})}$ *on* $\mathscr{H}$ *such that*

$$A_N J = J Q^{(\mathscr{S})} \tag{47}$$

and

$$[Q^{(\mathscr{S})}, H] = 0. \tag{48}$$

The proof is constructive and exhibits the specific form of $Q^{(\mathscr{S})}$.

The appropriately symmetrized scattering theory for systems with identical particles is thus obtained as follows. Construct the theory as if the particles were distinguishable. Then replace $\mathscr{H}_N$ by $A_N \mathscr{H}_N$, $\mathscr{H}$ by $Q^{(\mathscr{S})}\mathscr{H}$, and J by $JQ^{(\mathscr{S})}$. The resulting theory is the correct one.

It follows that the abstract structure of the scattering theory for systems with identical particles is the same as that for systems of distinguishable particles. The projection operators Π of the approximation theory must then be projection operators on $Q^{(\mathscr{S})}\mathscr{H}$, which is equivalent to the requirement that

$$[\Pi, Q^{(\mathscr{S})}] = 0. \tag{49}$$

That this is not a restrictive requirement is clear from the detailed considerations of [8]. With the addition of (49) to Assumption (Π) all of the previous considerations of approximation-solvability apply without change to systems with identical particles.

13. COULOMB INTERACTIONS

The formalism can include the long-range effects of repulsive Coulomb interactions [14]. The proof of this is based on the results of DOLLARD [22] and on the use of the chain rule [6, 7, 38].

When Coulomb interactions are present the cluster wave operators $\Omega_A^{c\pm}$ for a given clustering A have the form [22],

$$\Omega_A^{c\pm} \equiv \underset{t\to\pm\infty}{\text{s-lim}}\, e^{iH_N t} P_A e^{-iH_A t} U_A(t). \tag{50}$$

The projection operator P_A has the form given by (2) and the operator $U_A(t)$ is unitary.

Define now

$$H_A^c \equiv H_A + V_A^c, \tag{51}$$

where V_A^c is the Coulomb interaction that would take place between the fragments of A, were the fragments to be point particles. The wave operators

$$P_A^{\pm} \equiv \underset{t\to\pm\infty}{\text{s-lim}}\, e^{iH_A^c t} P_A e^{-iH_A t} U_A(t) \tag{52}$$

can then be shown to exist by methods of DOLLARD [22]. The important structure of $P_A^{\pm}$ is indicated by

$$P_A^{\pm} = \hat{P}_A \otimes \omega_A^{\pm}, \tag{53}$$

where $\omega_A^{\pm}$ are the wave operators corresponding to the pure Coulomb scattering of the fragments in A with the internal structure of the fragments being ignored. The operators $P_A^{\pm}$ have the intertwining property [21]

$$H_A^c P_A^{\pm} \supset P_A^{\pm} H_A, \tag{54}$$

with the result that $\Omega_A^{c\pm}$ can be written in the form

$$\Omega_A^{c\pm} = \underset{t\to\pm\infty}{\text{s-lim}}\, e^{iH_N t} P_A^{\pm} e^{-iH_A t}. \tag{55}$$

New injection operators $J^{\pm}: \mathscr{H}_N \to \mathscr{H}$ are now defined for all $\Phi \in \mathscr{H}$ by

$$J^{\pm}\Phi = J^{\pm} \oplus_A \phi_A \equiv \sum_A P_A^{\pm} \phi_A. \tag{56}$$

With this notation, the cluster wave operators $\Omega_A^{c\pm}$ can be combined into two-Hilbert-space wave operators

$$\Omega^{c\pm} \equiv \underset{t\to\pm\infty}{\text{s-lim}}\, e^{iH_N t} J^{\pm} e^{-iHt} \tag{57}$$

that have the same abstract structure as (11).

It is now necessary to assume:

Assumption (ω). The wave operators $\omega_A^{\pm}$ exist and are unitary.

This assumption seems reasonable if the Coulomb forces are all repulsive, as is the case in nuclear physics. Even in that case, however, it has not been proved for clusterings A with more than three fragments [21, 41-43].

If Assumption (ω) is satisfied, then Lemma 1 is also true for $J^{\pm}$. In addition,

$$J^- = J^+ S^{co} \tag{58}$$

where S^{co} is defined by

$$S^{co}\Phi = S^{co}\oplus_A \phi_A \equiv \oplus_A P_A^{+*} P_A^- \phi_A. \tag{59}$$

The operator S^{co} is the pure Coulomb scattering operator that one would obtain if the internal structures of the fragments in the clusterings were ignored.

One now has the following theorem, which is analogous to Theorem 1.

Theorem 12. *Let Assumption (ω) be satisfied and let* $S^c \equiv \Omega^{c+*}\Omega^{c-}$. *Then* $S^c - S^{co}$ *is given on* $\mathscr{H}$ *by*

$$S^c - S^{co} = \text{w-}\lim_{\epsilon_1 \to 0+} \text{s-}\lim_{\epsilon_2 \to 0+} (-2\pi i) \int\int dE(\mu)\delta_{\epsilon_1}(\lambda-\mu)\, T^c(\lambda+i\epsilon_2)\, dE(\lambda), \tag{60}$$

where δ_ϵ *is defined in (15) and where*

$$T^c(z) \equiv J^{+*}V^- + V^{+*}R_N(z)V^-, \tag{61}$$

$$V^{\pm} \equiv H_N J^{\pm} - J^{\pm} H, \tag{62}$$

$$V^{\pm *} \equiv J^{\pm *} H_N - H J^{\pm *}. \tag{63}$$

The operator $T^c(z)$ can be expressed in terms of an operator $M^c(z)$ in an analogous fashion to the procedure in the short-range case. Thus,

$$T^c(z) = R^{-1}(z) J^{+*} J^+ R(z) M^c(z), \tag{64}$$

where $M^c(z)$ is the unique solution of the equation

$$[J^{+*}J^+ - J^{+*}V^+ R] M^c = J^{+*}V^-. \tag{65}$$

The strong approximation-solvability of (65) is now proved in the same way as is Theorem 7. The only change in the analysis is that the operator J^+, instead of J, is to be used at every stage of the proof.

Equation (65) can be written in terms of the cluster matrix elements M_{BA}^c of M^c. These equations are

$$\sum_D P_B^{+*}(P_D^+ - \bar{V}_D^c P_D^+ R_D) M_{DA}^c = P_B^{+*} \bar{V}_A^c P_A^-, \tag{66}$$

where

$$\bar{V}_A^c \equiv H_N - H_A^c = H_N - H_A - V_A^c. \tag{67}$$

The major difference between (66) and the corresponding short-range equations (24) is the presence of the pure Coulomb wave operators $\omega_A^{\pm}$ in (53). This represents a formidable complication in that the many-body Coulomb scattering functions must be known. Calculation of these for two or three fragment clusterings is known to be difficult, but possible in principle [21, 41-43]. For more than three fragments, no calculational method is known.

The modifications needed to deal with the Coulomb interaction are compatible with those needed when some of the particles are identical. The operator $U(t)\colon \mathscr{H} \to \mathscr{H}$, defined for all $\Phi \in \mathscr{H}$ by

$$U(t)\Phi = U(t)\oplus_A \phi_A \equiv \oplus_A U_A(t)\phi_A, \tag{68}$$

satisfies the commutation relation [8]

$$[U(t), Q^{(\mathscr{I})}] = 0. \tag{69}$$

This leads in a straightforward way to the result

$$A_N J^{\pm} = J^{\pm} Q^{(\mathscr{I})}, \tag{70}$$

which, in turn, implies that the analysis of Sec. 12 can be applied.

Finally, it is noteworthy that although the wave operators $P_A^{\pm}$ were defined within the context of the Coulomb problem, a similar analysis can be applied to obtain a distorted-wave formalism in the short-range case. For that case, $U_A(t) \equiv 1$ and the distorting Hamiltonian K_A which replaces H_A^c is chosen to satisfy Assumption (ω) and otherwise to give easily calculable $\omega_A^{\pm}$. The operators $P_A^{\pm}$ are then expansions, not in terms of Coulomb scattering functions, but in terms of distorted waves which are the scattering functions of K_A.

14. AN EXAMPLE

We finally come to an example of how the projection operators $\Pi^{(n)}$ might be calculated [15].

For each channel α, operators $\rho_{\alpha j}: \mathscr{H}_\alpha \to L^2(\mathbb{R})$ are defined. Let p_A denote the momentum variables internal to the fragments of A, $\alpha \in A$, and let q_A denote the momentum variables of the fragments relative to each other. Let $\hat{\phi}_\alpha(p_A)$ denote the product of the bound state wave functions of the fragments of channel α. Then the operator $\rho_{\alpha j}$ is defined by

$$(\rho_{\alpha j}\psi_\alpha)(\lambda) \equiv \int dp_A\, dq_A \delta(\lambda - \epsilon_\alpha - T_A(q_A))\chi^*_{\alpha j}(q_A)\hat{\phi}^*_\alpha(p_A)\psi_\alpha(p_A, q_A). \tag{71}$$

Here, ϵ_α is the threshold energy of channel α and $T_A(q_A)$ is the kinetic energy of relative motion of the fragments of channel α. The functions $\chi_{\alpha j}$ are essentially arbitrary, but are assumed to satisfy three technical restrictions.

Assumption (χ). The functions $\chi_{\alpha j} \in L^2(\mathbb{R}^{3n_A-3})$ satisfy

$$\int dq_A\, \delta(\lambda - \epsilon_\alpha - T_A(q_A)\chi^*_{\alpha j}(q_A)\chi_{\alpha k}(q_A) = \delta_{jk}\Theta(\lambda - \epsilon_\alpha), \tag{72}$$

where δ_{jk} is the Kronecker delta and Θ is the step function. The $\chi_{\alpha j}$ also satisfy

$$\sup_{q_A} [T_A(q_A)]^{(3n_A-5)/2}\hat{J}(q_A/|q_A|)|\chi_{\alpha j}(q_A)|^2 < \infty, \tag{73a}$$

where the Jacobian $\hat{J}$ is defined by

$$dq_A = [T_A(q_A)]^{(3n_A-5)/2}\hat{J}(q_A/|q_A|)dT_A d(q_A/|q_A|). \tag{73b}$$

Finally, they have the property that for some $\epsilon > 0$ and $m > 0$,

$$||\bar{V}_A e^{-iH_A t}\rho^*_{\alpha j} f^{(m)}|| \leqslant k_{\alpha j}(1 + t^2)^{-\frac{1}{2}-\epsilon}, \tag{74a}$$

where $k_{\alpha j}$ is some constant and

$$f^{(m)}(\lambda) \equiv (i - \lambda)^{-m}. \tag{74b}$$

Equation (72) is a normalization condition and (74a) is essentially a smoothness condition. The inequality in (74a) is exactly of the type commonly encountered in the Cook's theorem approach to the existence of the cluster wave operators $\Omega_A^{\pm}$ [2, 21, 50]. From those calculations, it is apparent that (74a) will be satisfied, provided the potentials have short-range and the functions $\chi_{\alpha j}$ are at least three times continuously differentiable.

It is to be emphasized that the functions $\chi_{\alpha j}$ play much the same role in this scheme as do the trial functions in variational calculations. They are to be provided at the onset of a given calculation by the theorists' ingenuity. It is here, in particular, that physical intuition and experience is to be injected into the calculation.

As an example of how one might choose the functions $\chi_{\alpha j}$, consider a channel $\alpha \in A$ with two spin-zero fragments. Let $q_A \in \mathbb{R}^3$ be the relative momentum of the two fragments, and let μ_A be the reduced mass of the two fragments. Then one could choose

$$\chi_{\alpha j} = (\mu_A |q_A|)^{-1/2} Y_l^m (q_A/|q_A|), \tag{75}$$

where Y_l^m is some spherical harmonic with indices l,m depending on α, j.

As a second example consider a channel $\alpha \in A$ with three spinless fragments. Let $q_A^{(1)}$ and $q_A^{(2)} \in \mathbb{R}^3$ be a pair of Jacobi coordinate vectors for the channel, and let $\mu_A^{(1)}$ and $\mu_A^{(2)}$ be the corresponding reduced masses. Then one might choose

$$\chi_{\alpha j} = \left\{ \mu_A^{(1)} |q_A^{(1)}| \int dx |f_j(x)|^2 \Theta \left(\frac{|q_A^{(1)}|^2}{2\mu_A^{(1)}} + \frac{|q_A^{(2)}|^2}{2\mu_A^{(2)}} - \frac{|x|^2}{2\mu_A^{(2)}} \right) \right\}^{-1/2} f_j(q_A^{(2)}) \, Y_l^m(q_A^{(1)}/|q_A^{(1)}|), \tag{76}$$

where $f_j(q_A^{(2)}) \in L^2(\mathbb{R}^3)$, and the indicies l and m depend on α, j.

The projection operators $\Pi^{(n)}$ are now to be defined for all $\Phi \in \mathscr{H}$ by

$$\Pi^{(n)}\Phi = \Pi^{(n)} \oplus_A \phi_A \equiv \oplus_A \sum_{\alpha \in A} \sum_{j=1}^{r_\alpha} \rho^*_{\alpha j} \rho_{\alpha j} \phi_\alpha . \tag{77}$$

It is understood that the numbers r_α of trial functions $\chi_{\alpha j}$ satisfy

$$\sum_\alpha r_\alpha < \infty. \tag{78}$$

If $r_\alpha = 0$ then $\Pi^{(n)}$ annihilates functions from that channel. The important properties of the operators $\Pi^{(n)}$ are summarized in the following theorem.

Theorem 13. *Let the functions $\chi_{\alpha j}$ satisfy Assumption (χ). Then, the operators $\Pi^{(n)}$ defined in (77) satisfy Assumption (Π), and the operators*

$$C^{(n)} \equiv \Pi^{(n)} (J^*J - I) \Pi^{(n)} \tag{79}$$

are Hilbert-Schmidt.

If the functions $\chi_{\alpha j}$ are chosen to form a basis on the energy ellipsoid for channel α , then as all the $r_\alpha \to \infty$, the projection operators $\Pi^{(n)}$ tend monotonically to I, and the approximation scheme Γ defined by the $\Pi^{(n)}$ is a suitable approximation scheme. It follows that the $\Pi^{(n)}$ defined by (77) provide an example of how the approximation-solvability ideas can be implemented in the theory of N-particle systems.

The smoothness condition (74a) has extremely interesting further consequences.

Theorem 14. *Let Assumptions (A1) and (χ) be satisfied. Let $\Pi^{(n)}$ defined by (77) be such that the $\Pi^{(n)}$ tend strongly to I as $n \to \infty$. Then,*

(i) the assumptions of Theorems 9 and 10 are satisfied, with $E^{(n)}$ being the set of functions $\Psi = \oplus_A \psi_A$ in $\mathscr{H}$ with only a finite number of ψ_A nonzero and with ψ_A being infinitely differentiable and of compact support in the relative momentum variables of the fragments; and

(ii) $V\Pi^{(n)}E(\Delta)$ is trace class for all n and all finite intervals $\Delta \subset \mathbb{R}$.

It follows that the discussion of Sec. 11 can be applied to give an interpretation to the approximate equation (30). It further follows from Theorem 14 (ii) and the Kato-Birman-Pearson trace class theory [20, 37, 46] that the wave operators $\Omega^{(n)\pm}$ are asymptotically complete. From this comes the following important result.

Theorem 15. *Under the assumptions of Theorem 14, the approximate scattering operators $S^{(n)}$ are unitary on $\mathscr{H}^{(n)}$.*

Finally, we write the approximate equation (30) in a more detailed form which makes its attractive features more apparent. The cluster matrix elements $M_{BA}^{(n)}$ of $M^{(n)}|_{\mathscr{H}^{(n)}}$ have the form

$$M_{BA}^{(n)} = \sum_{\beta \in B} \sum_{\alpha \in A} \sum_{j=1}^{r_\beta} \sum_{k=1}^{r_\alpha} \rho^*_{\beta j} \rho_{\beta j} M_{BA}^{(n)} \rho^*_{\alpha k} \rho_{\alpha k}. \tag{80}$$

Let

$$M_{\beta j, \alpha k}(\lambda, \mu; z) \equiv \text{kernel of } \rho_{\beta j} M_{BA}^{(n)} \rho^*_{\alpha k}, \tag{81}$$

$$C_{\beta j, \alpha k}(\lambda, \mu) \equiv \text{kernel of } (1 - \delta_{\beta\alpha}) \rho_{\beta j} \rho^*_{\alpha k}, \tag{82}$$

$$B_{\beta j, \alpha k}(\lambda, \mu) \equiv \text{kernel of } \rho_{\beta j} \bar{V}_A \rho^*_{\alpha k}. \tag{83}$$

That the various operators are integral operators, and can hence be represented by kernels, follows from Theorems 13 and 14. Equation (30) then yields

$$M_{\beta j,\alpha k}(\lambda,\mu;z) = B_{\beta j,\alpha k}(\lambda,\mu) + \sum_{\gamma,n}\int_{-\infty}^{\infty}\left\{\frac{B_{\beta j,\gamma n}(\lambda,\eta)}{z-\eta} - C_{\beta j,\gamma n}(\lambda,\eta)\right\}M_{\gamma n,\alpha k}(\eta,\mu,z)\,d\eta. \tag{84}$$

This equation is an integral equation in one variable for a matrix-valued function. The driving terms are the overlap integrals $C_{\beta j,\alpha k}$ and the Born approximations $B_{\beta j,\alpha k}$.

We now recall the discussion of Secs. 12 and 13. Without altering the general form of (84), the effects of having identical particles and of having Coulomb interactions can be incorporated. Distorted waves other than those caused by Coulomb interactions can also be incorporated. In this latter case, the $C_{\beta j,\alpha k}$ are the overlap integrals of the distorted waves, and the $B_{\beta j,\alpha k}$ are the distorted-wave Born approximations.

15. A RELATED RESULT

Suppose the pair potentials $V_{ij}(x)$ satisfy

$$V_{ij} \in L^p(\mathbb{R}^3) \cap L^q(\mathbb{R}^3), \tag{85}$$

where $p \geqslant 2$ and $\frac{3}{2} > q \geqslant 1$. Define $\Pi = \oplus_A \Pi_A$ by

$$\Pi_A \equiv \begin{Bmatrix} P_A, A=2 - \text{fragment clusterings}, \\ 0, \text{ otherwise}. \end{Bmatrix} \tag{86}$$

Then the wave operators

$$\Omega^{\Pi\pm} \equiv \operatorname*{s-lim}_{t\to\pm\infty} e^{iP_\Pi H_N P_\Pi t} J\Pi e^{-iHt} \tag{87}$$

exist.

Theorem 16 [55]. *Assume that all subfragments of less than N particles can have only a finite number of bound states. Then $\Omega^{\Pi\pm}$ are asymptotically complete ($\Omega^{\Pi\pm}\Omega^{\Pi\pm *}$= orthogonal projection onto the absolutely continuous subspace of $P_\Pi H_N P_\Pi$).*

This result of TRUCANO [55] differs from the previous results of COMBES [19] and SIMON [52] in that the approximate Hamiltonian $P_\Pi H_N P_\Pi$ appears, rather than H_N itself. But the result is true for energies above the breakup threshold and the assumptions on the pair potentials are weaker. The proof is based on the techniques of [27, 31, 32] applied to a symmetrized version of (30), and is valid for an arbitrary number of particles.

C. C. gratefully acknowledges the generous support of the U. S.-German Fulbright Commission, the Minna-James-Heinemann-Stiftung in collaboration with the NATO Senior Scientists Programme, the German Academic Exchange Service (DAAD) and the University of Bonn during the beginning phases of this work. A. G. G. gratefully acknowledges the financial support of the DGRST and CNRS in France during the beginning stages of this work.

REFERENCES

[1] E.O. Alt, P. Grassberger, W. Sandhas: Nucl. Phys. **B2**, 167 (1967)
[2] W.O. Amrein, J.M. Jauch, K.B. Sinha: *Scattering Theory in Quantum Mechanics* (Benjamin, Reading, Massachusetts, 1977)
[3] A.L. Belopol'skii, M. S. Birman: Math. USSR Izv. **2**, 1117 (1968)
[4] Gy. Bencze: Nucl. Phys. **A210**, 568 (1973)
[5] Gy. Bencze: Phys. Lett. **72B**, 155 (1977)
[6] Gy. Bencze: Lett. Nuovo Cimento **17**, 91 (1976)
[7] Gy. Bencze, G. Cattapan, V. Vanzani: Lett. Nuovo Cimento **20**, 248 (1977)
[8] Gy. Bencze, C. Chandler: "On time dependent scattering theory for identical particles," CRIP Budapest, preprint KFKI-1979-14 (1979)
[9] F.A. Berezin: Soviet Math. Dokl. **6**, 997 (1965)
[10] G. Cattapan, V. Vanzani: Nuovo Cimento **41A**, 553 (1977)
[11] C. Chandler, A.G. Gibson: J. Math, Phys. **14**, 1328 (1973)

[12] C. Chandler, A.G. Gibson: J. Math. Phys. **18**, 2336 (1977)
[13] C. Chandler , A.G. Gibson: J. Math. Phys. **19**, 1610 (1978)
[14] C. Chandler, A.G. Gibson: in *Atomic Scattering Theory, Mathematical and Computational Aspects*, ed. by J. Nuttall (University of Western Ontario, London, Canada, 1978), p. 189
[15] C. Chandler, A.G. Gibson: in *Few Body Systems and Nuclear Forces I*, ed. by H. Zingl *et al.*, Lecture Notes in Physics, Vol. 82 (Springer, New York, 1978), p. 356
[16] F. Coester: Helv. Phys. Acta **38**, 7 (1965)
[17] F. Coester, L. Schlessinger: Ann. Phys. (N.Y.) **78**, 90 (1973)
[18] F. Coester: "Canonical scattering theory of relativistic particles," these proceedings
[19] J.M. Combes: Nuovo Cimento **A64**, 111 (1969)
[20] P. Deift: Ph.D. thesis, Princeton University (1976)
[21] J.D. Dollard: Rocky Mt. J. Math. **1**, 5 (1971)
[22] J.D. Dollard: J. Math Phys. **5**, 729 (1964)
[23] H. Ekstein: Phys. Rev. **101**, 880 (1956)
[24] D.E. Eyre, T.A. Osborn: "Cluster expansions of the three-body problem," University of Manitoba preprint (1979)
[25] L.D. Faddeev: in *The Three-Body Problem*, ed. by J.S.C. McKee, P. M. Rolph, (North-Holland, Amsterdam, 1970), p. 154
[26] L.D. Faddeev: *Mathematical Aspects of the Three-Body Problem in Quantum Scattering Theory* (Israel Program for Scientific Translations, Jerusalem, 1965)
[27] J. Ginibre, M. Moulin: Ann. Inst. Henri Poincaré **21**, 97 (1974)
[28] R. Goldflam, K.L. Kowalski, W. Tobocman: "Partition permuting array approach to few-body Hamiltonian models of nuclear reactions," Case Western Reserve University preprint (1979)
[29] P. Grassberger, W. Sandhas: Nucl. Phys. **B2**, 181 (1967)
[30] K. Gustafson, in *Operator Algebras, Ideals, and Their Applications in Theoretical Physics*, ed. by H. Baumgärtel *et al.* (Teubner-Texte, Leipzig, 1978), p. 335
[31] G.A. Hagedorn: Ph. D. thesis, Princeton University (1978)
[32] J. S. Howland: J. Functional Anal. **22**, 250 (1976)
[33] W. Hunziker, in *Lectures in Theoretical Physics*, ed. by A.O. Barut, W.E. Britten (Gordon and Breach, New York, 1968), Vol. X-A
[34] W. Hunziker: Helv. Phys. Acta **40**, 1052 (1967)
[35] J.M. Jauch: Helv. Phys. Acta **31**, 127, 661 (1958)
[36] B.R. Karlsson, E. M. Zeiger: Phys. Rev. D **11**, 939 (1975); D **16**, 2553 (1977)
[37] T. Kato: *Perturbation Theory for Linear Operators*, (Springer, New York, 1976), 2nd ed.
[38] T. Kato: J. Functional Anal. **1**, 342 (1967)
[39] D.J. Kouri, F. S. Levin: Nucl. Phys. **A250**, 127 (1975)
[40] H. Krüger, F. S. Levin: Phys. Lett. **65B**, 109 (1976)
[41] S.P. Merkuriev: Sov. J. Nucl. Phys. **24**, 150 (1976)
[42] S.P. Merkuriev: Theoret. Math. Phys. **32**, 680 (1977)
[43] S.P. Merkuriev: Teoret. Mat. Fiz. **38**, 201 (1979); Lett. Math. Phys. **3**, 141 (1979)
[44] I.M. Narodetskii, O.A. Yakubovskii: Sov. J. Nucl. Phys. **14**, 178 (1972)
[45] T.A. Osborn, K.L. Kowalski: Ann. Phys. (N. Y.) **68**, 36 (1971)
[46] D.B. Pearson: J. Functional Anal. **28**, 182 (1978)
[47] W.V. Petryshyn: Bull. Am. Math. Soc. **81**, 223 (1975); Proc. Sympos. Pure Math. **18**, part 1 (Am. Math. Soc., Providence, Rhode Island, 1970), p. 206
[48] W.N. Polyzou, E.F. Redish: Ann. Phys. (N.Y.), **119**, 1 (1979)
[49] E.F. Redish: Nucl. Phys. **A225**, 16 (1974)
[50] M. Reed, B. Simon: *Methods of Modern Mathematical Physics*, Vol. III (Academic, New York, 1979)
[51] I.M. Sigal: Comm. Math. Phys. **48**, 137 (1976)
[52] B. Simon: Comm. Math. Phys. **55**, 259 (1977)
[53] I.H. Sloan: Phys. Rev. C **6** , 1945 (1972)
[54] W. Tobocman: Phys. Rev. C **11** , 43 (1975)
[55] T.G. Trucano: Ph. D. thesis, University of New Mexico, in progress
[56] C. van Winter: Mat. Fys. Skr. Dan. Vid. Selsk. **2**, 1 (1964)
[57] V. Vanzani: in *Few-Body Nuclear Physics* (IAEA, Vienna, 1978), p. 57
[58] S. Weinberg: Phys. Rev. **133**, B232 (1964)
[59] C.H. Wilcox: J. Functional Anal. **12**, 257 (1973)
[60] O.A. Yakubovskii: Sov. J. Nucl. Phys. **5**, 937 (1967)
[61] W.W. Zachary: J. Math. Phys. **10**, 1098 (1969)

MATHEMATICAL QUESTIONS OF QUANTUM MECHANICS OF MANY-BODY SYSTEMS

I. M. Sigal

Department of Mathematics
Princeton University
Princeton, New Jersey 08540

INTRODUCTION

The purpose of this talk is to describe some mathematical methods which prove to be fruitful in the quantum many-body problem and, hopefully, are useful beyond the scope of this problem. Our discussion will be of a general character, and we will omit technical details and most of the proofs.

As a starting point, it is useful to analyze key points of the mathematical structure of the one-body problem, $H = H_o + V$, where $H_o = -\Delta$ and $Vf = V(x)f(x)$ on $L^2(\mathbb{R}^\nu)$, with V H_o-compact.

(1) The spectral theory of H is reduced to the study of the operator $V(H_o - z)^{-1}$. Indeed, the compactness of $V(H_o - z)^{-1}$ for $z \in \rho(H_o)$ gives information about the location of $\sigma_{ess}(H)$: $\sigma_{ess}(H) = \sigma(H_o)$. (Mathematical notations are listed at the end of the Introduction.) The behavior of $V(H_o - z)^{-1}$, considered on some Banach space, near $0 = \partial\sigma_{ess}(H)$, the only possible accumulation point of $\sigma_d(H)$, indicates whether $\sigma_d(H)$ is finite.

(2) In the time-independent approach, the abstract scattering theory of simple systems [35,36] reduces the scattering problem, i.e., the existence and completeness of the wave operators for the pair (H, H_o), to the limiting absorption principle, which in turn is reduced to the study of a behavior of the operator $V(H_o - z)^{-1}$ on certain Banach spaces as z approaches $\sigma(H_o)$.

(3) It is clear that the choice of the Banach spaces for the study of $V(H_o - z)^{-1}$ near $\sigma(H_o)$ is crucial. A very important tool in this respect is the factorization of V: let V be factorizable as $V = AB$, where A and B are, in general, unbounded operators on $L^2(\mathbb{R}^\nu)$ or bounded operators from a Hilbert space $\mathcal{K}$ into $L^2(\mathbb{R}^\nu)$ and from $L^2(\mathbb{R}^\nu)$ to $\mathcal{K}$, respectively. We define the Banach space $AL^2(\mathbb{R}^\nu)$. To consider $V(H_o - z)^{-1}$ on $AL^2(\mathbb{R}^\nu)$ is the same as considering $B(H_o - z)^{-1}A$ on the original Hilbert space $L^2(\mathbb{R}^\nu)$. The most popular factorization is $A = |V|^{1/2}$ and, therefore, $B = (\text{sign } V)|V|^{1/2}$.

In the following sections, we describe a generalization of the methods and results mentioned above for the case of many-body systems. Our generalization goes along the following lines:

(i) Generalization of the transition from H to $V(H_o - z)^{-1}$: the theory of regularizers for $H - z$. A regularizer for $H - z$ retains two main properties of $(H_o - z)^{-1}$ which played a crucial role in the conclusions above: that $(H - z)\ (H_o - z)^{-1} - I$ is compact for $z \in \rho(H_o)$ and well behaved for $z \to \sigma(H_o)$, and that $(H_o - z)^{-1}$, as an operator from $L^2(\mathbb{R}^\nu)$ into $D(H_o)$, is boundedly invertible.

(ii) Conditions on $V(H_o - z)^{-1}$ which ensure the desired properties of H are replaced by conditions on $V_l(H_o - z)^{-1}$ where V_l is a pair potential. In particular, the compactness of $V(H_o - z)^{-1}$ is replaced by the compactness of connected graphs $\Pi\,[V_l(H_o - z)^{-1}]$. The latter can be further reduced to the individual properties of V_l and some kinematical facts.

(iii) The factorization method is generalized as follows: Let each pair potential be factorizable as $V_l = A_l B_l$, where A_l, B_l have the same properties as A, B. Then the relevant Banach space is constructed as $\Sigma\, A_l L^2(\mathbb{R}^{\nu N})$ (N is the number of ν-dimensional particles). The operators of (ii) are replaced in this case by $B_l(H_o - z)^{-1}A_s$ and $\Pi[B_l(H_o - z)^{-1}A_s]$, respectively. Here l and s are any two pairs of indices and the product is taken with respect to such pairs with the restriction that the collection of all l's is connected.

(iv) The abstract scattering theory of simple systems is replaced with an abstract two-space scattering theory and with a theory of abstract multichannel systems.

In conclusion, we list some of the notations used in this article. $\sigma(A)$, $\rho(A)$, and $D(A)$ denote the spectrum, resolvent set, and domain of an operator A . $\sigma_{ess}(A)$, $\sigma_p(A)$ and $\sigma_d(A)$ denote the essential, point, and discrete spectra of A. If $\sigma(A) \subset \mathbb{R}$, we introduce $\delta_\epsilon(A-\lambda) = (2\pi i)^{-1}[(A-\lambda-i\epsilon)^1 - (A-\lambda+i\epsilon)^{-1}]$.

Let A be an operator on a Banach space X to another Banach space, then AX denotes the range $R(A)$, completed in the norm $||x|| = \inf\{||y||', Ay = x\}$, where $||\cdot||'$ is the norm in X. Note that $AX \approx X/\mathrm{Ker}\, A$. Let $\{X_k\}$ be a collection of Banach spaces which are subspaces of some linear space. The sum ΣX_k is a Banach space defined as $\Sigma X_k = j(\oplus X_k)$, $j(\oplus x_k) = \sum x_k$.

1. TWO-SPACE SCATTERING THEORY

a. This section is a condensed exposition of some results in the abstract two-space scattering theory. The proofs, details, and applications can be found in [27].

Since there is a simple criterion (see, e.g., [18], Theorem 3.7, p. 533) for the existence (and therefore isometry) of the time-dependent wave operators, we focus our attention on the problem of finding properties of a pair of self-adjoint operators which guarantee the completeness of the wave operators for this pair.

We use a stationary approach. Our main result shows, essentially, that if the resolvents of two self-adjoint operators, acting on different Hilbert spaces are proportional up to an operator-valued function which, considered between two appropriate Banach spaces, has strong boundary values on the real axis, then the corresponding wave operators exist and are complete.

b. Henceforth, H and $\hat{H}$ are self-adjoint operators on Hilbert spaces $\mathcal{H}$ and $\hat{\mathcal{H}}$ respectively, $E(\Delta)$, $R(z)$ and $\hat{E}(\Delta)$, $\hat{R}(z)$ are their spectral measures and resolvents, $E_p = E(\sigma_p(H))$ and $J \in L(\hat{\mathcal{H}}, \mathcal{H})$. In order to simplify the notation, we assume $\hat{H}$ to be absolutely continuous.

The main object of the two-space scattering theory is the strong limits $W^\pm = W^\pm(H,\hat{H},J) = \operatorname*{s-lim}_{t\to\pm\infty} e^{iHt}Je^{-i\hat{H}t}$, if they exist. These limits are called wave operators (for the triple $(H,\hat{H},J)$). Normally, one requires that the operator $Je^{-i\hat{H}t}$ be asymptotically isometric (AI): $\operatorname*{s-lim}_{t\to\pm\infty} ||Je^{-i\hat{H}t}\hat{u}|| = ||\hat{u}||$, $\hat{u} \in \hat{H}$. If $W^\pm$ exist, then they are intertwining for H and $\hat{H}$, $HW^\pm = W^\pm\hat{H}$, and $R(W^\pm) \subset R(I-E_p)$. Under the additional condition (AI), $W^\pm$ are isometric: $W^{\pm *}W^\pm = I$. $W^\pm$ are said to be (asymptotically) complete (AC) iff $R(W^\pm) = R(I-E_p)$, i.e., iff $W^\pm W^{\pm *} = I - E_p$. The main problem of the two-space scattering theory is to prove the existence and completeness of $W^\pm$.

c. The notion of wave operator can be generalized if we replace the usual strong limit in their definition by the strong Abelian limit. The new, generalized wave operators will be called the stationary wave operators. They may exist even when the nonstationary (usual) wave operators do not. When the latter exist, both definitions lead to the same operators. Since a proof of the existence of nonstationary wave operators usually is not difficult, one can recover their properties, namely completeness, from the corresponding properties of the stationary wave operators. The convenience of studying the stationary wave operators is based on the possibility of translating the expressions for them in terms of resolvents of H and $\hat{H}$, instead of evolution operators.

When one considers the stationary wave operators, it is also natural to replace the usual limits in (AI) by Abel limits.

d. The next step in modifying the original definition of the wave operators, in the hope of simplifying a proof of their properties, is to pass from the global operators H and $\hat{H}$ to the local ones $HE(\Delta)$ and $\hat{H}\hat{E}(\Delta)$, respectively. Here, Δ is a bounded interval of $\mathbb{R}$. The wave operators obtained in this way are called local. The stationary local wave operators for the triple $(H,\hat{H},J)$ and an interval $\Delta \subset \mathbb{R}$ can be written in the form $W^\pm(\Delta) = \operatorname*{s-lim}_{\epsilon\to\pm 0} W^{(\epsilon)}(\Delta)$, where $W^{(\epsilon)}(\Delta) = i\epsilon \int_\Delta R(\lambda+i\epsilon)J\hat{R}(\lambda-i\epsilon)\, d\lambda$, which will be taken henceforth as their definition. The condition of asymptotic isometry of $Je^{-i\hat{H}t}$ in the stationary, local case can be written as $\lim_{\epsilon\to 0} \int_\Delta ||J\hat{R}(\lambda+i\epsilon)\hat{u}||^2 d\lambda = ||E(\Delta)\hat{u}||^2$, $\forall \hat{u} \in \hat{\mathcal{H}}$.

Lemma 1.1 *If* $||J|| \leqslant 1$ *then the above condition is equivalent to the condition* $\operatorname*{s-lim}_{\epsilon\to 0} |\epsilon| \int_\Delta \hat{R}(\lambda-i\epsilon) J^*J\hat{R}(\lambda+i\epsilon)\, d\lambda = \hat{E}(\Delta)\,(SLAI)$.

The following lemma permits us to recover the existence and properties of the global wave operators from those of the local ones.

Lemma 1.2. *Let $W^{\pm}(\Delta)$ exist for all Δ's from a directed sequence $\Phi = \{\Delta_i\}$ of Borel subsets of $\mathbb{R}$. Futhermore for any $\Delta, \Delta' \in \Phi$, let $W^{\pm}(\Delta)^* W^{\pm}(\Delta') = \hat{E}(\Delta \cap \Delta')$ (respectively, $W^{\pm}(\Delta) W^{\pm}(\Delta')^* = E(\Delta \cap \Delta)$). Then $W^{\pm}(\bigcup_{\Delta \in \Phi} \Delta) = \text{s-lim}_{\Delta \to \cup \Delta} W_{\pm}(\Delta)$ (respectively, $W^{\pm}(\cup \Delta)^* = \text{s-lim}_{\Delta \to \cup \Delta} W^{\pm}(\Delta)^*$) exist. Moreover, if the Lebesgue measure of $\mathbb{R}/\bigcup_{\Delta \in \Phi} \Delta$ is zero, then $W^{\pm}(\bigcup_{\Delta \in \Phi} \Delta) = W^{\pm}(\mathbb{R})$ and are global wave operators.*

e. The following theorem states sufficient conditions on the resolvents of H and $\hat{H}$ in order for the stationary $W^{\pm}$ to be complete.

Theorem 1.1. *Let $\hat{H}$ and J satisfy (SLAI). Let the resolvents of H and $\hat{H}$ be connected by the equation*

$$R(z)(I - E_p) = J\hat{R}(z)Q(z), \tag{1.1}$$

where $Q(z)$: $\mathbb{C}/R \to \cup \, \text{Op}(\mathcal{H}, \hat{\mathcal{H}})$, and let there exist a collection Φ of Borel subsets of $\mathbb{R}$, a set $\{X_\Delta, \Delta \in \Phi\}$ of Banach spaces, with $X_\Delta \subset L^2(\Delta, \hat{\mathcal{H}})$ and with $X_\Delta \cap (L^2(\Delta) \otimes \hat{\mathcal{H}})$ dense in X, and a dense subset $X \subset (I - E_p)\mathcal{H}$ such that

(i) for any $\Delta \in \Phi$ and $x \in X_\Delta$, $|\epsilon| \int_\Delta ||\hat{R}(\lambda + i\epsilon)x(\lambda)||^2 \, d\lambda \leqslant M \, ||x||^2_{X_\Delta}$, i.e., $\delta_\epsilon(\hat{H} - .)$ is bounded from X_Δ to its dual X'_Δ, uniformly in $\epsilon \in \mathbb{R}^{\pm}$;

(ii) for any $f \in X$, $\epsilon \in \mathbb{R}^{\pm}$, $\Delta \in \Phi$, $Q(\cdot + i\epsilon)f \in X_\Delta$ and has strong limits in X_Δ as $\epsilon \to \pm 0$: $||(Q(\cdot + i\epsilon) - Q(\cdot + i\epsilon'))f||_{X_\Delta} \to 0$ $(\epsilon, \epsilon' \to \pm 0)$.

Then, (a) $\sigma_{s.c.}(H) \cap (\cup \Delta) = \phi$, (b) $W^{\pm}(\Delta)^* = \text{s-lim}_{\epsilon \to \pm 0} W^{(\epsilon)}(\Delta)^*$, exist for any $\Delta \in \Phi$ and equal $\text{s-lim}_{\epsilon \to \pm 0} W_1^{(\epsilon)}(\Delta)^*$, where $W_1^{(\epsilon)}(\Delta)^* = \int_\Delta \delta_\epsilon(\hat{H} - \lambda)Q(\lambda + i\epsilon) \, d\lambda$, (c) $W^{\pm}(\Delta) W^{\pm}(\Delta')^* = E(\Delta \cap \Delta')$, $\Delta, \Delta' \in \Phi$.

Corollary 1.1. *Under the conditions of the theorem, the stationary global wave operators exist and are complete.*

f. *Examples of Spaces X_Δ.* Given $\hat{H}$, Condition (i) of the theorem can be considered as a restriction on a space to which $Q(\lambda + i\epsilon)f$, $f \in X$, have to belong in order to insure the completeness of the wave operators $W^{\pm}$. It can be shown that the Sobolev space $H_s(\Delta, \hat{H})$, $s > \frac{1}{2}$, satisfies (i). We now give other examples of Banach spaces which satisfy (i).

First we formulate new conditions:

(α) $\delta_\epsilon(\hat{H} - \lambda)$ is bounded from a Banach space $\hat{\mathcal{B}}$ to its dual $\hat{\mathcal{B}}'$, uniformly in $\epsilon > 0$ and in λ from any compact subset of $\mathbb{R}$.

(β) For some bounded operators B_i from a Banach space $\mathcal{K}$ to $\hat{\mathcal{H}}$, $B_i^* \delta(\hat{H} - \lambda)B_i$ are bounded uniformly in $\epsilon > 0$ and in λ from any compact subset of $\mathbb{R}$.

Then (β) $\to$ (α) with $\hat{\mathcal{B}} = \Sigma B_i \mathcal{K}$ and (α) $\to$ (i) with $X_\Delta = L^2(\Delta, \hat{\mathcal{B}})$. Therefore, $L^2(\Delta, \Sigma B_i \mathcal{K})$, where B_i satisfy (β), obeys (i).

Remark 1.1 (β) holds if and only if each B_i is locally $\hat{H}$-smooth (so that it can be taken as a definition of local relative smoothness).

The following remark is useful in the N-body problem: Let T, self-adjoint on $\mathcal{H}$, have the form $T = T_1 \otimes I_2 + I_2 \otimes T_1$ on $\mathcal{H} = \mathcal{H}_1 \otimes \mathcal{H}_2$. Then $A_1 \otimes I_2$ is (locally) T-smooth if A is (locally) T-smooth.

Now we consider an example, which illustrates all essential properties of $\hat{\mathcal{H}}$ and $\hat{H}$, occurring in applications to quantum scattering. Let Δ be the self-adjoint extension on $L^2(\mathbb{R}^n)$ of the Laplacian on $C_o^2(\mathbb{R}^n)$.

Lemma 1.3. *Let $A_{fo\pi}$ be the operator from $L^2(\mathbb{R}^n)$ to $L^1_{loc}(\mathbb{R}^n)$ of multiplication by the function $fo\pi$, where $f \in L^p \cap L^q(\mathbb{R}^m)$, $p > m > q$, and π is a linear function from $\mathbb{R}^n$ to $\mathbb{R}^m$, $m \leqslant n$. Then*

$$||A_{fo\pi} R_\Delta(z) A_{\phi o\pi}|| \leqslant C ||f||_{L^p \cap L^q} ||\phi||_{L^p \cap L^q}, \quad p > m > q.$$

Let now π_i: $\mathbb{R}^n \to \mathbb{R}^m$, $m < n$, be linear functions and $f_i \in (L^p \cap L^q)(\mathbb{R}^m)$, $q < m < p$. Then Lemma 1.3 implies that the space $\Sigma_i A_{f_i o \pi_i} L^2(\mathbb{R}^n)$ and the operator Δ satisfy (α) and that, therefore, $L^2(\Delta, \Sigma A_{f_i o \pi_i} L^2(\mathbb{R}^n))$ and Δ satisfy condition (i) of the theorem.

2. ABSTRACT MULTICHANNEL SYSTEMS

Let H be a self-adjoint operator acting on a Hilbert space $\mathcal{H}$. Following W. Hunziker, we call the triple $\alpha = (\mathcal{H}_\alpha, H_\alpha, J_\alpha)$, where $\mathcal{H}_\alpha$ is a Hilbert space, H_α is a self-adjoint operator on $\mathcal{H}_\alpha$, and J_α is a bounded operator from $\mathcal{H}_\alpha$ into $\mathcal{H}$, a channel for H if it satisfies (AI) and $W^\pm(H, H_\alpha, J_\alpha)$ exist. $W_\alpha^\pm$ are called the (out- and in-) wave operators for the channel α (or simply channel wave operators). The system $\{\alpha\}$ of channels for H is said to be complete iff

$$\sum_\alpha W_\alpha^\pm W_\alpha^{\pm *} = I. \tag{2.1}$$

It is required in the multichannel scattering theory that all channels be mutually asymptotically orthogonal (independence of the channels). This means that for any pair α and β of different channels

$$\lim_{|t|\to\infty} (J_\alpha e^{-iH_\alpha t} f, J_\beta e^{-iH_\beta t}\phi) = 0,\ f \in \mathcal{H}_\alpha,\ \phi \in \mathcal{H}_\beta. \tag{2.2}$$

This implies that $R(W_\alpha^\pm) \perp R(W_\beta^\pm)$ if $\alpha \neq \beta$, or, equivalently, that $W_\alpha^{\pm *} W_\beta^\pm = 0$ $(\alpha \neq \beta)$. The latter together with the isometry of $W_\alpha^\pm$ gives $W_\alpha^{\pm *} W_\beta^\pm = \delta_{\alpha\beta}$.

We can define an abstract multichannel scattering system also in a weaker sense, replacing all time limits involved by Abel limits.

The abstract multichannel scattering system is a special case of the two-space scattering system. Indeed, define

$$\hat{\mathcal{H}} = \oplus\, \mathcal{H}_\alpha,\ \hat{H} = \oplus\, H_\alpha \text{ and } J(\oplus f_\alpha) = \sum J_\alpha f_\alpha. \tag{2.3}$$

Then $W^\pm = W^\pm(H, \hat{H}, J)$ can be expressed in terms of the channel wave operators $W^\pm = W^\pm$: $W^\pm(\oplus f_\alpha) = \sum W_\alpha^\pm f_\alpha$ and, conversely, $W_\alpha^\pm$ can be recovered from $W^\pm$: $W_\alpha^\pm = W^\pm \Pi_\alpha$, where Π_α is the projection on $\hat{\mathcal{H}}$ into $\hat{\mathcal{H}}_\alpha$.

(AI) for every α and (2.2) imply that (AI) is satisfied for the operators J and $\hat{H}$ defined in (2.3). Naturally, $W_\alpha^{\pm *} W_\beta^\pm = \delta_{\alpha\beta}$, written in the new language, is the isometry of $W^\pm$, which on the other hand is a direct consequence of (AI). Condition (2.1) of the completeness of the system of channels for H is translated into $W^\pm$-language as $W^\pm$-completeness.

3. REGULARIZERS

In this section, we give a definition of regularizers and present without proofs some results which illustrate their application.

Definition 3.1. We call an operator F from $\mathcal{H}$ to $D(T)$ a (right) regularizer for an operator T if and only if F is invertible and $TF - I$, raised to some power, is compact.

In the sequel, H and G denote, respectively, a self-adjoint operator (in general unbounded) acting on a Hilbert space $\mathcal{H}$, and an open set in $\mathbb{C}$ with a smooth boundary.

Theorem 3.1. *Let there exist a family $F(z)$ of regularizers for $H - z$, $z \in G$. Then $\sigma_{ess}(H) \subset \mathbb{C}\backslash G$.*

Theorem 3.2. *Let $G \subset \mathbb{C}\backslash\sigma_{ess}(H)$ and $F(z)$ be a regularizer for $H - z$ for all $z \in G$. Furthermore, let there exist Banach spaces $X \subset \mathcal{H}$ and $Y \supset D(H)$, such that*

(i) $F(z)$ is bounded from X into Y and is weakly continuous as $z \in G$ approaches possible accumulation points of $\sigma_d(H) \cap G$ and Ker *$F(z) = \{0\}$ for such points z.*

(ii) $(H - z)F(z) - I$ is bounded on X, strongly continuous as $z \in G$ approaches possible accumulation points of $\sigma_d(H) \cap G$, and, raised to some power, is compact on X and norm-continuous as $z \in G$ approaches the above points.

Then the number of eigenvalues of H inside G is finite.

Define $B_{\hat{\mathscr{H}},\mu} = \{f \in \hat{\mathscr{H}}: \|\Pi_{\lambda+h}f - \Pi_\lambda f\|_{\hat{\mathscr{H}}_\lambda} \leqslant C|h|^\mu\}$, where $\{\Pi_\lambda\}$ is a unitary operator from $\hat{\mathscr{H}}$ to a representation of $\hat{\mathscr{H}}$ as the fiber direct integral $\int^{\oplus} \mathscr{H}_\lambda \, d\mu(\lambda)$ with respect to $\hat{H}$.

Theorem 3.3. *Let $G \subset \mathbb{C}\backslash\sigma_{ess}(H)$ and $F(z)$ be a regularizer for $H - z$ for all $z \in G$. Let $\hat{H}$ and J be as in Sec. 1 and $F(z)$ be representable as $F(z) = J\hat{R}(z)\hat{F}(z)$. Furthermore, let there exist Banach spaces $X \subset \mathscr{H}$ and $\hat{X} \subset \hat{\mathscr{H}}$, such that*

(1) $\hat{F}(z)$ is bounded from X into $\hat{X}$ and strongly continuous as z approaches ∂G;

(2) $(H - z)F(z) - I$ is bounded on X and strongly continuous as $z \to \partial G$ and, raised to some power, is compact for all $z \in \bar{G}$.

Assume, in addition, that either (3) $\hat{X} \subset B_{\hat{\mathscr{H}},\mu}$, $\mu > \frac{1}{2}$, or

(3) there is a unitary representation, $\theta \to U(\theta)$, of $\mathbb{R}$ on $\mathscr{H}$ such that $U(\theta)HU(\theta)^{-1}$ and $U(\theta)F(z)\ U(\theta)^{-1}$ have analytic continuations to a domain $O \subset \mathbb{C}$, $O \cap \mathbb{R} \neq \phi$, for which $\sigma_{ess}(H(\theta)) \cap \bar{G} = \varnothing$ for $\operatorname{Im}\theta \neq 0$.

Then $R(z)$ is representable as $R(z)\,(I - E_{d,G}) = J\hat{R}Q(z)$, where $Q(z)(\in B(\mathscr{H}, \hat{\mathscr{H}}))$ is bounded from X to $\hat{X}$ and strongly continuous as $z \to \partial G$ and $E_{d,G}$ is the projection on the eigensubspace of $\sigma_p(H) \cap \bar{G}$.

4. HAMILTONIAN

In this section, we define the Schrödinger operator (Hamiltonian) of a many-body system and discuss some fundamental properties of the potentials.

Consider a system of N ν-dimensional particles of masses m_i, interacting via pair potentials $V_l(x_l)$. Here l labels pairs of indices and $x_l = x_i - x_j$ for $l = (ij)$. The configuration space of the system in the center-of-mass (CM) frame is defined as $R = \{x \in \mathbb{R}^{\nu N}, \Sigma m_i x_i = 0\}$, with the inner product $(x,\tilde{x}) = \Sigma m_i x_i \cdot \tilde{x}_i$. Denote by v_l and V_l the multiplication operators on $L^2(\mathbb{R}^\nu)$ and $L^2(R)$ with the functions $V_l(y)$ and $V_l(x_l)$, respectively.

We assume that v_l is Δ-compact, i.e., compact as an operator from the Sobolev space $H_2(\mathbb{R}^\nu)$ to $L^2(\mathbb{R}^\nu)$. Then the operator

$$H = H_o + \sum V_l, \quad H_o = -\frac{1}{2}\ (\text{Laplacian on } L^2(R)),$$

is defined on $L^2(R)$ and is self-adjoint there.

Remark 4.1. Potentials of the class $L^q(\mathbb{R}^\nu) + (L^\infty(\mathbb{R}^\nu))_\epsilon$, $q > \max(\nu/2, 2)$ if $\nu \neq 4$ and $q > 2$ if $\nu = 4$, where the subscript ϵ means that the L^∞-component can be taken arbitrarily small, are Δ-compact.

Definition 4.1. The monomials of the form $\Pi[V_l R_o(z)]$ and $\Pi[\operatorname{sign}(V_{l_i})|V_{l_i}|^{1/2} R_o(z)|V_{l_{i+1}}|^{1/2}]$, where $R_o(z) = (H_o - z)^{-1}$, are called graphs and modified graphs, respectively. A (modified) graph is called connected if and only if $\cup l = (1, \ldots, N)$ and the l's can be arranged in a sequence with any two neighboring pairs having at least one common index.

It is evident that (modified) graphs are bounded operators on $L^2(R)$.

Proposition 4.1. *If each v_l is Δ-compact, then any connected (modified) graph is compact.*

Before proceeding to the proof of this proposition we introduce some useful operators (see [7,8]). Let

$X \in C^\infty(\mathbb{R}^\nu)$, $X(x) = 0$ for $|x| \leqslant 1$ and $X(x) = 1$ for $|x| > 2$.

We define the multiplication operators $X_l^{(n)}$ on $L^2(R)$ by functions $X(x_l/n)$ and the operators $\bar{X}_l^{(n)} = I - X_l^{(n)}$. Note two important properties of these operators:

$$X_l^{(n)} \xrightarrow{s} 0\ (n \to \infty)\ (\text{also as an operator in } L^2(\mathbb{R}^\nu)) \tag{4.1}$$

and

$$[\bar{X}_l^{(n)}, H_o]\,(H_o - z)^{-1} = [X_l^{(n)}, H_0]\,(H_0 - z)^{-1} \overset{\text{norm}}{\longrightarrow} 0 \text{ as } n \to \infty$$

$$\text{(actually as } O(n^{-1})\text{)}. \tag{4.2}$$

Equation (4.1) (on $L^2(\mathbb{R}^\nu)$) and the relative compactness of v_l imply that

$$X_l^{(n)} V_l (H_o - z)^{-1} \overset{\text{norm}}{\longrightarrow} 0 \ (n \to \infty)\,. \tag{4.3}$$

Proof of Proposition 4.1. Consider for definiteness a graph. We write $V_l = (X_l^{(n)} + \bar{X}_l^{(n)})\,V_l$ and decompose the graph into the sum of the corresponding terms. Because of (4.3), each term, containing at least one $X_l^{(n)} V_l$, vanishes in norm as $n \to \infty$. In the summand with all $\bar{X}_l^{(n)} V_l$, we commute all $\bar{X}_l^{(n)}$ to the left, in front of the monomial. The terms, containing at least one commutator $([H_o, \bar{X}_l^{(n)}])$, vanish by (4.2) as $n \to \infty$. Thus, after all these operations, the only nonvanishing in the limit $n \to \infty$ term is of the form $\Pi \bar{X}_l^{(n)} \Pi [V_l R_o(z)]$. Of course, for any connected graph, an operator of the form $\Pi \bar{X}_l^{(n)} V_s R_o(z)$ is compact ($\Pi \bar{X}_l^{(n)} \in C_o^\infty(R)$ by the construction of $\bar{X}_l^{(n)}$ and the connectedness of the graph). Therefore, the nonvanishing term is also compact as a product of compact and bounded operators. Since this term approximates the graph in norm, the graph is compact as well (closeness in the uniform operator topology of the set of compact operators).

5. CONSTRUCTION OF REGULARIZERS. EXAMPLE

In this section, we give an example of regularizers which play an important role in the many-body problem.

Definition 5.1. Let A be a finite lattice and $\{H_a,\ a \in A\}$ a collection of operators on a Banach space $\mathcal{H}$. We define by induction on $a \in A$: $A_a(z) = I$, $a = \min A$,

$$A_a(z) = (H_a - z)\,(H_o - z)^{-1} \overrightarrow{\prod_{b \subset a}} A_b(z)^{-1}, \tag{5.1}$$

where $H_o = H_{\min A}$ and the arrow on the top of the product sign means that the order of the A^{-1}'s is such that if A_c^{-1} stands on the right of A_d^{-1} then $c \not\subset d$.

We set

$$F_a(z) = (H_o - z)^{-1} \overrightarrow{\prod_{b \subset a}} A_b(z)^{-1} \tag{5.2}$$

and $H = H_a$, $A = A_a$ and $F = F_a$ for $a = \max A$. Definition (5.2) implies that

$$(H_a - z)\,F_a(z) = A_a(z). \tag{5.3}$$

It follows immediately from the definition that $F_a(z)$ and $A_a(z)$ are bounded on $\mathcal{H}$ into $D(H_a)$ and on $\mathcal{H}$, respectively, and $F_a(z)$ is boundedly invertible. We will show later that, under certain restrictions on $\{H_a, a \in A\}$, $F(z)$ is a regularizer for $H - z$.

If we impose some additional restrictions on $\{H_a\}$, the operators $F_a(z)$ and $A_a(z)$ acquire certain useful properties. We assume now that the operators H_a are built as $H_a = T + \sum_{b \subseteq a} V_b$, where V_a, $a \in A$ are T-bounded operators with relative bound 0. In this case, the operators $F_a(z)$ and $A_a(z)$ have additional structure:

Lemma 5.1. *The operators $F_a(z)$ and $A_a(z) - I$ are finite linear combinations of monomials of the form*

$$R_o \Pi [V_c R_b],\ c, b \subset a, \ \textit{and}\ \Pi [V_c R_b],\ b \subset a,\ \cup c = a, \tag{5.4}$$

respectively.

The statement can easily be derived by induction. The details can be found in [25]. Note here only that, since V_a have T-bound 0, they are H_b-bounded as well. Therefore, monomials of form (5.6) are bounded and analytic in $z \in \cap\, \rho(H_b)$.

Lemma 5.2. *For z with* $\operatorname{dist}(z, \sigma(T))$ *sufficiently large, $A_a(z) - I$ is a norm-convergent series of a-connected graphs,*

$$\prod_{\cup c = a} [V_c (T - z)^{-1}].$$

Proof. The statement follows from Lemma 5.1 and the fact that for dist $(z,\sigma(T))$ large enough the following series is norm-convergent:

$$R_b(z) = (T-z)^{-1} \sum_{n=0}^{\infty} \left[\sum_{c \subseteq b} V_c (T-z)^{-1} \right]^n .$$

Indeed, $||A(T-z)^{-1}|| \to 0$ as dist $(z,\sigma(T)) \to \infty$ for any T-bounded operator A.

Now we proceed to the N-body systems.

Definition 5.2 (Lattice of Decompositions). Let $a = \{C_i\}$ be a decomposition of the set $\{1, \ldots, N\}$ into nonempty, disjoint subsets C_i, called clusters. Denote by $\mathfrak{A}$ the set of all such decompositions. $\mathfrak{A}$ can be given the structure of a lattice; namely, if b is a partition obtained by breaking up certain subsystems of a, we shall say that b is contained in the partition a, writing $b \subset a$. The smallest partition containing two partitions a and b will be denoted by $a \cup b$, i.e., $a \cup b = \sup(a,b)$. The largest partition contained in both a and b will be denoted by $a \cap b$: $a \cap b = \inf(a,b)$.

A pair l will be identified with the decomposition on $N-1$ clusters, one of which is the pair l itself and the others are free particles. Therefore, in the N-body case, $H_a = H_o + \sum_{l \subseteq a} V_l$.

Combining Lemma 5.2 and Proposition 4.1, we obtain

Corollary 5.1. *Let H be an N-body Hamiltonian as defined in Sec. 4 and $A(z)$ the operator constructed in (5.1) for it. Then $L(z) \equiv A(z) - I$ is compact for $z \in \cap \rho(H_a)$, i.e., $F(z)$ is a regularizer for $H-z$, $z \in \cap \rho(H_a)$.*

Corollary 5.1 and Theorem 3.1 imply

Corollary 5.2. *Let H be an N-body Hamiltonian in the CM frame, as defined in Sec. 4. Then $\sigma_{ess}(H) \subset \cap \sigma(H_a)$.*

Remarks 5.1. (1) The converse direction, $\sigma_{ess}(H) \supset \cup \sigma(H_a)$, is the easy one. It is usually proved [22] by explicitly constructing Weyl sequences for H. (2) Of course, one can allow in Corollary 5.1 for many-body potentials [31], and everything goes through in exactly the same way.

6. ASYMPTOTIC COMPLETENESS OF SINGLE-CHANNEL SYSTEMS

In this section, we illustrate methods of the previous sections with a proof of asymptotic completeness of short-range, single-channel systems. We remind the reader that single-channel, many-body systems are characterized by the condition: $\sigma_p(H^a) = \emptyset$ for all a with $1 < \#(a) < N$, where $\#(a)$ denotes the number of clusters in the decomposition a. Therefore, the collection of all channel Hamiltonians H_α is reduced to only one H_o.

As the first step in our analysis, we reformulate accordingly the results of Sec. 1. In the single-channel case, $\hat{\mathscr{H}} = \mathscr{H}$, $\hat{H} = H_o$, and $J = I$. Therefore, $Q(z) = (H_o - z)(H-z)^{-1}(I-E_p) = (I - V(H-z)^{-1})(I-E_p)$. Taking this into account in Theorem 1.1 and using in it the Banach spaces (suggested by the example in Sec. 1)

$$X_\Delta = L^2(\Delta, X), \; X = \sum |V_l|^{1/2} L^2(R), \tag{6.1}$$

we find the following

Theorem 6.1. *Let $|V_l|^{1/2}$ be H_o-smooth and let $|V_l|^{1/2} R(z)(I-E_p)|V_s|^{1/2}$ be defined on $L^2(R)$ for all l and s and strongly continuous as $\operatorname{Im} z \to 0$. Then ($\alpha$) $\sigma_{s.c.}(H) = \emptyset$, ($\beta$) the adjoint stationary wave operators $Z^\pm = \operatorname{s-lim}_{\epsilon \to \pm 0} i\epsilon \int R_o(\lambda - i\epsilon) R(\lambda - i\epsilon) d\lambda$ exist and are complete, i.e., $Z^{\pm *} Z^\pm = I - E_p$.*

Our next task is to reduce the second condition of Theorem 6.1 to a condition on the potentials V_l and free motion H_o. In order to simplify slightly the considerations, we assume that the potentials V_l are dilation-analytic [22]. We define

$R^a = \{x \in \mathbf{R} : \sum_{i \in C_k} m_i x_i = 0, \; \forall C_k \in a\}$ and $H^a = T^a + \sum_{l \subseteq a} V_l$.

Here, $T^a = -$ (1/2) (Laplacian on $L^2(R^a)$), V_l is the multiplication operator on $L^2(R^a)$ with the function $V_l(x_l)$ (note that $x_l = (\Pi^a x)_l$ for $l \subset a$, where Π^a is the projection from R onto R^a).

Our basic abstract restrictions on the potentials and free energy are

(I) (Strong H_o-smoothness of $\{|V_l|^{1/2}\}$) For all l and s, the operators $|V_l|^{1/2}R_o(z)|V_s|^{1/2}$ are bounded on $L^2(R^a)$ for each a with $l, s \subset a$ and strongly continuous in $z \in \mathbb{C}\setminus\mathbb{R}$ up to $\mathbb{R}$.

(II) (Compactness of (modified, a-)connected graphs) There is an $n < \infty$ such that a product of n operators of the form $\overrightarrow{\prod\limits_{\cup l_i = a}}$ $[\text{sign}(V_{l_i})|V_{l_i}|^{1/2}R_o(z)|V_{l_{i+1}}|^{1/2}]$ is compact on $L^2(R^a)$, $a \in A$.

Let $H^a(g) = T^a + \sum_{l \subset a} g_l V_l$ on $L^2(R^a)$, $R^a(z,g) = (H^a(g) - z)^{-1}$ and denote by G^a a connected subset of $\{g \in \mathbb{C}^{\#\text{ of } l \subset a}: \sigma_p(H^a(g)) = \emptyset,\ \forall_b \subset a,\ b \neq \min A\}$, containing 0.

Theorem 6.2. *Assume Conditions (I) and (II) are satisfied. Then for any $a \in \mathfrak{A}$ and $l, s \subset a$, $|V_l|^{1/2}R^a(z,g)|V_s|^{1/2}$ is a family of bounded operators on $L^2(R^a)$, strongly continuous in $z \in \mathbb{C}\setminus\mathbb{R}$ up to $\mathbb{R}$ and analytic in $g \in G^a$.*

Before we go over to the proof of Theorem 6.2, we first discuss Condition (II). Using the same arguments as in the proof of Proposition 4.1 (and a resolution of the identity containing a momentum cutoff function when momenta appear in the commutators) we obtain easily

Proposition 6.1. *Let F_l be multiplication operators on $L^2(R)$ by functions $F(x_l)$, where $F \in C_o^\infty(\mathbb{R}^\nu)$. Condition (II) with $n = 1$ follows from the following condition:*

(II′) For all l and s, the operators $B_l R_o(z) A_s$ are bounded on $L^2(R^a)$, $a \supset l, s$, and strongly continuous in $z \in \mathbb{C}\setminus\mathbb{R}$. Here A_l, $B_l = |V_l|^{1/2}$, $F_l\phi(p_l)$, with $\phi \in C_o^\infty(\mathbb{R})$ and p_l the relative momentum for l.

Thus, we can state

Theorem 6.3. *If Condition (II′) is satisfied, then the statement of Theorem 6.2 holds and, therefore, for $g \in G(\equiv G^a$ for $\#(a) = 1)$ $\sigma_{s_{s.c.}}(H(g)) = \emptyset$ and $Z^\pm(g) = W^\pm_{stat}(H_o, H(g))$ exist, are analytic in $g \in G$ and are complete, i.e., $Z^\pm(g)^* Z^\pm(g) = I - E_p(g)$.*

Using the relative smoothness estimates [14,15] or [25] one proves (see Lemma 1.3):

Proposition 6.2. *Let $V_l \in L^p \cap L^q(\mathbb{R}^\nu)$, $p > \nu/2 > q$. Then Condition (II′)* holds.

Corollary 6.1. *If the potentials satisfy the condition of Proposition 6.2, then the statement of Theorem 6.3 is true.*

Sketch of proof of Theorem 6.2. Looking at Theorem 3.3, we see what we have to prove about $F(z)$ and $A(z)$, defined in Sec. 5. We proceed by induction on $a \in A$. For $a = \min A$, the statement is trivial, $R^a = \{0\}$. Let the statement of Theorem 6.2 hold for all b, $b \subset a$, and let us prove it for a. Instead of studying $R^a(z)$ and then concluding about $|V_l|^{1/2}R^a(z)|V_s|^{1/2}$, we will investigate the latter directly. To this end, we first write the equation $R(z,g)A(z,g) = F(z,g)$ for the matrix $[|V_l|^{1/2}R^a(z,g)|V_s|^{1/2}]$, which follows from $R^a(z,g) \times A^a(z,g) = F^a(z,g)$. The boundedness of $A(z,g)$ and $F(z,g)$ on $\oplus L^2(\mathbb{R}^a)$ and analyticity in $g \in G^a$ for all $z \in \mathbb{C}\setminus\mathbb{R}$ up to $\mathbb{R}$ follows immediately from the inductive assumption and Condition (I). To see this, we represent the matrix elements in $L(z,g) \equiv A(z,g) - I$ as linear combinations of terms of the form

$$R^b(z,g) \prod_{b_i \subset a, l_i \subseteq a} \text{sign } V_{l_i}|V_{l_i}|^{1/2}R_{b_i}(z,g)|V_{l_{i+1}}|^{1/2}],$$

where $l_i, l_{i+1} \subseteq b_i$ if $\#(b_i \leqslant N$ (and similarly for $F(z,g)$). The only thing which needs to be proved here is that the estimates for $|V_l|^{1/2}R^b(z,g)|V_s|^{1/2}$ imply similar estimates for $|V_l|^{1/2}R^b(z,g)|V_s|^{1/2}$, i.e., that the CM motion of the clusters in the decomposition b can be introduced without trouble. The latter proof is rather straightforward.

To prove compactness of $L(z,g)$ for $g \in G^a$ and all $z \in \mathbb{C}\setminus\mathbb{R}$ up to $\mathbb{R}$ we note first that, since $L(z,g)$ is analytic in g for $g \in G^a$, it is enough to prove compactness for a small neighbourhood of $g = 0$. For g sufficiently small, each entry in $L(z,g)$ is, as follows from the previous paragraph, a norm-convergent series in powers of g whose coefficients are connected modified graphs. Since the latter graphs are compact by (II), each entry is itself compact by the theorem on the closedness of the set of compact operators in the uniform topology.

Applying a modification of Theorem 3.3 completes the proof.

ACKNOWLEDGMENT

The author is very grateful to Professors K. Hepp and W. Hunziker for their hospitality at the Institute for Theoretical Physics at ETH-Zürich, where this paper was completed. This work was partially supported by the National Science Foundation under Grant MCS-78-01885.

NOTES

1. Section 1 is a condensed version of [27,part I] (see also [28]). The two-space scattering theory was initiated by KATO [16] and its stationary formulation was developed by BIRMAN and SCHECHTER [1,2,20,23] for simple systems and by HOWLAND [11] and KATO [17] in the general case. It was applied to the study of three-body systems by HOWLAND [11], KATO [17], and YAJIMA [33].

Independently and on a more formal level, the two-space theory and its application to multiparticle scattering was studied by CHANDLER and GIBSON [3] (see also [4]) and by PRUGOVECKI [21].

The main definitions of this section follow KATO [16] and HUNZIKER [13]. Theorem 1.1 is close to a similar result of KATO [17] (actually, the manuscript of series [27] was completed long before the author learned about KATO's paper [17]). The Banach spaces $\Sigma A_i H$ of the example (see also Sec. 6) were introduced independently in [25] and [33] (see also [6]).

2. Abstract multichannel systems were introduced by HUNZIKER [13].

3. The results of Sec. 3 are due to SIGAL [26]. However, Theorem 3.3 was first proved with a restriction slightly stronger than (3). Theorem 3.3 with Condition (3) was proved by HAGEDORN [9].

Note that regularizers were first introduced in PDE's. Our definition differs from the one accepted in PDE's, where the invertibility of F is not required and, therefore, the condition that some power of $TF - I$ be compact is equivalent to the restriction that $TF - I$ is itself compact.

4. The space R and the inner product (x,y) of Sec. 4 were introduced in [31]. Relatively compact potentials were introduced in [5]. Compactness of connected graphs was proved by various authors [12,5], who assumed different restrictions on the potentials. Our proof is the simplest and allows for the most general systems. This is achieved by using the ENSS type operators $X_l^{(n)}$ [8] with the corresponding commutator estimates.

5. The regularizers of Sec. 5 come from equations introduced by BEREZIN [34] (see also [24-28]). The statements of this section, except for Corollary 5.2, were proved in [25,3]. Corollary 5.2 is the difficult part of the HVZ theorem [22].

6. The content of Sec. 6 is taken from an unpublished work of the author [29] (see also [30]). The completeness of the wave operators for single-channel systems was proved by HEPP [10] for smooth, finite-range potentials, by LAVINE [19] for repulsive potentials and by IORIO and O'CARROLL [14] for weak potentials (with the same restrictions on the decay and smoothness as in Corollary 6.1). Thus, the results of this section on the asymptotic completeness of short-range, single-channel systems are the most general.

Note that Condition (II′), actually, is not stronger than (I). There are other, direct proofs of Condition (II) with $n = 4N - 4$ [10] and $n = 3$ [25] for $\mathcal{S}(\mathbb{R}^\nu)$ potentials [10] and for a class of potentials for which (I) holds [25]. The proof in [25] is based on an approximation of the V's in a norm for which (I) holds and an estimation, in the momentum representation, of the kernels of the approximating graphs. The latter is done by the integration of the denominator (coming from R_0's) by parts, using Feynman's identity

$$\prod_1^s A_i^{-1} = \int \left[\sum_1^s \alpha_i A_i\right]^{-s} \delta\left(\sum_1^s \alpha_i - 1\right) d^s\alpha .$$

REFERENCES

[1] A.L. Belopol'skii, M.S. Birman: Math. USSR-Izv. **2**, 1117-1130 (1968)
[2] M.S. Birman: in Probl. of Math. Phys. **4**, 22-26, Leningrad University, Leningrad 1970
[3] C. Chandler, A.G. Gibson: J. Math. Phys. **14**, 1328-1335 (1973)
[4] C. Chandler, A.G. Gibson: J. Math. Phys. **18**, 2336-2346 (1977); **19**, 1610-1616 (1978)
[5] J.M. Combes: Comm. Math. Phys. **12**, 283-295 (1969)
[6] M. Combescure, J. Ginibre: Ann. Inst. Henri Poincaré **21**, 79-145 (1974)
[7] P. Deift, W. Hunziker, B. Simon, E. Vock: Comm. Math. Phys. **64**, 1-34 (1978)
[8] V. Enss: Comm. Math. Phys. **52**, 233-238 (1977)
[9] G.A. Hagedorn: Trans. Am. Math. Soc., in press
[10] K. Hepp: Helv. Phys. Acta **42**, 425-458 (1969)
[11] J.S. Howland: J. Functional Anal. **22**, 250-282 (1976)
[12] W. Hunziker: Helv. Phys. Acta **39**, 451-462 (1966)
[13] W. Hunziker: Helv. Phys. Acta **40**, 1052-1062 (1967)
[14] R.J. Iorio, Jr., M. O'Carroll: Comm. Math. Phys. **27**, 137-145 (1972)
[15] T. Kato: Math. Annalen **162**, 258-279 (1966)
[16] T. Kato: J. Functional Anal., **1**, 342-369 (1967)
[17] T. Kato: J. Fac. Sci. Univ. Tokyo, Sec. IA, **24**, 503-514 (1977)
[18] T. Kato: *Perturbation Theory for Linear Operators* (Springer, New York, 1966)
[19] R. Lavine: Comm. Math. Phys. **20**, 301-323 (1971)
[20] D.B. Pearson: "A generalization of the Birman trace theorem", J. Functional Anal. **28**, 182-186 (1978)
[21] E. Prugovečki: J. Math. Phys. **14**, 957-962 (1973)
[22] M. Reed, B. Simon: *Methods of Modern Mathematical Physics*, Vol. IV (Academic, New York, 1978)
[23] M. Schechter: J. Math. Pures Appl. (9) **57**, 373-396 (1974)
[24] I.M. Sigal: Comm. Math. Phys. **48**, 137-154 (1976); **48**, 155-164 (1976)
[25] I.M. Sigal: Mem. Am. Math. Soc. **209** (1978)
[26] I.M. Sigal: Bull. Am. Math. Soc. **84**, 152-154 (1978)
[27] I.M. Sigal: "Scattering theory for many particle systems," (I, II), preprint, ETH-Zürich (1977-1978); will appear in Lecture Notes in Mathematics
[28] I.M. Sigal: "On abstract scattering theory for two spaces," preprint, Princeton (1979)
[29] I.M. Sigal: (1972) unpublished
[30] I.M. Sigal: "Mathematical theory of single-channel systems," in preparation
[31] I.M. Sigal: "On the discrete spectrum," in preparation
[32] A.G. Sigalov and I.M. Sigal: Theor. and Math. Phys. **5**, 990-1005 (1970)
[33] K. Yajima: J. Fac. Sci. Univ. Tokyo, Sec. IA, **25**, 109-132 (1978)
[34] F.A. Berezin: Dokl. Akad. Nauk **163**, 795-798 (1965)
[35] T. Kato, S.T. Kuroda: Rocky Mt. J. Math. **1**, 127-171 (1971)
[36] S. Agmon: Ann. Scuola Norm. Sup. Pisa Cl. Sci. (4) **2**, 151-218 (1975)

SCATTERING FROM POINT INTERACTIONS

Lawrence E. Thomas

Department of Mathematics
University of Virginia
Charlottesville, Virginia 22903

INTRODUCTION

The simplest example of a Schrödinger operator with a point interaction in 3 dimensions is the Laplacian with a boundary condition at the origin

$$\lim_{r \downarrow 0} (\partial_r r - \alpha r)\psi = 0, \tag{1}$$

$r = |x|, \alpha$ a real constant. Two generalizations of this operator will be considered; (1) the 3-dimensional Laplacian with boundary conditions imposed at several points (quantum pin-ball machine), (2) a 3-particle operator given by the 9-dimensional Laplacian with boundary conditions analogous to (1) imposed along the particle-collision hypersurfaces.

These operators and relatives of theirs have a venerable history going back several decades. Originally, they were studied by BREIT, THOMAS, WIGNER, and others as nuclear models with interactions simulating potentials of short range [1]. They observed that potential scattering, with suitably scaled negative potential, converges in the low energy limit to scattering from a point interaction. In the late 50's and early 60's, HUANG, YANG, LEE, LUTTINGER, and WU studied multiparticle operators with pseudo-potential interactions (formally, $\sum_{i<j} \delta(x_i - x_j) \frac{\partial}{\partial r_{ij}} r_{ij})$ in low-order perturbation theory [2]. Their goals were to delineate spectral properties of the operators and to investigate the statistical mechanics of particles with these interactions. Because of the singular nature of the interactions, higher-order perturbation theory was difficult; no attempt was made to interpret the operators as self-adjoint.

Beginning in 1961, a series of papers by DANILOV, MINLOS, and FADDEEV appeared concerning the 3-body operator defined via boundary conditions [3]. The physical motivation was to compute the bound-state energy of tritium from 2-body scattering data. More will be said about this work in Sec. 2. There is a fine survey article by FLAMAND [4].

There is recent work on these operators as well. FRIEDMAN, ALONSO Y CORIA, and NELSON have considered the problem of operators with potentials of shrinking support [5]. (Some of this work has proceeded by techniques of nonstandard analysis.) Hoegh-Krohn and Grossmann have done work on the quantum pin-ball machine and have work in progress in which an infinite number of pins are placed in a periodic 3-dimensional array (solid-state model). The second section is a sketch of my own work on Birman-Schwinger bounds for the pin-ball machine [6].

It is perhaps appropriate to mention there is some vague "theoretical" justification for considering these interactions. DIMOCK has shown that the nonrelativistic limit of $P(\phi)_2$, quantum field theory in one space dimension, results in a δ-function potential between particles [7]. The boundary condition (1) is just the 3-dimensional analogue of the δ-function interaction in one dimension.

1. QUANTUM PIN-BALL MACHINE AND BIRMAN-SCHWINGER BOUNDS [6]

Let $-\Delta'$ be the operator acting in $L^2(\mathbb{R}^3)$ obtained from the Laplacian by imposing boundary conditions at N "pins" $x_1, x_2, \ldots, x_n \in \mathbb{R}^3$,

$$\lim_{r_i \downarrow 0}\left(\frac{\partial}{\partial r_i} r_i - \alpha r_i\right)\psi = 0, \; r_i = |x-x_i|, \; i = 1,2, \ldots, N. \tag{1.1}$$

As is well known, the resolvent kernel for the operator $-\Delta'$ is

$$(-\Delta' + k^2)^{-1}(\mathbf{x},\mathbf{y}) =$$

$$\frac{1}{4\pi}\frac{e^{-k|x-y|}}{|x-y|} - \frac{1}{(4\pi)^2}\sum_{i,j}\frac{e^{-k|x-x_i|}}{|x-x_i|}T_{ij}(k)\frac{e^{-k|y-x_j|}}{|y-x_j|}, \tag{1.2}$$

with $T(k)$ the matrix inverse of $A(k)$,

$$A_{ij}(k) = \begin{cases} \dfrac{1}{4\pi}\dfrac{e^{-k|x_i-x_j|}}{|x_i-x_j|}, & i \neq j, \\ \dfrac{-1}{4\pi}(k+\alpha), & i = j. \end{cases} \tag{1.3}$$

(The scattering amplitude for $-\Delta'$ is

$$f(k_{out},k_{in}) = -\frac{1}{4\pi}\sum_{i,j} e^{-ik_{out}\cdot x_i}T_{ij}(-ik)e^{ik_{in}\cdot x_j}.)$$

Let

$$k_N = \infty, \; k_{N-1} = \sup_{1 \leqslant i \leqslant N}\left(\sum_{j \neq i}^{N} |x_i-x_j|^{-2}\right)^{1/2}.$$

Assume, by relabeling the x_i's, if necessary, that the supremum is attained for $i = N$. Let

$$k_{N-2} = \sup_{i \leqslant N-1}\left(\sum_{j \neq i}^{N-1} |x_i-x_j|^{-2}\right)^{1/2}$$

and define $k_{N-3}, \ldots, k_1$ by continuing in this manner.

Theorem (Birman-Schwinger Bounds). *The eigenvalues* $\{e_i\}$ *of* $-\Delta'$ *satisfy*

$$\sum_{e_i < -\kappa^2} |e_i| \leqslant 2\kappa^2(N-1) + c\sum_{i<j}^{N}|x_i-x_j|^{-2} \tag{1.4}$$

with $\kappa = 0$ if $\alpha \geqslant 0$, $-\alpha$ if $\alpha < 0$, *and the constant c independent of* N, α. *For* $k > \kappa$, $N(k) \equiv$ *the number of eigenvalues, counting multiplicities* $\leqslant -k^2$, *satisfies*

$$N(k+\kappa) \leqslant 4\sum_{i<j}^{m} e^{-2(k+\kappa)|x_i-x_j|}\left(k^2|x_i-x_j|^2 + e^{-2(k+\kappa)|x_i-x_j|}\right)^{-1} + (N-m),$$

$$k_{m-1} \leqslant k < k_m, \; m = 1,2, \ldots, N. \tag{1.5}$$

The theorem has the following interpretation: suppose N non-self-interacting fermions were allowed to interact with N pins via the boundary conditions (1.1). In their ground state, the fermions would create an effective potential between pins which, if (1.3) were equality, would be $-r^{-2}$-like at small distances between pins. In a statistical mechanical context, (1.3) is an estimate on the potential between pins required for thermodynamic stability. (Note that if $\alpha = 0$, the eigenvalues e_i are homogeneous functions of degree -2 in the x_i's. In this sense, (1.4) gives the correct dependence on the x_i's.)

The proof of the theorem, in outline, runs as follows: $-\Delta'$ has an eigenvalue $-k^2$ if $A(k)$ has eigenvalue 0, or if $B(k) = C^{-1/2}(A(k)+C)C^{-1/2}$ has eigenvalue 1, where C is an arbitrary positive definite diagonal matrix. The derivative of $B(k)$ is negative definite, so that by first order perturbation theory an eigenvalue of $B(k)$ is decreasing in k. As a consequence, $N(k)$ is equal to the number of eigenvalues of $B(k)$ exceeding 1. An estimate for $N(k)$ is thus the square of the Hilbert-Schmidt norm of $B(k)$, $||B(k)||_2^2$.

In this argument, the matrix C is arbitrary; it can be chosen to optimize the estimate on $N(k)$. At this point, C acquires a dependence on k. Minimizing $||B(k)||_2^2$, we are led to a set of equations for the entries c_i of C,

$$c_i^{-1} + \sum_j^m L_{ij}(k) c_j^{-1} = 1/k, \ i = 1, 2, \ldots, m, \tag{1.6}$$

with

$$L_{ij}(k) = \begin{cases} \dfrac{e^{-2(k+\kappa)|x_i - x_j|}}{k^2 |x_i - x_j|^2}, & i \neq j, \\ 0 & , \quad i = j, \end{cases} \tag{1.7}$$

where we consider the equations as m-dimensional. The key fact is:

Lemma. *Equations (1.6) are solvable with each $c_i > 0$ if*

$$\sup_i \sum_{ij}^m L_{ij}(k) < 1.$$

The hypothesis of the lemma insures that the equations can be solved by a Neumann series, and further that the c_i's will be positive.

For $k_{m-1} \leqslant k < k_m$ the hypothesis of the lemma is satisfied. An estimate on $N(k)$ is thus $\|B(k)\|_2^2$, with $c_1, \ldots, c_m$ the solution to (1.6) and the remaining c_i's set to ∞ by a limiting argument. A simple estimate on the c_i's coming from (1.6) gives (1.5). This estimate on $N(k)$, plus an integration by parts on $\int k^2 dN(k)$, gives the estimate (1.4) of the theorem.

2. INTRODUCTION TO THE 3-BODY PROBLEM [4]

We first divide out by center-of-mass motion so that the relevant operator will be acting in $L^2(\mathbb{R}^6)$. Let (x,y) be the coordinates for $\mathbb{R}^6$ with x, y each 3-vectors. Let $S_1 = \{(x,y) \,|\, y = 0\}$, $S_2 = \{(x,y) \,|\, y - \sqrt{3}x = 0\}$, $S_3 = \{(x,y) \,|\, y + \sqrt{3}x = 0\}$ be the collision surfaces. Corresponding to the collision surface S_i, let $(x(i)),y(i)$ be coordinates of $\mathbb{R}^6$ so that $S_i = \{(x(i),y(i)) \,|\, y(i) = 0\}$. Then, the 3-body operator H is obtained from the 6-dimensional Laplacian by imposing boundary conditions along S_i,

$$\lim_{r_i \downarrow 0} \left(\frac{\partial}{\partial r_i} r_i - \alpha r_i \right) \Psi(x(i), y(i)) = 0 \quad r_i = |y(i)|, \quad i = 1, 2, 3. \tag{2.1}$$

By analogy with the quantum pin-ball machine, we attempt to write the resolvent as the free resolvent plus a correction consisting of the free resolvent integrated against a "charge" density χ supported on the collision surfaces. Imposing the boundary condition on this resolvent leads to a coupled set of integral equations for the density. Making the further assumption of Bose symmetry, we arrive at a single integral equation, the Skorniakov-Ter Martirosian equation for the density, which in a Fourier transform representation reads

$$A_\alpha(z)\hat{\chi}(p) \equiv (\alpha + \sqrt{p^2 - z})\hat{\chi}(p) - \frac{2}{\pi^2\sqrt{3}} \int_{\mathbb{R}^3} (p^2 - p \cdot p' + p'^2 - 3/4z)^{-1} \hat{\chi}(p') d^3p' = \hat{\chi}_0(p). \tag{2.2}$$

Here, $\hat{\chi}_0(p)$ is a known function.

The operator $A_\alpha(z)$ clearly commutes with rotations and can be reduced. The restriction of (2.2) to higher angular momentum channels presents no particular problems, but in the s-wave channel the situation is complicated. By an analysis involving the Mellin transform, MINLOS and FADDEEV [3] have shown that in this channel, (2.2) is solvable, but that the solution is not unique. In more technical terms, for z negative real, $A_\alpha(z)$ is not essentially self adjoint on, say, functions of rapid decay at ∞.

Further "boundary" conditions must be imposed on $A_\alpha(z)$. They take the form of specifying the asymptotics $p \to \infty$ of the solution to (2.2) in a manner reminiscent of a radiation condition. Moreover, the choice of asympotics should be z-independent if the first resolvent equation is to be satisfied by $(H-z)^{-1}$. (It is perhaps worthwhile to mention that ALONSO Y CORIA [5] found that if H' is the limit of a sequence of 3-body Schrödinger operators with shrinking potentials and H' is not the free Hamiltonian, then H' is not necessarily essentially self-adjoint on functions in the domain of H' supported away from the origin in $\mathbb{R}^6$.)

Then a peculiar phenomenon occurs. One can show there is an infinity of real negative z values, converging to $-\infty$, for which (2.2) has a homogeneous solution satisfying the asympotic condition. Consequently, regardless of the parameter α or the particular asympototic condition selected, H has a negative discrete point spectrum extending to $-\infty$. Intuitively, there is an infinity of eigenfunctions with supports more and more concentrated about the intersection of the collision surfaces. (One possible remedy to this situation is to replace the scalar α in the boundary condition (2.1) by an operator $\alpha_0 - \alpha_1\Delta_{x(i)}$ where $\Delta_x(i)$ is the "tangential" Laplacian. This leads to a semibounded operator while maintaining the point-like nature of the interaction. The physical interpretation of this new boundary condition is not clear, however.)

We conclude with remarks on the scattering theory for H. In preliminary studies, J.V. Ralston and I did show existence of the wave operators by a KUPSCH-SANDHAS argument [8]. We also established a local decay of the wave function ψ_t (and decay away from the collision surfaces) of the sort one might need in a geometrical approach to asympotic completeness.

REFERENCES

[1] G. Breit: Phys. Rev. **71** (1947), 215; J.M. Blatt, V.F. Weisskopf: *Theoretical Nuclear Physics* (Wiley, New York, 1952), p. 74; L.H. Thomas: Phys. Rev. **47** , 903 (1935)

[2] K. Huang, C.N. Yang: Phys. Rev. **105**, 767 (1957); K. Huang, C.N. Yang, J.M. Luttinger: Phys. Rev. **105**, 767 (1957); T.D. Lee, K. Huang, C.N. Yang: Phys. Rev. **106**, 1134 (1957); T.T. Wu, Phys. Rev. **115**, 1390 (1959); K. Huang, *Statistical Mechanics*, (Wiley, New York, 1963), p. 409

[3] G. S. Danilov: Sov. Phys.—JETP **13**, 349 (1961); L.D. Faddeev, R.A. Minlos: Sov. Phys.—JETP **14**, 1315 (1962)

[4] G. Flamand: In *Cargèse Lectures in Theoretical Physics*, ed. by F. Lurçat (Gordon and Breach, New York, 1965), p. 247

[5] C.N. Friedman: J. Functional Anal. **10**, 346 (1972); A. Alonso y Coria, "Shrinking potentials in the Schrodinger Equation," thesis, Princeton University (1978); E. Nelson, Bull. Amer. Math. Soc. **83**, 1165 (1977)

[6] L.E. Thomas: to appear in J. Math. Phys. (1979)

[7] J. Dimock: Commun. Math. Phys. **57**, 51 (1977)

[8] J. Kupsch, W. Sandhas: Commun. Math Phys. **2**, 147 (1966)

TWO PROBLEMS WITH TIME-DEPENDENT HAMILTONIANS

James S. Howland*

Department of Mathematics
University of Virginia
Charlottesville, Virginia 22903

INTRODUCTION

The subject of this talk is time-dependent Hamiltonians in quantum mechanics and a method of dealing with them. I hope to convince you of two things: first, that there are interesting problems with time-dependent Hamiltonians; and second, that the method is natural for dealing with them.

I will first describe the method, and then discuss its applications to two problems. I submit that the problems are of substantial physical interest, and that the method provides natural insights into them.

My handling of these subjects will be heuristic in the extreme, since full details are available elsewhere in printed form (or shortly will be).

1. THE METHOD.

The method arises from a procedure in classical mechanics. Consider Hamilton's equations of motion

$$\frac{dq}{dt} = \frac{\partial H}{\partial p}, \quad \frac{dp}{dt} = -\frac{\partial H}{\partial q},$$

where the Hamiltonian $H(p,q,t)$ depends on time. Bringing in the time t as a new coordinate, the energy E of outside sources as its conjugate momentum, and the new Hamiltonian

$$K(p,q,E,t) = E + H(p,q,t),$$

leads to the equivalent equations

$$\frac{dp}{d\sigma} = \frac{\partial H}{\partial p}, \quad \frac{dp}{d\sigma} = -\frac{\partial H}{\partial q},$$

$$\frac{dt}{d\sigma} = 1, \quad \frac{dE}{d\sigma} = -\frac{\partial H}{\partial t},$$

in which K is independent of the time σ. Thus, by a simple transformation, every temporally inhomogeneous system reduces to a conservative, homogeneous one.

In quantum mechanics, the state vector of a temporally inhomogeneous system satisfies a Schrodinger evolution equation

$$i\frac{\partial \psi}{\partial t} = H(t)\psi, \tag{1}$$

where ψ lies in a Hilbert space $\mathscr{H}$. The solution of such an equation is given, at least formally, by

$$\psi(t) = U(t,s)\psi(s),$$

where $U(t,s)$ is a unitary operatory called the *propagator* which satisfies $U(t,s) = I$ and $U(t,r)U(r,s) = U(t,s)$. If we proceed by analogy, we find that the corresponding conservative system has the equation

$$i\frac{d\Psi}{d\sigma} = K\Psi,$$

where Ψ is in $\mathscr{K} = L_2(-\infty, \infty; \mathscr{H})$ and

$$K = -i\frac{\partial}{\partial t} + H(t). \tag{2}$$

The theory of K should, therefore, be equivalent to the theory of the evolution equation (1), a fact which is surprisingly useful.

For one thing, it lets us generate propagators just as we generate semi-groups, by constructing a generator. The sort of generator we want is a self-adjoint operator K on $\mathscr{K}$ of the form (2), but what does that mean? One way to make it precise is to remark that, since the second term commutes with the scalar multiplication operator

$$Qf(t) = tf(t),$$

the operator K must have the same commutator with Q as the operator $P = -i\frac{\partial}{\partial t}$ does; that is, the CCR

$$KQ - QK = iI. \tag{3}$$

We shall define an *evolution generator* to be any self-adjoint operator K on $\mathscr{K}$ which satisfies the CCR (3) (in Weyl's integrated form, [1], p. 225). Formally, K and its propagator $U(t,s)$ are connected by the formula

$$e^{-i\sigma K}f(t) = U(t,t-\sigma)f(t-\sigma). \tag{4}$$

(To verify this, differentiate with respect to σ, use that U satisfies (1), and set σ equal to zero.) The connection is, in fact, rigorous:

Theorem. *For every evolution generator K, there is a unique measurable propagator satisfying (4).*

The proof is very near at hand. For K and Q are a canonical pair, so by von Neumann's theorem on representations of CCR, there is unitary U on $\mathscr{K}$, with

$$Q = UQU^* \tag{5}$$

and

$$e^{-i\sigma K} = UT_\sigma U^*, \tag{6}$$

where $T_\sigma f(t) = f(t-\sigma)$. (5) says that U commutes with Q and, hence, is a multiplication by a unitary $U(t)$. Substituting into (6) then gives (4), with $U(t,s) = U(t)U(s)^*$.

Of course, we would like to have $U(t,s)$ *strongly continuous* as well. This requires essentially that K also generate a contraction semigroup on $C_0(-\infty,\infty;\mathscr{K})$. In practice, when K is constructed by perturbation from a K_0 with a continuous propagator, it is usually easy to show that $U(t,s)$ is continuous.

2. THE AC STARK EFFECT

As our first application, we shall consider a hydrogen-like atom in a spatially uniform electric field which is oscillatory in time. The Hamiltonian in question is

$$H(t) = -\Delta + V(x) + \epsilon x_1 \cos \omega t \tag{7}$$

on $\mathscr{H} = L^2(\mathbb{R}^3)$, which has period $a = 2\pi\omega^{-1}$. To construct dynamics, we must show that

$$K = -i\frac{\partial}{\partial t} + H(t)$$

is essentially self-adjoint. Now, the ordinary Stark Hamiltonian

$$H_\epsilon = -\Delta + V(x) + \epsilon x_1 \tag{8}$$

is essentially self-adjoint by the commutator theorem [7, X.5] because it satisfies the inequality

$$\pm i[H_\epsilon,N] \leqslant CN,$$

where

$$N = H_\epsilon + cx^2 + b$$

As B. Simon has pointed out, the same proof works for K if we amend N to be

$$N' = -\frac{d^2}{dt^2} + H(t) + cx^2 + b.$$

The physical question to be considered is: What happens to the bound states of

$$H = -\Delta + V(x)$$

when the Stark field

$$\epsilon x_1 \cos\omega t$$

is turned on? In the ordinary (DC) Stark effect, we know that the bound states disappear, but that the eigenvalues reappear as "resonances"— poles of some sort of continuation of the resolvent of H_ϵ. The physical interpretation is that, for small ϵ, the bound state is unstable and decays at a rate given by the imaginary part of the pole. From the intuitive physical point of view, there does not seem to be a great deal of difference in the two cases, but mathematically, if one wants to study resonances, and perhaps calculate the decay rate, the immediate question is "what am I supposed to continue?". The present method leads quickly and naturally to the answer.

The point is that, because $H(t)$ is periodic in time, K has a symmetry: it commutes with the unitary translation

$$T_a f(t) = f(t-a).$$

In fact, since the method regards t as a spatial coordinate, the situation is exactly analogous to the spatially periodic "crystal" problem [9]. If $\mathcal{K}$ is decomposed as a direct integral

$$\mathcal{K} = \oplus\int_0^{2\pi} \mathcal{K}(\theta)\,d\theta$$

in which T_a is diagonal, then K will be an operator-valued multiplication:

$$K = \oplus\int_0^{2\pi} K(\theta)\,d\theta.$$

The operator T_a is easily diagonalized by the Fourier transform, and one finds that $\mathcal{K}(\theta)$ can be taken for all θ to be the space

$$\tilde{\mathcal{K}} = L_2([0,a];\mathcal{K})$$

and that

$$K(\theta) = \tilde{K} + \theta,$$

where $\tilde{K}$ is formally the same differential operator as K, but with the periodic boundary condition

$$u(0) = u(a).$$

The operator $\tilde{K}$ appears in the physical literature as the "Floquct Hamiltonian" [8] or the "quasi-energy" [10]. It is really quite analogous to an ordinary Schrödinger Hamiltonian. For example, by (6), K is unitarily equivalent to the generator $-i\frac{\partial}{\partial t}$ of T_σ, which is absolutely continuous, but $\tilde{K}$ can have point spectrum. A point λ is in $\sigma_p(\tilde{K})$ if

$$U(a,0)f = e^{-i\lambda a}f$$

for some nonzero f in $\mathcal{K}$. This means that the state f (considered as a *ray* in Hilbert space) is invariant under evolution through a period. Clearly, $\sigma_p(\tilde{K})$ is periodic with period $2\pi a^{-1} = \omega$. In the special case of constant $H(t)$,

$$\tilde{K} = -i\frac{\partial}{\partial t} + H,$$

for which $U(t,s) = e^{-iH(t-s)}$, one has that $\lambda \in \sigma_p(\tilde{K})$ *iff*

$$e^{-iaH}f = e^{-i\lambda a}f,$$

which means that $\gamma \in \sigma_p(H)$, modulo ω. Thus, eigenvalues of $\tilde{K}$ correspond to eigenvalues of H. Since eigenvalues correspond, it is easy to suppose that resonances also correspond. The resolvent of $\tilde{K}$ is what we want to continue.

The approach to resonances by complex scaling or dilation has become quite popular, especially for numerical work, where it lends itself to Rayleigh-Ritz calculations. The idea is that when the group of scale transformations

$$U(\phi)f(x) = e^{3\phi/2}f(e^{\phi}x) \tag{9}$$

is applied to a Hamiltonian H and ϕ is continued to complex values, the spectrum of

$$H(\phi) = U(\phi)HU(-\phi)$$

moves around in a tantalizing fashion, revealing portions of the second sheet where one can hunt for poles. The curious thing is that the *essential* spectrum moves, while the *discrete* spectrum stays put. The resonances are identified as the discrete eigenvalues of $H(\phi)$ in the unphysical region. For atomic Hamiltonians, the essential spectrum rotates about the thresholds ([7], Theorem XIII.10).

What happens to the spectrum in Stark fields? The question is important relative to the numerical work of REINHARDT *et. al.* (see [2]), who used a Rayleigh-Ritz method to compute eigenvalues of $H(\phi)$. For DC fields, the answer discovered by HERBST [2] is rather startling: the essential spectrum *disappears* for $\operatorname{Im}\phi \neq 0$, and the resolvent of $H(\phi)$ is meromorphic in the complex plane! This is quite satisfactory from the numerical standpoint.

For AC fields, we have to dilate $\tilde{K}$. CHU and REINHARDT [8] use the scale transformations

$$U(\phi)f(x,t) = e^{3\phi/2}f(e^{\phi}x,t). \tag{10}$$

For $\operatorname{Im}\phi \neq 0$, the essential spectrum of $H(\phi)$ then turns out to be the whole plane! This is initially rather discouraging, but the problem is merely one of gauge. Instead of (7), use the Hamiltonian

$$H_1(t) = (P_1 - \epsilon\omega^{-1}\sin\omega t)^2 + P_2^2 + P_3^2 + V(x),$$

where $P_k = -i\dfrac{\partial}{\partial x_k}$, in which the same field is described in a different gauge. Under the scaling (10), the essential spectrum of $\tilde{K}_1(\phi)$ is

$$\sigma_e(\tilde{K}_1(\phi)) = \omega Z + e^{-2i\operatorname{Im}\phi}\mathbb{R}^+ + \frac{\epsilon^2}{2}$$

where Z denotes the integers and $\mathbb{R}^+$ the nonnegative reals. Thus, again, the spectrum of $\tilde{K}_1$ rotates about thresholds (which here are the integral multiples of ω), and the entire spectrum shifts by $\epsilon^2/2$.

The Rayleigh-Ritz calculations are left in fine shape, because matrix elements of $\tilde{K}_1(\phi)$ are just those of $\tilde{K}(\phi)$ in another basis. In fact, since

$$K = GK_1G^*$$

for the unitary gauge transformation

$$Gf(x,t) = e^{i(\epsilon x/\omega)\sin\omega t}f(x,t),$$

one has, for real ϕ,

$$<K(\phi)f,g> = <K_1(\phi)G(\phi)f,G(\phi)g>,$$

where $G(\phi) = U(\phi)GU(-\phi)$. For suitable f, $G(\phi)f$ is well-behaved for complex ϕ.

It is also possible to treat the perturbation theory in ϵ of the bound states of $H = -\Delta + V(x)$ satisfactorily from this point of view, but time prevents us from going into that here. Instead, we turn to our second problem.

3. CHARGE TRANSFER

Our second problem is a multichannel scattering problem, one of the simplest imaginable. The picture is that of an ion bypassing an atom, and pulling off an electron. I had the name "charge transfer" from a colleague in physics.

To be precise, let $q_1(x)$ and $q_2(x)$ be two-body potentials on $\mathbb{R}^3$ for which the operators

$$h_j = -\Delta + q_j(x) \quad (j = 1,2)$$

each have a single bound state. The Hamiltonian

$$H(t) = -\Delta + q_1(x) + q_2(x - vt)$$

on $L_2(\mathbb{R}^3)$ then describes the interaction of the atom

$$h_1 = -\Delta + q_1(x)$$

with a center of force passing by on a straight-line path with vector velocity v. Thus, we are looking at a charged particle passing a one-electron atom in an approximation where the nucleus is fixed and the ion moves uniformly in a straight line. There are clearly three channels: the electron may come off bound either to the fixed or to the moving nucleus, or may be free of both. We want to do scattering theory. How do we go about it?

The first observation is that the scattering theory for a pair of evolution generators, K and K_0, is essentially the same as the scattering theory for their propagators, U and U_0. In fact, it follows from (4) that

$$e^{i\sigma K}e^{-i\sigma K_0}f(t) = U(t,t+\sigma)U_0(t+\sigma,t)f(t).$$

Letting σ tend to infinity, we find that the (K,K_0) wave operator is

$$W_+(K,K_0)f(t) = W_+(t)f(t)$$

where

$$W_+(t) = \operatorname*{s-lim}_{s\to\infty} U(t,s)U_0(s,t)$$

is the (U,U_0) wave operator, referred to initial time t. Thus, one can try all the standard tricks of scattering theory on the operators K and K_0. For example, one can use stationary scattering theory (Lippmann-Schwinger equations and the like) on K and K_0. The estimates are a bit different, but the formulas are the same [3,5].

From this point of view, the natural approach to the charge transfer problem is by Faddeev's equations! In fact, take

$$K_0 = -i\frac{\partial}{\partial t} - \Delta$$

and

$$K = K_0 + V_1 + V_2 + V_3,$$

where

$$V_1f(x,t) = q_1(x)f(x,t),$$

and

$$V_2f(x,t) = q_2(x - vt)f(x,t),$$

while $V_3 = 0$. Set $K_j = K_0 + V_j$ $(j = 1,2,3)$. The Faddeev formalism applies directly, and, suppressing the trivial V_3 terms, one obtains a formula for $(K - z)^{-1}$ in terms of the inverse of $I + F(z)$, where

$$F(z) = \begin{bmatrix} 0 & V_1R_1(z) \\ V_2R_2(z) & 0 \end{bmatrix}$$

and $R_j(z) = (K_j - z)^{-1}$. The usual sort of primary singularities arise. The channel projections are

$$P_j = \langle\cdot,\phi_j\rangle\phi_j \otimes I \quad (j = 1,2)$$

where $\otimes_1$ is the factorization for the coordinates (x,t), while $\otimes_2$ is for (x',t') with

$$x' = x - vt, \; t' = t.$$

The whole rigamarole works formally, and the only question is whether one can do the estimates proving compactness and the like.

The required estimates are not at all obvious. However, recently Kenji Yajima of Tokyo has informed me (private communication) that he has succeeded in proving the desired asymptotic completeness for K by a method similar to that outlined here. Details will be presented in a forthcoming publication.

4. NOTES AND REFERENCES

A number of authors not explicitly cited here have worked on time-dependent Hamiltonians, some using the method but most not. Among them are Davies, E.J.P.G. Schmidt, J. Goldstein, Monlezun, Hendrickson, Kuroda, Morita, and Yajima. I have referenced the papers I know in [5], to which the reader is referred. For the periodic case, there is a lot of physics literature (with which I am not familiar). Some of it is referenced by CHU and REINHARDT [8].

The results on the spectrum of the dilated AC Stark Hamiltonian are mine, and will appear in [6].

Finally, I wish to thank Barry Simon, Bill Reinhardt, K. Yajima, Hugh Kelly, Larry Thomas, and Ira Herbst for various entertaining and informative converstions.

REFERENCES

*Research partly supported by the National Science Foundation under Grant MCS-74-07313-A04

[1] G. Emch: *Algebraic Methods in Statistical Mechanics and Quantum Field Theory* (Interscience, New York 1972)
[2] I. Herbst: Comm. Math. Phys. **64**, 279 (1979)
[3] J. S. Howland: Math. Ann. **207,** 315 (1974)
[4] J. S. Howland: Functional Anal. **22**, 250 (1976)
[5] J. S. Howland: Indiana J. Math. (1979), to appear
[6] J. S. Howland: "Complex scaling of AC stark Hamiltonians" to appear
[7] M. Reed, B. Simon: *Methods of Modern Mathematical Physics,* Vol. II (Academic, New York, 1975); *ibid.*, Vol. IV (1978)
[8] S. I. Chu, W. P. Reinhardt: Phys. Rev. Lett. **39**, 1195 (1977)
[9] L. E. Thomas: Comm. Math. Phys. **33**, 335 (1973)
[10] Ya. B. Zeldovich: Sov. Phys.—Usp. **16**, 427 (1973)

TRANSLATION INVARIANCE OF N-PARTICLE SCHRODINGER OPERATORS IN HOMOGENEOUS MAGNETIC FIELDS

*I. W. Herbst**

Department of Mathematics
University of Virginia
Charlottesville, Virginia 22903

INTRODUCTION

This talk is based on work done some time ago with AVRON and SIMON which appeared in the Fall of 1978 in the Annals of Physics [1]. My reasons for wanting to talk about this subject are twofold: I find the subject interesting and have never talked about it before and, in addition, there may be some interesting physics here which deserves to be known. Previous work related to this subject appears in [2-6], where related work in solid state physics is discussed, and in [7], where part of the present results were obtained for Dirac Hamiltonians. Further references are given in [1] where the reader may also consult for details not discussed here.

Consider a classical particle of mass m and charge e in a constant magnetic field $\mathbf{B}$. We know from elementary physics that the particle will move in a spiral with axis parallel to $\mathbf{B}$. The frequency of revolution, the Larmor frequency, is given by

$$\boldsymbol{\omega} = -e\mathbf{B}/m.$$

The quantity

$$\mathbf{R} = \mathbf{v} \times \boldsymbol{\omega}/\omega^2,$$

where $\mathbf{v}$ is the velocity of the particle, describes the location of the particle with respect to the center of rotation (projected down into the plane with $\mathbf{x} = 0$). Thus, the quantity

$$\mathbf{c} = \mathbf{x}_\perp - \mathbf{R}$$

(with $\mathbf{x}_\perp = \mathbf{x} - (\mathbf{x} \cdot \mathbf{B})\,\mathbf{B}/B^2$) is the center of the classical Landau orbit (again projected). The quantity $\mathbf{c}$ is trivially conserved, $\frac{d\mathbf{c}}{dt} = \mathbf{0}$. It is the consequences of this obvious statement or, rather, its analog for many interacting particles in quantum mechanics, which I want to discuss.

1. THE PSEUDOMOMENTUM

Consider, again, a particle of mass m, charge e in a constant magnetic field $\mathbf{B}$, this time quantum mechanically.

The Hamiltonian is (in a particular gauge)

$$H_0 = (\mathbf{p} - e\mathbf{A})^2/2m, \quad \mathbf{p} = -i\nabla, \quad \mathbf{A} = \frac{1}{2}\mathbf{B} \times \mathbf{x}.$$

If we solve for the quantity $\mathbf{c}$ above, we find

$$\mathbf{c} = \mathbf{x}_\perp + m\mathbf{v} \times \mathbf{B}/eB^2, \quad \mathbf{v} = (\mathbf{p} - e\mathbf{A})/m.$$

The fact that $\mathbf{c}(t) = e^{iH_0t}\,\mathbf{c}\,e^{-iH_0t}$ is conserved follows easily from the Heisenberg equations of motion

$$m\frac{d\mathbf{v}(t)}{dt} = e\mathbf{v}(t) \times \mathbf{B}.$$

We now define a related quantity which reduces to the momentum when $\mathbf{B} = \mathbf{0}$:

$$\mathbf{k} = e\mathbf{c} \times \mathbf{B} + \mathbf{p} - \mathbf{p}_\perp.$$

Using our formula for $\mathbf{c}$, we find

$$\mathbf{k} = \mathbf{p} + \frac{1}{2}e\mathbf{B} \times \mathbf{x}. \tag{1}$$

This is the pseudomomentum. If we define $U(\boldsymbol{\alpha}) = \exp(i\boldsymbol{\alpha} \cdot \mathbf{k})$, we have

$$U(\boldsymbol{\alpha})^{-1}\mathbf{x}\, U(\boldsymbol{\alpha}) = \mathbf{x} - \boldsymbol{\alpha}, \tag{2}$$

$$U(\boldsymbol{\alpha})^{-1}\mathbf{v}\, U(\boldsymbol{\alpha}) = \mathbf{v}. \tag{3}$$

Equations (2) and (3) show that $U(\alpha)$ realizes the translation invariance of the system. (Note that H_0 does not commute with $\mathbf{p}$ even though there is an obvious translation invariance of the physics.)

The commutation relations of the components of $\mathbf{k}$ can be summarized by the formula

$$\mathbf{k} \times \mathbf{k} = -ie\mathbf{B}, \tag{4}$$

from which it follows that

$$U(\boldsymbol{\alpha})\, U(\boldsymbol{\beta}) = \omega(\boldsymbol{\alpha},\boldsymbol{\beta})\, U(\boldsymbol{\alpha} + \boldsymbol{\beta}),$$

where $\omega(\boldsymbol{\alpha},\boldsymbol{\beta}) = \exp\{ie[\boldsymbol{\alpha} \times \boldsymbol{\beta}]\cdot\mathbf{B}/2\}$. Thus, $\{U(\boldsymbol{\alpha}) : \boldsymbol{\alpha} \in \mathbf{R}^3\}$ is a projective representation of the translation group with a nontrivial multiplier ω. That ω is nontrivial means that $U(\boldsymbol{\alpha})$ cannot be multiplied by a phase factor $\eta(\boldsymbol{\alpha})$ so that $V(\boldsymbol{\alpha}) = \eta(\boldsymbol{\alpha})\, U(\boldsymbol{\alpha})$ satisfies $V(\boldsymbol{\alpha})\, V(\boldsymbol{\beta}) = V(\boldsymbol{\alpha} + \boldsymbol{\beta})$.

It is also an easily proved fact that if unitaries $V(\boldsymbol{\alpha})$ satisfy

$$V(\boldsymbol{\alpha})^{-1}\mathbf{x}\, V(\boldsymbol{\alpha}) = \mathbf{x} - \boldsymbol{\alpha}, \tag{5}$$

$$V(\boldsymbol{\alpha})^{-1} H_0 V(\boldsymbol{\alpha}) = H_0,$$

then $V(\boldsymbol{\alpha}) = \eta(\boldsymbol{\alpha})\, U(\boldsymbol{\alpha})$ for some phase factor η. Thus, V's satisfying (5) are essentially unique.

It is important to understand the structure of the Hilbert space with respect to H_0 and $\mathbf{k}$.

Let $\theta = e/|e|$ and suppose $\mathbf{B}$ points in the positive z-direction. Define $\beta = |e\mathbf{B}|/2$ and introduce

$$b = (\theta k_y + ik_x)/2\beta^{1/2}, \tag{6}$$

$$a = \frac{1}{2}\{\beta^{-1/2}(p_x + i\theta p_y) - i\beta^{1/2}(x + i\theta y)\}.$$

Then, a^* and b^* are independent creation operators, and we have

$$H_0 = (|e\mathbf{B}|/m)\left(a^*a + \frac{1}{2}\right) + p_z^2/2m.$$

We can write, with an obvious notation,

$$L^2(\mathbf{R}^3, d^3 x) = \mathscr{H} = \mathscr{H}_a \otimes \mathscr{H}_b \otimes L^2(\mathbf{R}, dz). \tag{7}$$

Thus, not only does $\mathbf{k}_\perp$ commute with H_0, but is completely independent of H_0, in the sense that H_0 operates on $\mathscr{H}_a \otimes L^2(\mathbf{R}, dz)$ and $\mathbf{k}_\perp$ on $\mathscr{H}_b$.

2. *N* PARTICLES

We consider a Hamiltonian of the form

$$H = \sum_{i=1}^{N}(\mathbf{p}_i - e_i\mathbf{A}_i)^2/2m_i + \sum_{i<j} V_{ij}(\mathbf{x}_i - \mathbf{x}_j),$$

with $\mathbf{A}_i = \frac{1}{2}\mathbf{B} \times \mathbf{x}_i$. The pseudomomentum is then defined by

$$\mathbf{k} = \sum_{i=1}^{N} \mathbf{k}_i.$$

We find the commutation relations ($Q = \sum_{i=1}^{N} e_i$)

$$\mathbf{k} \times \mathbf{k} = -iQ\mathbf{B},$$

$$[\mathbf{k}, H] = 0.$$

Thus, **k** is again a constant of the motion. A new possibility arises here which was not present with only one particle, namely, when $Q = 0$, the components of **k** commute. We discuss this case first.

3. $Q = 0$

Since the components of **k** commute with each other and with H, we can write $\mathcal{H} = L^2(\mathbb{R}^{3N})$ as a direct integral over the spectrum of **k** with H decomposable. It turns out that one can use a constant-fiber direct integral with Lebesgue measure:

$$\mathcal{H} = \int_{\mathbb{R}_3}^{\oplus} \mathcal{H}_{\mathbf{k}} d^3k, \ \mathcal{H}_{\mathbf{k}} \text{ constant},$$

$$H = \int_{\mathbb{R}^3}^{\oplus} H(\mathbf{k}) d^3k,$$

$$U(\boldsymbol{\alpha}) = \int_{\mathbb{R}^3}^{\oplus} e^{i\boldsymbol{\alpha}\cdot\mathbf{k}} d^3k.$$

Proof. We consider the case of N particles in two dimensions, with **B** in a fictitious third dimension. The third dimension can be dealt with as in the case $\mathbf{B} = 0$. For simplicity, we also assume all particles are charged.

In the notation of Sec. 1, we have

$$\mathcal{H} = L^2(\mathbb{R}^{2N}, dx^{2N}) = \mathcal{H}_{a_1} \otimes \cdots \otimes \mathcal{H}_{a_N} \otimes \mathcal{H}_{b_1} \otimes \cdots \otimes \mathcal{H}_{b_N},$$

where

$$a_j = \frac{1}{2}\{\beta_j^{-1/2}(p_{jx} + i\theta_j p_{jy}) - i\beta_j^{1/2}(x_j + i\theta_j y_j)\},$$

$$b_j = (\theta_j k_{jy} + ik_{jx})/2\beta_j^{1/2}.$$

Since $\mathbf{k}_j$ is linear in b_j and b_j^*, $\mathbf{k}_j$ acts in $\mathcal{H}_{b_j}$. Let $\mathcal{H}' = \mathcal{H}_{b_1} \otimes \cdots \otimes \mathcal{H}_{b_n}$, $\xi_x = (e_1B)^{-1}k_{1y}$, $\xi_y = -(e_2B)^{-1}k_{2x}$. One easily verifies the commutation relations

$$[\xi_x, \xi_y] = [\xi_x, k_y] = [\xi_y, k_x] = 0,$$

$$[\xi_x, k_x] = [\xi_y, k_y] = i.$$

Thus, the vectors $\boldsymbol{\xi}$ and $\mathbf{k} = \mathbf{k}_1 + \cdots + \mathbf{k}_N$ satisfy the canonical commutation relations, with $\boldsymbol{\xi}$ the position and **k** the momentum. By von Neumann's theorem [8], we have

$$\mathcal{H}' = \mathcal{H}'' \otimes L^2(\mathbb{R}^2, d^2k)$$

where $\boldsymbol{\xi} = I \otimes i\nabla_{\mathbf{k}}$, $\mathbf{k} = I \oplus M_{\mathbf{k}}$ and here $M_{\mathbf{k}}$ means multiplication by the components of **k**. We let $\mathcal{H}_{\mathbf{k}} = \mathcal{H}_{a_1} \otimes \cdots \otimes \mathcal{H}_{a_1} \otimes \mathcal{H}''$ so that

$$\mathcal{H} = \int_{\mathbb{R}^2}^{\oplus} \mathcal{H}_{\mathbf{k}} d^2k, \ U(\boldsymbol{\alpha}) = \int_{\mathbb{R}^2}^{\oplus} e^{i\mathbf{k}\cdot\boldsymbol{\alpha}} d^2k, \ H = \int_{\mathbb{R}^2}^{\oplus} H(\mathbf{k}) d^2k.$$

Remark. Note that $H_0 = \sum_{i=1}^{N} (\mathbf{p}_i - e_i\mathbf{A}_i)^2/2m_i$ acts in $\mathcal{H}_{a1} \otimes \cdots \otimes \mathcal{H}_{a_N}$ so that $H_0(\mathbf{k})$ is independent of **k** (this depends on the fact that we have removed the third dimension). The **k** dependence of $H(\mathbf{k})$ arises from the potentials V_{ij}, i.e., from the dependence of $\mathbf{x}_i - \mathbf{x}_j$ on **k**.

As a concrete example, we look at the two-body problem. We take $e_1 = -e_2 = 1$. Then $H(\mathbf{k})$ is unitarily equivalent to

$$\tilde{H}(\mathbf{k}) = (\mathbf{p} - \mathbf{A})^2/2m_1 + (\mathbf{p} + \mathbf{A})^2/2m_2 + k_{\parallel}^2/2m + V(\mathbf{r} - \boldsymbol{\beta}) \tag{8}$$

where $\tilde{H}(\mathbf{k})$ operates in $L^2(\mathbb{R}^3, d^3r)$, $\mathbf{p} = -i\nabla$, $\mathbf{A} = \frac{1}{2}\mathbf{B} \times \mathbf{r}$, $M = m_1 + m_2$, $k_{\parallel}$ is the component of **k** parallel to **B** and $\boldsymbol{\beta} = \mathbf{k} \times \mathbf{B}/B^2$.

There are two spectral results which easily follow from (8).

Proposition. *Suppose* $V \in (L^2 + L^\infty)_\epsilon$, *and let* $E(\mathbf{k}) = \inf \sigma(H(\mathbf{k}))$ *and*

$$\mu = m_1 m_2/(m_1 + m_2).$$

Then

(*a*) $\sigma_{ess}(H(\mathbf{k})) = [|\mathbf{B}|/2\mu + k_{\|}^2/2M, \infty)$,

(*b*) $\lim_{\mathbf{k}_\perp \to \infty} E(\mathbf{k}) = |\mathbf{B}|/2\mu + k_{\|}^2/2M$.

These results follow easily once the Hamiltonian $H_0(\mathbf{k}) = (2m_1)^{-1}(\mathbf{p}-\mathbf{A})^2 + (2m_2)^{-1}(\mathbf{p}+\mathbf{A})^2 + k_{\|}^2/2M$ is analyzed. We have $\sigma_{ess}(H_0(\mathbf{k})) = [|\mathbf{B}|/2\mu + k_{\|}^2/2M, \infty)$ and $\inf \sigma(H_0(\mathbf{k})) = |\mathbf{B}|/2\mu + k_{\|}^2/2M$. Thus, (a) is a standard compactness result while (b) follows from the fact that as $\mathbf{k}_\perp \to \infty, V(\mathbf{x}-\boldsymbol{\beta}) \to 0$ in some sense.

Part (b) contrasts sharply with the case where $\mathbf{B} = \mathbf{0}$ where $E(\mathbf{k}) = \text{const} + k_{\|}^2/2M + k_\perp^2/2M$.

We now go on to analyze the case where $Q \neq 0$.

4. $Q \neq 0$

The one-body problem with $V = 0$, i.e., the case discussed in Sec. 1, has an infinite degeneracy associated with the fact that with $\mathcal{H} = (\mathcal{H}_a \otimes L^2(dz)) \otimes \mathcal{H}_b$ we have

$$H_0 = h_0 \otimes I.$$

We will show that this result is of general validity for $N > 1$ particles if $Q \neq 0$. As usual, let $\mathbf{k} = \sum_{i=1}^{N} \mathbf{k}_i$ and $\mathbf{B}$ in the positive z-direction. Then, $[k_x, k_y] = -iQB$. If $\xi = -QBk_x$, $\eta = k_y$, we have $[\xi, \eta] = i$; thus, ξ and η satisfy the CCR. By von Neumann's theorem, $\mathcal{H} = \mathcal{H}_0 \otimes L^2(\mathbb{R}, d\xi)$, where ξ is multiplication by ξ in $L^2(\mathbb{R}, d\xi)$ and $\eta = -i\frac{d}{d\xi}$. Since e^{itH} commutes with all operators of the form $I \otimes B$, with B an arbitrary bounded operator, we must have $e^{itH} = e^{ith} \otimes I$ for some h. Thus, $H = h \otimes I$ and, again, we have an infinite degeneracy. This result still holds after the center of mass in the z-direction is separated out, i.e., for the operator $H - k_{\|}^2/2M$, as a bit of thought shows.

5. THE HVZ THEOREM

In this section, we discuss an analog of the HVZ theorem [9-11] locating the bottom of the essential spectrum for multiparticle Hamiltonians. The first thing one wants to do is find suitable "reduced" Hamiltonians which will in general have some discrete spectrum. That is, we must do the analog of removing the center-of-mass motion in the case $\mathbf{B} \neq \mathbf{0}$. For $Q = 0$, a suitable reduced Hamiltonian is the operator $H(\mathbf{k})$. In the case of nonzero total charge, we must be more explicit in removing the center of mass motion in the $\mathbf{B}$-direction. Thus, consider the operator $H - k_{\|}^2/2M$. $\mathcal{H}$ can be written as $\mathcal{H}_1 \otimes L^2(\mathbb{R}, dk_{\|})$ so that $H - k_{\|}^2/2M = \tilde{H} \otimes I$. We now apply the same considerations as in Sec. 4 to $\tilde{H}$ and write $\mathcal{H}_1 = \mathcal{H}_0 \otimes L^2(\mathbb{R}, d\xi)$ so that

$$\tilde{H} = H' \otimes I.$$

It is H' which is the right operator to consider when looking for discrete spectrum.

Given a partition of $\{1, \cdots, N\}$ into two nonempty clusters C_1 and C_2 we define

$$H_{C_1, C_2} = H - V_{C_1, C_2},$$

where V_{C_1, C_2} is the sum of all V_{ij} which connect C_1 and C_2.

In the following, we assume all V_{ij} to be in $(L^2 + L^\infty)_\epsilon$. For $Q = 0$, the operators $H_{C_1, C_2}(\mathbf{k})$ are defined just as $H(\mathbf{k})$.

Theorem. *Suppose* $Q = 0$. *Then*

$$\inf \sigma_{ess}(H(\mathbf{k})) = \inf_\alpha \{\inf \sigma(H_{C_1^\alpha, C_2^\alpha}(\mathbf{k}))\},$$

where the infimum is taken over all partitions of $\{1, \cdots, N\}$ *into nonempty clusters* C_1^α and C_2^α.

In the case $Q \neq 0$, we have

Theorem. $\inf \sigma_{ess}(H') = \inf_\alpha \{\inf \sigma(H_{C_1^\alpha, C_2^\alpha})\}$

I will indicate how to prove the inequalities $\inf \sigma_{ess} \geq \inf_{\alpha} \{\inf \sigma(H_{C_1^\alpha, C_2^\alpha})\}$. I will make use of the following proposition, which may be interesting for its own sake, for the case $Q = 0$:

Proposition. *Suppose* $A = \int_{\mathbf{R}^n}^{\oplus} A(\mathbf{p}) d^n p$ *on* $\mathscr{H} = \int_{\mathbf{R}^n}^{\oplus} \mathscr{H}_{\mathbf{p}} \, d^n p$ *where* $\mathscr{H}_{\mathbf{p}} = \mathscr{H}_0$ *is independent of* $\mathbf{p}$. *Suppose in addition that*

(a) for all $\phi_j \in C_0^\infty(\mathbf{R}^n)$, $M_{\phi_2} A \phi_1(\mathbf{x})$ *is compact (here,* $\mathbf{x} = i\nabla_{\mathbf{p}}$ *and* $(M_{\phi_2} f)(\mathbf{p}) = \phi_2(\mathbf{p}) f(\mathbf{p})$*);*

(b) $A(\mathbf{p})$ *is norm-continuous in* $\mathbf{p}$.

Then $A(\mathbf{p})$ *is compact for all* $\mathbf{p}$.

Sketch of Proof. Suppose $\eta \in S(\mathbf{R}^n)$, with $\|\eta\|_2 = 1$. Define the projection P_η on $\mathscr{H}$ by

$$P_\eta f(\mathbf{p}) = (\eta, f(\cdot))\eta(\mathbf{p}),$$

where (,) is the $L^2(\mathbf{R}^n, d^n p)$ inner product, so that $(\eta, f(\cdot)) \in \mathscr{H}_0$. Without loss of generality, we can assume $A\phi(\mathbf{x})$ is compact for all $\phi \in C_0^\infty$. Let $A_\eta = AP_\eta$. We first show that A_η is compact. Thus, if $\phi \in C_0^\infty$ with $\phi(\mathbf{x}) = 1$ for $|\mathbf{x}| \leq 1$, let $\phi_j(\mathbf{x}) = \phi(\mathbf{x}/j)$. We have

$$B_j = A\phi_j(\mathbf{x})P_\eta$$

is compact for each j and, since $\|(\phi_j(\mathbf{x}) - 1)\eta\|_2 \to 0$ as $j \to \infty$,

$$\lim_{j \to \infty} \|B_j - A_\eta\| = 0,$$

and, hence, A_η is compact.

Now choose η with $\eta(\mathbf{p}_0) > 0$ and suppose $\{P_m'\}$ is a sequence of finite-dimensional orthogonal projections in $\mathscr{H}_0$ with $P_m' \uparrow I$; thinking of $\mathscr{H}$ as $\mathscr{H}_0 \otimes L^2(\mathbf{R}^n, d^n p)$, let $P_m = P_m' \otimes I$.

I now want to show that, as $m \to \infty$,

$$\|A(\mathbf{p}_0)(1 - P_m')\| \to 0,$$

using the fact that $\|A_\eta(1 - P_m)\| \to 0$. If B is a closed ball centered at $\mathbf{p}_0$ where $\eta(\mathbf{p})$ is strictly positive and $f \in \mathscr{H}$ is such that $f(\mathbf{p})$ equals f_0 in B and vanishes outside B, we have $(\eta, f(\cdot)) = \left(\int_B \eta(\mathbf{p}) d^n \mathbf{p}\right) f_0 = \gamma f_0$ and

$$\|A_\eta(1 - P_m)f\|^2 = \int_B \|A(\mathbf{p})(I - P_m')f_0\|_{\mathscr{H}_0}^2 \gamma^2 |\eta(\mathbf{p})|^2 d^n p \geq c \int_B \|A(\mathbf{p})(1 - P_m')f_0\|_{\mathscr{H}_0}^2 d^n p,$$

with $c > 0$. From this inequality, a bit of fooling with subsequences and an explicit use of the norm continuity of $A(\mathbf{p})$ easily gives $\|A(\mathbf{p}_0)(1 - P_m')\| \to 0$. Since $A(\mathbf{p}_0)P_m'$ is finite rank, the proof is complete.

In discussing the proof of the HVZ theorem for $Q = 0$, for simplicity we restrict to the case of N 2-dimensional particles in discussing how to show $\inf \sigma_{ess}(H(\mathbf{k})) \geq \inf_{\alpha} \{\inf \sigma(H_{C_1^\alpha, C_2^\alpha}(\mathbf{k}))\}$. In proving the result, we can assume all $V_{ij} \in C_0^\infty$ by a limiting argument. In the Weinberg-Van winter-equation approach [9,10], the major element in the proof in the case $\mathbf{B} = \mathbf{0}$ is to show that an operator of the form

$$A = V_1(z - H_0)^{-1} V_2(z - H_0)^{-1} \cdots\cdots V_m(z - H_0)^{-1}$$

is compact when A is "connected" in the sense that

$$\prod_{i=1}^{m} V_i = 0 \text{ for } \sum_{i \leq j} |\mathbf{x}_i - \mathbf{x}_j| \text{ large.}$$

In our case, we need the same result when A is restricted to a fiber $\mathscr{H}_{\mathbf{k}} = \mathscr{H}_0$, i.e., for $A(\mathbf{k})$. Hypothesis (b) of the proposition can be shown to be satisfied and, thus, in the notation of Sec. 3, we need only show that $A\phi(\boldsymbol{\xi})$ is compact for all $\phi \in C_0^\infty(\mathbf{R}^2)$, as an operator on $\mathscr{H}$. The function ϕ introduces the necessary fall-off in the nontranslation invariant variables to give compactness. We leave the details to the reader.

In the case $Q \neq 0$, a similar but less complicated procedure gives the same type of result (for details, see [1]).

6. CONCLUDING REMARKS

The pseudomomentum is a rather elusive object and it is hoped that some clever experimentalist will come up with a way of measuring it. In closing, I would like to remark that in radiative transitions, it is the pseudomomentum plus photon momentum which is conserved. This can be shown explicitly in a theory where the Hamiltonian is to the form

$$H = \sum_i (\mathbf{p}_i - e_i \mathbf{A}_i - \mathbf{A}(\mathbf{x}_i))^2/2m_i + \sum_{i<j} V_{ij}(\mathbf{x}_i - \mathbf{x}_j) + H_{0,\mathrm{rad}}$$

where $\mathbf{A}(\mathbf{x})$ is the quantized radiation field and $H_{0,\mathrm{rad}}$ is the free energy of the radiation field. Again see [1] for details.

The Proposition in Sec. 5 has a converse. Namely, if $A(\cdot)$ is norm-continuous, then $M_{\phi_2} A \phi_1(\mathbf{x})$ is compact for all $\phi_1, \phi_2 \in C_0^\infty(\mathbb{R}^n)$.

REFERENCES

*Partially supported by the National Science Foundation Grant MCS 78-00101

[1] J. Avron, I. Herbst, B. Simon: Ann. Phys. (N.Y.) **114**, 431 (1978)
[2] M. Boon: J. Math. Phys. **13**, 1268 (1972)
[3] E. Brown: in *Solid State Physics,* ed. by E. Ehrenreich, F. Seitz (Academic, New York, 1968)
[4] A. Grossman: in *Statistical Mechanics and Field Theory,* ed. by R. N. Sen, C. Weil (Halsted, New York, 1972)
[5] J. Zak: *Solid State Physics,* ed. by E. Ehrenreich, F. Seitz (Academic, New York, 1972), Vol. 27
[6] J. Zak: Phys. Rev. **134**, 1602 (1964); **135**, 776 (1964)
[7] H. Grotsch, R. A. Hegstrom: Phys. Rev. A **4**, 59 (1971)
[8] J. von Neumann: Math. Ann. **104**, 570 (1931)
[9] W. Hunziker: Helv. Phys. Acta **39**, 451 (1966)
[10] C. Van Winter: Mat.-Fys. Skr. Danske Vid. Selsk. **1** (8), 1 (1964)
[11] G. Zhislin: Tr. Mosk. Mat. Obs. **9**, 81 (1960)

WAVE OPERATORS FOR MULTICHANNEL LONG-RANGE SCATTERING

A. W. Sáenz

Naval Research Laboratory
Washington, District of Columbia 20375

ABSTRACT

We consider a multichannel scattering system composed of $N \geqslant 2$ spinless, distinguishable, nonrelativistic particles, each with configuration space $\mathbf{R}^3$ and interacting pairwise by local potentials $V_{ij}(1 \leqslant i < j \leqslant N)$ consisting of long-range and short-range parts. Each V_{ij} can be chosen, roughly, with the same degree of generality as in ALSHOLM [3], when the configuration space in the latter reference is taken as $\mathbf{R}^3$. Using techniques similar to those of ALSHOLM [2,3], we have proved for the class of potentials considered that suitable modified wave operators $\Omega_\alpha^\pm$ exist and have a generalized intertwining property for each channel α such that the corresponding bound states have a mild decay property of infinity. For the present class of V_{ij}'s, this property is known to be possessed by those bound states corresponding to eigenvalues of the discrete spectrum of the pertinent cluster Hamiltonian, or even to arbitrary nonthreshold eigenvalues if in addition the V_{ij}'s are dilatation analytic. We have also proved that the usual range-orthogonality property of the $\Omega_\alpha^\pm$'s holds under the conditions stated below. The results of this paper can be readily generalized to the case when the single-particle configuration space is $\mathbf{R}^\nu(\nu \geqslant 1)$.

INTRODUCTION

It is well known that DOLLARD [1] was the first to define and prove the existence of modified Møller wave operators for both single-channel and multichannel scattering by certain Coulombic potentials, and also the first to prove asymptotic completeness for single-channel scattering by pure Coulomb potentials. For single-channel scattering by local long-range potentials, ALSHOLM [2,3] obtained the most general results on the existence of modified wave operators defined by iteration, and even more general results were obtained by HÖRMANDER [4] on the existence of *some* kind of modified wave operators [5]. The existence of such operators for a class of long-range momentum-dependent potentials has been proved in a single-channel context [6]. Furthermore, asymptotic completeness has been established in this context for certain local potentials of long range [7].

Our main purpose is to prove the existence of multichannel wave operators for a wide class of local pair potentials. DOLLARD's proof [1] of the existence of such operators explicitly assumed, for channels other than the free channel, a certain property of the pertinent k-body $(2 \leqslant k < N)$ bound-state wave functions, related to their decay at large distances [8,9]. A similar assumption is made in the present paper.

A published proof [10] of the existence of modified wave operators for multichannel scattering systems whose particles interact by a rather general type of long-range local potentials makes no assumptions about bound states. Unfortunately, this proof is incorrect.

In this paper, we consider $N \geqslant 2$ distinguishable, spinless, nonrelativistic particles, each having configuration space $\mathbf{R}^3$ and interacting pairwise by local potentials $V_{ij}(1 \leqslant i < j \leqslant N)$, each of which is the sum of a long-range and a short-range part. Our assumptions on the V_{ij}'s are roughly the same as those made by ALSHOLM [3] when the configuration space in the latter reference is specialized to $\mathbf{R}^3$ [11]. In particular, given $\gamma \in (0, 1]$, each $V_{ij}(\mathbf{x})$ can be $O(|\mathbf{x}|)^{-\gamma}$ at infinity, and pure Coulomb potentials are allowed. In Sec. 1, we define modified channel wave operators $\Omega_\alpha^\pm$. Our main results, stated in Sec. 1, are as follows: (1) the existence of $\Omega_\alpha^\pm$ for the assumed class of potentials, for each channel α whose bound states satisfy a certain mild decay property at infinity; (2) a generalized intertwining property of the $\Omega_\alpha^\pm$'s, holding under the same conditions assumed for existence; (3) the usual range-orthogonality property of the $\Omega_\alpha^\pm$'s, under slightly more restrictive conditions on the long-range parts of the V_{ij}'s.

Proofs of these results are sketched in Sec. 2.

1. DEFINITION OF THE MODIFIED WAVE OPERATORS AND STATEMENT OF THE MAIN RESULTS

Consider $N \geqslant 2$ distinguishable spinless particles whose Hamiltonian operator is given formally by

$$-\sum_{i=1}^{N} (2m_i)^{-1}\Delta_{x_i} + \sum_{1 \leqslant i<j \leqslant N} V_{ij}, \tag{1.1}$$

where each $V_{ij}(1 \leqslant i < j \leqslant N)$ is a multiplication operator in $L^2(\mathbb{R}^3)$ by $V_{ij}(\mathbf{x}_i - \mathbf{x}_j)$, with $\mathbf{x}_k \in \mathbb{R}^3$ the position vector of the kth particle. $V_{ij}(\cdot)(1 \leqslant i < j \leqslant N)$, denoted by V_{ij} henceforth, is a real-valued function on the individual-particle configuration space $\mathbb{R}^3$ of the form

$$V_{ij} = V_{ij}^L + V_{ij}^S. \tag{1.2}$$

For each $1 \leqslant i < j \leqslant N$, the long-range and short-range portions, V_{ij}^L and V_{ij}^S, of V_{ij} are real-valued functions having the respective properties (L) and (S) stated below.

(L) for some positive integer m,

$$V_{ij}^L \in C^m(\mathbb{R}^3) \tag{1.3a}$$

and

$$\left.\begin{aligned} |D^k V_{ij}^L(\mathbf{x})| \leqslant \text{const.}(1 + |\mathbf{x}|)^{-k-\gamma_k}, \\ k = 1, \dots, m, \end{aligned}\right\} \tag{1.3b}$$

at each $\mathbf{x} \in \mathbb{R}^3$. The first line of (1.3b) applies to all the partial derivatives D^k of the specified orders k and the "decay exponents" γ_k are constants obeying the inequalities

$$0 < \gamma_1 < 1,\ m\gamma_1 + \gamma_m > 1. \tag{1.3c}$$

Furthermore, if $\frac{1}{2} < \gamma_1 < 1$, they also obey the inequalities

$$\gamma_k - \gamma_{k+1} \leqslant \gamma_1,\quad k = 1, 2, \dots, m-1, \tag{1.3d}$$

when $m \geqslant 2$.

(S) V_{ij}^S admits the decomposition

$$V_{ij}^S = V_{ij}^{(1)} + V_{ij}^{(2)}, \tag{1.3e}$$

where

$$(1 + |\mathbf{x}|)^{1+\epsilon} V_{ij}^{(1)} \in L^2(\mathbb{R}^3) + L^\infty(\mathbb{R}^3), \tag{1.3f}$$

$$V_{ij}^{(2)} \in L^s(\mathbb{R}^3), \tag{1.3g}$$

for some constants $\epsilon > 0$ and $2 \leqslant s < 3\gamma_1$, provided that $2 < 3\gamma_1$. If $2 \geqslant 3\gamma_1$, then $V_{ij}^{(2)} = 0$.

Since each V_{ij} in (1.1) is such that (1.2), (L), and (S) obtain, it is well known that the differential operator (1.1) on $C_o^\infty(\mathbb{R}^{3N})$ has a unique self-adjoint extension in $\mathcal{H} = L^2(\mathbb{R}^{3N})$. We denote this extension by H.

Consider a fixed decomposition $D = \{C_1, \dots, C_n\}$ of the N particles into $n \geqslant 2$ clusters $C_1, \dots, C_n$. Define a linear mapping $x = (\mathbf{x}_1, \dots, \mathbf{x}_N) \to (X_D, \xi)$ of $\mathbb{R}^{3n}$ onto itself, with Jacobian of absolute value unity, where $X_D = (\mathbf{X}_1, \dots, \mathbf{X}_n) \in \mathbb{R}^{3n}$ specifies the center-of-mass vector $\mathbf{X}_l = M_l^{-1} \sum_{j \in C_l} m_j \mathbf{x}_j$ of the n_l particles in C_l $(l = 1, \dots, n)$, with $M_l = \sum_{j \in C_l} m_j$. For $n < N$, $\xi = (\boldsymbol{\xi}_1, \dots, \boldsymbol{\xi}_{N-n}) \in \mathbb{R}^{3(N-n)}$ specifies the positions of the particles in each cluster with respect to their center of mass, every $\boldsymbol{\xi}_r (r = 1, \dots, N-n)$ being a linear combination of differences $\mathbf{x}_j - \mathbf{x}_k$ such that j and k are in the same cluster. For $n_l \geqslant 2$, we write $\xi^{(l)}$ for the vector $(\boldsymbol{\xi}_r) \in \mathbb{R}^{3(n_l-1)}$ whose components $\boldsymbol{\xi}_r$ involve only differences of this kind with $j,k \in C_l$.

Let H_D be the "free" Hamiltonian corresponding to the decomposition D, i.e., the unique self-adjoint operator in $\mathcal{H}$ which is the extension of the operator on $C_o^\infty(\mathbb{R}^{3N})$ obtained from (1.1) by omitting all operators V_{ij} with i and j in different clusters of D. We write $h_l (l = 1, \dots, n)$ for the "internal" Hamiltonian of C_l, i.e., a self-adjoint operator in $L^2(\mathbb{R}^{3(n_l-1)})$ formally obtained from the operator of type (1.1) involving only $i,j \in C_l$ by eliminating the center-of-mass motion.

Let D be such that either $n = N$ or that $n < N$ and that each h_l with $n_l \geq 2$ has a nonempty point spectrum. Let α be a channel consistent [12] with D, i.e., a pair $\alpha = \alpha_D = (b_D, D)$, where b_D is empty for $n = N$ and is otherwise a set $\{\psi_l, n_l \geq 2\}$ of mutually orthogonal eigenstates of h_l, one for each l with $n_l \geq 2$. The corresponding channel subspace $\mathscr{H}_\alpha$ is the set of all $f_\alpha \in \mathscr{H}$ of the form

$$f_\alpha(x) = F(X_D) \prod_{\substack{1 \leq l \leq n, \\ n_l \geq 2}} \psi_l(\xi^{(l)}), \tag{1.4}$$

where $F \in L^2(\mathbb{R}^{3n})$ and where the product is replaced by unity for $n = N$.

In analogy with the single-channel case, we define modified channel wave operators $\Omega_\alpha^\pm$ as restrictions to $\mathscr{H}_\alpha$ of the respective operators

$$\operatorname*{s-lim}_{t \to \pm\infty} \exp(itH)\exp(-itH_D - iG_{D,t}^{(m)})P_\alpha$$

when they exist, where P_α is the orthogonal projection with domain $\mathscr{H}$ and range $\mathscr{H}_\alpha$. If they exist, the operators $G_{D,t}^{(r)}$ $(r = 0, 1, 2, \ldots)$ have the form

$$G_{D,t}^{(r)} = \Gamma_{D,t}^{(r)} \otimes I \tag{1.5}$$

when $\mathscr{H}$ is written as

$$\mathscr{H} = \mathscr{H}_{cm} \otimes \mathscr{H}_{rel},$$

$\mathscr{H}_{cm} = L^2(\mathbb{R}^{3n})$ being the space of center-of-mass motion and $\mathscr{H}_{rel} = L^2(\mathbb{R}^{3(N-n)})$ the space of relative motion of the particles in the various clusters with respect to the appropriate centers of mass ($\mathscr{H} \equiv \mathscr{H}_{cm}$ if $n = N$). In (1.5), I is the identity operator in $\mathscr{H}_{rel}$ and, when it exists, $\Gamma_{D,t}^{(r)}$ $(r = 0, 1, 2, \ldots)$ is a bounded self-adjoint operator of multiplication by $\Gamma_{D,t}^{(r)}(\cdot): \mathbb{R}^{3n} \to \mathbb{R}$ in the momentum-space representation of $\mathscr{H}_{cm}$. Each function $\Gamma_{D,t}^{(r)}(\cdot)$ is defined recursively by

$$\begin{aligned} &\Gamma_{D,t}^{(0)}(P) = 0, \\ &\Gamma_{D,t}^{(r)}(P) = \int_o^t V(sM^{-1}P + \nabla\Gamma_{D,t}^{(r-1)}(P))\,ds, \\ &r \geq 1, \end{aligned} \tag{1.6}$$

for $P = (\mathbf{P}_1, \ldots, \mathbf{P}_n) \in \mathbb{R}^{3n}$. Here, $M^{-1}P = (M_i^{-1}\mathbf{P}_1, \ldots, M_n^{-1}\mathbf{P}_n)$ and $V: \mathbb{R}^{3n} \to \mathbb{R}$ is defined by

$$V(Z) = \sum_{r \neq s} \sum_{i \in C_r, i < j \in C_s} V_{ij}^L(\mathbf{Z}_r - \mathbf{Z}_s) \tag{1.7}$$

for $Z = (\mathbf{Z}_1, \ldots, \mathbf{Z}_n) \in \mathbb{R}^{3n}$. For a given $r,s = 1, \ldots, n$ with $r \neq s$, the rightmost sum in (1.7) runs over the set of all those i,j with $1 \leq i < j \leq N$ such that $i \in C_r$, $j \in C_s$, and is taken to be zero if this set is empty. The remaining sum in (1.7) runs over all such pairs r,s.

The definitions (1.4)-(1.7) were motivated by heuristic arguments, analogous to those used for similar purposes in the single-channel case [2,13] and in the multichannel case with Coulombic interactions [1]. In particular, $\Gamma_{D,t}^{(r)}$ is, formally speaking, the rth iterate of a solution of the equation

$$\frac{\partial\Gamma_{D,t}(P)}{\partial t} = V(sM^{-1}P + \nabla\Gamma_{D,t}(P)),$$

whose analogue in single-channel scattering was solved exactly by HORMÄNDER [4].

We close this section by stating our principal results.

Theorem 1. *Let each function V_{ij} $(1 \leq i < j \leq N)$ be of the form (1.2), with V_{ij}^L and V_{ij}^S obeying* (**L**) *and* (**S**), *respectively. Then $\Omega_\alpha^\pm$ exist for each channel α for which the bound states satisfy the condition*

$$\begin{aligned} &\int_{\mathbb{R}^{3n_l}} |\boldsymbol{\eta}_i|^{2+\delta} |\psi_l(\xi^{(l)})|^2 d\xi^{(l)} < \infty, \\ &i \in C_l,\ l = 1, \ldots, n,\ n_l \geq 2, \end{aligned} \tag{1.8}$$

for some $\delta > 0$, where $\boldsymbol{\eta}_i = \mathbf{X}_l - \mathbf{x}_i$ for $i \in C_l$ and $d\xi^{(l)}$ denotes Lebesgue measure in $\mathbb{R}^{3n_l}$.

Theorem 2. *Under the hypotheses on the V_{ij}'s in Theorem 1, the operators $\Omega_\alpha^\pm$ have the intertwining properties*

$$H\Omega_\alpha^\pm \subset \Omega_\alpha^\pm(H_\alpha + W_\alpha) \tag{1.9}$$

for each channel α for which they exist, where H_α is the restriction of H_D to $\mathscr{H}_\alpha$ and W_α is a bounded self-adjoint operator of multiplication in the momentum-space representation of $\mathscr{H}_{cm}$. Every W_α vanishes when the additional condition that $\lim_{|\mathbf{x}|\to\infty} V_{ij}(\mathbf{x}) = 0$ for each $1 \leqslant i < j \leqslant N$ is imposed.

Remarks. (1) Theorems 1 and 2 are multichannel analogues of, e.g., Theorems 1 and 3, respectively, of [3].

(2) Under the hypotheses on the V_{ij}'s in Theorem 1, each $\psi_l(n_l \geqslant 2)$ belonging to an eigenvalue of h_l in $\sigma_{\text{disc}}(h_l)$ decays exponentially in the L^2 sense, thus obeying the condition (1.9) for all $\delta > 0$ [14]. If each V_{ij} is as stated in the latter theorem and is in addition dilatation analytic (e.g., if $V_{ij}(\mathbf{x}) =$ const $|\mathbf{x}|^{-\gamma}(1 \leqslant i < j \leqslant N)$ for some $0 < \gamma \leqslant 1$ independent of i,j [15]) then each $\psi_l(n_l \geqslant 2)$ pertaining to a nonthreshold eigenvalue of h_l has this decay property [16].

(3) The definition of W_α in (1.9) is too lengthy to give here.

Proposition 1. *For each $1 \leqslant i < j \leqslant N$, let V_{ij}^L obey* (**L**) *and, in addition, have the property $\lim_{|\mathbf{x}|\to\infty} V_{ij}^L(\mathbf{x}) = 0$. Then if α and β are two different channels such that $\Omega_\alpha^\pm$ and $\Omega_\beta^\pm$ exist, the ranges of Ω_α^+ (respectively, Ω_α^-) and Ω_β^+ (respectively Ω_β^-) are orthogonal.*

A. Proof of Theorem 1

To prove the existence of each pair $\Omega_\alpha^\pm$ of modified channel wave operators under the hypotheses of Theorem 1, we adopt the following strategy:

(a) As in [2,3], we prove their existence under conditions stronger, as far as the V_{ij}'s are concerned, than those of Theorem 1. It can be shown that the existence of $\Omega_\alpha^\pm$ for any α under the stronger conditions entails their existence under those of the theorem.

(b) We reduce the existence problem for each D with $n < N$ to one no more complicated than for the case when $n = N$ (free channel).

The stronger conditions in question on the V_{ij}'s are as follows: For each $1 \leqslant i < j \leqslant N$, the function V_{ij} admits the decomposition (1.2), where V_{ij}^S obeys (**S**), but where now V_{ij}^L obeys the condition

(**L'**) For all $1 \leqslant i < j \leqslant N$,

$$V_{ij}^L \in C_o^\infty(\mathbf{R}^3) \tag{2.1a}$$

and

$$\left.\begin{array}{l}|D^k V_{ij}^L(\mathbf{x})| \leqslant \text{const}(1 + |\mathbf{x}|)^{-k-\gamma_k},\\ \quad k = 1,2,3,\ldots.\end{array}\right\} \tag{2.1b}$$

The infinite set $\{\gamma_k\}_{k=1}^\infty$ of decay exponents in (2.1b) is nonincreasing and concave, and satisfies both (1.3c) and

$$\gamma_k - \gamma_{k+1} \leqslant \gamma_1, \quad k = 1,2,3,\ldots. \tag{2.1c}$$

We proceed to sketch how to accomplish (b) in the typical case of Ω_α^+, where now and henceforth in this subsection α denotes a fixed channel consistent with a cluster decompostion D, with $2 \leqslant n < N$. Ω_α^+ will be shown to exist by proving that

$$Z_\alpha(t) = \left\| \frac{d}{dt}[\exp(itH)\exp(-itH_D - iG_{D,t}^{(m)})f_\alpha] \right\|_{\mathscr{H}} \tag{2.2}$$

is in $L^1(1,\infty)$, as a function of t. In (2.2), f_α is an arbitrary, but fixed element of $\mathscr{H}$ of the form (1.4), but with $F_\alpha \in S_D$. Here, S_D is the set of all $h \in \mathscr{H}_{cm}$ having Fourier transforms $\hat{h} \in C_o^\infty(\mathbf{R}^{3n})$ such that supp $\hat{h} \subset \{P = (\mathbf{P}_1, \ldots, \mathbf{P}_n) \in \mathbf{R}^{3n} | M_i^{-1}\mathbf{P}_i \neq M_j^{-1}\mathbf{P}_j, 1 \leqslant i < j \leqslant N\}$. (It is well known that S_D is dense in $\mathscr{H}_{cm}$.)

From now on, we write $X_D = X$ and drop the channel subscript in f_α, F_α, and Z_α. Standard arguments yield the inequality

$$Z(t) \leqslant Z_L(t) + Z_S(t) + Z'_S(t), \tag{2.4}$$

where

$$Z_L(t) = \|[V(X) - V(tM^{-1}(1/i)\nabla_X + \nabla\Gamma_{D,t}^{(m-1)})]\exp(-iY_t)F\|, \tag{2.5a}$$

$$Z_S(t) = \sum_{r\neq s}{}' \sum_{i\in C_r, i<j\in C_s} \|V_{ij}^S(\mathbf{X}_{rs} + \eta_{ij})(\exp(-iY_t)F)\phi\|_H, \tag{2.5b}$$

$$Z'_S(t) = \sum_{r\neq s}{}' \sum_{i\in C_r, i<j\in C_s} \|[V_{ij}^L(\mathbf{X}_{rs} + \eta_{ij}) - V_{ij}^L(\mathbf{X}_{ij})](\exp(-iY_t)F)\phi\|_{L^2(\mathbf{R}^{3N})}. \tag{2.5c}$$

Here and henceforth, $\|\ \|$ denotes the norm in $\mathscr{H}_{cm}$; $V(tM^{-1}P + \nabla\Gamma_{D,t}^{(m-1)})$ is multiplication by $V(tM^{-1}P + \nabla\Gamma_{D,t}^{(m-1)}(P))$ in the momentum-space representation of $\mathscr{H}_{cm}$; Y_t is the self-adjoint operator

$$Y_t = tK_D + \Gamma_{D,t}^{(m)}$$

in $\mathscr{H}_{cm}$, where K_D is the unique self-adjoint extension of $-\sum_{l=1}^{n} M_l^{-1}\Delta_{X_l}$ on $C_o^\infty(\mathbf{R}^{3n})$; and ϕ is the product $\prod_{1\leqslant l\leqslant n, n_l\leqslant 2} \psi_l$ of bound-state wave functions for the channel α considered, with $\|\phi\|_{L^2(\mathbf{R}^{3(N-n)})} = 1$. We have also written $\mathbf{X}_{rs} = \mathbf{X}_r - \mathbf{X}_s$ and $\mathbf{x}_i - \mathbf{x}_j = \mathbf{X}_{rs} + \eta_{ij}$ for $i \in C_r$, $j \in C_s$.

We now sketch the arguments used to prove that Z_L, Z_S, and Z'_S, are in $L^1(1,\infty)$, and, hence, that Z has this same property.

By a train of reasoning similar, but more complicated than that in [2,3] it follows that

$$Z_L(t) \leqslant \sum_{|\kappa|=0}^{m-1} (A_\kappa(t) + B_\kappa(t) + C_\kappa(t)) + D(t) + E(t) + R(t), \tag{2.6}$$

where A_κ, B_κ, C_κ, D, E, and R are nonnegative functions analogous to a_k, b_k, d_k, e, c, and r, respectively, in inequality (5.29), p. 39 of [2]. Here, κ is a multiindex $(\kappa_1, \ldots, \kappa_{3n})$ of $3n$ nonnegative integers and, as usual, $|\kappa| = \sum_{i=1}^{3n} \kappa_i$. For example,

$$A_\kappa(t) = \sum_{r\neq s} A_\kappa^{(r,s)}(t), \quad |\kappa| = 0, 1, \ldots, m-1, \tag{2.7}$$

where the sum runs over all C_r, C_s with $r \neq s (r,s = 1, \ldots, n)$. In particular,

$$A_\kappa^{(1,2)}(t) = \text{const}\,|t| \sum_{\kappa_1+\kappa_2=\kappa} \int_0^1 (1 + \lambda|\mathbf{r}|)^{-(k+2+\gamma_{k+2})} \exp(-i\,\mathscr{Y}_{t,\lambda})H_t^{(\kappa_1,\kappa_2)}\|d\lambda, \tag{2.8}$$

where $k = |\kappa| = 0, \ldots, m-1$. We are using the following notation for the relative and center-of-mass variables of the pair C_1, C_2:

$$\mathbf{r} = \mathbf{X}_{12},\ \mathbf{R} = (M_1 + M_2)^{-1}(M_1\mathbf{X}_1 + M_2\mathbf{X}_2),$$

$$X' = (\mathbf{R}, \mathbf{X}_3, \ldots, \mathbf{X}_n),$$

$$\mathbf{p} = (M_1 + M_2)^{-1}(M_2\mathbf{P}_1 - M_1\mathbf{P}_2),\ \mathbf{P} = \mathbf{P}_1 + \mathbf{P}_2,$$

$$P' = (\mathbf{P}, \mathbf{P}_3, \ldots, \mathbf{P}_n).$$

In (2.8), $H_t^{(\kappa_1,\kappa_2)}(\mathbf{r},X') = ((\nabla_1\Gamma_{D,t}^{(m)})^{\kappa_1}(\nabla_2\Gamma_{D,t}^{(m)})^{\kappa_2}F)(X)$, $(\nabla_i\Gamma_{D,t}^{(m)})^{\kappa_i}(i = 1,2)$ being multiplication by $(\nabla_{P_i}\Gamma_{D,t}^{(m)}(P))^{\kappa_i}$ in the momentum-space representation of $\mathscr{H}_{cm}$. The summation in (2.8) runs over all multiindices κ_1 and κ_2 such that $\kappa = \kappa_1 + \kappa_2$, in an obvious notation. $\mathscr{Y}_{t,\lambda}$ is multiplication by $(t/2\mu\lambda)|\mathbf{p}|^2 + \mathscr{G}_t(\mathbf{p},P')$ in this momentum-space representation, with $\mu = M_1M_2/(M_1 + M_2)$ and $\mathscr{G}_t(\mathbf{p},P') = \Gamma_{D,t}^{(m)}(P)$. The remaining functions $A_\kappa^{(r,s)}$ in (2.7) are given by formulas analogous to (2.8) in terms of the relative and center-of-mass variables of the relevant pair of clusters.

We claim that each $A_\kappa^{(r,s)}$ in (2.7) is in $L^2(1,\infty)$, so that each A_κ in (2.6) also has this property. Without loss of generality, it suffices to consider $r = 1, s = 2$. To prove that $A_\kappa^{(1,2)} \in L^2(1,\infty)$, one uses estimates of

$|D^{\gamma}\Gamma_{D,t}^{(r)}(P)|$ and $\|(1+\lambda|\mathbf{r}|)^{l}\exp(-i\mathcal{Y}_{t,\lambda})\|$ similar to those in [2,3], for suitable multiindices γ and suitable positive integers r,l, together with interpolation. Here, D^{γ} is a partial derivative of order $|\gamma|$ with respect to the argument P and $h(\mathbf{r},X')$ a function which, when expressed in terms of $\mathbf{X}_1,\ldots,\mathbf{X}_n$, is in S_D.

Similar arguments show that the remaining functions $B_\kappa,\ldots,R$ in (2.6) are also in $L^2(1,\infty)$. Hence, $Z_L \in L^2(1,\infty)$.

To show that $Z_S \in L^2(1,\infty)$, it suffices to prove that each summand $\|\,.\,\|_{L^2(\mathbb{R}^{3n})}$ in (2.5b) has this property. Without loss of generality, only the summand with $r=i=1, s=j=2$ need be considered (renumbering the particles, if necessary). Therefore, using (2.5b) and (S), we see that it suffices to show that

$$\int_0^1 \|v_i(\mathbf{r}+\eta_{12})(\exp(-iY_{t,1})\mathcal{F})\cdot\phi\|_{\mathcal{H}}dt < \infty, \tag{2.9}$$

where

$$\mathcal{F}(\mathbf{r},X') = F(X) \in S_D$$

and

$$(1+|\mathbf{x}|)^{1+\epsilon}|v_1(\mathbf{x})| \leqslant \text{const},$$
$$(1+|\mathbf{x}|)^{1+\epsilon}v_2 \in L^2(\mathbb{R}^3),$$
$$v_3 \in L^s(\mathbb{R}^3),$$

for some $\epsilon > 0$ and some $s \in (2, 3\gamma_1)$.

One can prove (2.9) for $i=1$, and also that $Z'_S \in L^1(1,\infty)$, by procedures similar, but simpler than those employed to show that $A_\kappa \in L^2(1,\infty)$. Property (2.9) can be proved for $i=2$ by using, in particular, the following elementary estimate. For all $\eta \in (0,1]$, $\lambda \in (0,1]$, and $t \geqslant 1$, there exists $\zeta \in (0,\eta]$ such that

$$\|(\exp(-i\mathcal{Y}_{t,\lambda})\mathcal{F})(\mathbf{r},\cdot)\|_{L^2(\mathbb{R}^{3(n-1)})} \leqslant \text{const}\ t^{-(1+\zeta)}(1+\lambda|\mathbf{r}|)^{1+\eta},$$

where $\mathcal{F}$ is as above. Finally, (2.9) can be proved for $i=3$ by using Hölder's inequality, the usual L_p estimate for $\exp it\Delta$, and an estimate for $\|(1+|\mathbf{r}|)^k\exp(-i\mathcal{Y}_{t,1})\mathcal{F}\|$ for $k>0$. This completes our sketch of the proof of Theorem 1.

B. Proof of Theorem 2 and Proposition 1

Theorem 2 can be proved by an approach analogous to that followed in [2,3] (see also [10]).

Proposition 1 can be established by a method similar to that used in [10].

ACKNOWLEDGMENTS

I particularly wish to thank Dr. W.W. Zachary for many valuable conversations and letters during the course of this work, which began while I was on sabbatical leave at Princeton University (1976-7). It is also a pleasure to thank Prof. P. Alsholm and Prof. W.O. Amrein for their valuable advice.

NOTES AND REFERENCES

[1] J.D. Dollard: Ph. D. thesis, Princeton University, 1963; J. Math. Phys. **5**, 729 (1964); Rocky Mt. J. Math. **1**, 5 (1971) [Erratum: Rocky Mt. J. of Math. **2**, 217 (1972)]
[2] P. Alsholm: "Wave operators for long-range scattering," Ph. D. thesis, University of California, Berkeley, 1972
[3] P. Alsholm: J. Math. Anal. Appl. **59**, 550 (1977)
[4] L. Hormänder, Math. Zeits. **146**, 69 (1976)
[5] For a review of investigations on modified wave operators for single-channel scattering by long-range potentials, see, e.g., W.O. Amrein: in *Scattering Theory in Mathematical Physics*, ed. by J. LaVita and J.-P. Marchand (Reidel, Dordrecht, Holland, 1974). See also [7]
[6] A.M. Berthier, P. Collet: Ann. Inst. Henri Poincaré **26**, 279 (1977)
[7] M. Reed, B. Simon: *Methods of Modern Mathematical Physics*, Vol. III (Academic, New York, 1979), Sec. XI. 9 and Notes to Sec. XI. 9

[8] A similar assumption has been made by J. H. Hendrickson, J. Math. Phys. **17**, 729 (1976) to establish the existence of modified wave operators for multichannel scattering systems with long-range time-dependent potentials
[9] For rigorous results on the decay of bound-state wave functions of multiparticle systems and a review of the literature, see M. Reed, B. Simon: *Methods of Modern Mathematical Physics,* Vol. IV (Academic, New York, 1978), Sec. XIII.11 and Notes to Sec. XIII.11
[10] W.W. Zachary: J. Math. Phys. **17**, 1056 (1976) [Erratum: J. Math. Phys. **18**, 536 (1977)]
[11] In this paper, the $\mathbb{R}^3$ case is considered for simplicity. The present work is easily generalized to the case of a single-particle configuration space $\mathbb{R}^\nu(\nu \geqslant 1)$ under conditions on the long-range parts of the V_{ij}'s roughly the same as in [3].
[12] W. Hunziker: J. Math. Phys. **6**, 6 (1965)
[13] V. Buslaev, V.B. Matveev: Teor. Mat. Fiz. **2**, 367 (1970) [Theoret. Math. Phys. **2**, 266 (1970)]
[14] See, e.g., Theorem XIII.39 of [9]
[15] See p. 234 of [9]
[16] See, e.g., Theorem XIII.41 of [9]

ASYMPTOTIC COMPLETENESS
IN CLASSICAL
THREE-BODY SCATTERING

W. W. Zachary

Naval Research Laboratory
Washington, District of Columbia 20375

INTRODUCTION

The present work is motivated by the paper of DEIFT and SIMON [1] in which the problem of proving asymptotic completeness in N-body quantum scattering is reduced to the proof of existence of certain strong limits closely related to the adjoints of the wave operators and certain spectral information about the M-body subsystems with $2 \leqslant M \leqslant N-1$. This "geometric" approach is time-dependent in character as contrasted with the time-independent character of the more customary approach involving resolvents. DEIFT and SIMON describe the difficulties involved in the application of the latter approach and we refer to their paper for these details as well as for references to recent work.

As of this writing, the existence of the limits proposed by Deift and Simon has not been proved for any N-body quantum system for all energies for $N \geqslant 3$, although similar geometric methods have been used to give new proofs of completeness in N-body systems below the lowest three-body threshold [2,3].

In the present work we will use an analogue of the Deift-Simon approach to sketch a proof of asymptotic completeness of three-body classical scattering systems for a certain class of short-range spherically symmetric pair potentials. It is hoped that this work will provide some insight into the corresponding quantum problem.

The only previous results on asymptotic completeness in N-body classical systems with $N \geqslant 3$ are those of HUNZIKER [4,5] and these are restricted to the case of finite range forces. It seems to be characteristic of the many-body classical scattering problem that one has to make strong assumptions—much stronger than in the corresponding quantum problem — in order to prove asymptotic completeness. We will also have to make strong assumptions, but we will be able to handle some forces with infinite range in the case $N = 3$. The difficulties arise in the treatment of channels with bound fragments. Such problems do not occur in the case $N = 2$ for which PROSSER [6] and SIMON [7,8] could prove asymptotic completeness for forces falling off at infinity faster than the Coulomb force. See also ANTONEČ [9].

Our approach consists of a generalization of the method of SIMON [7,8] to the case $N = 3$. We encounter a difficulty with bound systems which effectively limits the results to this value of N and spherically symmetric potentials.

Asymptotic completeness can be proved for three-body classical scattering by consideration of the following steps:

(A) Prove existence and uniqueness of solutions of Newton's equations with specified asymptotic conditions at $|t| = \infty$.

(B) Prove existence and other properties of the Møller wave transformations.

(C) Establish classical analogues of the results in [1] relating asymptotic completeness to the existence of certain limits closely related to the inverses of the Møller transformations.

(D) Prove asymptotic completeness by establishing the existence of the limits defined in (C).

1. PRELIMINARY CONSIDERATIONS AND DEFINITIONS

We first consider the kinematics. Denote by $\mathbf{r}_i$, $\boldsymbol{\kappa}_i$, and m_i (i = 1,2,3) the positions, momenta, and masses respectively of the three-body system. We will consider the three particles with the kinetic energy of their center-of-mass removed. Thus let (i,j,k) be in cyclic order, introduce pair notation $\alpha = (ij), \beta = (jk), \gamma = (ki)$, and define $\mathbf{x}_\alpha$ as the relative coordinates of the pair α, and $\mathbf{y}_\alpha$ as the position of the third particle relative to this pair. The momenta $\boldsymbol{\kappa}_\alpha$ and $\mathbf{p}_\alpha$ are conjugate to $\mathbf{x}_\alpha$ and $\mathbf{y}_\alpha$ respectively.

We define the free Hamiltonian,

$$H_o = \sum_{i=1}^{3} \frac{\mathbf{k}_i^2}{2m_i} = \frac{\boldsymbol{\kappa}_\alpha^2}{2\mu_\alpha} + \frac{\mathbf{p}_\alpha^2}{2n_\alpha},$$

the full Hamiltonian

$$H = H_o + \sum_{\alpha=1}^{3} \mathrm{V}_\alpha(\mathbf{x}_\alpha), \tag{1.1}$$

and the channel Hamiltonian H_α corresponding to the pair α

$$H_\alpha = H_o + \mathrm{V}_\alpha,$$

where μ_α and n_α denote reduced masses.

Using (1.1) we find the following expressions for Hamilton's equations relating to the coordinates $\mathbf{y}_\alpha, \mathbf{p}_\alpha, \mathbf{x}_\alpha$, and $\boldsymbol{\kappa}_\alpha$:

$$\begin{aligned}
&n_\alpha \dot{\mathbf{y}}_\alpha(t) = \mathbf{p}_\alpha(t)\\
&\dot{\mathbf{p}}_\alpha(t) = -\frac{\partial}{\partial \mathbf{y}_\alpha} \sum_{\beta\neq\alpha} V_\beta(x_\beta(t)) \equiv \sum_{\beta\neq\alpha} \mathbf{F}_{\beta\alpha}(x_\beta(t))\\
&\mu_\alpha \dot{\mathbf{x}}_\alpha(t) = \boldsymbol{\kappa}_\alpha(t)\\
&\dot{\boldsymbol{\kappa}}_\alpha(t) = \frac{-\partial}{\partial \mathbf{x}_\alpha} \sum_{\beta=1}^{3} V_\beta(\mathbf{x}_\beta(t)) \equiv \sum_{\beta=1}^{3} \mathbf{G}_{\beta\alpha}(\mathbf{x}_\beta(t))
\end{aligned} \tag{1.2}$$

where $\mathbf{x}_\beta, \beta\neq\alpha$, are to be expressed in terms of $\mathbf{x}_\alpha$ and $\mathbf{y}_\alpha$, and the dots denote differentiation with respect to time.

We will consider solutions of (1.2) which are asymptotic to "free" solutions in the limits $t\rightarrow\pm\infty$ For the case in which all particles are asymptotically free, i.e., for which no stable bound systems exist in these limits, the corresponding solutions are

$$\begin{aligned}
&\mathbf{y}_\alpha^{(0)}(t) = \mathbf{y}_\alpha^{(0)}(0) + \frac{\mathbf{p}_\alpha^{(0)}}{n_\alpha} t\\
&\mathbf{x}_\alpha^{0}(t) = \mathbf{x}_\alpha^{(0)}(0) + \frac{\boldsymbol{\kappa}_\alpha^{(0)}}{\mu_\alpha} t, \dot{\mathbf{p}}_\alpha^{(0)} = 0 = \dot{\boldsymbol{\kappa}}_\alpha^{(0)}.
\end{aligned} \tag{1.3}$$

SIMON [7] considered this case for $N \geqslant 2$.

We will generalize this approach to $N = 3$ where not all particles are asymptotically free. Thus, assume that the pair α is bound and that the third particle does not interact with it. The corresponding solution is

$$\begin{aligned}
&\mathbf{y}_\alpha^{(0)}(t) = \mathbf{y}_\alpha^{(0)}(0) + \frac{\mathbf{p}_\alpha^{(0)}}{n_\alpha} t, \dot{\mathbf{p}}_\alpha^{(0)} = 0,\\
&\mu_\alpha \ddot{\mathbf{x}}_\alpha^{(0)}(t) = -\frac{\partial}{\partial \mathbf{x}_\alpha^{(0)}} V_\alpha(\mathbf{x}_\alpha^{(0)}(t)) \equiv \mathbf{G}_{\alpha\alpha}(\mathbf{x}_\alpha^{(0)}(t))\\
&\boldsymbol{\kappa}_\alpha^{(0)}(t) = \mu_\alpha \dot{\mathbf{x}}_\alpha^{(0)}(t).
\end{aligned} \tag{1.4}$$

With our assumptions on the potentials and forces (see below), the phase space of the system is $\Gamma = \mathbb{R}^{12}$ and we will also consider the subset

$$\Gamma' = \{(\mathbf{y},\mathbf{p},\mathbf{x},\boldsymbol{\kappa})\in\Gamma : \mathbf{p}_\alpha \neq 0, \boldsymbol{\kappa}_\alpha \neq 0 \text{ for some } \alpha\}.$$

It is clear that Γ and Γ' differ by a set of Lebesgue measure zero.

The concept of bound system will be defined by linking it to boundedness of trajectories in configuration space. This is possible because it follows from (1.1), our assumptions on the potentials stated below, and conservation of energy that the momenta are bounded functions of time.

Thus, let $B_\pm$ denote the measurable subsets of Γ such that

$$\sup_{t \gtrless 0}(|\mathbf{y}_\alpha(t)| + |\mathbf{x}_\alpha(t)|) < \infty$$

respectively, for all α where $(\mathbf{y}_\alpha(t), \mathbf{x}_\alpha(t))$ are solutions of (1.2). The subset $B = B_+ \cap B_-$ is defined to be the region of phase space corresponding to bound systems. For N = 2 the corresponding subsets $b_\pm$ of the two-body phase space are defined by

$$b_\pm = \sup_{t \gtrless 0}|\mathbf{x}_\alpha(t)| < \infty \tag{1.5}$$

respectively.

Solutions of (1.2) define a self map F (t) of Γ:

$$F(t):(\mathbf{y}(0), \mathbf{p}(0), \mathbf{x}(0), \boldsymbol{\kappa}(0)) \rightarrow (\mathbf{y}(t), \mathbf{p}(t), \mathbf{x}(t), \boldsymbol{\kappa}(t)) \tag{1.6}$$

called the Hamiltonian flow. It is desirable that this flow be complete, i.e., that it be defined for all $t \in (-\infty, \infty)$. In this case F(t) is a one-parameter group of diffeomorphisms of Γ and the map (1.6) is a canonical transformation. The completeness of the flow is the classical analogue of the essential self-adjointness of the corresponding Hamiltonian in quantum mechanics. For scattering solutions this completeness follows from the existence and uniqueness of solutions of Newton's equations discussed in Sec. 2. The completeness of the flow for two-body bound systems can be proved with our assumptions on the forces stated below. See also the remarks on this point in [4].

We will consider the following class of potentials and forces.

(I_α) V_α is a continuously differentiable spherically symmetric bounded function which approaches zero as $|\mathbf{x}_\alpha| \rightarrow \infty$.

(II_α) The forces $\mathbf{F}_{\beta\alpha}$ ($\beta \neq \alpha$) and $\mathbf{G}_{\beta\alpha}$ are conservative as in (1.2). They are locally Lipschitz on $\mathbb{R}^3$ with

$$\sup_{\mathbf{x} \in \mathbb{R}^3}|\mathbf{F}_{\beta\alpha}(\mathbf{x})| < \infty, \sup_{\mathbf{x} \in \mathbb{R}^3}|\mathbf{G}_{\beta\alpha}(\mathbf{x})| < \infty,$$

i.e., given a compact subset $D \subset \mathbb{R}^3$ there exist positive constants $D'_{\beta\alpha}, D_{\beta\alpha}$ (depending upon D) such that

$$|\mathbf{F}_{\beta\alpha}(\mathbf{x}) - \mathbf{F}_{\beta\alpha}(\mathbf{y})| \leqslant D'_{\beta\alpha}|\mathbf{x}-\mathbf{y}|, \beta \neq \alpha,$$

$$|\mathbf{G}_{\beta\alpha}(\mathbf{x}) - \mathbf{G}_{\beta\alpha}(\mathbf{y})| \leqslant D_{\beta\alpha}|\mathbf{x} - \mathbf{y}|$$

for all $\mathbf{x}, \mathbf{y} \in D$.

(III_α) For some $R_0 > 0$ and positive constants $A'_{\beta\alpha}, A_{\beta\alpha}, B_{\beta\alpha}, B'_{\beta\alpha}$ with $\beta \neq \alpha$,

$$|\mathbf{F}_{\beta\alpha}(\mathbf{x}_\beta)| \leqslant A'_{\beta\alpha}\exp(-B'_{\beta\alpha}|\mathbf{x}_\beta|)$$

$$|\mathbf{G}_{\beta\alpha}(\mathbf{x}_\beta)| \leqslant A_{\beta\alpha}\exp(-B_{\beta\alpha}|\mathbf{x}_\beta|) \text{ if } |\mathbf{x}_\beta| \geqslant R_0.$$

(IV_α) For some $R_1 > 0$ and positive constants $C'_{\beta\alpha}, C_{\beta\alpha}, K_{\beta\alpha}, K'_{\beta\alpha}$ with $\beta \neq \alpha$,

$$|\mathbf{F}_{\beta\alpha}(\mathbf{x}) - \mathbf{F}_{\beta\alpha}(\mathbf{y})| \leqslant C'_{\beta\alpha}\exp(-K'_{\beta\alpha}r)|\mathbf{x} - \mathbf{y}|,$$

$$|\mathbf{G}_{\beta\alpha}(\mathbf{x}) - \mathbf{G}_{\beta\alpha}(\mathbf{y})| \leqslant C_{\beta\alpha}\exp(-K_{\beta\alpha}r)|\mathbf{x} - \mathbf{y}|, \text{ if } |\mathbf{x}|, |\mathbf{y}| \geqslant r \geqslant R_1.$$

Conditions (II_α)-(IV_α) are similar to those used by SIMON [7,8] except that we have imposed exponential fall-off at infinity. This strong decay at infinity is not needed in the two-body case.

Spherical symmetry is imposed because our proof of asymptotic completeness requires that the two-body bound systems be completely integrable. Two-body systems with spherically symmetric potentials are the most important class of three-dimensional integrable systems.

2. EXISTENCE AND UNIQUENESS OF SCATTERING SOLUTIONS

Theorem 2.1. *Assume (I_α)-(IV_α) for all α. Then for given solutions $(\mathbf{y}_\alpha^{(0)}(t),\ \mathbf{x}_\alpha^{(0)}(t))$ satisfying (1.3) or (1.4), there exist unique functions $(\mathbf{y}_\alpha(t),\ \mathbf{x}_\alpha(t))$ satisfying (1.2) with*

$$\lim_{t\to\pm\infty} |t|^n\{|\mathbf{u}_\alpha(t)| + |\dot{\mathbf{u}}_\alpha(t)| + |\mathbf{v}_\alpha(t)| + |\dot{\mathbf{v}}_\alpha(t)|\} = 0 \tag{2.1}$$

for any non-negative integer n, where

$$\left.\begin{aligned} \mathbf{u}_\alpha(t) &= \mathbf{y}_\alpha(t) - \mathbf{y}_\alpha^{(0)}(t), \\ \mathbf{v}_\alpha(t) &= \mathbf{x}_\alpha(t) - \mathbf{x}_\alpha^{(0)}(t). \end{aligned}\right\} \tag{2.2}$$

A proof for the asymptotic conditions (1.3) and $n = 0$ is discussed in [7]. The only essential difference in the method of proof is that we eliminate the kinetic energy of the center-of-mass whereas SIMON does not. With our more restrictive assumptions we obtain (2.1) for the stated range of n.

Our proof in the case of the asymptotic conditions (1.4) makes use of the usual existence and uniqueness theorems for ordinary differential equation and of fixed point, theorems to show that the mapping (in the case of the limit $t\to-\infty$)

$$\mathcal{F}\begin{bmatrix}\mathbf{u}_\alpha(t)\\ \mathbf{v}_\alpha(t)\end{bmatrix} = \int_{-\infty}^{t} ds \int_{-\infty}^{s} d\tau \begin{bmatrix} \dfrac{1}{n_\alpha}\sum\limits_{\beta\neq\alpha} \mathbf{F}_{\beta\alpha}(\mathbf{x}_\beta(\tau)) \\ \dfrac{1}{\mu_\alpha}\left[\mathbf{G}_{\alpha\alpha}(\mathbf{x}_\alpha^{(0)}(t) + \mathbf{v}_\alpha(t)) - \mathbf{G}_{\alpha\alpha}(\mathbf{x}_\alpha^{(0)}(t)) + \sum\limits_{\beta\neq\alpha} \mathbf{G}_{\beta\alpha}(\mathbf{x}_\beta(t))\right]\end{bmatrix} \tag{2.3}$$

has a unique fixed point. In addition to the stated conditions(I_α)-(IV_α), our proof of existence and uniqueness of fixed points requires the additional condition that $D_{\alpha\alpha}$ be small in comparision with $B'_{\beta\alpha}$ and $B_{\beta\alpha}$ for $\beta\neq\alpha$. If $\mathbf{G}_{\alpha\alpha}$ is differentiable, this is seen to be a restriction on the magnitude of $\dfrac{\partial}{\partial\mathbf{x}_\alpha}\mathbf{G}_{\alpha\alpha}$ as compared with the asymptotic rate of decay of $\mathbf{F}_{\beta\alpha}$ and $\mathbf{G}_{\beta\alpha}$. Unfortunately, we do not have the space to give a precise statement of this condition here. A similar expression with different limits of integration is used for the study of the limit $t\to+\infty$. In the proof we use a different metric space than in the corresponding two-body problem [7,8,11] or for $N > 2$ with the asymptotic conditions (1.3) [7].

3. EXISTENCE OF WAVE TRANSFORMATIONS

The existence of the Møller wave transformations express the fact that $F(t)$ can be approximated by simpler flows in the limits $t\to\pm\infty$. Thus, just as the flow $F(t)$ is associated with solutions of (1.2), the "free flows" $F_o(t)$ and $F_\alpha(t)$ are associated with the solutions (1.3) and (1.4) respectively. In particular, $F_\alpha(t)$ propagates the pair α according to the internal evolution of the bound system while the third particle travels freely. It is clear that $F_o(t)$ is complete whereas the completeness of $F_\alpha(t)$ can be proved as we have indicated in Sec. 1.

We define the following mappings for given $\mathbf{a}_\alpha$, $\mathbf{b}_\alpha'$, $\mathbf{c}_\alpha, \mathbf{d}_\alpha \neq 0$, $\mathbf{x}_\alpha^{(0)}(t)$, and $\boldsymbol{\kappa}_\alpha^{(0)}(t)$:

$$\Omega_0^+(\mathbf{a}_\alpha, n_\alpha\mathbf{b}_\alpha, \mathbf{c}_\alpha, \mu_\alpha\mathbf{d}_\alpha) = (\mathbf{y}_\alpha(0), \mathbf{p}_\alpha(0), \mathbf{x}_\alpha(0), \boldsymbol{\kappa}_\alpha(0))$$

$$\Omega_\alpha^+(\mathbf{a}_\alpha, n_\alpha\mathbf{b}_\alpha, \mathbf{x}_\alpha^{(0)}0), \boldsymbol{\kappa}_\alpha^{(0)}(0)) = (\mathbf{y}_\alpha(0), \mathbf{p}_\alpha(0), \mathbf{x}_\alpha(0), \boldsymbol{\kappa}_\alpha(0)),$$

i.e., mappings from the asymptotic data at $t = -\infty$ to the interacting solutions evaluated at zero time. Similar mappings $\Omega_0^-, \Omega_\alpha^-$ can be defined for asymptotic data specified at $t = +\infty$.

Let $\Gamma_\alpha\subset\Gamma'$ denote the product of the phase spaces of the bound pair and the free particle (with non-vanishing momenta). We have

Theorem 3.1. *Assume $(I_\alpha)-(IV_\alpha)$ for all α as well as the restriction on $D_{\alpha\alpha}$ noted in Sec. 2. Then:*

$$(a)\ \Omega_0^\pm = \lim_{t\to\mp\infty} F(-t)F_0(t) \tag{3.1}$$

uniformly on compact subset of Γ',

$$\Omega_\alpha^\pm = \lim_{t\to\mp\infty} F(t)F_\alpha(t), \alpha = 1,2,3, \tag{3.2}$$

uniformly on compact subsets of Γ_α.

(b) $F(t)\Omega_\alpha^\pm = \Omega_\alpha^\pm F_\alpha(t)$, $\alpha = 0,1,2,3$,

for all finite t.

(c) $\Omega_\alpha^\pm, \alpha = 0,1,2,3,$ *are measure preserving.*

(d) *The ranges of* Ω_α^+ and Ω_β^+ *(resp.* Ω_α^- and Ω_β^-*) are disjoint* for $\alpha \neq \beta$, $\alpha,\beta = 0,1,2,3$,

The existence of the limits for Ω_0^+ has been proved in [7] under different assumptions. The existence of $\Omega_\alpha^\pm$ for $\alpha \neq 0$ is proved in a similar fashion but is more complicated in its technical details. The idea is that

$$\Omega_\alpha(t) \equiv F(-t)F_\alpha(t)D_\alpha,$$

where D_α denotes a compact subset of Γ_α, can be written in terms of an integral mapping $F(t)$ which has the same integrand as (2.3) but different integration limits. (See [7] or [8] for the corresponding procedure in the two-body case.) $\mathcal{F}(t)$ has the property that it is a contraction on our metric space for finite t and converges to $\mathcal{F}$ uniformly as $t \to -\infty$. The existence of Ω_α^+ follows essentially from these properties and (2.1).

The assertions (b)-(d) can be proved from (a), Theorem 2.1, and some well-known properties of the flows.

4. CLASSICAL VERSION OF THE DEIFT-SIMON APPROACH

We now describe a classical version of the DEIFT-SIMON analysis [1] which reduces the proof of asymptotic completeness to the proof of existence of certain limits closely related to the inverses of the Møller transformations (3.1) and (3.2). A sketch of an existence proof of these limits will be given in Sec. 5.

We define

$$Q_\alpha(m,R) = \{(\mathbf{y}_\alpha, \mathbf{x}_\alpha): |\mathbf{x}_\alpha| \leqslant m|\mathbf{y}_\alpha|^{1/3},\ |\mathbf{y}_\alpha| \geqslant m^{-1}R\} \tag{4.1}$$

and consider smooth functions $J_\alpha \in C^\infty(\mathbf{R}^6)$ satisfying

$$0 \leqslant J_\alpha \leqslant 1$$

$$\operatorname{supp} J_\alpha \subset Q_\alpha(2,R)$$

$$J_\alpha = 1 \text{ on } Q_\alpha(1,R) \tag{4.2}$$

for $\alpha \neq 0$. The corresponding function for $\alpha = 0$ is given by

$$J_0 = 1 - \sum_{\alpha=1}^{3} J_\alpha. \tag{4.3}$$

The relations (4.1)-(4.3) are the same definitions as in [1]. We do not change the momentum part of phase space. Intuitively, J_α is a function that is large in regions where the particles comprising the pair α are close together compared with their distances from the third particle.

Lemma 4.1. *Assume* (I_α) *and* (II_α). *For* $\alpha \neq 0$ *we can find* $T > 0$ *such that*

(a) $J_\beta F_\alpha(t) = 0 = (1 - J_\alpha)F_\alpha(t)$

for $|t| \geqslant T$ *on compact subsets of* Γ_α *where* $\beta \neq \alpha$ *with* $\beta = 0$ *allowed.*

(b) $J_\alpha F_0(t) = 0 = (1 - J_0)F_0(t)$

for $|t| \geqslant T$ *on compact subsets of* Γ'.

The proof uses the definition of the flows, of J_α, and (1.5).

Note that in the presently considered classical case the limits in Lemma 4.1 become zero in finite time. This does not happen in the quantum case because of the spreading of wave packets.

Using the fact that the ranges of the wave transformations are disjoint by Theorem 3.1 (d) we say that the scattering system is *asymptotically complete* if

$$\Gamma \approx B\bigcup_{\alpha=0}^{3} \text{Ran } \Omega_\alpha^+ \approx B\bigcup_{\alpha=0}^{3} \text{Ran } \Omega_\alpha^-$$

where we use the notation $X \approx Y$ for subsets of Γ such that $X = Y$ up to sets of Lebesgue measure zero. (More precisely, the symmetric difference $X \bigcup Y - X \bigcap Y$ has measure zero.)

The next result relates this concept to the existence of certain limits $W_\alpha^\pm$.

Theorem 4.1. *Assume that asymptotic completeness holds. Then the limits*

$$W_\alpha^\pm = \lim_{t \to \mp\infty} F_\alpha(-t) J_\alpha F(t), \alpha = 0, 1, 2, 3,$$

exist in the topology of uniform convergence on compact subsets of $\Gamma' - B$. *Also,*

$$W_\alpha^\pm \, \Omega_\beta^\pm \Gamma_\beta = \delta_{\alpha\beta} \Gamma_\alpha \tag{4.4}$$

where Γ_0 *is defined to be* Γ'.

The proof of existence of $W_\alpha^\pm$ uses Lemma 4.1 and is similar to that for the quantum case [1]. The relations (4.4) then follow from the group property of $F(t)$ and Lemma 4.1.

In the other direction we have

Theorem 4.2. *Assume that the limits* $W_\alpha^\pm$ *exist for* $\alpha = 0, 1, 2, 3$ *and suppose that* (I_α)-(IV_α) hold for all α. *Then the three-body system is asymptotically complete.*

The proof goes essentially as in the quantum case. The assumptions (I_α)-(IV_α) are imposed so that we can use SIMON's result that the two-body subsystems are asymptotically complete. Our assumptions are stronger than necessary for this task, and we could just as well impose SIMON's conditions.

In the quantum analogue of Theorem 4.2 proved in [1] one finds the additional condition that the Hamiltonian of each two-body subsystem must have no singular continuous spectrum. A similar assumption is not needed in the classical case because such phenomena occur on sets of measure zero.

5. ASYMPTOTIC COMPLETENESS

We now sketch a proof of asymptotic completeness of classical three-body systems by discussing the existence of the limits $W_\alpha^\pm$.

Theorem 5.1. *Assume* (I_α)-(IV_α) *for all pairs* α *as well as the restriction on* $D_{\alpha\alpha}$ *noted in Sec. 2. Then the three-body system is asymptotically complete.*

One sees from Theorems 4.1 and 4.2 that if W_α^+ exists then it is the inverse of Ω_α^+, so that it is sufficient to consider solutions of Newton's equations with asymptotic conditions corresponding to the flow $F_\alpha(t)$.

Let D denote a compact subset of $\Gamma' - B$. In order to prove the existence of W_0^+ it is sufficient to prove the existence of the limits

$$\lim_{t \to -\infty} F_0(-t) F(t) D \tag{5.1}$$

$$\lim_{t \to -\infty} F_0(-t) J_\alpha F(t) D, \alpha = 1, 2, 3. \tag{5.2}$$

The appropriate solution of Newton's equations is

$$\begin{aligned} \mathbf{y}_\alpha(t) &= \mathbf{a}_\alpha + \mathbf{b}_\alpha t + \mathbf{u}_\alpha(t) \\ \mathbf{p}_\alpha(t) &= n_\alpha(\mathbf{b}_\alpha + \dot{\mathbf{u}}_\alpha(t)) \\ \mathbf{x}_\alpha(t) &= \mathbf{c}_\alpha + \mathbf{d}_\alpha t + \mathbf{v}_\alpha(t) \\ \boldsymbol{\kappa}_\alpha(t) &= \mu_\alpha(\mathbf{d}_\alpha + \dot{\mathbf{v}}_\alpha(t)) \end{aligned} \tag{5.3}$$

with $\mathbf{a}_\alpha, \mathbf{b}_\alpha, \mathbf{c}_\alpha, \mathbf{d}_\alpha \neq 0$ where $\mathbf{u}_\alpha$ and v_α satisfy (2.1) for the limit $t \to -\infty$. Using the property (1.3) of the flow $F_0(t)$ and (2.1) we find

$$\lim_{t\to-\infty} F_0(-t)F(t)(\mathbf{y}_\alpha(0),\mathbf{p}_\alpha(0),\mathbf{x}_\alpha(0),\boldsymbol{\kappa}_\alpha(0)) = (\mathbf{a}_\alpha, n_\alpha\mathbf{b}_\alpha, \mathbf{c}_\alpha, \mu_\alpha\mathbf{d}_\alpha).$$

The result for the limits (5.2),

$$\lim_{t\to-\infty} F_0(-t)J_\alpha F(t)D = 0,$$

arises from the fact that $J_\alpha = 0$ unless

$$|\mathbf{x}_\alpha(t)| \leqslant 2|\mathbf{y}_\alpha(t)|^{1/3}$$

and

$$|\mathbf{y}_\alpha(t)| \geqslant 1/2\ R \text{ for some } R>0.$$

One shows from (2.1) and (2.2) that the first inequality fails for sufficiently large $|t|$.

For W_α^+ with $\alpha\neq 0$ we consider the solution

$$\begin{aligned} \mathbf{y}_\alpha(t) &= \mathbf{a}'_\alpha + \mathbf{b}'_\alpha t + \mathbf{u}_\alpha(t) \\ \mathbf{p}_\alpha(t) &= \mathbf{n}_\alpha(\mathbf{b}'_\alpha + \dot{\mathbf{u}}_\alpha(t)) \\ \mathbf{x}_\alpha(t) &= \mathbf{x}_\alpha^0(t) + \mathrm{v}_\alpha(t) \\ \boldsymbol{\kappa}_\alpha(t) &= \mu_\alpha(\dot{\mathbf{x}}_\alpha^{(0)}(t) + \dot{\mathrm{v}}_\alpha(t)). \end{aligned} \tag{5.4}$$

From (1.5) and (2.1) we see that the inequalities

$$|\mathbf{x}_\alpha^{(0)}(t) + \mathrm{v}_\alpha(t)| \leqslant |\mathbf{a}'_\alpha + \mathbf{b}'_\alpha t + \mathbf{u}_\alpha(t)|^{1/3}$$

and

$$|\mathbf{a}'_\alpha + \mathbf{b}'_\alpha t + \mathbf{u}_\alpha(t)| \geqslant R \text{ for some } R>0$$

are valid for sufficiently large $|t|$, and consequently $J_\alpha = 1$ for such t. Thus, we only have to consider the existence of the limit.

$$\lim_{t\to-\infty} F_\alpha(-t)F(t)D. \tag{5.5}$$

We have previously alluded to the fact that the flow $F_\alpha(t)$ factorizes as the product of two simpler flows — that of a free particle and that of the internal flow of a two-body bound system. The convergence of that part of (5.5) involving the flow of the free particle is proved in the same manner as that for the limit (5.1) with the solution (5.3). For the bound part we use (5.4) to express this part of the flow as the sum of two terms — the first involving the solutions $\mathbf{x}_\alpha^{(0)}(t)$ and $\boldsymbol{\kappa}_\alpha^{(0)}(t)$ of the bound system and the second depending upon the quantities $\mathrm{v}_\alpha(t)$ and $\dot{\mathrm{v}}_\alpha(t)$ defined by (2.2). For the first set of terms it is seen that we have to have control over the internal flow of the two-body bound systems. In order to achieve this we use the fact that a two-body bound system with a spherically symmetric potential is completely integrable. (This is the main reason for assuming that the potentials are spherically symmetric in the present work.) We then make a canonical transformation from the coordinates $(\mathbf{x}_\alpha^{(0)}, \boldsymbol{\kappa}_\alpha^{(0)})$ to the action-angle variables $(\mathbf{I}_\alpha, \mathbf{w}_\alpha)$ in terms of which the flow of a bound system is

$$(\mathbf{I}_\alpha, \mathbf{w}_\alpha) \to (\mathbf{I}_\alpha, \mathbf{w}_\alpha + \boldsymbol{\nu}_\alpha t)$$

where $\boldsymbol{\nu}_\alpha$ are the associated frequencies. After the flow is evaluated in these variables we transform back to the original ones.

By collecting the preceding results one finds

$$W_\alpha^+(\mathbf{y}_\alpha(0),\mathbf{p}_\alpha(0),\mathbf{x}_\alpha(0),\boldsymbol{\kappa}_\alpha(0)) = F_\alpha(T)^{-1}(\mathbf{a}'_\alpha + \mathbf{b}'_\alpha T, n_\alpha\mathbf{b}'_\alpha, \mathbf{x}_\alpha^{(0)}(T), \boldsymbol{\kappa}_\alpha^{(0)}(T)).$$

6. NON-EXTENSION OF THE RESULTS TO N > 3

In conclusion we make a remark concerning the lack of extendability of our results to N-body systems with $N > 3$.

Note that the proof breaks down for channels which contain bound systems of three or more particles. The reason is, of course, that we make essential use of the fact that two-body bound systems with spherically symmetric potentials are completely integrable. For bound systems of three or more particles, integrable systems in three dimensions are harder to find. Stated more precisely, integrable Hamiltonian systems are not generic [12].

It is interesting to note that our proof breaks down for the same types of situations for which HUNZIKER [4] has trouble. That is, the bound system stability assumption formulated by Hunziker is satisfied for two-body systems with spherically symmetric potentials [4]. In fact, the details of our proofs of Theorems 3.1 and 5.1 require the use of this stability condition in an explicit manner. For bound systems of three or more particles the validity of Hunziker's stability condition remains an open question for forces with infinite range.

ACKNOWLEDGMENTS

I wish to thank A. W. Sáenz, C. Chandler, and I. W. Herbst for discussions concerning the results reported here.

REFERENCES

[1] P. Deift, B. Simon: Comm. Pure Appl. Math. **30,** 573 (1977)
[2] B. Simon: Comm. Math. Phys. **55,** 259 (1977); **58,** 205 (1978)
[3] V. Enss: Comm. Math. Phys. **65** , 151 (1979)
[4] W. Hunziker: Comm. Math. Phys. **8,** 282 (1968)
[5] W. Hunziker: "Scattering in Classical Mechanics," in *Scattering Theory in Mathematical Physics,* ed. by J.A. La Vita J.-P. Marchand (D. Reidel, Dordrecht-Holland, 1974)
[6] R.T. Prosser: J. Math. Phys. **13,** 186 (1972)
[7] B. Simon: Comm. Math.Phys. **23,** 37 (1971)
[8] M. Reed, B. Simon: *Methods of Modern Mathematical Physics. III. Scattering Theory* (Academic, New York, 1979)
[9] M.A. Antoneč: Funk. Anal. i Priložen **12,** 68 (1978)
[10] I.W. Herbst: Comm. Math. Phys. **35,** 193 (1974)
[11] L. Markus, K.R. Meyer: "Generic Hamiltonian Dynamical Systems are neither Integrable nor Ergodic," Mem. Am. Math. Soc., No. 144 (1974)

CANONICAL SCATTERING THEORY FOR RELATIVISTIC PARTICLES

F. Coester

Argonne National Laboratory
*Argonne, Illinois 60439**

ABSTRACT

A multichannel relativistic scattering theory can be formulated in a manner similar to the nonrelativistic multichannel theory. The mass operator plays the role of the Hamiltonian of the nonrelativistic theory. The problems of existence and completeness of wave operators are formally the same. Poincaré invariance and cluster separability of the S operator impose nontrivial restrictions. The purpose of this paper is to review the known solutions for the case where the number of constituent particles is bounded and explicit representations for the physical one-particle states exist.

The formal framework of a relativistic multi-channel scattering theory is similar to that of the nonrelativistic theory. The formulations of EKSTEIN [1] and HAAG [2] point to a two-Hilbert space description [3-7] in which the identification operator — introduced into the mathematical theory by KATO [4] — is realized in terms of bound state wave functions, [1,3] or equivalently in terms of creation operators of physical particles [2,8]. Let $\mathscr{H}$ be the Hilbert space of the states of the interacting system and $\mathscr{H}_f$ the Fock space of free particles that occur in the initial and final states. In a relativistic theory, we require unitary representations of the Poincaré group $U(a,\Lambda)$ and $U_f(a,\Lambda)$ in $\mathscr{H}$ and $\mathscr{H}_f$ respectively. The generators of the infinitesimal transformations are self-adjoint operators $H,\mathbf{P}$ for time and space translations, $\mathbf{J}$ for rotations, and $\mathbf{K}$ for Lorentz boosts. The generators $H,\mathbf{P}$, and $\mathbf{J}$ have the physical significance of energy, momentum, and angular momentum. The mass operator (rest energy) M is defined by

$$M:=\sqrt{H^2-\mathbf{P}^2}. \tag{1}$$

I will assume in the following that M is positive and has a bounded inverse. Zero-mass particles are thus excluded. The commutation relations of the generators are

$$[P_i,P_j]=[P_i,H]=0, \tag{2}$$

$$[J_p,J_q]=i\sum_r \epsilon_{pqr}J_r, \tag{3}$$

$$[J_p,P_q]=i\sum_r \epsilon_{pqr}P_r,\ [J_i,H]=0, \tag{4}$$

$$[J_p,K_q]=i\sum_r \epsilon_{pqr}K_r, \tag{5}$$

$$[K_p,K_q]=-i\sum_r \epsilon_{pqr}J_r, \tag{6}$$

$$[K_p,P_q]=i\delta_{pq}H,\ [K_r,H]=iP_r. \tag{7}$$

We will also need the NEWTON-WIGNER position operator $\mathbf{X}$, defined as a function of the generators by [9]

$$\mathbf{X}:=\frac{1}{2}(H^{-1}\mathbf{K}+\mathbf{K}H^{-1})-\frac{\mathbf{P}\times(H\mathbf{J}+\mathbf{P}\times\mathbf{K})}{MH(M+H)}, \tag{8}$$

which satisfies the canonical commutation relations

$$[X_r,X_s]=0,\ [X_r,P_s]=i\delta_{rs}, \tag{9}$$

as a consequence of (2)-(7). It follows that $\mathbf{K}$ can be written as a function of $\mathbf{P},\mathbf{X},M$, and $\mathbf{j}$,

$$\mathbf{j}:=\mathbf{J}-\mathbf{X}\times\mathbf{P}, \tag{10}$$

$$\mathbf{K}=\frac{1}{2}(H\mathbf{X}+\mathbf{X}H)-\mathbf{j}\times\mathbf{P}(M+H)^{-1}. \tag{11}$$

The eigenvectors of M define a unitary mapping Φ_1 of the one particle subspace $\mathscr{H}_{f1} \subset \mathscr{H}_f$ onto the one-particle subspace $\mathscr{H}_1 \subset \mathscr{H}$. The identification operator Φ needed for the formulation of the initial condition in the time dependent scattering theory,

$$\lim_{\tau \to -\infty} ||\Psi(\tau) - \Phi e^{-iM_f \tau} \chi|| = 0, \tag{12}$$

can be defined by tensor products of the Φ_1's such that

$$\text{w-lim}_{\tau \to \pm\infty} e^{iM_f \tau} \Phi^\dagger \Phi \, e^{-iM_f \tau} = 1. \tag{13}$$

The wave operators $\Omega_\pm$ are then defined by

$$\Omega_\pm(M, \Phi, M_f) := \text{s-lim}_{\tau \to \pm\infty} e^{iM\tau} \Phi \, e^{-iM_f \tau} \tag{14}$$

and the scattering operator S is

$$S := \Omega_+^\dagger \Omega_-. \tag{15}$$

The mathematical problems of existence and asymptotic completeness of the wave operators are similar to those of the nonrelativistic theory. The new features are related to relativistic invariance of S,

$$U_f(a, \Lambda) S U_f^{-1}(a, \Lambda) = S, \tag{16}$$

and the cluster-separability requirement which assures the independence of spatially separated scattering events. Neither property follows automatically from the existence of the wave operators (14). The dynamics of the system is specified by the generators, H, $\mathbf{P}$, $\mathbf{J}$, $\mathbf{K}$, and the identification operator Φ. The problem is to satisfy sufficient conditions to be imposed on these operators such that S is Poincaré-invariant and satisfies the cluster-separability requirement. The purpose of this paper is to review the known solutions for directly interacting particles [3,10], as well as the suggestion that no solution exists [11].

The following equivalence relations [3,4] play a key role.

Definition. Two dynamical systems $\{M, \Phi\}$ and $\{M', \Phi'\}$ are called scattering-equivalent if $S(M, \Phi, M_f) = S(M', \Phi', M_f)$.

Lemma 1. *Let $\Phi \in L(\mathscr{H}_f, \mathscr{H})$ and $\Phi' \in L(\mathscr{H}_f, \mathscr{H})$ be two identification operators. The limiting relation,*

$$\text{s-lim}_{\tau \to \pm\infty} (\Phi - \Phi') e^{iM_f \tau} = 0, \tag{17}$$

is necessary and sufficient for the identity of the wave operators:

$$\Omega_\pm(M, \Phi, M_f) = \Omega_\pm(M, \Phi', M_f).$$

Lemma 2. *For any two unitary operators $A \in L(\mathscr{H}_i, \mathscr{H}_i)$ and $B_f \in L(\mathscr{H}_f, \mathscr{H}_f)$,*

$$\Omega_\pm(M, A\Phi B_f^\dagger, M_f) = A\,\Omega_\pm(A^\dagger M A, \Phi, B_f^\dagger M_f B_f) B_f^\dagger. \tag{18}$$

Corollary. *If $[B_f, M_f] = 0$, and $[S, B_f] = 0$ then $\{M, A\Phi B_f^\dagger\}$ and $\{A^\dagger M A, \Phi\}$ are scattering-equivalent.*

Lemma 3. *If $\{M', \Phi'\}$ and $\{M, \Phi\}$ are scattering-equivalent and the wave operators $\Omega'_\pm := \Omega_\pm(M', \Phi', M_f)$ and $\Omega_\pm := \Omega_\pm(M, \Phi, M_f)$ are complete, then $W := \Omega'_+ \Omega_+^\dagger = \Omega'_- \Omega_-^\dagger$ is unitary and the operators $W\Phi$ and Φ satisfy the equivalence relation (17).*

Lemma 4. *The wave operators $\Omega_\pm(M, \Phi, M_f)$ are Poincaré-invariant,*

$$U(a, \Lambda) \Omega_\pm = \Omega_\pm U_f(a, \Lambda), \tag{19}$$

if and only if

$$\text{s-lim}_{\tau \to \pm\infty} (U(a, \Lambda) \Phi U_f^{-1}(a, \Lambda) - \Phi) e^{-iM_f \tau} = 0. \tag{20}$$

On physical grounds, we should expect

$$\Omega_\pm(H, \Phi, H_f) := \text{s-lim}_{t \to \pm\infty} e^{iHt} \Phi \, e^{-iH_f t} = \Omega_\pm(M, \Phi, M_f). \tag{21}$$

Theorem 1. *If $\Omega_\pm(M,\Phi,M_f)$ (or $\Omega_\pm(H,\Phi,H_f)$) exists and is Poincaré-invariant then $\Omega_\pm$ (H,Φ,H_f) (or $\Omega_\pm$ (M,Φ,M_f)) also exists and they are equal,*

$$\Omega_\pm(H,\Phi,H_f) = \Omega_\pm(M,\Phi,M_f).$$

From the invariance of $\Omega_\pm(M,\Phi,M_f)$, it follows that

$$H\Omega_\pm(M,\Phi,M_f) = \Omega_\pm(M,\Phi,M_f)H_f. \tag{22}$$

It can be shown that the limit

$$\operatorname*{s-lim}_{t\to\pm\infty} (\Phi - \Omega_\pm)e^{-iH_f t} = 0 \tag{23}$$

follows from

$$\operatorname*{s-lim}_{\tau\to\pm\infty} (\Phi - \Omega_\pm)e^{-iM_f \tau} = 0, \tag{24}$$

using $H_f = \sqrt{1+\mathbf{Q}_f^2}\,M_f, \mathbf{Q}_f := \mathbf{P}_f M_f^{-1}$, and the dominated convergence theorem [12].

The existence of the wave operators $\Omega_\pm(H,\Phi,H_f)$ can be proved by an obvious generalization of Cook's method [13].

Theorem 2. *Let*

$$V := H\Phi - \Phi H_f. \tag{25}$$

If there is a dense set $D \subset D(H_f)$ such that for $t > t_0$ and $\chi \in D$ the conditions

$$\left.\begin{aligned} &\Phi e^{\pm iH_f t}\chi \in D(H)\cap \Phi D(H_f), \\ &\int_0^\infty \| Ve^{\pm iH_f t}\chi\| dt < \infty, \end{aligned}\right\} \tag{26}$$

are satisfied, then $\Omega_\pm(H,\Phi,H_f)$ exist.

It might be tempting to assure the invariance of the wave operators by requiring Poincaré invariance of Φ,

$$U(a,\Lambda)\Phi U_f^{-1}(a,\Lambda) = \Phi.$$

But this is possible only for trivial theories, since then $M\Phi = \Phi M_f, \Omega_\pm = \Phi$, and $S = 1$. From (11), it follows that (19) is satisfied if

$$\mathbf{P}\Phi = \Phi\mathbf{P}_f, \tag{27}$$

$$\mathbf{J}\Phi = \Phi\mathbf{J}_f, \tag{28}$$

and

$$\mathbf{X}\Phi = \Phi\mathbf{X}_f. \tag{29}$$

Theorem 3 (SOKOLOV [14]). *Define*

$$\mathbf{Q} := \mathbf{P}M^{-1}, \; \mathbf{R} := \mathbf{X}M, \tag{30}$$

$$\zeta := \exp\{\frac{1}{2}(\mathbf{X}\cdot\mathbf{P} + \mathbf{P}\cdot\mathbf{X})\ln M\} \tag{31}$$

and

$$\Phi_s := \zeta^\dagger \Phi \zeta_f. \tag{32}$$

Then Φ_s and Φ are scattering-equivalent. If Φ satisfies (27)-(29), then it follows that

$$\mathbf{Q}\Phi_s = \Phi_s\mathbf{Q}_f, \tag{33}$$

$$\mathbf{R}\Phi_s = \Phi_s\mathbf{R}_f, \tag{34}$$

$$\mathbf{K}\Phi_s = \Phi_s\mathbf{K}_f, \tag{35}$$

$$\mathbf{J}\Phi_s = \Phi_s\mathbf{J}_f. \tag{36}$$

The scattering equivalence of Φ_s and Φ follows from Lemma 2. Equations (33) and (34) follow from

$$\mathbf{P}\zeta = \zeta \mathbf{P} M^{-1}, \quad \mathbf{X}\zeta = \zeta \mathbf{X} M, \tag{37}$$

(27), and (29). Equation (35) follows from (11), (30), (33), and (34).

For a mathematical formulation of cluster separability, consider two channels α and β. Let D be a partition of the particles in α and β into two disjoint subsets $\{\alpha_1, \alpha_2\}$ and $\{\beta_1, \beta_2\}$ respectively. ($\alpha_1 \cup \alpha_2 = \alpha$, $\alpha_1 \cap \alpha_2 = \varnothing$, $\beta_1 \cup \beta_2 = \beta$, $\beta_1 \cap \beta_2 = \varnothing$). By definition of H_f, the channel subspaces $\mathscr{H}_{f\alpha}$ and $\mathscr{H}_{f\beta}$ are tensor products

$$\mathscr{H}_{f\alpha} = \mathscr{H}_{f\alpha_1} \otimes \mathscr{H}_{f\alpha_2},$$
$$\mathscr{H}_{f\beta} = \mathscr{H}_{f\beta_1} \otimes \mathscr{H}_{f\beta_2}, \tag{38}$$

and the momentum in each channel is additive in the clusters,

$$\mathbf{P}_{f\alpha} = \mathbf{P}_{f\alpha_1} + \mathbf{P}_{f\alpha_2},$$
$$\mathbf{P}_{f\beta} = \mathbf{P}_{f\beta_1} + \mathbf{P}_{f\beta_2}. \tag{39}$$

Let

$$S_{\beta\alpha} := \mathscr{P}_{f\beta} S \mathscr{P}_{f\alpha} \tag{40}$$

where $\mathscr{P}_{f\alpha}$ is the projection onto $\mathscr{H}_{f\alpha}$ and define

$$T_{\alpha D}(\mathbf{a}_1, \mathbf{a}_2) = e^{-i(\mathbf{P}_{f\alpha_1}\cdot \mathbf{a}_1 + \mathbf{P}_{f\alpha_2}\cdot \mathbf{a}_2)}. \tag{41}$$

The separability requirement for S can be stated in the form

$$\operatorname*{w\text{-}lim}_{|\mathbf{a}_1 - \mathbf{a}_2| \to \infty} T_{\beta D}(\mathbf{a}_1, \mathbf{a}_2) S_{\beta\alpha} T_{\alpha D}^{-1}(\mathbf{a}_1, \mathbf{a}_2) = S_{\beta_1\alpha_1} \otimes S_{\beta_2\alpha_2} \tag{42}$$

for all channels α and β and all cluster decompositions D. The strong limit

$$\operatorname*{s\text{-}lim}_{|\mathbf{a}_1 - \mathbf{a}_2| \to \infty} (S\mathscr{P}_{f\alpha} - S\mathscr{P}_{f\alpha_1} \otimes S\mathscr{P}_{f\alpha_2}) T_{\alpha D}(\mathbf{a}_1, \mathbf{a}_2) = 0 \tag{43}$$

follows from (42) [3]. The dynamical history of the system is specified by

$$\Phi(t) := e^{iHt} \Phi e^{-iH_f t}. \tag{44}$$

We are, therefore, seeking a condition on $\Phi(t)$ which is sufficient for the validity of (42). Such a condition is

$$\operatorname*{w\text{-}lim}_{|\mathbf{a}_1 - \mathbf{a}_2| \to \infty} T_{\beta D}(\mathbf{a}_1, \mathbf{a}_2) (\Phi^\dagger(t')\Phi(t))_{\beta\alpha} T_{\alpha D}^{-1}(\mathbf{a}_1, \mathbf{a}_2) = (\Phi^\dagger(t')\Phi(t))_{\beta_1\alpha_1} \otimes (\Phi^\dagger(t')\Phi(t))_{\beta_2\alpha_2}, \tag{45}$$

uniformly in t' and t for all D, α, β. It follows that

$$\operatorname*{w\text{-}lim}_{|\mathbf{a}_1 - \mathbf{a}_2| \to \infty} T_{\beta D}(\mathbf{a}_1, \mathbf{a}_2) (\Phi^\dagger(t')\Omega_\pm)_{\beta\alpha} T_{\alpha D}^{-1}(\mathbf{a}_1, \mathbf{a}_2) = (\Phi^\dagger(t')\Omega_\pm)_{\beta_1\alpha_1} \otimes (\Phi^\dagger(t')\Omega_\pm)_{\beta_2\alpha_2}. \tag{46}$$

Equation (42) follows from (46).

For N particles, $\mathscr{H} = \mathscr{H}_f$, and $\Phi = 1$, MUTZE [11] has shown, assuming (27)-(29), that the right hand side of (42) is unity. The conditions imposed are obviously too strong and they are not necessary. The key to satifying both the apparently imcompatible relations (29) and (45) is found in Lemmas 1-4. It would be sufficient to show that for every cluster decomposition D there is a scattering-equivalent Φ_D which satisfies (45).

For three particles in $\mathscr{H} = \bigotimes_{i=1}^{3} \mathscr{H}_1^{(i)}$, it is possible to satisfy (27)-(29) and (42) by the following construction [2]. Assume the mass operator M_{12} on $\mathscr{H}_1^{(1)} \otimes \mathscr{H}_1^{(2)}$, for particles 1 and 2, is known. It is then possible to construct a mass operator $M_{12,3}$ on H for the three particle system in which particle 3 is a noninteracting spectator such that

$$S_{12,3} = S_{12} \otimes 1 \tag{47}$$

while

$$\Omega_{\pm 12,3} \neq \Omega_{\pm 12} \otimes 1 \tag{48}$$

Let M_o be the mass operator of the noninteracting particles and define

$$\begin{aligned} V_{12} &:= M_{12,3} - M_o, \\ V_{13} &:= M_{13,2} - M_o, \\ V_{23} &:= M_{23,1} - M_o. \end{aligned} \tag{49}$$

For

$$M = M_o + V_{12} + V_{13} + V_{23}, \tag{50}$$

one obtains an S operator which satisfies (42). It is possible to construct operators Φ_D, equivalent to Φ by Lemma 2, for which (45) holds. This procedure could not be generalized to more than three particles.

For N particles, SOKOLOV [10] has proposed a dynamics in which the identification operator is Lorentz-invariant and the interactions are introduced into the energy-momentum vector in such a manner that the Lorentz invariance and separability are manifest. The commutativity of the components of P^μ requires n-body interactions for all $n \leqslant N$. They are generated recursively for successively increasing N.

Consider for $\mathcal{H}$ the Hilbert space of N particles $\mathcal{H}_N := \bigotimes_{i=1}^{N} \mathcal{H}_1^{(i)}$ and assume that $\mathbf{K}$ and $\mathbf{J}$ do not depend on the interaction.

$$\mathbf{K} = \mathbf{K}_o := \sum_{i=1}^{N} \mathbf{K}_i \otimes 1, \tag{51}$$

$$\mathbf{J} = \mathbf{J}_o := \sum_{i=1}^{N} \mathbf{J}_i \otimes 1. \tag{52}$$

For the energy-momentum vector P^μ we assume a form that is additive for separated clusters,

$$P^\mu = P_o^\mu + \sum_{n=2}^{N} U_n^\mu, \tag{53}$$

where

$$P_o^\mu = \sum_{i=1}^{N} P_i^\mu \otimes 1, \tag{54}$$

and each n-body interaction term U_n^μ has the form

$$U_n^\mu = \sum_{i_1 < i_2 \dots i_n} V_{i_1 \dots i_n}^\mu \otimes 1, \tag{55}$$

where $V_{i_1 \dots i_n}^\mu$ is a short-range n-body interaction defined on $\bigotimes_{i=1}^{n} \mathcal{H}_1^{(i)}$ independent of N. By definition, it vanishes for any division of the cluster $i_1, \dots, i_n$ into separated fragments. Let D_k be a partition of the N particles into k disjoint subsets $\{\sigma_1, \dots, \sigma_k\}$ and define

$$U_n^\mu(D_k) := \sum_{j=1}^{k} \sum_{i_1 < \dots i_n \in \sigma_j} V_{i_1 \dots i_n}^\mu \otimes 1. \tag{56}$$

Obviously,

$$P^\mu(D_k) = P_o^\mu + \sum_n U_n^\mu(D_k) = \sum_{j=1}^{k} P_j^\mu, \tag{57}$$

where P_j is the energy-momentum vector of the jth cluster. Each $V_{i_1 \dots i_n}^\mu$ transforms as a four-vector. With

$$T_{D_k}(\mathbf{d}_1, \dots, \mathbf{d}_k) := e^{i \sum_{j=1}^{k} \mathbf{P}_j \cdot \mathbf{d}_j}, \tag{58}$$

we have

$$\underset{\min |\mathbf{d}_i - \mathbf{d}_j| \to \infty}{\text{s-lim}} T_{D_k}(d_1, \dots, d_k) e^{iP^\mu a_\mu} T_{D_k}^{-1}(\mathbf{d}_1, \dots, \mathbf{d}_k) = e^{iP^\mu(D_k) a_\mu}. \tag{59}$$

For $\mathbf{K}, \mathbf{J}$, and P^μ defined by (51)-(53), the identification operator Φ defined by tensor products of Φ_1's is Lorentz-invariant,

$$\mathbf{K}\Phi = \Phi \mathbf{K}_f, \ \mathbf{J}\Phi = \Phi \mathbf{J}_f. \tag{60}$$

The $V^{\mu}_{i_1 \ldots i_n}$ are related to each other by the requirement that different components of P commute,

$$[P^{\mu},P^{\nu}] = 0. \tag{61}$$

For $N = 2$, it is easy to satisfy (61) by

$$V^{\mu 12} = Q_o^{\mu}\, v_2,$$

where v_2 is Lorentz invariant and commutes with Q_0. Assume that the $V^{\mu}_{i_1 \ldots i_n}$ have been determined for all $n < N$ such that P satisfies (61) for all $N' < N$. The problem is to construct $V^{\mu}_{1 \ldots N}$ such that (61) is satisfied for N particles. It follows from (59) that

$$P^{\mu} = \sum_{K=2}^{N} (-1)k \sum_{D_k} P^{\mu}(D_k) + U^{\mu}_N. \tag{62}$$

If A is a Lorentz-invariant unitary operator on $\mathscr{H}_N$ which transforms $\mathbf{Q}$ into $\mathbf{Q}_o$,

$$A^{-1}\mathbf{Q}A = \mathbf{Q}_o, \tag{63}$$

and

$$A(D_k) := \underset{\min|\mathbf{d}_i-\mathbf{d}_j|\to\infty}{\text{s-lim}} T_{D_K}(\mathbf{d}_1,\ldots,\mathbf{d}_k)AT^{-1}_{D_k}(\mathbf{d}_1,\ldots,\mathbf{d}_k), \tag{64}$$

then

$$A^{-1}(D_k)\mathbf{Q}(D_k)A(D_k) = \mathbf{Q}_o, \tag{65}$$

and from (62)

$$P^{\mu} = AQ^{\mu}_o\{\sum_k (-1)^k(k-1)!\sum_{D_k} A^{-1}(D_k)M(D_k)A(D_k) + v_N\}A^{-1}, \tag{66}$$

where

$$v_N := A^{-1}MA - \sum_k (-1)^k(k-1)!\sum_{D_k} A^{-1}(D_k)M(D_k)A(D_k) \tag{67}$$

is Lorentz-invariant, commutes with Q_o, and vanishes for any cluster separation,

$$[v_N,\mathbf{Q}_o] = 0, \quad v_N(D_k) = 0. \tag{68}$$

A construction of A satisfying (64) and (65) can be based on the following Lemma.

Lemma 5. *Let* $A(D_k) = e^{i\alpha(D_k)}$ *be a set of unitary operators, defined for all* D_k, *such that*

$$\underset{\min|\mathbf{d}_i-\mathbf{d}_j|\to\infty}{\text{s-lim}} T_{D_r}(\mathbf{d}_1,\ldots,\mathbf{d}_r)A(D_k)T^{-1}_{D_r}(\mathbf{d}_1,\ldots,\mathbf{d}_r) = A(D_k \times D_r) \tag{69}$$

and $A(D_k)$ *satisfies (65). Then*

$$A := e^{i\sum_{k=1}^{N}(-1)^k(k-1)!\sum_{D_k}\alpha(D_k)} \tag{70}$$

satisfies (64).

Assuming $P^{\mu}(D_k)$ is known SOKOLOV [10] gives a prescription for constructing operators $A(D_k)$ which satisfy (65) and (69). IF A is then constructed according to (70) and v_N is an arbitrary Lorentz-invariant operator satisfying (68), then P^{μ} and hence U^{μ}_N is determined by the right-hand side of (66).

The restriction to a fixed number of particles is not an essential feature of the dynamical systems we have considered. It is easy to incorporate particle creation [3,10,15] as long as there are Poincaré-invariant sectors of the Hilbert space $\mathscr{H}$ in which the particle number is bounded. The relativistic Lee model [15] is an example.

Canonical field theories, on the other hand, are radically different in that locality and infinitely many degrees of freedom are essential features of the relativistic invariance. This is evident from the construction of the Poincaré generators as integrals over the energy-momentum tensor $T^{\mu\nu}(x)$:

$$P^{\mu} = \int T^{o\mu}(x)d^3\mathbf{x}, \tag{71}$$

$$\mathbf{K} = \int \mathbf{x}\,T^{oo}(x)d^3\mathbf{x},$$

$$J_i = \frac{1}{2} \sum_{mn} \epsilon_{imn} \int [x^m T^{on}(x) - x^n T^{om}(\mathbf{x})] d^3\mathbf{x}. \tag{72}$$

The commutation relations (2)-(8) are satisfied if and only if the energy density $T^{oo}(x)$ satisfies the local SCHWINGER commutation relations [16]

$$i[T^{oo}(x), T^{oo}(x')] = - T^{ok}(x')\partial'_k \delta(\mathbf{x} - \mathbf{x}') + T^{ok}(x)\partial_k \delta(\mathbf{x} = \mathbf{x}') + \sigma(\mathbf{x}, \mathbf{x}'),$$

where σ must be antisymmetric in $\mathbf{x}$ and $\mathbf{x}'$, $(\sigma(\mathbf{x},\mathbf{x}') = -\sigma(\mathbf{x}',\mathbf{x}))$ and satisfy the relations

$$\int \sigma(\mathbf{x},\mathbf{x}') d^3\mathbf{x} = 0, \int \mathbf{x}\sigma(\mathbf{x},\mathbf{x}') d^3\mathbf{x} = 0. \tag{73}$$

$\sigma = 0$ is sufficient but not necessary.

The main results of Sokolov's construction is to demonstrate that Poincaré-invariance and cluster separability of a nontrivial S operator can be realized for a finite number of particles with direct interactions.

NOTES AND REFERENCES

*Work performed under the auspices of the U.S. Department of Energy

[1] H. Ekstein: Phys. Rev. **101**, 880 (1956)
[2] R. Haag: Phys. Rev. **112**, 669 (1958)
[3] F. Coester: Helv. Phys. Acta **38**, 7 (1965)
[4] T. Kato: J. Functional Anal. **1**, 342 (1967)
[5] M. Reed and B. Simon: *Methods of Modern Mathematical Physics*, (Academic, New York, 1979), Vol. III, Sec. XI.3
[6] I. M. Sigal: "Scattering theory for many-particle systems", preprint
[7] C. Chandler and A. G. Gibson: proceedings of this conference
[8] See [5], p. 321
[9] M. H. L. Pryce: Proc. Roy. Soc. **A195**, 62 (1949), definition (e); T. D. Newton, E. P. Wigner: Revs. Mod. Phys. **21**, 400 (1949)
[10] S. N. Sokolov: Dokl Akad. Nauk. SSSR **233**, 575 (1977)
[11] U. Mutze: J. Math. Phys. **19**, 231 (1978)
[12] T. Kato: *Perturbation Theory for Linear Operators* (Springer, New York, 1966), Theorem 3.7, p. 533
[13] [5], Vol. I, Theorem I.11
[14] S. N. Sokolov: Teor. Mat. Fiz. **24**, 236 (1975)
[15] B. de Dormale: J. Math. Phys. **20**, 1229 (1979)
[16] J. Schwinger: Phys. Rev. **127**, 324 (1962)

CONSISTENT MODELS OF SPIN 0 AND 1/2 EXTENDED PARTICLES SCATTERING IN EXTERNAL FIELDS

S. Twareque Ali

Department of Mathematics
University of Prince Edward Island
Charlottetown, Prince Edward Island, Canada C1A 4P3

Eduard Prugovečki

Department of Mathematics
University of Toronto
Toronto, Canada M5S 1A1

ABSTRACT

By replacing sharp with stochastic localizability, positive-definite and covariant probability densities yielding conserved and covariant probability currents can be introduced in relativistic quantum mechanics. The resulting stochastic phase-space formalism can be used to construct covariant models of extended spin 0 and 1/2 particles, whose interaction with an external electromagnetic field leaves the space of positive-energy wave functions invariant.

INTRODUCTION

Historically, the two outstanding problems of relativistic quantum mechanics had been [1]: (a) the search for a positive definite one-particle probability density with the right covariance and conservation properties; (b) the construction of consistent relativistic one-particle models in which the interaction of the particle with an external field does not cause spontaneous transitions to negative-energy states. It has been evident from the very inception of modern quantum theory that the Klein-Gordon theory could not lead in the spin zero case to a satisfactory solution of either of these problems, but initially it appeared that in case of spin 1/2 particles the Dirac theory provided a satisfactory answer to the first problem in terms of the current $j^\nu = \bar{\psi}\gamma^\nu\psi$. However , it eventually became clear that, after all, $j^0(\mathbf{x})$ could not be interpreted as a particle-position probability density since, in the words of WIGHTMAN [2], "multiplication by $\mathbf{x}$ does not carry positive energy solutions (of the Dirac equation) into positive energy wave functions, (and therefore) in this respect the Klein-Gordon equation is neither better nor worse than the Dirac equation".

In this report, we shall survey results which show that a satisfactory answer to both of the above problems becomes feasible as soon as the assumption of perfectly sharp localizability in space-time is replaced by a physically more realistic notion of extended stochastic localizability. In other words, instead of resorting to the extreme idealizations of point-like particles whose location in relation to a classical inertial frame can be ascertained at any given instant with arbitrary precision, we shall acknowledge from the very start the fact that, first of all, actual measurements are of finite precision and, second, that all hadronic matter (and probably even the leptonic world [3,4]) consists of extended particles.

Our approach to the notion of extended particle is not based on any of the standard models (harmonic oscillator quark models, bag or string models, QCD, etc.) although there are some clear-cut mathematical connections with some of these models (see Sec. 4). Rather, our interpretation originates in the more fundamental level of a theory of measurement based on space-time symmetries derived from phase-space representations [5,6] of the Poincaré group, i.e., it is in the spirit of HEISENBERG's ideas [7] on the close interrelation of the notion of fundamental symmetry and elementary particle. Indeed, in a series of earlier articles (see [8] for a review), we have studied the physical interpretation of representations of the Galilei group on $L^2(\Gamma)$ (Γ = phase space), and we have shown [9,10] (as briefly surveyed in Sec. 1) that certain of the invariant subspaces of $L^2(\Gamma)$ consist of wave functions which can be interpreted as probability amplitudes for nonsharp measurements of position and momentum. The key idea leading to the present relativistic results was to substitute in that nonrelativistic approach the Poincaré group in place of the Galilei group (see Sec. 2), and thus develop a new method of quantization of classical relativistic models [11]. This procedure has been then applied [11-13] to standard relativistic Hamiltonians describing relativistic particles possessing in general intrinsic electric and magnetic dipole moments [14] (see Sec. 3), thus arriving at relativistically covariant and gauge-invariant models of spin 0 and ½ extended particles. These models possess a consistent probabilistic interpretation, and display no spontaneous transitions to negative energy states.

1. NONRELATIVISTIC STOCHASTIC PHASE SPACES

In the Hilbert spaces $H^s(\Gamma) = l^2(s) \otimes L^2(\Gamma)$, $s = 0, 1/2, 1, \ldots$, we consider [11] the ray representation

$$(U(b,\mathbf{a},\mathbf{v},R)\psi)(\mathbf{q},\mathbf{p};t) = \exp\left\{\frac{i}{\hbar}\left[-\frac{m|\mathbf{v}|^2}{2}(t-b) + m\mathbf{v}\cdot(\mathbf{q}-\mathbf{a})\right]\right\}$$
$$\times D^s(R^{-1})\psi(R^{-1}[\mathbf{q}-\mathbf{v}(t-b)-\mathbf{a}], R^{-1}(\mathbf{p}-m\mathbf{v}); t-b) \tag{1.1}$$

of the proper Galilei group $G_+^\dagger$, [15] and denote by $\mathscr{H}^s(\Gamma_\xi)$ the invariant subspaces that possess a (unique) resolution generator $\xi(\mathbf{q}',\mathbf{p}')$ which, by definition, is a rotationally invariant function that has the following two properties:

$$\psi(\mathbf{q},\mathbf{p}) = \langle\xi_{\mathbf{q},\mathbf{p}}|\psi\rangle, \forall\, \psi \in \mathscr{H}^s(\Gamma), \tag{1.2}$$

$$\mathbf{P}(\Gamma_\xi) = \int_\Gamma |\xi_{\mathbf{q},\mathbf{p}}\rangle d\mathbf{q}\, d\mathbf{p} \langle\xi_{\mathbf{q},\mathbf{p}}|, \tag{1.3}$$

In (1.2) $\langle\cdot|\cdot\rangle$ denotes the inner product in $\mathscr{H}^s(\Gamma)$, in (1.3) $\mathbf{P}(\Gamma_\xi)$ is the orthogonal projector onto $\mathscr{H}^s(\Gamma_\xi)$, and

$$\xi_{\mathbf{q},\mathbf{p}} = U(0,\mathbf{q},\frac{\mathbf{p}}{m},I)\xi \otimes 1_s, \tag{1.4}$$

where 1_s is the identity operator in $l^2(s)$. The space and velocity infinitesimal generators of $U(b,\mathbf{a},\mathbf{v},R)$ provide a representation

$$\mathbf{X} = \mathbf{q} + i\hbar\nabla_{\mathbf{p}},\ \mathbf{P} = -i\hbar\nabla_{\mathbf{q}}, \tag{1.5}$$

of the canonical commutation relations, which is irreducible on each $\mathscr{H}^0(\Gamma_\xi) = L^2(\Gamma_\xi)$. The transition to the configuration and momentum representations can be executed if ξ has represenatives $\xi(\mathbf{x}) \in l^2(s) \otimes L^2(\mathbf{R}^3)$ and

$$\tilde{\xi}(\mathbf{k}) = h^{-3/2}\int \exp\left(-\frac{i}{\hbar}\mathbf{k}\cdot\mathbf{x}\right)\xi(\mathbf{x})d\mathbf{x} \tag{1.6}$$

in these representations, so that

$$\xi(\mathbf{q},\mathbf{p}) = \int \exp\left[\frac{i}{\hbar}\mathbf{p}\cdot(\mathbf{x}-\mathbf{q})\right]\xi^*(\mathbf{x})\xi(\mathbf{x}-q)d\mathbf{x}, \tag{1.7}$$

where the asterisk denotes complex conjugation. Then it turns out [9] that, for any $\psi(\mathbf{q},\mathbf{p}) \in \mathscr{H}^s(\Gamma_\xi)$,

$$\int |\psi(\mathbf{q},\mathbf{p})|^2 d\mathbf{q} = \int \chi_{\mathbf{q}}^\xi(x)|\psi(\mathbf{x})|^2 d\mathbf{x}, \tag{1.8}$$

$$\int |\psi(\mathbf{q},\mathbf{p})|^2 d\mathbf{p} = \int \hat{\chi}_{\mathbf{p}}^\xi(\mathbf{k})|\tilde{\psi}(\mathbf{k})|^2 d\mathbf{k}, \tag{1.9}$$

$$\chi_{\mathbf{q}}^\xi(\mathbf{x}) = |\xi(\mathbf{x}-\mathbf{q})|^2,\ \hat{\chi}_{\mathbf{p}}^\xi(\mathbf{k}) = |\tilde{\xi}(\mathbf{k}-\mathbf{p})|^2, \tag{1.10}$$

where $\psi(\mathbf{x})$ and $\tilde{\psi}(\mathbf{k})$ are, respectively, the configuration and momentum representatives [11] of $\psi(\mathbf{q},\mathbf{p})$, e.g.,

$$\psi(\mathbf{q},\mathbf{p}) = \int \tilde{\xi}_{\mathbf{q},\mathbf{p}}^*(\mathbf{k})\tilde{\psi}(\mathbf{k})d\mathbf{k}. \tag{1.11}$$

Consequently, $|\psi(\mathbf{q},\mathbf{p})|^2 = \psi^\dagger(\mathbf{q},\mathbf{p})\psi(\mathbf{q},\mathbf{p})$ ($\psi^\dagger$ = matrix adjoint of ψ) can be interpreted as the probability density of obtaining the stochastic value [8] $(q,\chi_{\mathbf{q}}^\xi) \times (p,\hat{\chi}_{\mathbf{p}}^\xi) \in \Gamma_\xi$ for the simultaneous measurement of position and momentum — the uncertainty principle being automatically satisfied by the confidence functions [9,16] $\chi_{\mathbf{q}}^\xi$ and $\hat{\chi}_{\mathbf{p}}^\xi$ on account of (1.10).

The consistency of this interpretation is confirmed by the existence of the Galilei-covariant probability 3-current

$$\mathbf{j}^{\xi}(\mathbf{q},t) = \int \frac{\mathbf{p}}{m} |\psi(\mathbf{q},\mathbf{p};t)|^2 d\mathbf{p}, \tag{1.12}$$

which, in conjunction with $\rho^{\xi}(\mathbf{q},t) = \int |\psi(\mathbf{q},\mathbf{p};t)|^2 d\mathbf{p}$, obeys the usual probability conservation equation. Furthermore, for the optimal case [9] of

$$\xi^{(r)}(\mathbf{x}) = (2\pi^{3/2}\hbar r)^{-3/2} \exp\left(-\frac{\mathbf{x}^2}{2r^2}\right), \quad 0 < r < \infty, \tag{1.13}$$

we easily obtain [17] that, in the sharp-point limit $r \to +0$, $\rho^{\xi}(\mathbf{q})$ and $\mathbf{j}^{\xi}(\mathbf{q})$ converge to their conventional counterparts $|\psi(\mathbf{x})|^2$ and $\frac{i}{2m}\psi^*(\mathbf{x})\overleftrightarrow{\nabla}_x (\mathbf{x})\overleftrightarrow{\nabla}_x\psi(\mathbf{x})$, respectively.

The conventional nonrelativistic quantum theory of point-like particles moving in an external electromagnetic field (and/or mutually interacting via local potentials) can be recast ([10], [17]-[19]) into the present framework, with the ensuing advantage of the above outlined stochastic phase-space interpretation becoming readily available. However, the *dynamically* interesting new features emerge when we consider interactions

$$A_{\xi}^{\nu}(ct) = \mathbf{P}(\Gamma_{\xi})A^{\nu}(ct,\mathbf{Q})\mathbf{P}(\Gamma_{\xi}), \quad \nu = 0,1,2,3, \tag{1.14}$$

with an external 4-potential $A^{\nu}(x) = A^{\nu}(ct,\mathbf{x})$ by introducing, e.g., in the $s = 0$ case, Hamiltonians of the form

$$H_{\xi}(t) = \frac{1}{2m}\, |\mathbf{p} - \frac{e}{c}\mathbf{A}_{\xi}(ct)|^2 + eA_{\xi}^{0}(ct) \tag{1.15}$$

acting on $\mathscr{H}^0(\Gamma_{\xi})$. The appropriate physical interpretation of such interactions reveals itself as soon as we make the transition to the configuration representation, when we discover that [11]

$$(A_{\xi}^{\nu}(ct)\psi)(\mathbf{x}) = \int \chi_{\mathbf{q}}^{\xi}(\mathbf{x})A_{\xi}^{\nu}(ct,\mathbf{q})\psi(\mathbf{x})d\mathbf{q}; \tag{1.16}$$

and, therefore, $\chi_{\mathbf{q}}^{\xi}(\mathbf{x})$ reveals itself as a configuration-space form factor (with its Fourier transform thus appearing as usual [20] in the first Born approximation as the ratio of the extended to the point particle cross section when the field is electrostatic). Hence, the interpretation of (1.15) is obvious: $H_{\xi}(t)$ governs the motion of a particle that is extended in a stochastic sense (see Sec. 4) in the configuration space and moves under the influence of $A^{\nu}(x)$.

The free propagator in $L^2(\Gamma_{\xi})$,

$$K_{\xi}(\mathbf{q}'',\mathbf{p}'',t'';\mathbf{q}',\mathbf{p}\,t') = \langle U(t'',\mathbf{q}'',\frac{\mathbf{p}''}{m},I)\xi\,|\,U(t',\mathbf{q}',\frac{\mathbf{p}'}{m},I)\xi\rangle, \tag{1.17}$$

describes stochastically (i.e., in the role of a probability amplitude) the motion of this extended particle when it is free. Indeed, the formal analogies with Brownian motion are striking, and so are the ensuing path integral formulas when compared to the Wiener integrals for a Brownian particle [12]. The propagator

$$K_{\xi}'(\mathbf{q}'',\mathbf{p}'',t'';\mathbf{q}',\mathbf{p}',t') = \langle \xi_{\mathbf{q}'',\mathbf{p}''}|\,T\exp\left[-\frac{i}{\hbar}\int_{t'}^{t''} H_{\xi}(t)\,dt\right]\xi_{\mathbf{q}',\mathbf{p}'}\rangle$$

for the interacting case can also be expressed in terms of path integrals which manifestly exhibit the Galilei-covariance of the model [12].

2. RELATIVISTIC STOCHASTIC PHASE SPACE KINEMATICS

Conventional relativistic quantum mechanics has a consistent interpretation in the momentum representation, but not in the configuration representation, so that we cannot fall back on it to test our ideas on relativistic extended particles and stochastic localization, as we could do in Sec. 1. Therefore, we shall extrapolate the key features of the nonrelativistic stochastic theory of the preceding section to the relativistic domain, instead of using conventional approaches as guidelines.

In the relativistic phase space , $M_m = \hat{\mathbb{R}}^4 \times V_m$ ($\hat{\mathbb{R}}^4$ = Minkowski space-time; V_m = forward mass-m hyperboloid) we introduce a time-ordered [12] family $S^{\dagger}$ of space-like hyperplanes along which measurements are performed by all relativistic observers and adopt some $\sigma \in S^{\dagger}$ as the initial-data surface. On the $\Sigma_m = \sigma \times V_m$ hypersurface in M_m, we introduce the invariant surface measure

$$d\Sigma_m(q,p) = 2p^\nu d\sigma_\nu(q)\delta(p^2 - m^2c^2)d^4p, \tag{2.1}$$

which assumes the form $d\mathbf{q}d\mathbf{p}$ in the inertial frames where σ is the q^0 = const hyperplane. The Hilbert space $\mathscr{H}^s(\Sigma_m) = l^2(s) \otimes L^2(\Sigma_m)$ of one-column matrix-valued functions (with $2s+1$ components), which extrapolates $\mathscr{H}^s(\Gamma)$ to the relativistic regime, has the inner product

$$(\psi_1|\psi_2) = \int_{\Sigma_m} \psi_1^\dagger(q;p)\psi_2(q,p)d\Sigma_m(q,p). \tag{2.2}$$

Upon introducing, in the frame where σ is the $q^0 = 0$ hypersurface, the free time-evolution

$$\psi(q,p) = \left\{\exp\left(-\frac{i}{\hbar}P^0 q_0\right)\psi\right\}(0,\mathbf{q},p), \tag{2.3}$$

with

$$P^0 = (\mathbf{P}^2 + m^2c^2)^{1/2}, \tag{2.4}$$

(where we have retained the 3-momentum operators in (1.5)), we consider the representations [6, 11, 21, 22]

$$(U(a,\Lambda)\psi)(q,p) = D^s(\Lambda_p^{-1}\Lambda\Lambda_{\Lambda^{-1}p})\psi(\Lambda^{-1}(q-a),\Lambda^{-1}p) \tag{2.5}$$

of the proper Poincaré group $P_+^\uparrow$, where Λ_p denotes a Lorentz boost to the 4-velocity p/m .

We denote by $\mathscr{H}^s(\Sigma_m^\eta)$ the invariant subspaces with a resolution generator $\eta(q,p)$, $(q,p) \in \Sigma_m$, for which, by definition and by analogy with (1.2)-(1.4),

$$\psi(q,p) = (\eta_{q,p}|\psi), \forall\psi \in H^s(\Sigma_m^\eta), \tag{2.6}$$

$$\mathbf{P}(\Sigma_m^\eta) = \int_{\Sigma_m} |\eta_{q,p})d\Sigma_m(q,p)(\eta_{q,p}|, \tag{2.7}$$

$$\eta_{q,p} = U(q,\Lambda_p)\eta. \tag{2.8}$$

In fact, to each $L^2(\Gamma_\xi)$ we can set in correspondence an $L^2(\Sigma_m^\eta)$ with η having the momentum space representative

$$\eta(k) = (mc)^{1/2}\tilde{\xi}(\mathbf{k}),\ k = (k^0,\mathbf{k}) \in V_m, \tag{2.9}$$

so that if we write, by analogy with (1.7) and (1.11),

$$\eta(q,p) = \int \exp(-\frac{i}{\hbar}q\cdot k)\eta^*(\Lambda_p^{-1}k)\eta(k)\ \frac{d\mathbf{k}}{2k^0}, \tag{2.10}$$

$$\psi(q,p) = \int \eta_{q,p}^*(k)\tilde{\psi}(k)\frac{d\mathbf{k}}{2k^0}, \tag{2.11}$$

we shall obtain

$$|\psi(q,p)|^2 = |\psi(\mathbf{q},\mathbf{p};t)|^2 + O\left(\frac{\mathbf{p}^2}{m^2c^2}\right), \tag{2.12}$$

as $|\mathbf{p}| \to 0$, i.e., the relativistic theory indeed merges with the nonrelativistic one at small velocities. It turns out that $|\psi(q,p)|^2$ is properly normalized over Σ_m, that the 4-current

$$j^\nu(q) = \int \frac{p^\nu}{m}|\psi(q,p)|^2\frac{d\mathbf{p}}{p^0} \tag{2.13}$$

is conserved, and that $|\psi(q,p)|^2$ transforms as a scalar and $j^\nu(q)$ as a 4-vector under (2.5). Hence, the nonrelativistic interpretation can be consistently transferred to the present case; albeit that marginality conditions (1.8)-(1.9) lose their significance in the absence of a probabilistically consistent conventional configuration representation and of a group property for Lorentz boosts (as opposed to Galilean boosts).

The relativistic free propagator in $\mathscr{H}^s(\Sigma_m^\eta)$ is

$$K_\eta(q'',p'';q',p') = (\eta_{q'',p''}|\eta_{q',p'}), \tag{2.14}$$

and it indeed has the property that [11-12]

$$\psi(q,p) = \int_{\Sigma_m} K_\eta(q,p;q',p')\psi(q',p')d\Sigma_m(q',p'). \tag{2.15}$$

Furthermore, this free propagator is manifestly covariant.

For the $s = 1/2$ case the present quantum kinematics constitutes a stochastic phase-space counterpart of the Foldy-Wouthuysen kinematics for free spin 1/2 particles. The transition to a Dirac γ-matrix representation can also be performed, but the minimal coupling of an extended particle to an external field has not been proven to leave the space of positive energy solutions invariant [21]. Consequently, we shall treat both cases, $s = 0, 1/2$, simultaneously, by quantizing classical relativistic theories describing particles with or without internal degrees of freedom.

3. CONSISTENT EXTERNAL FIELD DYNAMICS

The classical Hamiltonian of a particle with intrinsic electric and magnetic dipole moments $\mathbf{d}$ and $\boldsymbol{\mu}$, respectively, is [14]

$$H^{cl} = eA^0 + c\left[|\mathbf{p} - \frac{e}{c}\mathbf{A}|^2 + \left(mc + \frac{1}{2c}\sigma^{cl}_{\mu\nu}F^{\mu\nu}\right)^2\right]^{1/2}, \tag{3.1}$$

where $\sigma^{cl}_{0j} = - d^j$, $j = 1,2,3$, $\sigma^{cl}_{ij} = \epsilon_{ijk}\mu^k$, $i,j = 1,2,3$, and $\sigma^{cl}_{\mu\nu} = - \sigma^{cl}_{\nu\mu}$. To quantize [13] this model on $\mathscr{H}^s(\Sigma^{\eta}_m)$, we set

$$\mathbf{d} = f\frac{e\hbar}{2mc}\boldsymbol{\sigma},\quad \boldsymbol{\mu} = g\frac{e\hbar}{2mc}\boldsymbol{\sigma}, \tag{3.2}$$

(where $\boldsymbol{\sigma}$ has the Pauli matrices as components) and replace the classical potential $A^\nu(q)$ by the self-adjoint operators

$$A^\nu_\eta(q^0) = \mathbf{P}(\Sigma^{\eta}_m)A^\nu(q^0,\mathbf{Q})\mathbf{P}(\Sigma^{\eta}_m), \tag{3.3}$$

$$(A^\nu(q^0,\mathbf{Q})\psi)(q,p) = A^\nu(q)\psi(q,p), \tag{3.4}$$

in the frame where the initial data surface σ is the $q^0 =$ const. hyperplane. The resulting Hamiltonian

$$H_\eta(q^0) = eA^0_\eta + c\left[|\mathbf{p} - \frac{e}{c}\mathbf{A}_\eta|^2 + \left(mc + \frac{1}{2c}\sigma_{\mu\nu}F^{\mu\nu}_\eta\right)^2\right]^{1/2} \tag{3.5}$$

is self-adjoint on $\mathscr{H}^s(\Sigma^{\eta}_m)$. The fact that $H_\eta(q^0)$ transforms indeed as the time-like component of a 4-vector is established [11-13] by expressing A^ν_η as integral operators and noting that $\sigma^{\mu\nu}$ can be written in a manifestly covariant form in terms of the metric tensor $g^{\mu\nu}$ and the totally antisymmetric tensor $\epsilon^{\mu\nu\lambda\kappa}$ as follows,

$$\sigma^{\mu\nu} = \frac{i}{2}(\mu g^{\mu\lambda}g^{\nu\kappa} + d\epsilon^{\mu\nu\lambda\kappa})(\sigma_\kappa\sigma_\lambda - \sigma_\lambda\sigma_\kappa), \tag{3.6}$$

and recalling that σ^ν transforms as a 4-vector. The case of spin $s = 0$ is obtained by setting $f = g = 0$ and working on $\mathscr{H}^0(\Sigma^{\eta}_m)$ instead of $\mathscr{H}^{1/2}(\Sigma^{\eta}_m)$.

$H_\eta(q^0)$ is clearly self-adjoint in $\mathscr{H}^s(\Sigma^{\eta}_m)$, so that the time-evolution operator

$$U_\eta(q''_0,q'_0) = T\exp\left(-\frac{i}{\hbar c}\int_{q'_0}^{q''_0} H_\eta(q^0)dq^0\right) \tag{3.7}$$

does not give rise to spontaneous transitions to negative energy states. It is well known that although the classical Hamiltonian (3.1) is not gauge invariant, the theory as a whole does possess gauge invariance, and the same remains true in the quantum case [11 - 13].

To prove the covariance of the quantum theory as a whole one has to establish the covariance of the propagator for the interesting case, which assumes the form

$$K'_\eta(q'',p'';q',p') = (\eta_{0,\mathbf{q}'',\mathbf{p}''}|U_\eta(q''_0,q'_0)\eta_{0,\mathbf{q}',\mathbf{p}'}) \tag{3.8}$$

in the frame where the initial-data surface σ is the $q^0 = 0$ hyperplane. This can be generally achieved by expressing (3.8) as a path integral, which turns out to be manifestly covariant [12]. The same goal can be also achieved by means of a perturbation series [11], which is also term by term manifestly covariant. This series is of special interest in the weak-coupling limit since it makes apparent the presence of charge renormalization when a comparison is carried out between the classical and quantum Lagrangians [12].

To obtain a manifestly covariant form for the S-martrix, we write the interaction-picture evolution operator from the reference surface $\sigma' \in S^\uparrow$ to the reference surface $\sigma'' \in S^\uparrow$ in the form

$$\tilde{U}_\eta(\sigma'',\sigma') = \exp\left[\frac{i}{\hbar}P\cdot q(\sigma'')\right]U_\eta(|q(\sigma'')|,|q(\sigma'|)\exp\left[-\frac{i}{\hbar}P\cdot q(\sigma')\right] \tag{3.9}$$

where $q(\sigma')$ is the (time-like) 4-vector perpendicular to σ' from the origin of a coordinate system where σ is the $q^0 = 0$ hyperplane and $|q(\sigma')|$ is its invariant length. Then

$$(k''|S|k') = \lim_{\substack{q^0(\sigma'')\to+\infty \\ q^0(\sigma')\to-\infty}} (k''|U_\eta(\sigma'',\sigma')|k'), \tag{3.10}$$

where $|k)$ is the element of $\mathscr{H}^s(\Sigma_m^\eta)$ obtained by considering $\eta_{q,p}(k)$ in (2.11) as a function of (q,p) at fixed $k \in V_m$.

4. DISCUSSION

The most notable departure of the present models from the conventional external-field models is that the dynamics is not governed by an invariant equation, but rather by a Schrödinger equation whose Hamiltonian is the total energy operator and therefore transforms as the time-like component of a 4-vector. Invariant equations retain their significance in the present stochastic approach, but only as *field* equations for second-quantized theories [23].

The second notable feature is that the present quantum models for extended particles are obtained by quantizing classical models of point-like rather than extended particles. Indeed, the latter kind of models [3,24] run into serious instability problems already at the classical level, whereas in our case the extended feature of the particle appears only at the quantum level as an expression of a (consistent) extrapolation of the uncertainty principle to the quantum regime. Naturally, each resolution generator $\eta(k)$ in the momentum representation could be viewed as the internal-motion wave function for a quark-antiquark pair and, in fact, $\eta^{(r)}(k)$, corresponding to $\xi^{(r)}(\mathbf{x})$ in (1.13), coincides with the wave-functions obtained in relativistic harmonic oscillator models [25,26]. We believe, however, that a purely stochastic interpretation of η as a proper wave function for an extended particle is preferable, since, even if quarks were proven to exist as *bona fide* particles, we would again be running into difficulties with providing a consistent probabilistic interpretation in the configuration representation if we treated them as (perfectly) point-like particles (as is done in the above models). Instead, an analysis [23] of basic measurement-theoretical aspects at a second-quantization level reveals that a purely stochastic approach to the very notion of space-time is not only feasible, but in fact mandatory if a totally self-consistent combination of the uncertainty and relativity principles is desired.

REFERENCES

[1] A. Pais: in *Aspects of Quantum Theory,* ed. by A. Salam and E.P. Wigner (Cambridge University Press, Cambridge, 1972)
[2] A.S. Wightman: [1], p. 98
[3] P.A.M. Dirac: Proc. Roy. Soc. (London) **A268**, 57 (1962)
[4] G. Rosen: Int. J. Theor. Phys. **17,** 1 (1978)
[5] E. Prugovečki: J. Math. Phys. **19,** 2260 (1978)
[6] S.T. Ali: J. Math. Phys. **20** , 1385 (1979)
[7] W. Heisenberg: Am. J. Phys. **43,** 389 (1975)
[8] E. Prugovečki: Found. Phys. **9,** 575 (1979)
[9] S.T. Ali, E. Prugovečki: J. Math. Phys. **18,** 219 (1977)
[10] S.T. Ali, E. Prugovečki: Physica **89A,** 501 (1977)
[11] E. Prugovečki: Phys. Rev. D **18,** 3655 (1978)
[12] E. Prugovečki: "Quantum action principle and functional integration over paths in stochastic phase space," to appear
[13] S.T. Ali, E. Prugovečki: "Self-consistent relativistic models for extended spin models 1/2 particles in external fields," to appear
[14] A.O. Barut: *Electrodynamics and Classical Theory of Fields and Particles* (MacMillan, New York, 1964)
[15] J.M. Lévy-Leblond: *Group Theory and Its Applications*, Vol. II, ed. by E.M. Loebl (Academic, New York, 1971)
[16] C.F. Dietrich: *Uncertainty, Calibration, and Probability* (Wiley, New York, 1973)
[17] E. Prugovečki: Ann. Phys. (N.Y.) **110,** 201 (1978)
[18] S.T. Ali, E. Prugovečki: Int. J. Theor. Phys. **16,** 689 (1977)
[19] E. Prugovečki: Physica **91A,** 202 (1978)
[20] R. Hofstadter: Ann. Rev. Nucl. Sci. **7,** 231 (1956)
[21] E. Prugovečki: Rep. Math. Phys. (1979-80), to appear
[22] S.T. Ali: "On some representations of the Poincaré group on phase space II," to appear
[23] E. Prugovečki: "A self-consistent approach to quantum field theory for extended particles," to appear

[24] P. Gnadig, Z. Kunszt, P. Hasenfratz, G. Koti: Ann. Phys. (N.Y.) **116**, 380 (1978)
[25] R.P. Feynman, M. Kislinger, F. Ravndal: Phys. Rev. D **3**, 2706 (1971)
[26] Y.S. Kim, M.E. Noz: Phys. Rev. D **15**, 335 (1977)

CURVED-SPACE SCATTERING

Jeffrey M. Cohen and Michael W. Kearney

Department of Physics
University of Pennsylvania
Philadelphia, Pennsylvania, 19104

Lawrence S. Kegeles

Department of Physics
Stevens Institute of Technology
Hoboken, New Jersey 07030

ABSTRACT

Problems involving curvilinear coordinates and accelerating reference frames can be treated using powerful methods applicable to Maxwell's equations in curved space. A method developed by Cohen and Kegeles reduces the curved-space problem to that of solving a single complex linear scalar wave equation. Gravitational perturbations and neutrino fields can be treated using the same curved-space method. When applied to curved-space scattering problems, the method yields results in a straightforward manner.

INTRODUCTION

A major source of difficulty in integrating the electromagnetic field equations in a given curved space-time is the coupled structure of the Maxwell system, which consists of eight partial differential equations in six unknowns. Standard devices for the flat-space treatment which successfully decoupled the equations fail in curved space, since space-time curvature leads to a more strongly coupled system. One such device, however, has not been fully exploited in the context of curved spaces: the Debye or two-component Hertz-potential formalism. It is the purpose of this paper to show that the Hertz formalism can be extended to all curved space-times, and that the Debye formalism can be extended to a wide and astrophysically interesting class of spaces, in each of which the potential obeys one (decoupled) linear scalar wave equation. Included are, for example, the Friedmann cosmological models, the Kerr and Schwarzschild solutions of black holes and neutron stars, the Gödel universe, Taub-NUT space, the Bondi and Kantowski-Sachs universes, and other universes of various Bianchi types. In fact, the results of ELLIS [1] and WAINWRIGHT [2] show that this method applies to every perfect-fluid model with local rotational symmetry. Mathematically, the class of space-times to which the scalar Debye formalism has been extended is the generalized GOLDBERG-SACHS [3-5] class: every algebraically special geometry, in the sense of PETROV [6], which admits a shear-free congruence of null geodesics along the repeated principal direction of the Weyl tensor. (One must admit from among these the spaces with strong background electromagnetic fields, as required in the test-field approximation.)

In Sec. 1, we summarize the Hertz and Debye potential theory in flat space and formulate the problem of generalizing it to curved spaces. Section 2 couches the theory in the covariant language of differential forms; the notation and powerful theorems of this formalism provide the desired generalization of the Hertzian scheme. Section 3 translates these results into the standard tensor notation, both to assist the reader unfamiliar with forms, and also to facilitate the eventual use of the fully explicit NEWMAN-PENROSE (NP) formalism [4]. In Sec. 4 of [7a], translation of the above results into the concise and explicit NP formalism enables us to contruct a decoupled linear wave equation for the scalar Debye potential in the generalized Goldberg-Sachs class of space-times. Examples for important spaces are given in Sec. 5 of [7a]. The aim of Appendix A of [7a] is to illustrate the procedure of explicitly writing differential form equations in a definite Cartan frame, which is of use at several points in the text. Appendix B of [7a] is intended to enable the reader unfamiliar with the spinor or the NP spin-coefficient formalisms to understand the latter from a purely tetrad or Cartan frame viewpoint, and to calculate spin coefficients, necessary in the applications, by standard Cartan methods.

These methods provide a straightforward procedure for computing spin coefficients with a minimum of calculation.

1. HERTZ AND DEBYE POTENTIALS IN FLAT SPACE

This section is a brief summary of the flat-space theory and is largely based on the paper of NISBET [8]; the reader is referred there for a more detailed discussion. We mention only those results necessary for the subsequent generalization.

HERTZ [9] introduced a potential for the Maxwell field while investigating electric dipole fields; the true covariant bivector nature of this potential was noted considerably later by LAPORTE and UHLENBECK [10]. This type of bivector (antisymmetric second-rank tensor) potential is related by second derivatives to the physical field, hence by first derivatives to the familiar four-vector potential. In fact,

$$\phi = -\nabla\cdot\mathbf{P}_E,\ \mathbf{A} = \dot{\mathbf{P}}_E + \nabla\times\mathbf{P}_M, \tag{1.1}$$

and

$$\begin{aligned}\mathbf{E} &= \nabla(\nabla\cdot\mathbf{P}_E) - \mathbf{P}_E - \nabla\times\ddot{\mathbf{P}}_M = -\nabla\times\dot{\mathbf{P}}_M + \nabla\times(\nabla\times\mathbf{P}_E),\\ \mathbf{B} &= \nabla\times\dot{\mathbf{P}}_E + \nabla\times(\nabla\times\mathbf{P}_M) = \nabla\,\nabla\cdot\ddot{P}_M - \mathbf{P}_M + \nabla\times\dot{\mathbf{P}}_E,\end{aligned} \tag{1.2}$$

are the relations in question. The notation is standard; we choose $c = 1$ and denote the Hertz bivector by $\mathbf{P}_E$ and $\mathbf{P}_M$, according to the natural electric and magnetic labeling of components. The conditions imposed upon the Hertz potential (1.2) and the Maxwell equations

$$\begin{aligned}&\nabla\times\mathbf{E} + \dot{\mathbf{B}} = 0,\ \nabla\cdot\mathbf{B} = 0,\\ &\nabla\times B - \dot{E} = 0,\ \nabla\cdot\mathbf{E} = 0,\end{aligned} \tag{1.3}$$

are just

$$\Box\mathbf{P}_E = 0,\ \Box\mathbf{P}_M = 0 \tag{1.4}$$

in the source-free vacuum case. Here, $\Box = \partial^2/\partial t^2 - \nabla^2$ is the d'Alembertian operator.

A new type of gauge freedom, termed by Nisbet "gauge transformations of the third kind," is associated with the Hertzian potentials. Here we consider gauge transformations of the sources, that is, those gauge terms which may appear as sources in (1.4) while preserving the source-free property of the Maxwell field itself. These turn out to be bivectors of the form

$$\begin{aligned}&\mathbf{Q}_E = \nabla\times\mathbf{G},\ \mathbf{Q}_M = -\dot{\mathbf{G}} - \nabla g,\\ &\mathbf{R}_E = -\mathbf{L} - \nabla l,\ \mathbf{R}_M = -\nabla\times\mathbf{L},\end{aligned} \tag{1.5}$$

where $(\mathbf{G}, g)$ and $(\mathbf{L}, l)$ are arbitrary four-vectors. In this scheme, the wave equations (1.4) for the potentials are modified to become

$$\Box\mathbf{P}_E = \mathbf{Q}_E + \mathbf{R}_E,\ \Box\mathbf{P}_M = \mathbf{Q}_M + \mathbf{R}_M. \tag{1.6}$$

The new fields given by

$$\begin{aligned}\mathbf{E} &= \mathbf{R}_E + \nabla(\nabla\cdot\mathbf{P}_E) - \ddot{\mathbf{P}}_E - \nabla\times\dot{\mathbf{P}}_M = -\nabla\times\dot{\mathbf{P}}_M + \nabla\times(\nabla\times\mathbf{P}_E),\\ \mathbf{B} &= -\ddot{\mathbf{R}}_M + \nabla\times\dot{\mathbf{P}}_E + \nabla\times(\nabla\times\mathbf{P}_M) = \mathbf{Q}_M + \nabla(\nabla\cdot\mathbf{P}_M) - \ddot{\mathbf{P}}_M + \nabla\times\dot{\mathbf{P}}_E,\end{aligned} \tag{1.7}$$

may be verified to remain source-free by substitution into the Maxwell equations (1.3).

NISBET's [8] reduction of the Hertz bivector to two purely radial vectors (DEBYE [11] potentials) utilizes the gauge transformations just discussed. In this representation, the potential is given by $\mathbf{P}_E = \hat{\mathbf{r}}P_E, \mathbf{P}_M = \hat{\mathbf{r}}\mathbf{P}_m$ (with $\hat{\mathbf{r}}$ the unit radial vector), and the gauge bivectors are obtained from (1.5) with $\mathbf{G} = \mathbf{L} = 0$ and $g = 2P_E/r,\ l = 2P_M/r$. The functions P_E and P_M are the Debye potentials and (1.6) implies that they each obey the wave equation (which differs in the radial operator from the scalar d'Alembertian wave equation)

$$-\frac{\partial^2\psi}{\partial t^2} + \frac{\partial^2\psi}{\partial r^2} + \frac{1}{r^2}\left[\frac{1}{\sin\theta}\frac{\partial}{\partial\theta}\sin\theta\frac{\partial\psi}{\partial\theta} + \frac{1}{\sin^2\theta}\frac{\partial^2\psi}{\partial\phi^2}\right] = 0. \tag{1.8}$$

The solutions, which are of the form $\psi = e^{-ikt}rz_l(kr)\,Y_l^m(\theta, \phi)$, with $z_l(kr)$ a spherical Bessel function and $Y_l^m(\theta, \phi)$ a spherical harmonic, give rise to the static $(k = 0)$ and dynamic electric $(P_M = 0)$ and magnetic

($P_E = 0$) multipoles of order l when inserted into the prescription (1.7) for the electromagnetic field [12]. Only the monopole field is missing in this scheme, since the differential operations (1.7) kill the $l = 0$ solution of (1.8).

We emphasize the essential role played in the treatment sketched above by the gauge terms g and l. For, it the d'Alembertian operators of (1.6) are computed explicitly upon the Debye choice $\mathbf{P}_E = \hat{\mathbf{r}}P_E$, $\mathbf{P}_M = \hat{\mathbf{r}}P_M$ of Hertzian vectors, the resultant expressions each contain three components; only by adding the specified gauge terms to the right-hand side does one reduce two components of each equation to identities and the third component to (1.8).

A remarkable economy is achieved by the Debye potentials; the arbitrary source-free Maxwell field is specified by two scalar functions which obey a single separable second-order wave equation. Roughly speaking, one might expect that since a zero-rest-mass field possesses two degrees of freedom, no more economical representation of the Maxwell field is possible.

With the intention of formulating a covariant generalization of the Debye potentials, one may consider the two-potential representation from the following viewpoint. By a suitable choice of bivector direction in space-time $1-m$ the *rt* direction (and its dual $\theta\phi$ direction) $1-m$ one has succeeded in "diagonalizing" the Hertz potential in the sense of SYNGE [13]. That is, one has found the principal directions (and values) of the Hertz vector. Of course, one may in this sense "diagonalize" any bivector, with the resultant principal directions algebraically dependent upon the bivector itself. Remarkably, the Debye scheme shows that all source-free Maxwell fields may be represented by Hertzian bivector potentials with the *same* principal directions, indendent of the Maxwell field, and determined *a priori.*

The problem of covariant generalization of the Debye scheme may thus be viewed as the search for a special bivector direction in space-time, determined *a priori* and presumably geometrically, independently of the details of any particular Maxwell field.

How the generalized GOLDBERG-SACHS theorem [3-5] provides special directions of the required sort in a wide class of space-times is shown in [7,14].

2. DIFFERENTIAL FORMS

The reader totally unfamiliar with the language of differential forms may omit this section; the chief loss will be a certain lack of motivation for some of the formulas of the next section, which is provided here.

Once the results of Sec. 1 are translated from the three-space vector notation into the notation of differential forms [15-20], the framework will be provided for investigation of the curved-space problem, since the latter is a covariant notation. Comparison of this section with Sec. 3 may convince the reader of the superior adaptation of the present notation over standard tensor analysis for problems of this sort (where antisymmetric tensors play a central role).

We make use of the operators *, the Hodge dual; d, the exterior derivative; $\delta = {}^*d^*$, the coderivative; and $\Delta \equiv d\delta + \delta d = d^*d^* + {}^*d^*d$, the harmonic operator. The operator Δ has the property of reducing in Minkowski space (or flat three-space) to the d'Alembertian (or Laplacian) operator.

The flat-space equations of Sec. 1 are now presented in the differential-forms notation; once an equation is written in this formalism, it is fully covariant. That the translations are correct may be verified by explicitly writing out the equations below in some Cartan frame. This procedure is illustrated in Appendix A of [7a] for selected equations.

In terms of the Maxwell 2-form $f = \frac{1}{2}f_{\mu\nu}\omega^{\mu} \wedge \omega^{\nu}$, where $f_{\mu\nu}$ is the Maxwell tensor and ω^{α}, $\alpha = 0, 1, 2, 3$, are the basis forms, the Maxwell equations (1.3) become

$$df = 0,$$
$$\delta f = 0. \tag{2.1}$$

The equations (1.1) relating the Hertz bivector (2-form) P to the four-vector (1-form) potential A becomes

$$A = \delta P. \tag{2.2}$$

The analogue of (1.2) giving f in terms of P is

$$f = d\delta P = -\delta dP, \tag{2.3}$$

the equality of the last two expressions requiring

$$\Delta P = 0, \tag{2.4}$$

the analogue of (1.4). Now, the fact that an f given by (2.3) is Maxwell field (i.e., satisfies (2.1)) is a trivial consequence of the identity $d^2 \equiv 0$— the exterior derivative applied twice kills any form. That is, we have

$$df = d(d\delta P) \equiv 0,$$
$$\delta f = \delta(-\delta dP) \equiv 0, \tag{2.5}$$

where we have used the corollary $\delta^2 \equiv 0$ (a consequence of $*^2 = \pm$ the identity, the sign depending on the dimension of the form, so that $\delta^2 = (*d*)(*d*) = \pm *d* \equiv 0$).

The 2-form gauge terms (1.5) are

$$Q = dG,$$
$$R = *dL, \tag{2.6}$$

where G and L are arbitrary 1-forms. The wave equation (1.6) with gauge terms is therefore

$$\Delta P = dG + *dL, \tag{2.7}$$

so that the gauge-transformed field (1.7) becomes

$$f = d\delta P - dG$$
$$= *dL - \delta dP. \tag{2.8}$$

That the transformed fields (2.8) still obey the Maxwell equations (2.1) is again a trivial application of $d^2 \equiv \delta^2 \equiv 0$.

Equations (2.7) and (2.8) represent a *fully covariant generalization of the Hertz potential scheme to all curved space-times.*

The Debye two-component reduction of this formalism in flat space may now be summarized as follows. We use the spherical orthonormal Cartan frame $\omega^0 = dT$, $\omega^1 = dr$, $\omega^2 = r\, d\theta$, $\omega^3 = r \sin\theta\, d\phi$; we choose the Hertz 2-form to be of the form $P = P_E\omega^0 \wedge \omega^1 + P_M\omega^2 \wedge \omega^3$, and select gauge 1-forms $G = (2P_E/r)\omega^0$, $L = (2P_M/r)\omega^0$. Then, just as in Sec. 1, (2.7) results in the wave equation (1.8) for both P_E and P_M (see Appendix A of [7a]), and (2.8) yields the standard electromagnetic multipoles.

It should be remarked that the entire Hertzian scheme as generalized above to curved spaces would fail if the world were Riemannian (as opposed to pseudo-Riemannian). For, if space-time were a compact Riemann space, a result in Hodge theory [16, 17] would say that $\Delta P = 0$ if and only if dP and δP vanish. But then the prescriptions (2.3) for the Maxwell field would be identically zero.

3. TENSOR NOTATION

There is no unique translation of the formulas of Sec. 1 into tensor notation; in particular, many second-order tensor operators reduce in flat space to the wave operator of (1.4) (two examples are the contracted second covariant derivative operator and the operator which we, in fact, adopt below). There is, however, a unique translation of the formulas of Sec. 2 into tensor notation, since the forms language is covariant; we choose this unique prescription. In effect, we are allowing the operators defined on forms to make the choice for us. The reader who has noted the elegance of the formulas of the previous section (based on powerful properties of the operators in the theory of differential forms) will see the motivation for this choice.

For relations between the operators defined on forms and the covariant derivative operator of conventional tensor analysis, [17] is especially recommended.

In the present notation, the Maxwell equations (1.3) or (2.1) become

$$\nabla_\mu f_{\nu\lambda} + \nabla_\nu f_{\lambda\mu} + \nabla_\lambda f_{\mu\nu} = 0,$$
$$\nabla^\mu f_{\mu\nu} = 0, \tag{3.1}$$

where ∇_μ denotes the covariant derivative. The Hertzian bivector potential $P_{\mu\nu}$ is related to the four-vector potential A_μ (see (1.1) or (2.2)) by

$$A_\mu = -\nabla^\lambda P_{\lambda\mu}. \tag{3.2}$$

Equations (1.2) or (2.3), giving $f_{\mu\nu}$ in terms of $P_{\mu\nu}$, become

$$\begin{aligned} f_{\mu\nu} &= -\nabla_\mu \nabla^\lambda P_{\lambda\nu} + \nabla_\nu \nabla^\lambda P_{\lambda\mu} \\ &= \nabla_\lambda \nabla^\lambda P_{\mu\nu} - \nabla^\lambda \nabla_\mu P_{\lambda\nu} + \nabla^\lambda \nabla_\nu P_{\lambda\mu}. \end{aligned} \tag{3.3}$$

The last equality specifies the wave equation analogous to (1.4) or (2.4) to be

$$-\nabla_\lambda \nabla^\lambda P_{\mu\nu} + (\nabla^\lambda \nabla_\mu - \nabla_\mu \nabla^\lambda) P_{\lambda\nu} + (\nabla_\nu \nabla^\lambda - \nabla^\lambda \nabla_\nu) P_{\lambda\mu} = 0. \tag{3.4}$$

It is readily seen via the Ricci identities that the last two terms are proportional to the Riemann tensor, so that the operator of equation (3.4) could no be preferred over the contracted second covariant derivation (the first term alone) purely from the standpoint of a generalization from flat space. Thus, an investigator working in the tensor formalism might not have succeeded in finding the operator of equation (3.4).

The fact that the field tensor (3.3) obeys (3.1) is no longer proved by inspection as in the notation of forms. We present a proof.

For the first set of Maxwell equations (the first of (3.1)) we choose the first equation (3.3) for $f_{\mu\nu}$ and express it in terms of A_μ ((3.2)), so that $f_{\mu\nu} = \nabla_\mu A_\nu - \nabla_\nu A_\mu$. Then

$$\begin{aligned} \nabla_\mu f_{\nu\lambda} + \nabla_\nu f_{\lambda\mu} + \nabla_\lambda f_{\mu\nu} &= \nabla_\mu \nabla_\nu A_\lambda - \nabla_\mu \nabla_\lambda + A_\nu \nabla_\nu \nabla_\lambda A_\mu - \nabla_\nu \nabla_\mu A_\lambda + \nabla_\lambda \nabla_\mu A_\nu - \nabla_\lambda \nabla_\nu A_\mu \\ &= (\nabla_\mu \nabla_\nu - \nabla_\nu \nabla_\mu) A_\lambda + (\nabla_\lambda \nabla_\mu - \nabla_\mu \nabla_\lambda) A_\nu + (\nabla_\nu \nabla_\lambda - \nabla_\lambda \nabla_\nu) A_\mu \\ &= R^\sigma_{\lambda\nu\mu} A_\sigma + R^\sigma_{\nu\mu\lambda} A_\sigma + R^\sigma_{\mu\lambda\nu} A_\sigma \text{ (by the Ricci identities)} \\ &= 3R^\sigma_{[\lambda\nu\mu]} A_\sigma = 0 \text{ (by the cyclic symmetry of the Riemann tensor).} \end{aligned}$$

Similarly, for the second equation (3.1) we express $f_{\mu\nu}$ by the second equation (3.3) and introduce a potential $B_{\lambda\mu\nu}$ whose four-divergence (as opposed to A_μ whose curl) gives $f_{\mu\nu}$. Thus,

$$B_{\lambda\mu\nu} = \nabla_\lambda P_{\mu\nu} - \nabla_\mu P_{\lambda\nu} + \nabla_\nu P_{\lambda\mu} \quad \text{and} \quad f_{\mu\nu} = \nabla^\lambda B_{\lambda\mu\nu}$$

together yield the second equation (3.3) (note the total anti-symmetry of $B_{\lambda\mu\nu}$). Now, the second Maxwell equation gives

$$\begin{aligned} \nabla^\mu f_{\mu\nu} &= \nabla^\mu \nabla^\lambda B_{\lambda\mu\nu} \\ &= \frac{1}{2}(\nabla^\mu \nabla^\lambda - \nabla^\lambda \nabla^\mu) B_{\lambda\mu\nu} \\ &= \frac{1}{2}(R^{\sigma\lambda\mu}_{\lambda} B_{\sigma\mu\nu} + R^{\sigma\lambda\mu}_{\mu} B_{\lambda\sigma\nu} + R^{\sigma\lambda\mu}_{\nu} B_{\lambda\mu\sigma}), \end{aligned}$$

where the second line follows by the antisymmetry of $B_{\lambda\mu\nu}$ and the third line by the Ricci identities. The first and second terms inside the brackets in the third line vanish, since the first factor of these terms is symmetric and the second antisymmetric in σ, μ (first term) or σ, λ (second term). The third term equals

$$-\frac{1}{2} R_{\nu\sigma\lambda\mu} B^{\sigma\lambda\mu} = -\frac{1}{2} R_{\nu[\sigma\lambda\mu]} B^{\sigma\lambda\mu},$$

by the antisymmetry of $B^{\sigma\lambda\mu}$, and also vanishes by the symmetry property $R_{\nu[\sigma\lambda\mu]} = 0$ of the Riemann tensor.

The gauge terms of (1.5) and (2.6) are, in tensor notation,

$$\begin{aligned} Q_{\mu\nu} &= \nabla_\mu G_\nu - \nabla_\nu G_\mu, \\ R_{\mu\nu} &= -\nabla^\lambda L_{\lambda\mu\nu}, \end{aligned} \tag{3.5}$$

where G_μ and $L_{\lambda\mu\nu}$ are arbitrary tensors, except that $L_{\lambda\mu\nu}$ is totally antisymmetric (L is essentially the dual of the arbitrary 4-vector or 1-form with the same kernel letter of the previous sections).

With gauge terms included, the wave equation for the potential becomes ((1.6), (2.7))

$$-\nabla_\lambda \nabla^\lambda P_{\mu\nu} + (\nabla^\lambda \nabla_\mu - \nabla_\mu \nabla^\lambda) P_{\lambda\nu} + (\nabla_\nu \nabla^\lambda - \nabla^\lambda \nabla_\nu) P_{\lambda\mu} = \nabla_\mu G_\nu - \nabla_\nu G_\mu - \nabla^\lambda L_{\lambda\mu\nu} \tag{3.6}$$

and the gauge-transformed field tensor ((1.7), (2.8)) is given by

$$\begin{aligned} f_{\mu\nu} &= -\nabla_\mu \nabla^\lambda P_{\lambda\nu} + \nabla_\nu \nabla^\lambda P_{\lambda\mu} - \nabla_\mu G_\nu + \nabla_\nu G_\mu \\ &= \nabla_\lambda \nabla^\lambda P_{\mu\nu} - \nabla^\lambda \nabla_\mu P_{\lambda\nu} + \nabla^\lambda \nabla_\nu P_{\lambda\mu} - \nabla^\lambda L_{\lambda\mu\nu}. \end{aligned} \tag{3.7}$$

Using a proof similar to that given above for the field (3.3), one can show that the transformed field still obeys the source-free Maxwell equations.

Equations (3.6) and (3.7) comprise a covariant generalization of the Hertz potential formalism to all space-times.

The flat-space Debye two-component reduction of the potential is now given in tensor notation. In the natural spherical coordinate basis for tensors, we choose $P_{tr} = -P_{rt} = P_E$; $P_{\theta\phi} = -P_{\phi\theta} = r^2 \sin\theta\, P_M$; all other components vanish. For gauge terms we take $G_t = 2P_E/r$, the other components being zero; and $L_{r\theta\phi} = 2r^2 \sin\theta P_M/r$, other components given anti-symmetrically by permuting indices, or else vanish. Again, the statement is that with these choices, (3.6) yields the wave equation (1.8) for P_E and P_M, and that (3.7) gives the standard electromagnetic multipole fields.

4. DEBYE POTENTIALS IN CURVED SPACES

With the covariant machinery for the Hertzian potentials set up in the last two sections, it is possible [7] to formulate a two-component (or one-complex-component) reduction of the potential analogous to the Debye scheme in flat space. The problem consists of finding special bivector directions in space-time so that (3.6) or (2.7) yields decoupled wave equations for the corresponding components of the potential for some choice of gauge terms (3.5) or (2.6). In [7], we showed that in a class of space-times, the Weyl tensor provides such special bivectors through its principal directions. These are, as required, defined geometrically by the space-time itself and independently of the Maxwell fields to be computed.

5. DISCUSSION

The flat-space method of electromagnetic Hertz potentials has been generalized to all curved space-times. The covariant formulation of this procedure has provided the framework for an extension of the Debye potential scheme to an astrophysically interesting class of spaces, where it gives a new, direct, and practical method for constructing Maxwell fields by solving one decoupled linear scalar wave equation. This formulation allows realistic problems in relativistic astrophysics associated with neutron stars, pulsars, black holes, and global (cosmological) phenomena to be investigated by direct computation.

Results strictly annlogous to the spin-1 results of this paper [7] have been obtained for zero-rest-mass fields with other physically interesting values of spin and have been presented elsewhere [14].

When applied to curved-space scattering problems, the methods yield results in a straightforward manner [21]. An example involving black hole scattering can be found in [21].

FOOTNOTES AND REFERENCES

*Supported in part by Air Force Office of Scientific Research Grant AF0SR-78-3608
**Supported in part by NSF Grant PH No. Y77-28356

[1] G.F.R. Ellis: J. Math. Phys. **8**, 1171 (1967)
[2] J. Wainwright: Commun. Math. Phys. **17**, 42 (1970)
[3] J.N. Goldberg, R.K. Sachs: Acta Physica Polonica **22**, 13 (1962)
[4] E. Newman, R. Penrose: J. Math. Phys. **3**, 566 (1962)
[5] W. Kundt, A.H. Thompson: C.R. Acad. Sci. (Paris) **254**, 4257 (1962); I. Robinson, A. Schild: J. Math. Phys. **4**, 484 (1962)
[6] A.Z. Petrov: Sci. Not. Kazan State Univ. **114**, 55 (1954)
[7] J. M. Cohen, L.S. Kegeles: Phys. Rev. D **10**, 1070 (1974); Phys. Lett. **47A**, 261 (1974)
[8] A. Nisbet: Proc. Roy. Soc. (London) **A231**, 250 (1955)
[9] H. Hertz: Ann. Phys. (Leipz.) **36,** 1 (1889)
[10] O. Laporte, G.E. Uhlenbeck: Phys. Rev. **37**, 1380 (1931)
[11] P. Debye: Ann. Phys. (Leipz.) **30**, 57 (1909)
[12] W.K.H. Panofsky, M. Phillips: *Classical Electricity and Magnetism* (Addison-Wesley, Cambridge, Massachusetts, 1955), Chap. 13
[13] J.L. Synge: *Relativity: The Special Theory* (North-Holland, Amsterdam, 1965), 2nd ed., p. 335
[14] J. M. Cohen, L.S. Kegeles: Phys. Lett. **54A**, 5 (1975); Phys. Rev. D **19**, 1641 (1979)
[15] E. Cartan: *Les Systémes Différentiels extérieurs et leurs applications géométriques,* Exposés de Géometrie **14** (Hermann, Paris, 1945)

[16] W.V.D. Hodge: *The Theory and Applications of Harmonic Integrals* (Cambridge, 1952)
[17] G. de Rham: *Variétés différentiables* (Hermann, Paris, 1960)
[18] H. Flanders: *Differential Forms with Applications to the Physical Sciences* (Academic, New York, 1963)
[19] C.W. Misner and J.A. Wheeler: Ann. Phys. N.Y. **2**, 525 (1957)
[20] D.R. Brill, J.M. Cohen: J. Math. Phys. **7**, 238 (1966)
[21] M.W. Kearney, L.S Kegeles, J.M. Cohen: Astrophys. and Space Sci. **56**, 129 (1978)

LEVINSON'S THEOREMS AND THE QUANTUM-MECHANICAL PARTITION FUNCTION FOR PLASMAS

D. Bollé

Instituut voor Theoretische Fysica
Universiteit Leuven, B-3030 Leuven
Belgium

INTRODUCTION

It is perfectly clear by now that the well-known Levinson's theorem not only provides some deep theoretical insight into the scattering process but also serves as a basis for many applications in a wide variety of mathematical and physical problems. Therefore, it is still worthwhile to extend this Levinson's theorem to more general scattering situations.

In the first part of this talk, we discuss a new set of Levinson-type theorems for two-body scattering with potentials belonging to $L^1 \cap L^2$. These theorems are expressed in the form of trace formulas for the two body time delay operator. In that respect, this part can be viewed as a contribution to the general study of time delay in scattering.

In the second part, we want to give an idea how these time delay aspects of modern scattering theory can be used successfully to discuss some properties of the quantum-mechanical partition function in statistical mechanics. In particular, we study a high-temperature expansion of the two-body interaction part of the partition function for a gas of Boltzmann particles. We clearly see how a cancellation between bound-state contributions and scattering contributions to this expansion comes about. This cancellation is important, e.g., in chemical equilibrium calculations in plasmas. Up to now, this cancellation has only been shown in WKB approximation. Here we report a full quantum-mechanical proof for smooth potentials.

1. TWO-BODY TIME DELAY

We start with a very short discussion of the known features of two-body time delay [1] which we employ in this analysis.

The scattering system we look at is characterized by an interacting Hamiltonian h and an asymptotic Hamiltonian h_o. These operators act on a Hilbert space $H = L^2(\mathbf{R}^3)$. Here and in the following we systematically remove the center-of-mass motion from the problem.

To define the time delay, we consider a family of projection operators $P(R)$ onto a sphere of radius R in position space around the center-of-mass point. Given an incident wave packet $f \in H$ and a certain radius R, the time delay is determined by

$$T(f,R) = \int_{-\infty}^{+\infty} [\|P(R)V_t\Omega_- f\|^2 - \|P(R)U_t f\|^2]dt, \tag{1}$$

where $V_t = e^{-iht}$ is the total evolution, $U_t = e^{-ih_ot}$ the free evolution, and $\Omega_\pm = \underset{t\to\pm\infty}{\text{s-lim}}\; V_t U_t$ the Møller wave operators. The physical meaning of $T(f,R)$ may be read off from the right-hand side of (1). It is the difference between the total time spent in the sphere R by the scattering state $\Omega_- f$ evolving under the total evolution and the initial state f evolving freely. In order to obtain a quantity which is independent of the radius of the sphere, we take the limit $R \to \infty$ of (1). We assume that this limit exists on a dense set D of vectors f for the class of potentials $V \in L^1 \cap L^2$ and is associated with a certain operator q which can be written in terms of the scattering operator and whose expectation values on D coincide with this limit. For rigorous work in this connection we refer to [2-4].

Let us now describe the relation between the time delay operator q and the scattering matrix, $S = \Omega_+^*\Omega_-$ [5]. It is known that the two-body S-operator is decomposable in the spectral representation of h_o. Furthermore, we shall assume that S is simply multiplication by a function $s(E)$ in this spectral representation (S is "diagonalizable," which is always verified if h_o has a simple spectrum) [6]. In physical terms, this means that S is an on-shell operator and that we want to subtract out the energy conserving δ-function by defining a family of reduced energy-dependent S-matrices $s(E)$ that act on $\hat{H} = L^2(\theta,\phi)$. In the same way, the theory of time delay allows us to construct a family of operators $q(E)$ acting on $\hat{H}$ [2,5]. In terms of $s(E)$, the operator $q(E)$ may be expressed as [3,5]

$$q(E) = -is^*(E)\frac{d}{dE}s(E). \tag{2}$$

We remark that this relation has also been generalized to many-particle scattering [7].

A last feature of time delay we need is known as the spectral property. Let $r(z) = (h - z)^{-1}$ and $r_o(z) = (h_o - z)^{-1}$ be the resolvents of h and h_o defined for complex energy z. Then [2,8],

$$\operatorname{Im}\operatorname{Tr}[r(E+i\eta) - r_o(E+i\eta)]$$
$$= \sum_{j=1}^{N_b} \frac{\eta}{|\chi_j^2 + E + i\eta|^2} + \frac{1}{2\pi}\int_0^\infty \frac{\eta}{(E-E')^2+\eta^2}\operatorname{tr}q(E')dE' \tag{3}$$

where Tr is the trace on H, tr the trace on $\hat{H}$ and the $-\chi_j^2$ are the bound-state energies; the index j runs over all bound states. In the limit $\eta \to 0$, this property has a simple interpretation: the change of state density produced by the interaction V becomes the trace of the time delay operator.

2. A SET OF TWO-BODY LEVINSON TYPE THEOREMS

To derive the new set of two-body Levinson's theorems in terms of the energy moments of the trace of the time-delay operator, we need to know the analytic behavior of $\operatorname{Tr}[r(z) - r_o(z)]$ in the complex plane (see (3)).

Let Π_η be defined as the set of points in the z-plane a distance η or greater away from the positive real axis. Then [9],

Lemma 1. *Let $V \in L^1 \cap L^2$. For all positive integers n, the operators $r_o(z)[Vr_o(z)]^n$ are trace class for $z \in \Pi_\eta$. The function* $\operatorname{Tr}\{r_o(z)[Vr_o(z)]^n\}$ *is an analytic function of z in the Π_η domain.*

Lemma 2. *There exist finite k_r and k_i such that for all $|z| > (k_r^2 + k_i^2)/2\mu$ the Born series expansion*

$$\operatorname{Tr}[r(z) - r_o(z)] = \sum_{n=1}^{\infty} (-1)^n \operatorname{Tr}\{r_o(z)[Vr_o(z)]^n\}$$

is valid. The series is absolutely convergent in z.

Lemma 3. *Set $z = k^2/2\mu$. For all integers $n \geq 2$,* $\operatorname{Tr}[Vr_o(z)]^n$ *satisfies*

$$\lim_{|\operatorname{Re}k|\to\infty} \operatorname{Tr}[Vr_o(z)]^n = 0 \qquad \text{for all } \operatorname{Im}k \geq 0.$$

The next step is to consider the function $Q_N(z)$ defined by

$$Q_N(z) = z^N \operatorname{Tr}\{r(z) - r_o(z) - \sum_{n=1}^{N+1} (-1)^n r_o(z)[Vr_o(z)]^n\}$$

for $N = 0,1,2,\ldots$ In this formula we have introduced $(N+1)$ "correction terms", anticipating the high-energy behavior of the trace. From the foregoing lemmas, it is clear that this function $Q_N(z)$ is analytic in Π_η. Therefore, its contour integral around the spectrum of h is equal to zero. This contour can be split up in the segments C_j, C_Γ, and C_η. The contours C_j encircle the P distinct eigenvalues of the Hamiltonian h. The segment C_Γ is a circle at a distance Γ from the origin. The contour C_η "runs around" the positive real axis at a distance η. It is straightforward to verify that the C_j contributes an amount $2\pi i \sum_{j=1}^{N_b} (-\chi_j^2)^N$. For C_η, we can show that [10]

Lemma 4. *Let $V \in L^1 \cap L^2$ and assume that the Hamiltonian h has no zero energy eigenstates. Then*

$$\lim_{\Gamma\to\infty}\lim_{\eta\to 0}\int_{C_\eta} Q_N(z)\,dz$$

$$= i\int_0^\infty E^N\{\mathrm{tr}\,q(E) - 2\,\mathrm{Im}\,\mathrm{Tr}\sum_{n=1}^{N+1}(-1)^n r_o(E)[Vr_o(E)]^n\}\,dE.$$

Finally, the C_Γ integral does not contribute.

Lemma 5. *For potentials $V \in L^1 \cap L^2$, we have that*

$$\lim_{\Gamma\to\infty}\int_{C_\Gamma} Q_N(z)\,dz = 0.$$

These conclusions can be combined to give us the following set of Levinson's theorems [10].

Theorem. *Let $V \in L^1 \cap L^2$. Assume that h has no zero energy resonances or bound states, and no bound states in the continuum. Then*

$$\int_0^\infty E^N\{\mathrm{tr}\,q(E) - 2\,\mathrm{Im}\,\mathrm{Tr}\sum_{n=1}^{N+1}(-1)^n r_o(E)[Vr_o(E)]^n\}\,dE$$

$$= -2\pi\sum_{j=1}^{N_b}(-\chi_j^2)^N,$$

for $N = 0,1,2,3,\ldots$.

The justification for calling these trace relations Levinson's theorems is based on the fact that the time delay operator in a given partial wave is just the derivative of the phase shift with respect to energy, as can be shown from (2). We remark that alternative discussions of the first $N = 0$ relation were given recently by NEWTON [11] and DREYFUS [12].

Finally, we discuss the correction terms appearing in these theorems. Ideally, one would like to have an easy way to obtain these terms as functions of the potential describing the scattering. So far, we know that they stem from the high-energy behaviour of the time delay operator, or, equivalently, from the high-energy behaviour of the full resolvent. For smooth potentials ($V \in S$) we can show [10] that these correction terms are given by $\sum_{k=1}^{N} c_k E^{-k+1/2}$, where the c_k can be calculated easily by using the elegant recurrence relations of PERELOMOV [13]. Because of the basic relationship between the Schrödinger equation and the Korteweg-de Vries equation [14], these c_k are generalizations of the polynomial conserved densitities of the latter. For explicit forms of the c_k we refer to the literature [14]. So for $V \in S$ we arrive at

$$\int_0^\infty dE_{N+1}\int_{E_{N+1}}^\infty dE_N \ldots \int_{E_2}^\infty dE_1\Big[\mathrm{tr}\,q(E_1) - 2c_1E_1^{-1/2} - 4c_2E_2^{-1/2}\ldots$$

$$\ldots - \frac{2^{N+1}}{(2N-1)!!}\,c_{N+1}E_{N+1}^{-1/2}\Big] = \frac{-2\pi}{N!}\sum_{j=1}^{N_b}(-\chi_j^2)^N. \tag{4}$$

We remark that these relations, which are a special case of the relations given in the theorem, have been discussed by BUSLAEV [15] under differing circumstances. In Buslaev's version the time delay is replaced by the logarithmic S-matrix derivative form according to (2). In our derivation, we do not need the S-matrix. Furthermore, in the results reported by Buslaev there is no recognition that the correction terms are Korteweg-de Vries type invariants.

3. THE QUANTUM-MECHANICAL PARTITION FUNCTION FOR PLASMAS

In this section, we want to give an idea how these time delay results can be applied to understand certain features of the quantum-mechanical partition function [10].

We start by writing the partition function for a gas of Boltzmann particles in terms of the time-delay operator. Using the cluster-expansion representation of the quantum-mechanical grand canonical partition function, we write for the second virial coefficient [16]

$$a_2 = -\sqrt{2}\,\lambda^3\,\mathrm{Tr}\,(e^{-\beta h} - e^{-\beta h_o}), \tag{5}$$

where λ is the thermal wavelength, $\lambda = (2\pi\hbar^2/\mu kT)^{1/2}$. We then introduce the Watson transform [17], which connects the statistical operator with the resolvent via a contour integral over the spectrum of h. Evaluating that integral we obtain

$$a_2 = -\sqrt{2}\,\lambda^3 \left\{ \sum_{j=1}^{N_b} e^{\beta\chi_j^2} + \frac{1}{2\pi}\int_0^\infty e^{-\beta E}\,\mathrm{tr}\,q(E)\,dE \right\}. \tag{6}$$

It is straightforward to show that this formula is a generalization of the well-known Beth-Uhlenbeck result [16]. One advantage of (6), is that it remains defined even when angular momentum is not conserved and the phase shift is undefined. Furthermore a formula of the type (6) remains valid for the third and higher virial coefficients [8].

On the basis of this representation (6), we are able to derive a high-temperature expansion for a_2 in the case of smooth potentials [10,18] using the set of Levinson's theorems (4). To do this, let us concentrate on the continuum contribution in (6). We successively perform the following steps: add and subtract a term $2c_1E^{-1/2}$ under the integral, do a partial integration with respect to energy, and use the 0th order ($N = 0$) Levinson's theorem, to rewrite the result; we again add and substract a term $-\ 4c_2E_1^{-1/2}$, do a partial integration, use the $N = 1$ Levinson's theorem, etc. We then arrive at the following series:

$$a_2 = -\sqrt{2}\,\lambda^3 \left\{ \sum_{j=1}^{N_b} e^{\beta\chi_j^2} + \frac{c_1}{(\pi\beta)^{1/2}} - N_b - 2c_2\left(\frac{\beta}{\pi}\right)^{1/2} - \beta\sum_{j=1}^{N_b}\chi_j^2 \right.$$
$$\left. + \frac{4}{3}\,\frac{c_3}{\pi^{1/2}}\,\beta^{3/2} - \frac{\beta^2}{2}\sum_{j=1}^{N_b}\chi_j^4 - \frac{8}{15}\,\frac{c_4}{\pi^{1/2}}\,\beta^{5/2} - \frac{\beta^3}{3}\sum_{j=1}^{N_b}\chi_j^6 + \ldots \right\}, \tag{7}$$

or

$$a_2 = -\ 4\pi\left(\frac{\hbar^2}{2\mu}\right)^{3/2} \sum_{k=1}^{\infty} (-1)^{k+1}\,\frac{2^{k-1}}{(2k-3)!!}\,c_k\beta^k. \tag{8}$$

The result (8) for the high-temperature expansion of the two-body interacting part of the quantum mechanical partition function is nothing but the Kirkwood-Wigner expansion [16], which can be derived in many other ways. Only recently, however [19], it was remarked that the coefficient of β^k is the kth conserved charge density for the Korteweg-de Vries equation.

The advantage of the method of derivation explained in this talk is apparent in (7). There we see an explicit cancellation between the bound state and continuum contributions to a_2. Such a cancellation is important, e.g., in chemical equilibrium calculations in plasmas, where the bound state contributions are taken as the appropriate low-order term (= Saha equation) to calculate the internal partition function. Since the effect of the scattering states only appears in the higher order terms, it is essential to properly include compensation with the continuum in the definition of that internal partition function. In that respect, this compensation has been studied already numerically in the literature [20] in the framework of partial wave scattering. Consequently, one employs the following form for the effective internal partition function

$$Z_1 = -\sqrt{2}\,\lambda^3 \sum_{j=1}^{\infty}\sum_{l=0}^{j-1} (2l+1)(e^{\beta\chi_{jl}^2} - 1 - \beta\chi_{jl}^2). \tag{9}$$

Furthermore, it was also shown that the -1 compensation with the continuum depends only on the analytic properties of the phase shift [21]. Only recently, ROGERS [22] realized that the exact compensation of the $\beta\chi_{jl}^2$ term, if it applied to general potentials, would imply a higher-order Levinson-type theorem. In the absence of such a theorem, he studied the compensation in WKB approximation.

We have used precisely a set of higher-order Levinson's theorems to provide a proof and an explanation of the total cancellation of the bound state contributions for smooth potentials on a full quantum mechanical level. Moreover, our method does not use a partial wave projection so that it is valid in more general cases like, e.g., nonspherically symmetric interactions. Finally, it also clearly shows how this cancellation comes into play. Indeed, the basic ingredients of our method are the Levinson's theorems in terms of time delay. It is exactly those relations that couple the time delay, describing the scattering continuum, with the bound states and the high-energy behavior of the resolvent. In this way, the derivation of the high-temperature expansion and this cancellation are naturally interconnected.

4. CONCLUSION

To conclude this talk, we first remark that the general method described here is, in principle, applicable to higher virial coefficients, written down as functions of the two basic scattering parameters, namely, the bound state energy and the many-body time delay. It would certainly be interesting to see, e.g., what the contribution of the third virial coefficient is to the high-temperature behavior of a plasma.

Secondly, we recall that we have studied the behavior of the partition function for smooth potentials. In the case of more realistic potentials, e.g., Yukawa or Coulomb (long range!), which have a $1/r$ singularity at the origin, it is known [23] that nonanalytic terms appear in the series (8) for a_2. We are presently investigating if our methods can be modified to incorporate this effect.

FOOTNOTES AND REFERENCES

*Bevoegdverklaard Navorser NFWO, Belgium

[1] D. Bollé, T.A. Osborn: Phys. Rev. D **13**, 299 (1976) and references cited therein
[2] J.M. Jauch, K.B. Sinha, B.N. Misra: Helv. Phys. Acta **45**, 398 (1972)
[3] Ph.A. Martin: Comm. Math. Phys. **47**, 221 (1976)
[4] R. Lavine: in *Scattering Theory in Mathematical Physics*, ed. by J.A. LaVita, J.-P. Marchand (D. Reidel, Dordrecht, Holland, 1974), p. 141
[5] J.M. Jauch, J.-P. Marchand: Helv. Phys. Acta **40**, 217 (1967)
[6] W.O. Amrein, J.M. Jauch, K.B. Sinha: *Scattering Theory in Quantum Mechanics* (Benjamin, Reading, Massachusetts, 1977)
[7] D. Bollé, T.A. Osborn: J. Math. Phys. **20**, 1121 (1979)
[8] T.A. Osborn, T.Y. Tsang: Ann. Phys. (N.Y.) **101**, 119 (1976)
[9] T.A. Osborn, D. Bollé: J. Math. Phys. **18**, 432 (1977)
[10] D. Bollé: to be published in Ann. Phys. (N.Y.)
[11] R.G. Newton: J. Math. Phys. **18**, 1348 (1977)
[12] T. Dreyfus: Helv. Phys. Acta **51**, 321 (1978)
[13] A.M. Perelomov: Ann. Inst. H. Poincaré **24**, 161 (1976)
[14] R.M. Miura, C.S. Gardner, M.D. Kruskal: J. Math. Phys. **9**, 1204 (1968); C.S. Gardner, J.M. Greene, M.D. Kruskal, R.M. Miura: Phys. Rev. Lett. **19**, 1095 (1967)
[15] V.S. Buslaev: in *Topics in Mathematical Physics*, ed. by M. Sh. Birman (Consultants Bureau, New York), Vol. I, p. 69
[16] K. Huang: *Statistical Mechanics* (Wiley, New York, 1964)
[17] K.M. Watson, Phys. Rev. **103**, 489 (1956)
[18] D. Bollé, H. Smeesters: Phys. Lett. **62A**, 290 (1977)
[19] M. Kac and P. Van Moerbeke: Proc. Nat. Acad. Sci. USA **71**, 2350 (1974); H.P. McKean, P. Van Moerbeke, Invent. Math. **30**, 217 (1975)
[20] A.I. Larkin: Sov. Phys.-JETP **11**, 1363 (1960); Y.G. Krasnikov: Sov. Phys.-JETP **26**, 1252 (1968); V.P. Kopyshev: Sov. Phys.-JETP **28**, 648 (1969); W. Ebeling, Ann. Physik **22**, 33, 383, 392 (1969); W. Ebeling, R. Sandig; Ann. Physik **28**, 289 (1973); W. Ebeling, Physica **73**, 573 (1974)
[21] F.J. Rogers, H.C. Graboske, Jr., H.E. DeWitt: Phys. Lett. **34A**, 127 (1971); A.G. Petchek: Phys. Lett. **34A**, 411 (1971); W.G. Gibson: Phys. Lett. **36A**, 403 (1971); D. Kremp, W.D. Kraeft: Phys. Lett. **38A**, 167 (1972)
[22] F.J. Rogers: Phys. Lett. **61A**, 358 (1977)
[23] J.E. Avron: Ann. Phys. (N.Y.) **108**, 448 (1977); H.E. DeWitt: J. Math. Phys. **3**, 1003 (1962)

A CLUSTER EXPANSION AND EFFECTIVE-POTENTIAL REPRESENTATION OF THE THREE-BODY PROBLEM

T. A. Osborn and D. Eyre

Department of Physics, University of Manitoba
Winnipeg R3T 2N2, Canada

INTRODUCTION

The objective of this paper is to describe cluster expansions of the three-body problem and to assess, by computation, the utility of both the exact and approximate descriptions of the three-body scattering theory that emerge from the cluster approach. In part, this work is motivated by the success in nuclear collisions of the resonating group theory to approximately describe the elastic scattering of two tightly bound clusters. However, the resonating group theory treatment of the scattering process [1] is not an exact description of the N-body scattering process and a number of attempts to bring it within the framework of a complete and exact scattering formalism have been made [2]. The basic goal of any cluster expansion is to arrive at a simple approximate description of the scattering process that is accurate when clustering dominates the underlying physical states of the system. In this paper, we obtain a set of integral equations that feature in a natural way two cluster states and at the same time remain exact descriptions of three-body scattering. The approximate solutions of these equations obtained by low order iterations defines the cluster expansion of the three-body problem and enables us to quantitatively investigate the accuracy of this description and its rate of convergence to the exact three-body scattering solutions.

Our approach to cluster representations is set in the context of the KARLSSON-ZEIGER (hereafter KZ) integral equations [3]. Generally, integral equations for the three-body problem require a choice of complete basis in momentum space. In the approach of FADDEEV [4], the basis is taken to be plane waves of three free particles, whereas in the KZ equations the basis is taken as one free particle adjoined to an interacting pair that is either in a scattering state or a bound state. The resultant KZ equations are structurally different from those found by Faddeev. In particular, the complex three-body energy z appears only in the propagators of the equations. The simplicity of the propagators in the KZ equations allows us the decompose these Green functions into a sum of two propagators — one contains only a two-cluster state and the other the orthogonal three-particle continuum state. This decomposition, in turn, allows us to decouple the integral equations, so that the continuum and the two-cluster contributions can be treated separately.

The method outlined above for decoupling the KZ equations has been investigated at a formal level by BOLLÉ [5] and given an explicit kernel form using a three-body scattering wave functions approach by KUZMICHEV [6]. The equations outlined in the next two sections in part consolidate and extend the results of these two authors. A feature of considerable interest is the emergence of an effective channel potential that arises when the three-body degrees of freedom are decoupled from the problem. This effective intercluster potential is given as the solution of an integral equation.

1. DECOUPLED KARLSSON-ZEIGER EQUATIONS

In this section, we recall the form of the KZ equation, introduce the cluster decomposition of the propagators, and give the operator form of the cluster expansion. We adhere to the choice of three-body momentum variables given by KARLSSON and ZEIGER [3], namely the two independent Jacobi momentum variables in the three-body c.m. system are denoted by $(\mathbf{p}_\alpha, \mathbf{q}_\alpha)$. The momentum of particle α having mass m_α is $\mathbf{p}_\alpha$ and the relative momentum of the pair β and γ is $\mathbf{q}_\alpha$. The reduced mass of pair $\beta\gamma$ is $\mu_\alpha = m_\beta m_\gamma (m_\beta + m_\gamma)^{-1}$ and the reduced mass of cluster $\beta\gamma$ relative to particle α is $n_\alpha = m_\alpha (m_\beta + m_\gamma)(m_\alpha + m_\beta + m_\gamma)^{-1}$. We often will represent the pair of vectors $(\mathbf{p}_\alpha, \mathbf{q}_\alpha)$ by a single six dimensional vector $\mathbf{p}_0$ and associate with $\mathbf{p}_0$ a reduced mass $n_0^2 = n_\alpha \mu_\alpha$. In this case, the three-body Hamiltonian for kinetic energy, H_0, may be represented by

$$H_0 = \frac{p_\alpha^2}{2n_\alpha} + \frac{q_\alpha^2}{2\mu_\alpha} = \frac{p_0^2}{2n_0}. \tag{1.1}$$

The scattering problem is assumed to be initiated in channel α. Particle α, with initial momentum $\mathbf{p}_\alpha'$, is incident on the pair $\beta\gamma$. The amplitudes for elastic and rearrangement scattering are denoted by $H_{\beta\alpha}^+(\mathbf{p}_\beta, \mathbf{p}_\alpha')$, where $\beta = 1,2,3$ and labels the final channel. The breakup amplitudes will be the sum over β of the three functions $E_{\beta\alpha}^+(\mathbf{p}_0, \mathbf{p}_\alpha')$. As KARLSSON and ZEIGER have shown [3], these six functions satisfy a set of coupled integral equations, whose kernels are given by the set of energy-independent rearrangement potentials,

$$V_{\beta\alpha}^{bb}(\mathbf{p}_\beta, \mathbf{p}_\alpha') = -\bar{\delta}_{\beta\alpha}\phi_\beta(\mathbf{q}_\beta^{(1)})\psi_\alpha(\mathbf{q}_\alpha^{(2)}), \quad V_{\beta\alpha}^{bc}(\mathbf{p}_\beta, \mathbf{p}_0') - \bar{\delta}_{\beta\alpha}\phi_\beta(\mathbf{q}_\beta^{(1)})\Omega_\alpha^-(\mathbf{q}_\alpha^{(2)}, \mathbf{q}_\alpha'),$$
$$V_{\beta\alpha}^{cb}(\mathbf{p}_0, \mathbf{p}_\alpha') = \bar{\delta}_{\beta\alpha}t_\beta^+(\mathbf{q}_\beta^{(1)}, \mathbf{q}_\beta)\psi_\alpha(\mathbf{q}_\alpha^{(2)}), V_{\beta\alpha}^{cc}(\mathbf{p}_0, \mathbf{p}_0') = \bar{\delta}_{\beta\alpha}t_\beta^+(\mathbf{q}_\beta^{(1)}, \mathbf{q}_\beta)\Omega_\alpha^-(\mathbf{q}_\alpha^{(2)}, \mathbf{q}_\alpha'), \tag{1.2}$$

where $\bar{\delta}_{\beta\alpha} = 1 - \delta_{\beta\alpha}$. The momenta $\mathbf{q}_\beta^{(1)}$ and $\mathbf{q}_\alpha^{(2)}$ are

$$\mathbf{q}_\beta^{(1)} = \mathbf{p}_\alpha' + \frac{\mu_\beta}{m_\gamma}\mathbf{p}_\beta, \quad \mathbf{q}_\alpha^{(2)} = -\frac{\mu_\alpha}{m_\gamma}\mathbf{p}_\alpha' - \mathbf{p}_\beta. \tag{1.3}$$

The functions $\Omega_\beta^\pm$, $t_\beta^\pm$, ψ_β, and ϕ_β are all constructed from solution of the α-γ two-body problem. If v_β is the two-body interaction and $h_\beta^0 = q_\beta^2/2\mu_\beta$ the kinetic energy operator, then $\psi_\beta(q)$ is the bound state eigenfunction with binding energy ϵ_β, viz.,

$$h_\beta\psi_\beta = (h_\beta^0 + v_\beta)\psi_\beta = -\epsilon_\beta\psi_\beta. \tag{1.4}$$

The vertex function ϕ_β is defined by multiplying $\psi_\beta(\mathbf{q})$ by $(q^2/2\mu_\beta + \epsilon_\beta)$.

The scattering wave function solutions of h_β are the momentum-space matrix elements of the wave operator defined by the strong limits

$$\Omega_\beta^\pm = \operatorname*{s-lim}_{-t\to\pm\infty} e^{iht}e^{-ih_\beta^0 t}. \tag{1.5}$$

In terms of $\Omega_\beta^\pm$, the half-on-shell t-matrices are the kernels

$$t_\beta^\pm(\mathbf{q}, \mathbf{q}') = \langle\mathbf{q}|v_\beta\Omega_\beta^\pm|\mathbf{q}'\rangle = \langle\mathbf{q}'|\Omega_\beta^\pm{}^\dagger v_\beta|\mathbf{q}\rangle, \tag{1.6}$$

where $\dagger$ denotes operator adjoint.

For fixed values of α, the six functions $H_{\beta\alpha}^+$ and $E_{\beta\alpha}^+$ satisfy the coupled integral equations

$$H_{\beta\alpha}^+(\mathbf{p}_\beta, \mathbf{p}_\alpha) = V_{\beta\alpha}^{bb}(\mathbf{p}_\beta, \mathbf{p}_\alpha) - \sum_{\gamma>0} 2n_\gamma\int \frac{V_{\beta\gamma}^{bb}(\mathbf{p}_\beta, \mathbf{p}_\gamma)H_{\gamma\alpha}^+(\mathbf{p}_\gamma'', \mathbf{p}_\alpha')}{p_\gamma''^2 - k_\gamma^2 - i0}\,d\mathbf{p}_\gamma''$$
$$- \sum_{\gamma>0} 2n_0\int \frac{V_{\beta\gamma}^{bc}(\mathbf{p}_\beta, \mathbf{p}_0'')E_{\gamma\alpha}^+(\mathbf{p}_0'', \mathbf{p}_\alpha')}{p_0''^2 - k_0^2 - i0}\,d\mathbf{p}_0'', \tag{1.7}$$

$$E_{\beta\alpha}^+(\mathbf{p}_0, \mathbf{p}_\alpha') = V_{\beta\alpha}^{cb}(\mathbf{p}_0, \mathbf{p}_\alpha') - \sum_{\gamma>0} 2n_\gamma\int \frac{V_{\beta\gamma}^{cb}(\mathbf{p}_0, \mathbf{p}_\gamma'')H_{\gamma\alpha}^+(\mathbf{p}_\gamma'', \mathbf{p}_\alpha')}{p_\gamma''^2 - k_\gamma^2 - i0}\,d\mathbf{p}_\gamma''$$
$$- \sum_{\gamma>0} 2n_0\int \frac{V_{\beta\gamma}^{cc}(\mathbf{p}_0, \mathbf{p}_0'')E_{\gamma\alpha}^+(\mathbf{p}_0'', \mathbf{p}_\alpha')}{p_0''^2 - k_0^2 - i0}\,d\mathbf{p}_0''. \tag{1.8}$$

In describing an arbitrary final state channel γ, it is useful to know the magnitude of the corresponding momentum allowed by energy conservation. We have reserved the symbol k_γ for this momentum, which is determined by $\mathbf{p}_\alpha'$ and the eigenvalues ϵ_γ of (1.4), and where by convention we set $\epsilon_0 = 0$.

$$k_\gamma = (2n_\gamma)^{1/2}\left|\frac{p_\alpha'^2}{2n_\alpha} - \epsilon_\alpha + \epsilon_\gamma\right|^{1/2}, \quad \gamma = 0,1,2,3. \tag{1.9}$$

Examining (1.7) and (1.8) reveals the basic structure of the KZ equations. The equations form a set of connected scattering equations for the physical amplitudes $H_{\beta\alpha}^+$ and $E_{\beta\alpha}^+$. The kernels depend on simple energy independent potential-like functions given in (1.2). These potentials utilize only half-off shell two-body t-matrix information determined at positive two-body scattering energies. In (1.7) and (1.8), the singular denominator $(p_\gamma''^2 - k_\gamma^2 - i0)^{-1}$ describes propagation of the cluster ψ_γ; the denominator $(p_0''^2 - k_0^2 - i0)^{-1}$ gives the propagation of an intermediate state that is composed of a free three-particle continuum.

We now turn our attention to the problem of decoupling these equations. In the process of solving this problem, we will automatically construct a cluster expansion of the three-body problem. Define for complex z

the Green functions $G(z)$ and $G_\alpha(z)$ by $(H-z)^{-1}$ and $(H_\alpha - z)^{-1}$ respectively. The transition operator $U_{\beta\alpha}(z)$ of ALT, GRASSBERGER, and SANDHAS [7] is determined by

$$G(z) = \delta_{\alpha\beta} G_\beta(z) - G_\beta(z) U_{\beta\alpha}(z) G_\alpha(z). \tag{1.10}$$

Use $U_{\beta\alpha}(z)$ to define a related operator $T_{\beta\alpha}(z)$ by

$$T_{\beta\alpha}(z) = V_\beta G_0(z) U_{\alpha\beta}(z) G_0(z) V_\alpha, \quad \alpha > 0,\ \beta > 0. \tag{1.11}$$

It is a simple matter to show that $T_{\beta\alpha}(z)$ satifies the equation

$$T_{\beta\alpha}(z) = -\bar{\delta}_{\beta\alpha} V_\beta G_0(z) V_\alpha - \sum_\gamma \bar{\delta}_{\beta\gamma} V_\beta G_\gamma(z) T_{\gamma\alpha}(z), \tag{1.12}$$

where γ is summed over 1,2,3. It is well known [3,8] that the interacting matrix elements of $T_{\beta\alpha}(z)$ give representations of the physical transition amplitudes in momentum space. Specifically, if $s_\alpha = p_\alpha'^2/2n_\alpha - \epsilon_\alpha$ is the available incident energy, then

$$H^+_{\beta\alpha}(\mathbf{p}_\beta, \mathbf{p}'_\alpha) = <\mathbf{p}_\beta \psi_\beta | T_{\beta\alpha}(s_\alpha + i0) | \mathbf{p}'_\alpha \psi_\alpha>, \quad E^+_{\beta\alpha}(\mathbf{p}_0, \mathbf{p}'_\alpha) = <\mathbf{p}_\beta \Omega^-_\beta(\cdot, \mathbf{q}_\beta) | T_{\beta\alpha}(s_\alpha + i0) | \mathbf{p}'_\alpha \psi_\alpha>. \tag{1.13}$$

The half-off-shell function $H^+_{\beta\alpha}$ is identical to the function appearing in Faddeev's description [4] of the elastic and rearrangement amplitude. Whereas $E^+_{\beta\alpha}$ — the β component of the breakup amplitude — differs from Faddeev's breakup representation and the two are only equal for on-shell kinematics.

To proceed with the decoupling of the KZ equations, it is convenient to write (1.12) in a matrix form. Let V, $\tilde{V}$, $R_0(z)$, $R(z)$ and $T(z)$ denote the 3 × 3 matrices with operator elements $\delta_{\alpha\beta} V_\alpha$, $V_\alpha \bar{\delta}_{\alpha\beta}$, $\delta_{\alpha\beta} G_0(z)$, $\delta_{\alpha\beta} G_\alpha(z)$, and $T_{\alpha\beta}(z)$, respectively. Then, (1.12) can be written

$$T(z) = \tilde{V}(-R_0(z) V) - \tilde{V} R(z) T(z). \tag{1.14}$$

Now expand $G_\alpha(z)$ into cluster and continuum parts. Let the projection operators P^b_α and P^c_α be defined by

$$<\mathbf{p}_\alpha, \mathbf{q}_\alpha | P^b_\alpha | \mathbf{p}''_\alpha, \mathbf{q}''_\alpha> = \delta(\mathbf{p}_\alpha - \mathbf{p}''_\alpha) \psi_\alpha(\mathbf{q}_\alpha) \psi_\alpha(\mathbf{q}''_\alpha)^*, \tag{1.15}$$

$$<\mathbf{p}_\alpha, \mathbf{q}_\alpha | P^c_\alpha | \mathbf{p}''_\alpha, q''_\alpha> = \delta(\mathbf{p}_\alpha - \mathbf{p}''_\alpha) <\mathbf{q}_\alpha | \Omega^-_\alpha \Omega^{-\dagger}_\alpha | \mathbf{q}''_\alpha>, \tag{1.16}$$

where * denotes complex conjugation. Because the two-body problem is taken to be asymptotically complete, the projectors P^b_α and P^c_α have orthogonal ranges, diagonalize H_α, and span the three-body Hilbert space. Thus, one can write

$$G_\alpha(z) = G^b_\alpha(z) + G^c_\alpha(z), \quad G^b_\alpha(z) = P^b_\alpha G_\alpha(z) P^b_\alpha, G^c_\alpha(z) = P^c_\alpha G_\alpha(z) P^c_\alpha. \tag{1.17}$$

Note that the product $G^b_\alpha(z) G^c_\alpha(z) = 0$. Using this decomposition to write $R(z)$ as the sum of a cluster state and a continuum state gives us

$$R(z) = R^b(z) + R^c(z), \tag{1.18}$$

where $[R^{b,c}(z)]_{\alpha\beta} = \delta_{\alpha\beta} G^{b,c}_\alpha(z)$.

The next stage is to decouple the equations for $H^+_{\beta\alpha}$ and $E^+_{\beta\alpha}$. We assume that $(1+VR^c(z))^{-1}$ exists. This is very probable, since the operator $\tilde{V}R^c(z)$ is completely connected. In this circumstance, the equation

$$T^c(z) = \tilde{V} - \tilde{V} R^c(z) T^c(z), \tag{1.19}$$

has the formal solution

$$T^c(z) = (1 + \tilde{V} R^c(z))^{-1} \tilde{V}. \tag{1.20}$$

Using (1.18) in (1.14) leads to

$$(1 + \tilde{V} R^c(z)) T(z) = \tilde{V}(-R_0(z) V) - \tilde{V} R^b(z) T(z). \tag{1.21}$$

Applying the inverse of the left factor gives

$$T(z) = T^c(z)(-R_0(z) V) - T_c(z) R^b(z) T(z). \tag{1.22}$$

At a formal level, (1.19) and (1.22) constitute a decoupling of the scattering problem. In (1.19), the only intermediate states that can propagate are continuum states. Given knowledge of $T^c(z)$, then (1.22) essentially turns on the cluster features of the scattering problem.

2. INTERACTING REPRESENTATIONS

We investigate now the consequence of the identities (1.19) and (1.22). In component form, these two equations read

$$T^c_{\beta\alpha}(z) = \tilde{\delta}_{\beta\alpha} V_\beta - \sum_\gamma \tilde{\delta}_{\beta\gamma} V_\beta G^c_\gamma(z) T^c_{\gamma\alpha}(z), \tag{2.1}$$

$$T_{\beta\alpha}(z) = T^c_{\beta\alpha}(z)[-G_0(z) V_\alpha] - \sum_\gamma T^c_{\beta\gamma}(z) G^b_\gamma(z) T_{\gamma\alpha}(z). \tag{2.2}$$

After appropriate matrix elements are formed and the completeness of the two-body wave operators Ω^-_γ is utilized, the operator identities (2.1) and (2.2) will give us four independent equations — two integral equations and two quadrature relations. We define a quasi-breakup amplitude $E^c_{\beta\alpha}$ from the operator $T^c_{\beta\alpha}(z)$ by

$$E^c_{\beta\alpha}(\mathbf{p}_0, \mathbf{p}'_\alpha) = \langle \mathbf{p}_\beta \Omega^-_\beta(\cdot, \mathbf{q}_\beta) | T^c_{\beta\alpha}(s_\alpha + i0) | \mathbf{p}'_\alpha \psi_\alpha \rangle. \tag{2.3}$$

This definition clearly parallels that of (1.13) for the β component of the physical breakup amplitude. Now multiply (2.1) from the left by $\langle \mathbf{p}_\beta \Omega^-_\beta(\cdot, \mathbf{q}_\beta)|$ and from the right by $|\mathbf{p}'_\alpha \psi_\alpha\rangle$, and set $z = s_\alpha + i0$. It is readily found that

$$E^c_{\beta\alpha}(\mathbf{p}_0, \mathbf{p}'_\alpha) = V^{cb}_{\beta\alpha}(\mathbf{p}_0, \mathbf{p}'_\alpha) - \sum_{\gamma>0} 2n_0 \int \frac{V^{cc}_{\beta\gamma}(\mathbf{p}_0, \mathbf{p}''_0) E^c_{\gamma\alpha}(\mathbf{p}''_0, \mathbf{p}'_\alpha)}{p''^2_0 - k^2_0 - i0} d\mathbf{p}''_0. \tag{2.4}$$

This is an integral equation for $E^c_{\beta\alpha}$. The kernel and driving terms are constructed from the KZ potentials $V^{cc}_{\beta\alpha}$ and $V^{cb}_{\beta\alpha}$.

The second relation to be extracted from (2.1) is obtained by using states $\langle \mathbf{p}_\beta \psi_\beta|$ and $|\mathbf{p}'_\alpha \psi_\alpha\rangle$ to construct matrix elements. In terms of the two-cluster-like amplitude

$$H^c_{\beta\alpha}(\mathbf{p}_\beta, \mathbf{p}'_\alpha) = \langle \mathbf{p}_\beta \psi_\beta | T^c_{\beta\alpha}(s_\alpha + i0) | \mathbf{p}'_\alpha \psi_\alpha \rangle, \tag{2.5}$$

(3.1) becomes

$$H^c_{\beta\alpha}(\mathbf{p}_\beta, \mathbf{p}'_\alpha) = V^{bb}_{\beta\alpha}(\mathbf{p}_\beta, \mathbf{p}'_\alpha) - \sum_{\gamma>0} 2n_0 \int \frac{V^{bc}_{\beta\gamma}(\mathbf{p}_\beta, \mathbf{p}''_0) E^c_{\gamma\alpha}(\mathbf{p}''_0, \mathbf{p}'_\alpha)}{p''^2_0 - k^2_0 - i0} d\mathbf{p}''_0. \tag{2.6}$$

Given $E^c_{\gamma\alpha}$, (2.6) is a quadrature relation for $H^c_{\beta\alpha}$. Finally, turn to (2.2). Its two-cluster matrix elements are obviously

$$H^+_{\beta\alpha}(\mathbf{p}_\beta, \mathbf{p}'_\alpha) = H^c_{\beta\alpha}(\mathbf{p}_\beta, \mathbf{p}'_\alpha) - \sum_{\gamma>0} 2n_0 \int \frac{H^c_{\beta\gamma}(\mathbf{p}_\beta, \mathbf{p}''_\gamma) H^+_{\gamma\alpha}(\mathbf{p}''_\gamma, \mathbf{p}'_\alpha)}{p''^2_\gamma - k^2 - i0} d\mathbf{p}''_\gamma. \tag{2.7}$$

Equations (2.4), (2.6), and (2.7) form a solvable system. In (2.4), we solve a three-body problem that allows only continuum intermediate states. Then, via (2.6), we take $E^c_{\beta\alpha}$ and by integration determine an effective potential function $H^c_{\beta\alpha}$. Lastly, this effective potential is used in the coupled cluster structure given by (2.7) to find the exact elastic and rearrangement amplitudes $H^+_{\beta\alpha}$. Taken together, this system of three equations provides a cluster description of three-body scattering which is exact. The effect of the intermediate three particle continuum states is summed up into the form of the effective two-cluster potentials $H^c_{\beta\alpha}$. Iterative approximations to (2.4), will automatically define a cluster expansion of the three-body problem wherein one can take into account successively more accurate, but approximate effects of the three-body continuum. At each stage of the iteration of (2.4) one is able to construct a better approximation to the exact effective potential $H^c_{\beta\alpha}$.

One additional equation is needed to make the system (2.4)-(2.7) complete. We must find the breakup function $E^+_{\beta\alpha}$. This is given by the matrix element of (2.2) with respect to $\langle \mathbf{p}_\beta \Omega^-_\beta(\cdot, \mathbf{q}_\beta)|$ and $|\mathbf{p}'_\alpha \psi_\alpha\rangle$. The result is

$$E^+_{\beta\alpha}(\mathbf{p}_0, \mathbf{p}'_\alpha) = E^c_{\beta\alpha}(\mathbf{p}_0, \mathbf{p}'_\alpha) - \sum_{\gamma>0} 2n_\gamma \int \frac{E^c_{\beta\gamma}(\mathbf{p}_0, \mathbf{p}''_\gamma) H^+_{\gamma\alpha}(\mathbf{p}''_\gamma, \mathbf{p}'_\alpha)}{p''^2_\gamma - k^2_\gamma - i0} d\mathbf{p}'_\gamma. \tag{2.8}$$

This is a quadrature relation for $E^+_{\beta\alpha}$. In order to use it, we must first have solved for $H^+_{\gamma\alpha}$ and $E^c_{\beta\alpha}$. Essentially, the integral on the right of (2.8) computes for us the difference between the quasi-breakup component $E^c_{\beta\alpha}$ and the exact breakup function $E^+_{\beta\alpha}$. This difference is just due to the effect of including the influence of the cluster structure on the breakup amplitude.

Equations (2.4), (2.6)-(2.8) have the feature of unlinking the continuum from the two-particle states. This structure then gives us the opportunity to treat the continuum intermediate states in a perturbation expansion. An expansion of this sort is attractive because it has a well defined physical meaning.

Finally, recall that the solvability of our decoupled scheme requires that (2.4) have a unique solution. In a previous study of the KZ equations, we have shown how to introduce an algebraic decomposition of the amplitudes that transforms the KZ equations into a Fredholm integral equation [9]. Similarly, (2.4) can be brought into a fredholm form. Thus, (2.4) obeys the Fredholm alternative and has a unique solution provided that the corresponding homogeneous equation has no solution. A complete mathematical proof of uniqueness would now show that this homogeneous equation can have no solution or only has a solution in exceptional situations. We have dealt with this problem by solving the numerical equivalent of (2.4). For three-body energies in the bound-state and two-cluster scattering sector (i.e., below breakup), we have always found unique solutions.

3. EXPANSION APPROXIMATIONS AND NUMERICAL SOLUTIONS

This section will develop the iterative expansion of the effective potential representation (EPR) system of equations. Iterative approximations to (2.4) lead to approximations to the effective channel potentials $H^c_{\beta\alpha}(\mathbf{p}_\beta, \mathbf{p}_\alpha')$. We start by defining the zeroth iterate of $E^c_{\beta\alpha}$,

$$E^{c,0}_{\beta\alpha}(\mathbf{p}_0, \mathbf{p}_\alpha') = 0. \tag{3.1}$$

The ith iterate of (2.4) is defined by the recursion relation

$$E^{c,i+1}_{\beta\alpha}(\mathbf{p}_0, \mathbf{p}_\alpha') = V^{cb}_{\beta\alpha}(\mathbf{p}_0, \mathbf{p}_\alpha') - \sum_{\gamma>0} 2n_0 \int \frac{V^{cc}_{\beta\gamma}(\mathbf{p}_0'', \mathbf{p}_0) E^{c,i}_{\gamma\alpha}(\mathbf{p}_0'', \mathbf{p}_\alpha')}{p_0''^2 - k_0^2 - i0} d\mathbf{p}_0''. \tag{3.2}$$

In terms of the ith iterate for the pseudo-breakup amplitude $E^{c,i}_{\beta\alpha}$, there is a corresponding effective channel potential given by

$$H^{c,i}_{\beta\alpha}(\mathbf{p}_\beta, \mathbf{p}_\alpha') = V^{bb}_{\beta\alpha}(\mathbf{p}_\beta, \mathbf{p}_\alpha') - \sum_{\gamma>0} 2n_0 \int \frac{V^{bc}_{\beta\gamma}(\mathbf{p}_\beta, \mathbf{p}_0'') E^{c,i}_{\gamma\alpha}(\mathbf{p}_0'', \mathbf{p}_\alpha')}{p''^2 - k_0^2 - i0} d\mathbf{p}_0''. \tag{3.3}$$

The last step in obtaining an approximation for the physical scattering amplitude is to use the approximate potential $H^{c,i}_{\beta\alpha}$ in place of $H^c_{\beta\alpha}$ in (2.7). The corresponding solution of Eqn. (2.7) we will denote $H^{+,i}_{\beta\alpha}$. In all the examples studied numerically in this paper,

$$H^{c,i}_{\beta\alpha} \to H^c_{\beta\alpha}, \quad H^{+,i}_{\beta\alpha} \to H^+_{\beta\alpha} \ \text{as} \ i \to \infty. \tag{3.4}$$

The convergence property reflects the fact that (2.4) may be solved by iteration. This is characteristic of many physical circumstances where it occurs that the coupling of the 3-particle continuum to itself is weak. Technically, this means that the operator $\tilde{V}G^c(z)$ has all of its eigenvalues inside the unit circle of the complex plane. This will not always be the case and alternate approximate methods can be devised to deal with this situation. The last stage of obtaining a solution is to solve the coupled channel problem given in (2.7). Essentially, (2.7) is a kinematic copy of the two-body problem and is always easy to solve whatever the strength of the coupling.

It is interesting to know whether or not the approximate solutions obtained by iteration are self-consistent with the impulse approximation. The standard expression for the impulse approximation given by

$$H^{imp}_{\alpha\alpha}(\mathbf{p}_\alpha, \mathbf{p}_\alpha') = \sum_{\gamma\neq\alpha} \int \frac{\phi_\alpha(\mathbf{q}_\alpha^{(1)}) \langle \mathbf{q}_\gamma^{(2)} | t_\gamma(\tilde{p}_\alpha'^2 - \epsilon_\alpha - \tilde{p}_\gamma^2 + i0) | \mathbf{q}_\gamma^{(1)} \rangle \psi_\alpha(\mathbf{q}_\alpha^{(2)})}{\tilde{q}_\gamma^{(2)2} + \tilde{p}_\gamma^2 - \tilde{p}_\alpha'^2 + \epsilon_\alpha - i0} d\mathbf{p}_\gamma, \tag{3.5}$$

where $\tilde{p}_\gamma^2 = p_\gamma^2/2n_\gamma$ and $\tilde{q}_\gamma^2 = q_\gamma^2/2\mu_\gamma$. For large values of p_α', one expects the impulse approximation to become increasingly accurate. We find that approximation $i = 0$ does not have the right structure to form $H^{imp}_{\alpha\alpha}$, whereas once $i \geqslant 1$ the impulse approximation is reproduced correctly. This in part explains why the $i = 0$ cluster expansion never yields reasonable three-body phase shifts.

We now report the computational results we have obtained using the decoupling scheme. We compare the approximate solutions found by iterative expansions for the effective potentials with those determined from an exact solution of Faddeev's equations. We consider the case where the two-particle interaction is a separable (rank one) potential in momentum space. In this special situation, the original Faddeev integral equations for the three-body problem reduce [10] to coupled integral equations in one continuous variable and are thus easy to solve numerically. Although this problem represents a somewhat simple version of the three-body problem, it can give us a good idea as to the rate of convergence implicit in the effective potential approximations.

The separable interaction is chosen to be a Yamaguchi potential, viz.,

$$v(\mathbf{q},\mathbf{q}') = \frac{\lambda}{(q^2+\beta^2)(q'^2+\beta^2)}. \tag{3.6}$$

The physical system is taken to be either three identical bosons or three fermions. The range parameter β is set to 1.44401 fm^{-1} and λ adjusted to reproduce a two-particle binding energy of the deuteron, $\epsilon = 0.053695$ fm^{-2}. We state our solutions in terms of the cotangent of the phase shift for the lowest angular momentum state of our system. This phase shift parameterization of the on-shell scattering solution is

$$H^{+,i}(k,k) = -\frac{3}{4k} e^{i\delta(k)} \sin\delta(k). \tag{3.7}$$

The first set of solutions correspond to newtron-deuteron scattering in the spin-quartet state. The curve in Fig. 1 labeled EXACT is the numerical solution of the Faddeev equations throughout the elastic scattering sector. The value of k required to breakup the bound state subsystem is denoted by k_B. Also shown on Fig. 1 are the first four approximate solutions, namely the value of $k \cot\delta$ determined from $H^{+,i}$ where $i = 0,1,2,3$. When $i = 3$, the approximate solutions is indistinguishable from the exact result on the graph. Note that the approximation for $i = 0$ is quite bad. This in accord with the fact that only for $i \geqslant 1$ do our approximations become sufficiently self-consistent to contain the impulse approximation.

Fig. 2 shows the behavior of $k \cot\delta$ in the elastic scattering sector for three identical bosons in the $J = 0$ state. This is a more strongly interacting system than the three fermion system because of the different spin and isospin recoupling coefficients. Again, for $i \geqslant 1$ there is rapid convergence of the approximate solutions. Fig. 3 show the values of the Fredholm determinant in the boson system for negative three-body energies, $-E$. The dotted curve gives the Fredholm determinant defined by the Faddeev equations. The solid curve is the Fredholm determinant for the negative energy version of integral equation (2.7). This negative energy form of the cluster representation is obtained by replacing the variable k_γ^2 in (2.7) by $2n_\gamma(-E+\epsilon_\gamma)$. It is seen that, although the Fredholm determinants are different functions of E, both have the same zeros and thus predict the same three-body bound-state energies.

REFERENCES

[1] K. Wildermuth, Y.C. Tang: *A Unified Theorv of the Nucleus* (Vieweg, Braunschweig, 1977)
[2] J. Schwager: Ann. Phys. (N.Y.) **98**, 14 (1976); R. Raphael, P.C. Tandy, W. Tobocman: Phys. Rev. C **14**, 1355 (1976); S.F.J. Wilk, T.A. Osborn: Ann. Phys. (N.Y.) **114**, 410 (1978); E.W. Schmidt, M. L'Huillier: "N-cluster dynamics and effective interaction of composite particle. I. The dynamical equation", Tübingen preprint (1979)
[3] B.R. Karlsson, E.M. Zeiger: Phys. Rev. D **11**, 939 (1975)
[4] L.D. Faddeev: *Mathematical Aspects of the Three-Body Problem in Quantum Scattering Theory* (Davey, New York, 1965)
[5] D. Bollé: Phys. Rev. D **13**, 1809 (1976)
[6] V.E. Kuzmichev: "New many-channel Lippmann-Schwinger-type equation for three particles", Kiev preprint (1977), and "On the three-particle system low energy properties", Kiev preprint (1977)
[7] E.O. Alt, P. Grassberger, W. Sandhas: Nucl. Phys. **B2**, 167 (1967)
[8] T.A. Osborn, K.L. Kowalski: Ann. Phys. (N.Y.) **68**, 361 (1971)
[9] D. Eyre and T.A. Osborn: Nucl. Phys. **A299**, 301 (1978)
[10] C. Lovelace: Phys. Rev. **135B**, 1225 (1964); R.D. Amado: Annu. Rev. Nucl. Sci. **19**, 61 (1969)

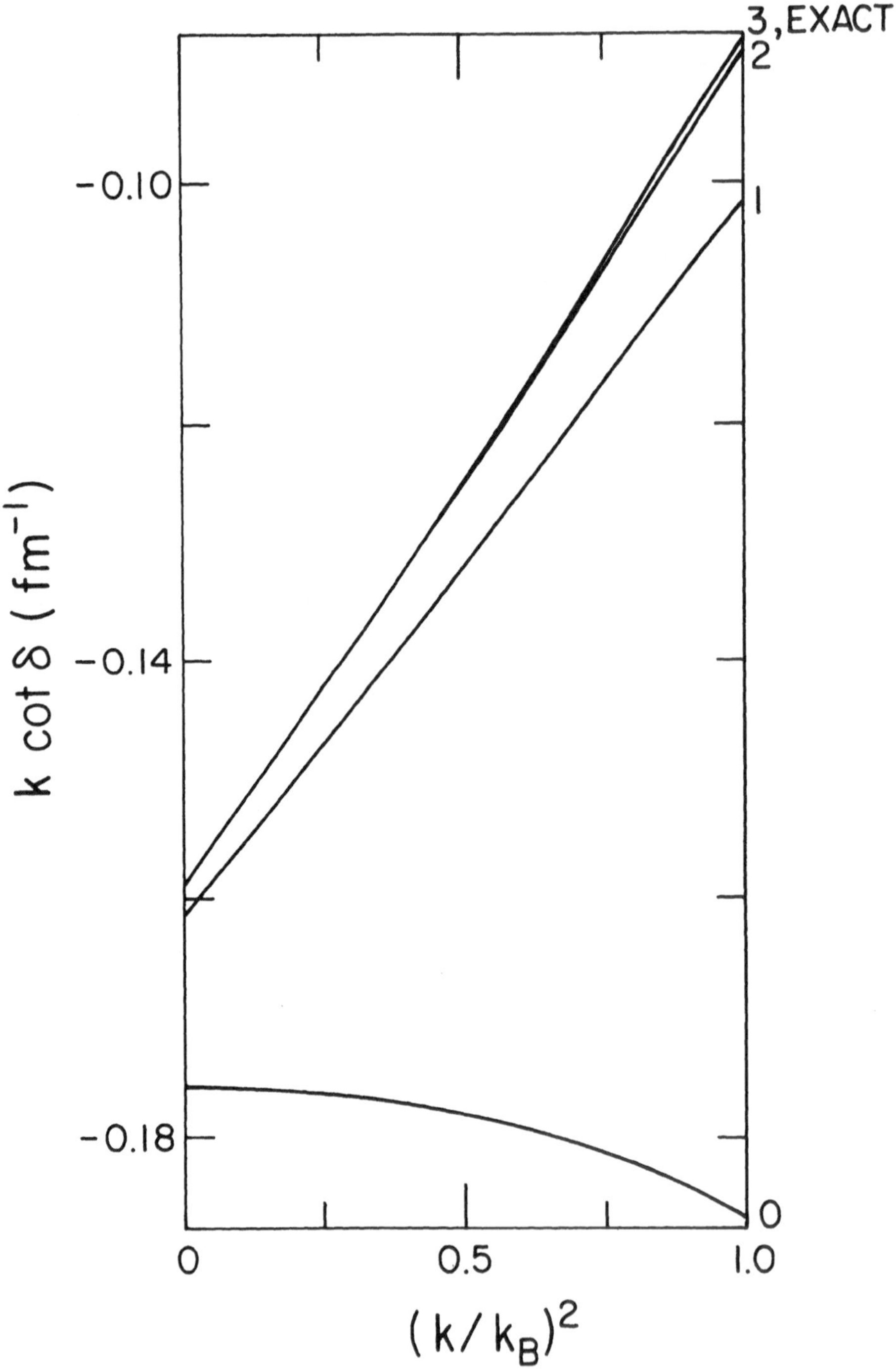

Fig. 1. Solution of the EPR equations for a system of fermions corresponding to n-d scattering in the ^4S state. Each curve is labeled on the right by the ith iterate of the EPR expansion.

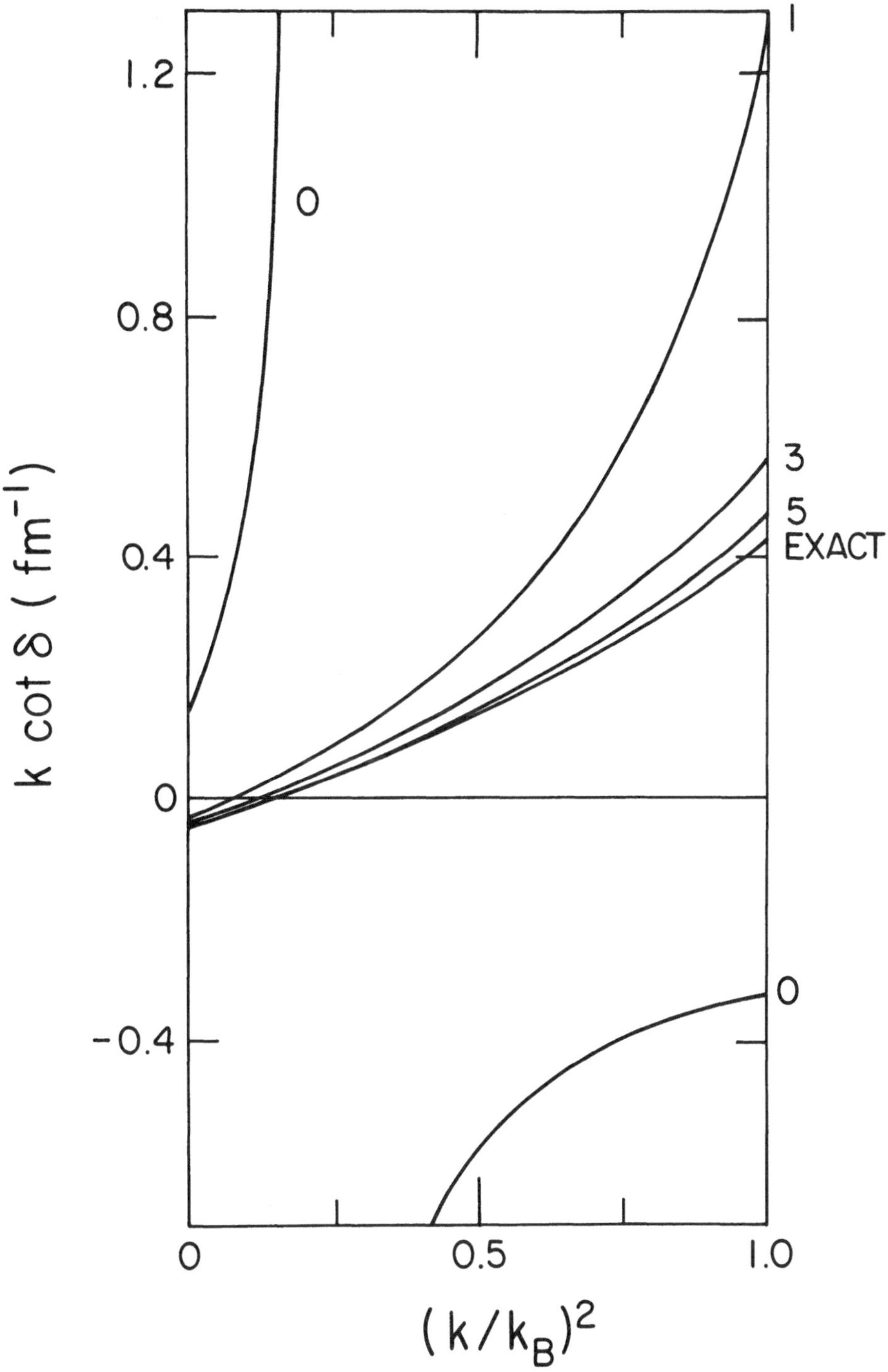

Fig. 2. Solution of the EPR equations for identical spin-0 and isospin-0 particles. Curves are labeled on the right by the ith iterate of the EPR expansion.

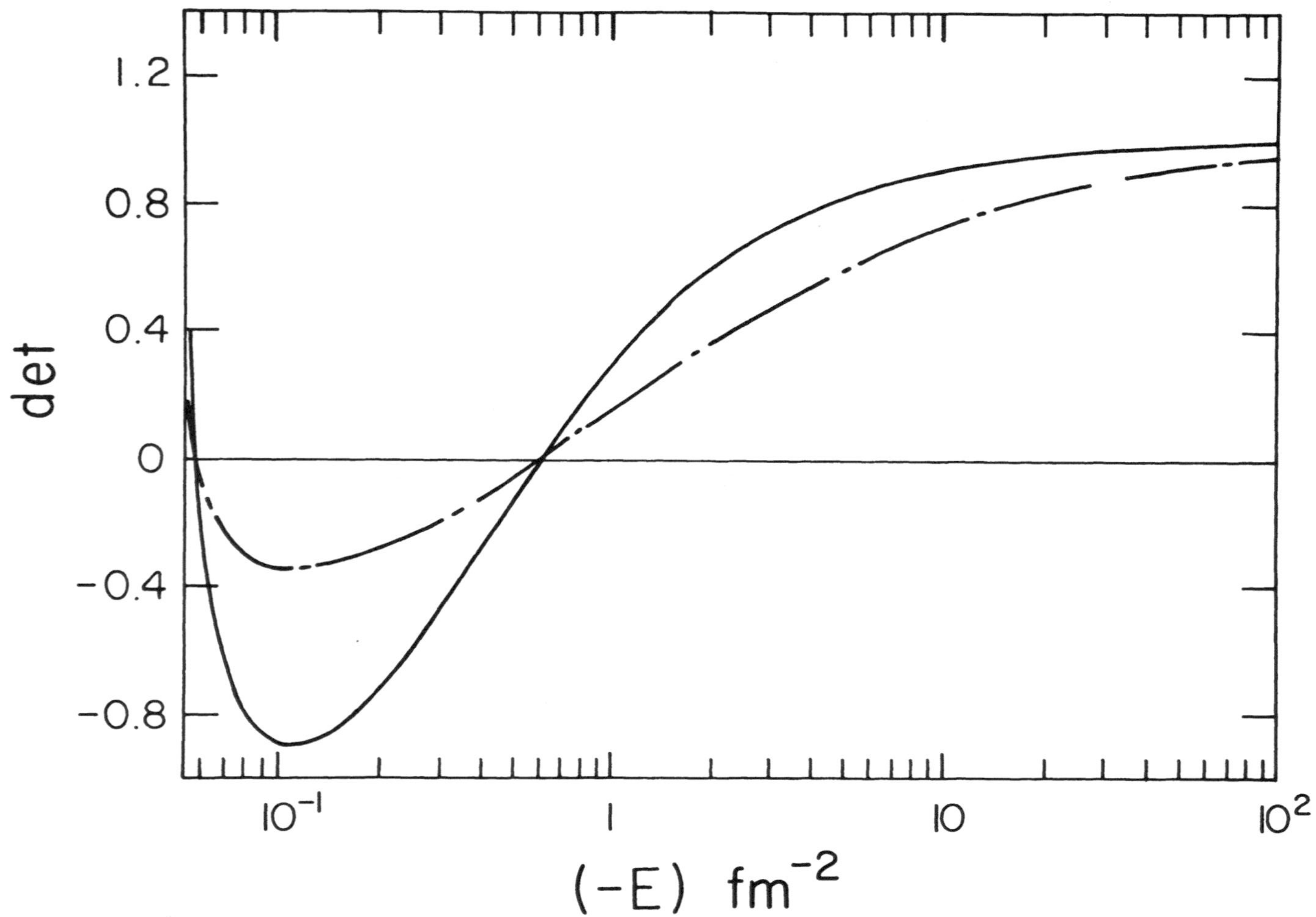

Fig. 3. Fredholm determinant at negative energies. The continuous curve is obtained from an exact EPR and the dotted curve from the Faddeev equation.

QUANTUM-STATISTICAL MECHANICS AND SCATTERING THEORY ON HOMOGENEOUS SPACES OF FINITE VOLUME

Norman E. Hurt

MRJ, Inc.
10400 Eaton Place
Fairfax, Virginia 22030

INTRODUCTION

Quantum field theorists have recently been interested in calculations of the following type. Let M be a multiply connected Riemannian manifold. Assume the Hamiltonian H is intrinsically defined on M. Let

$$Z(\beta) = Tr(e^{-\beta H})$$

be the partition function on M. Then the effective Lagrangian density $L^{(1)}$ on M is determined by

$$L^{(1)} = \frac{1}{2}\int_o^\infty \frac{d\beta}{\beta} Z(\beta)e^{-\beta m^2}/\mathrm{Vol}(M).$$

Of course, the partition function is of interest in its own right from the point of view of quantum-statistical mechanics on M. This paper will concern itself with a review of three important concepts: partition functions, asymptotics, and scattering theory. The calculation of the partition functions in simple cases for compact spaces in quantum field theory and quantum statistical mechanics is demonstrated in Secs. 4 and 5. The general result is presented in Sec. 6 for compact quotients of noncompact symmetric spaces of rank one. The traces are evaluated in terms of the density matrices.

When the spaces are no longer compact, the density matrix is no longer of trace class. As we motivate in Sec. 2 from scattering theory, the "noncompact" part of the density matrix must be subtracted off to obtain a well-defined partition function. This is precisely what happens in the finite-volume theory presented in Sec. 7. The noncompact term in that case is expressed in terms of the theory of Eisenstein series and this is where scattering theory plays the crucial role. The class of spaces that we are ultimately led to consider are Riemannian locally symmetric spaces with nonpositive curvature. Thus $M = \Gamma\backslash G/K$ where G is a real connected noncompact semisimple Lie group with finite center (e.g. $SO_0(n,1)$), K is a maximal compact subgroup and Γ is a discrete subgroup of G which we shall assume is without torsion and $\Gamma\backslash G$ compact in Sec. 6 and such that Vol $(\Gamma\backslash G)$ is finite in Sec. 7.

1. PARTITION AND DENSITY FUNCTIONS

The partition function is determined by the density matrix ρ which is the solution of the Bloch equation on M:

$$\frac{\partial\rho}{\partial\beta} + H\rho = 0,$$

where $\rho(x'',x',0) = \delta(x''-x')$. The density matrix on M and its simply connected covering space $\tilde{M}$ are related by

$$\rho(x'',x',\beta) = \sum_{\gamma\in\Gamma} a(\gamma)\tilde{\rho}(x'',x',\beta),$$

where a is a unitary one-dimensional representation of Γ.

If $\tilde{H}$ has a discrete set of eigenvalues $\tilde{\lambda}_n$,

$$\tilde{H}|\tilde{n}> = \tilde{\lambda}_n|\tilde{n}>,$$

then formally we have

$$\tilde{\rho} = e^{-\beta \tilde{H}} = \sum_n e^{-\beta \tilde{\lambda}_n} |\tilde{n}><\tilde{n}|$$

and

$$Z(\beta) = \mathrm{Tr}(e^{-\beta H}) = \sum_n e^{-\beta \tilde{\lambda}_n}.$$

2. SCATTERING PHASE SHIFT

For scattering theory problems involving a central potential $V(r)$, the renormalized partition function

$$Z_L(\beta) = \mathrm{Tr}(e^{-\beta H_L} - e^{-\beta H_L^o})$$

is related to the phase shift $\eta_L(k)$ by

$$Z_L(\beta) = \frac{1}{\pi} \int_o^\infty e^{-\beta \lambda_k} \frac{d\eta_L(k)}{dk} \, dk,$$

where $H_L = H_L^o + V(r)$ and $H_L^o = \frac{\hbar^2}{2m} \left\{ \frac{d^2}{dr^2} - L(L+1) r^{-2} \right\}$.

The scattering length A_L is given by the "low temperature" asymptotics on $Z_L(\beta)$. Here, $\eta_L(k) \sim -A_L k^{2L+1} (\mathrm{mod}\ \pi)$ as $k \to 0$. and $A_l = (-2m/\hbar^2) \lim_{\beta \to \infty} [\beta^{L+\frac{1}{2}} Z_L(\beta)/a(L)]$, where $a(L)$ is a known constant.

3. ZETA FUNCTION RENORMALIZATION

Zeta-function renormalization in quantum field theory treats the problem of finding the effective energy momentum tensor $< T_{\mu\nu} >$ or the total energy $E = -\int_M L^{(1)}$ (or the integrated form of $< T_o^o >$), in terms of the zeta functions $\zeta(s)$. The connection between $\zeta(s)$ and the density matrices $\rho(\beta)$ is given by the Mellin transform

$$\zeta(s,x'',x') = \frac{1}{\Gamma(s)} \int_o^\infty \beta^{s-1} \rho(\beta,x'',x') \, d\beta.$$

One has

$$\zeta(s) = \int_M \zeta(s,x',x') dx' = \sum_{\tilde{\lambda}} n_{a\tilde{\lambda}} \tilde{\lambda}^{-s}$$

in the case of a discrete spectrum, where $n_{a\tilde{\lambda}}$ is the multiplicity of the representation a in the $\tilde{\lambda}$-representation. One of the standard results of zeta function renormalization is that the one-loop effective action $W^{(1)}$ is given by

$$W^{(1)} = -\lim_{s \to 1} \frac{i}{2} \left[\frac{\zeta(0)}{s-1} + \zeta'(0) \right].$$

4. PLANAR RIGID ROTATOR

The partition function of the planar rigid rotator ($H = -\frac{h^2}{2m} \left(\frac{2\pi}{L} \right)^2 \frac{\partial^2}{\partial \theta^2}$, $M = S^1$) is

$$Z(\beta) = 1 + 2 \sum_{m=1}^\infty \exp\left[\frac{-\beta \hbar^2}{2m} \left(\frac{2\pi}{L} \right)^2 m^2 \right],$$

which is twice the classical theta function. The Mellin transform of $[Z(\beta) - 1/2]/2$ is the Riemann zeta function

$$\zeta(2s) = \sum_{n=1}^\infty n^{-2s}.$$

The theory of high-temperature asymptotics is a very interesting field. Mathematicians have called this area geometric asymptotics since the high temperature limit of $Z(\beta)$ is an asymptotic expansion whose coefficients are Riemannian or spectral invariants of the underlying manifold. This theory first arose in spectroscopy (as defined by Sir Arthur Schuster) and the first examples were developed in early quantum-statistical mechanics. For more details, see [8].

Quantum field theory of scalar fields $\phi^{(\alpha)}$ on S^1 can be described in terms of the zeta-function approach. Viz., the vacuum average of the Hamiltonian $E = \langle H \rangle$ for the $\phi^{(\alpha)}$ fields is given as follows. Let $M = S^1$, $\tilde{M} = R$, and $\Gamma = \mathbb{Z}$. Representation a is given by $a(\gamma_n) = \exp(2\pi i n\alpha)$, $0 \leqslant \alpha \leqslant \frac{1}{2}$. The eigenvalues of $H = d^2/d\phi^2$ are $-k^2$ where $k = n - \alpha$. Finally,

$$E = \zeta(-1/2)_{S^1}.$$

This can be evaluated, using the Hurwitz-Lerch-Hermite zeta functions, to be

$$E = -\pi^2 \sum_{n=1}^{\infty} n^{-2}\cos(2\pi n\alpha) = \alpha - \alpha^2 - \frac{1}{6}.$$

5. SCALAR FIELDS ON EINSTEIN UNIVERSE

Another example is the case of conformally coupled scalar fields on Einstein universe $T \otimes S^3(a)$. The eigenvalues and degeneracies are well known in this case to be $\tilde{\zeta}_n = (n/a)^2$ and $d_n = n^2$, $n = 1, 2, 3, \ldots$, where a is the radius of S^3. Thus,

$$\zeta(s)_{S^3} = \sum d_n \tilde{\lambda}_n^{-s} = \sum_{n=1}^{\infty} n^2(n^2/a^2)^{-s} = a^{2s}\zeta(2s-2).$$

Thus, the total energy for real fields is

$$E = \frac{1}{2}\zeta(-1/2)_{S^3} = \frac{1}{240a}.$$

As stated in the Introduction, there is considerable interest among quantum field theorists to study examples of modified Einstein universes of the form $T \otimes \tilde{M}/\Gamma$. To calculate the total energy or energy-momentum tensor, we have seen that it is necessary to know the partition function on the space of interest. We turn our attention to this problem.

6. PARTITION FUNCTIONS ON COMPACT QUOTIENTS OF SYMMETRIC SPACES OF NONCOMPACT TYPE OF RANK 1

For spaces of this type, the partition function is determined by the Selberg trace formula which is stated as follows. Let $\tilde{p}(\beta)$ be the fundamental solution of the Bloch equation on G/K. Let U be the representation of G induced by the identity representation of Γ. Then certain spherical representations $\{U_j, j \geqslant 0\}$ of G occur in U with multiplicities n_j. The partition function is given by

$$\zeta(\beta) = \mathrm{Vol}(\Gamma\backslash G)\tilde{p}(1) + (4\pi\beta)^{-\frac{1}{2}} \sum_{\gamma} |u_\gamma| j(\gamma)^{-1} C(h(\gamma)) \exp(-u_\gamma^2/4\beta),$$

where $\tilde{p}(1) = \int_{-\infty}^{\infty} \exp(-r^2\beta) c(r)^{-2} dr$. Here $c(r)$ is the Harish-Chandra c-function (which determines the Plancherel measure of G/K), $j(\gamma)$ is a positive integer (such that $\gamma = \sigma^{j(\gamma)}$ where σ is primitive), and $C(\cdot)$ is a known positive function.

7. PARTITION FUNCTIONS AND SCATTERING THEORY

Scattering theory arises naturally when the space $\Gamma\backslash G/K$ considered in Sec. 6 is no longer compact but has only finite volume. Selberg himself used concepts from scattering theory when he originally treated this case. Basically, the problem is that the left-regular representation of G on $L^2(G/\Gamma)$ in this case is no longer a discrete direct sum of irreducible unitary representation as occurred in Sec. 6. Instead $L^2(G/\Gamma) = L^2_{dis} \oplus L^2_{Eis}$ where $U|L^2_{Eis}$ is a direct integral of unitary principal series. The space $L^2_{Eis}(K\backslash G/\Gamma)$ is the closure of the subspace spanned by wave packets formed with Eisenstein series of type 1. The scattering matrix S arises naturally in the theory of Eisenstein series.

In the simple case of $SO(2)\backslash SL(2,R)/\Gamma$, $S(z)$ is the scattering matrix for the automorphic wave equation

$$u_{tt} = y^2\Delta u + \frac{1}{4}u.$$

This relationship is explored in detail in LAX and PHILLIPS [10], LANG [9], and FADDEEV *et al.* [3,4].

The result we need is that the partition function is now given by

$$Z(\beta) = \left[z(\Gamma)\right] \mathrm{Vol}(G/\Gamma) \frac{1}{4\pi} \int_{-\infty}^{\infty} \exp(-r^2\beta) c(r)^{-1} c(-r)^{-1} dr$$

$$+ (4\pi\beta)^{-\frac{1}{2}} \sum_{\gamma} |u_\gamma| j(\gamma)^{-1} C(h(\gamma)) \exp(-u_\gamma^2/4\beta)$$

$$+ \frac{1}{4\pi} \exp(-r^2\beta) \mathrm{Tr}(S'(ir) S(ir)^*) dr$$

$$+ \frac{1}{4}(d - \mathrm{Tr}(S(0)) - \frac{d}{2\pi} \int_{-\infty}^{\infty} \exp(-r^2\beta) [\Gamma'(1+ir)/\Gamma(1+ir)] dr$$

$$+ c_1 \int_{-\infty}^{\infty} \exp(-r^2\beta) dr$$

$$+ c_2 \int_{-\infty}^{\infty} \exp(-r^2\beta) J(r) dr,$$

where c_1 and c_2 are known constants, $J(r)$ is a known function, and $z(\Gamma)$ is the center of Γ.

APPENDIX

Let G be a connected, noncompact, simple Lie group with finite center K, a maximal compact subgroup. We assume that the rank $(G/K) = 1$. Let $\mathbf{g} = \mathbf{k} + \mathbf{p}$ be the Cartan decomposition with Cartan involution θ and let $\mathfrak{a}_\mathbf{p}$ be a maximal abelian subspace of $\mathbf{p}$. Let Φ^+ be the set of positive roots, let $P_+ = \{\alpha \in \Phi^+ | \alpha \neq 0 \text{ on } \mathfrak{a}_\mathbf{p}\}$, and let $\rho = \frac{1}{2} \sum_{\alpha \in P_+} \alpha$.

For $\sum = P_+|_{\mathfrak{a}_\mathbf{p}}$, there is an element $\beta \in \sum$ such that 2β is the only other possible element of $\sum$. p,q are the number of elements of P_+ which restrict to β and 2β, respectively. Let H_o be the element of $\mathfrak{a}_\mathbf{p}$ for which $\beta(H_o) = 1$. Then $c_o = (2p + 8q)^{\frac{1}{2}}$.

The dual space Λ of $\mathfrak{a}_\mathbf{p}$ is identified with $\mathbb{R}$ and for h in $A_\mathbf{p}$ we put $u(h) = \beta(\log h)$. For the discrete subgroup Γ of G, we assume Γ has no elements of finite order other than those in its center $z(\Gamma)$. Thus every element of Γ is conjugate in G to an element of the Cartan subgroup $A_\mathbf{p}$. We write $A = A_\mathbf{k} A_\mathbf{p}$ and choose an element $h(\gamma)$ of A to which γ is conjugate. Let $h(\gamma) = h_\mathbf{k}(\gamma) h_\mathbf{p}(\gamma)$. Then $u_\gamma = \beta(\log h_\mathbf{p}(\gamma))$. The norm of u_γ is essentially the length of the shortest geodesic in the free homotopy class associated to γ on $\Gamma\ G/K$.

Let U be the representation of G induced by a representation a of Γ. If $\Gamma\backslash G$ is compact or on $L^2_{dis}(G/\Gamma)$, U is discretely decomposable with finite multiplicites. Let $U_1, U_2, \ldots$ be a complete list of inequivalent irreducible unitary representations of class one which occur in $L^2_{dis}(G/\Gamma)$ with multiplicities n_j. Each n_j is finite. Since each U_j is class one, it corresponds to a unique elementary positive definite spherical function $\phi_{\nu_j}, \nu_j \in \Lambda^{\mathbb{C}}$.

Finally, d is the number of equivalence classes of Γ-cuspidal minimal parabolic subgroups of G, the equivalence relation being conjugacy by an element of Γ.

REFERENCES

[1] J. S. Dowker, R. Critchley: "Scalar effective lagrangian in de Sitter Space," University of Manchester preprint

[2] J. S. Dowker, R. Banach: "Quantum field theory on Clifford-Klein space-times," University of Manchester preprint
[3] L. D. Faddeev, B. S. Pavlov: Proc. Steklov Math. Inst. **27**, 161 (1972)
[4] L. D. Faddeev, V. L. Kalinin, A. B. Venkov: Proc. Steklov Math. Inst. **37**, 5 (1973)
[5] R. Gangolli: Acta Math. **121**, 151 (1968)
[6] R. Gangolli: Ill. J. Math. **21**, 1 (1977)
[7] R. Gangolli, G. Warner: "Zeta functions of Selberg's type for some noncompact quotients of symmetric spaces of rank one," University of Washington preprint
[8] R. Hermann, N. Hurt: *Quantum Statistical Mechancis and Lie Group Harmonic Analysis* (Math Sci Press Brookline, Massachusetts, to be published)
[9] S. Lang: *SL(2,R)* (Addison-Wesley, Reading, Massachusetts, 1975)
[10] P. D. Lax, R. S. Phillips: *Scattering Theory for Automorphic Functions* (Princeton University Press, Princeton, New Jersey, 1977)
[11] A. Selberg: J. Indian Math. Soc. **20**, 47 (1956)
[12] A. Venkov: Proc. Steklov Math. Inst. **125**, 6 (1973)

INVERSE SCATTERING THEORY

WHAT DO WE KNOW ABOUT THE GEOMETRIC NATURE OF EQUATIONS WHICH CAN BE SOLVED USING THE INVERSE SCATTERING TECHNIQUE?

Robert Hermann*

Division of Applied Sciences
Harvard University
Cambridge, Massachusetts 02138

EULER-ARNOLD-LAX VECTOR FIELDS ON LIE ALGEBRAS

It is well-known that certain *nonlinear* partial differential equations can be solved in terms of *linear* analysis by what has been called the Inverse Scattering Technique [1-4]. There has been much speculation among workers in this field that there might be a unified way of looking at equations that are amenable to this method, but there are only so far isolated, but tantalizing clues to what such a unified theory might be.

Work of F. Estabrook and H. Wahlquist and the author has suggested that there are close links to the theory of Lie groups, and that solutions to the differential equations in question determine certain families of submanifolds (curves and surfaces) of Lie groups [5-17]. The aim of this paper is to tentatively suggest a possible classification scheme based on the ideas presented in [9,11,12,16]. Here is a brief introduction.

Let $\mathcal{G}$ be a Lie algebra with the real numbers as field of scalars. Assume, for the moment, that $\mathcal{G}$ is finite-dimensional. (Ultimately, it will be essential to extend the ideas to infinite dimensional Lie algebras.) Let $V: \mathcal{G} \to \mathcal{G}$ be a vector field over $\mathcal{G}$. It determines a set of curves in $\mathcal{G}$, the solutions of the differential equation

$$\frac{dA}{dt} = V(A(t)). \tag{1}$$

Let $(g,A) \to \operatorname{Ad} g(A)$ be the map $G \times \mathcal{G} \to \mathcal{G}$ which determines the *adjoint representation* of G and $\mathcal{G}$. (Thus, if G is a group of matrices,

$$\operatorname{Ad} g(A) = gAg^{-1},$$

$$g \in G,\ A \in \mathcal{G}.)$$

The action of G on $\mathcal{G}$ partitions $\mathcal{G}$ into submanifolds, the *orbits* of the action.

Definition. The vector field V on $\mathcal{G}$ is said to be of *Euler-Arnold-Lax* type if each solution of (1) lies on a single orbit of Ad G.

Remark. This is a Lie-group theoretic version of what the analysts now call an *isospectral flow.*

Here is one construction of such vector fields V. Let

$$B: \mathcal{G} \to \mathcal{G}$$

be any map. Set

$$V(A) = [B(A),A],$$

$$\text{for } A \in \mathcal{G}. \tag{2}$$

Equation (1) then takes the form

$$\frac{dA}{dt} = [B(A(t)),A(t)]. \tag{3}$$

For example, the equations for the geodesics of a left-invariant Riemannian metric on G take the form (3) [17] with $B: \mathcal{G} \to \mathcal{G}$ a linear map. This often provides a "Hamiltonian" structure for (1.3), as described in [9].

1. EULER-ARNOLD-LAX VECTOR FIELDS WHICH ARE REDUCED BY SUBGROUPS

Let $\mathcal{G}$, G, V be as in the previous section. Suppose H is a Lie subgroup of G and let S be a submanifold of $\mathcal{G}$. The pair (S,H) is said to reduce V if the following conditions are satisfied:

(a) V is tangent to S. (1.1)

(b) The solution of (1) which lie on S also lie on orbits of H.

In [9], it is shown how such reductions, with H a *solvable* Lie group, can lead to "complete integrability" or "solvability by quadratures" of differential equations.

It may also happen that on certain structures there are two such groups H and K, with both (S,H) and (S,K) reducing V. For example, as we shall show below, for the Toda lattice, $G = GL(n,R)$, H = natural solvable subgroup, K = maximal compact subgroup, S = set of Jacobi matrices of the Lie algebra $\mathcal{G}$.

For example, suppose V is of the form (3). Then, two mappings B, B': $\mathcal{G} \to \mathcal{G}$ provide the same solutions *on the submanifold S* providing that the following condition is satisfied:

$$B(A) - B'(A) \in \text{centralizer of } A \text{ in } \mathcal{G},\ A \in S. \tag{1.2}$$

This provides the algebraic freedom to change from one group to another.

Now, we examine the "Toda lattice" from this general point of view.

2. THE GENERALIZED TODA LATTICE AND GRADED ASSOCIATIVE ALGEBRAS

In addition to its Lie algebra structure, suppose that $\mathcal{G}$ has an *associative algebra structure*

$$(A_1,A_2) \to A_1A_2$$

with the Lie algebra structure given by the commutator:

$$[A_1,A_2] = A_1A_2 - A_2A_1.$$

In addition, we suppose that $\mathcal{G}$ has a *graded structure*, i.e., $\mathcal{G}$ as a vector space is a direct sum

$$\mathcal{G} = \sum_{n=1}^{\infty} \mathcal{G}^n$$

of linear subspaces. Finally, we suppose that

$$\mathcal{G}^n \mathcal{G}^m \subset \mathcal{G}^{n+m}. \tag{2.1}$$

Set

$$\mathcal{J} = \mathcal{G}^{-1} \oplus \mathcal{G}^0 \oplus \mathcal{G}^1. \tag{2.2}$$

The elements of $\mathcal{J}$ are called the *Jacobi elements* of $\mathcal{G}$. Let B: $\mathcal{G} \to \mathcal{G}$ be the linear mapping defined by the following formula

$$B(A_{-m} + \ldots + A_0 + \ldots + A_m) = A_{-m} + \ldots + A_{-1} - A_1 \ldots - A_m. \tag{2.3}$$

Set

$$V_B(A) = [B(A),A],\ A \in \mathcal{J}. \tag{2.4}$$

Theorem 2.1. *V_B, considered as a vector field on $\mathcal{G}$, is tangent to $\mathcal{J}$.*

Proof. Suppose $A = A_{-1} + A_0 + A_1 \in \mathcal{J}$, with $A_{-1} \in \mathcal{G}^{-1}$; $A_0 \in \mathcal{G}^0$; $A_1 \in \mathcal{G}^1$. Then,

$$B(A) = A_{-1} - A_1.$$

Hence,

$$\begin{aligned} V_B(A) &= [A_{-1} - A_1, A_{-1} + A_0 + A_1] \\ &= [A_{-1},A_0] + [A_{-1},A_1] - [a_{-1},A_1] - [A_1,A_0] \in \mathcal{J}. \end{aligned}$$

Q.E.D.

Thus, the orbits of V_B which lie in $\mathcal{J}$ are precisely the solutions of the differential equations, *with constraints*:

$$\frac{dA}{dt} = [B(A),A], \; A(t) \in \mathcal{G}. \tag{2.5}$$

Now set

$$\mathcal{G}^+ = \mathcal{G}^0 + \mathcal{G}^1 + \mathcal{G}^2 + \dots. \tag{2.6}$$

Because of the relations (2.1), $\mathcal{G}^+$ is a Lie subalgebra of $\mathcal{G}$. Note that

$$B(A) - A \in \mathcal{G}^+, \; A \in \mathcal{G}. \tag{2.7}$$

Hence,

$$V_B(A) = [B(A) - A, A]. \tag{2.8}$$

Assume for the moment that $\mathcal{G}$ is finite dimensional. Let G be its Lie group. (The "Toda lattice" corresponds to the case $G = GL(n,R)$.) Let G^+ be the connected Lie group of G corresponding to the Lie subalgebra $\mathcal{G}^+$ of $\mathcal{G}$.

Theorem 2.2. *The vector field V_B is tangent to the orbits of G^+.*

Note also that G^+ is *solvable* if $\mathcal{G}^0$ is abelian. In this case, the "integrability by quadratures" of (2.5) (which is the "Toda lattice" when $G = GL(n,R)$) is related, as explained in [9], to the classic 19th century work relating solvable Lie groups and ordinary differential equations which are "integrable by quadratures".

There are two directions of generalization inherent in this simple algebraic formalism which have not yet been pursued in detail: the infinite dimensional $\mathcal{G}$'s and the case where $\mathcal{G}^0$ is *not* abelian, so the G^+ is not solvable.

There is a certain intuition underlying this condition which is related to the Gel'fand-Levitan formula of Inverse Scattering theory. I will now leave the domain of precise Lie theory in order to suggest certain formal, algebraic directions which might prove useful in further research.

3. INTUITIVE BACKGROUND OF GEL'FAND-LEVITAN

Let X be an orientable manifold with volume element differential form dx. Let $F(X)$ denote the C^∞, real-valued functions on X. Let $D(X)$ denote the linear differential operators

$$\Delta: F(X) \to F(X).$$

$D(X)$ forms an associative algebra. Let $F_0(X)$ denote the subspace of $F(X)$ consisting of the *compactly supported* functions, i.e., those which vanish outside of a compact subset of X. Then $\Delta \in D(X)$ also maps $F_0(X)$ into itself.

One can now—by imposing boundary conditions and norms—complete $F_0(X)$ to a Hilbert space H, with operators Δ defined in suitable domains within H. Given two operators Δ, Δ', one can often construct *unitary* (or isometric) operators $U: H \to H$ which intertwine Δ and Δ'.

$$U\Delta = \Delta' U$$

Typically, such operators U involve the processes of *scattering theory*. Differential-geometrically, these involve "global" properties of the differential operators.

In parallel, one can construct *integral* operators

$$\alpha: F_0(X) \to F(X)$$

which intertwine differential operators. For example, suppose α is of the following form:

$$\alpha(f)(y) = \int_X K(y,x) f(x) dx. \tag{3.1}$$

Here, "dx" is a volume element differential form on X. For each y: $x \to K(y,x)$ is a *generalized function* on X. For example, it might be a smooth function on a certain open dense subset of X, with certain singularities, *which move with y*, of this subset. The goal is to set up the following conditions:

(a) differential equations for the function $(x,y) \to K(x,y)$ on the open subset where it is smooth;

(b) relations between the singularities and the differential equations in (a);

so that α intertwines Δ and Δ'. The Gel'fand-Levitan operators are the prototypes for this scheme, which will now be explained.

Let

$$X = \mathbb{R},$$
$$\Delta = d_x^2 + V(x), \tag{3.2}$$
$$\Delta' = d_x^2 + V'(x).$$

($d_x = d/dx$). Thus, Δ, Δ' are Sturm-Liouville operators with potentials V and V'. dx = Lebesgue measure on X. Look for α of the following form:

$$\alpha(f)(y) = \int_{\mathbb{R}} K(y,x)f(x)dx, \tag{3.3}$$

where

$$K(y,x) = 0, \quad \text{if } x > y. \tag{3.4}$$

Thus, $K(x,y)$ is smooth on $\mathbb{R}^2$, minus the diagonal $x = y$, and is zero in the region (3.4). One can then find integral operators α which intertwine Δ and Δ'. Now,

$$\alpha(f)(y) = \int_{-\infty}^{y} K(y,x)f(x)dx. \tag{3.5}$$

Remark. This indicates that α is a *causal* operator, i.e., if x,y are physically interpreted as "times", then $\alpha(f)$, at a given time y, only depends on the values of f at times $\leqslant y$. Part of the problem involved in extending the inverse scattering technique to higher dimensions is to discover the appropriate notion of "causality".

We will not be concerned here with any further details about the way the integral operator (3.5) is determined. What is important for group theoretical purposes is that the operators of form (3.1) again form a group.

Abstractly, we have the following set-up. A Lie algebra $\mathcal{G}$ of operators, together with two groups L and H acting as automorphisms of $\mathcal{G}$. We are—as in the toy theory constructed in the previous section—then motivated to look for mappings B: $\mathcal{G} \to \mathcal{G}$, together with curves $t \to A(t)$ in $\mathcal{G}$ which satisfy the equations

$$\frac{dA}{dt} = [B(A),A], \tag{3.6}$$

together with the possibility that dA/dt is tangent to the orbits of *both* H and L. Again, we might do this bv requiring that the following conditions be satisfied:

$B(A) \in \mathcal{H} \equiv$ Lie algebra of H.
For some function $F(A)$ of the operator A,
$F(A) + A \in \mathcal{L} \equiv$ Lie algebra of L.

Now, the Gel'fand-Levitan situation (together with the later work by GEL'FAND and DIKKI [18,19], leads us to suspect that $F(A)$ will not be a *polynomial* function of A, but something more general—a "pseudodifferential operator" of some form. Thus we are led to the program of extending the Lie algebra of differential operators by adding certain "pseudodifferential operators".

Notice that, for group-theoretic purposes, it is not crucial that $\mathcal{G}$ and its possible extensions in which $F(A)$ is to lie, be *operators*. What is important are the Lie-algebraic properties. Now, in quantum mechanics one is faced with a similar problem, i.e., that certain physical calculations involve only algebraic properties of operators, not necessarily the representations in which they are encountered. (In 1920's vintage quantum mechanics, this is called the "matrix method.") The best formalism for dealing with such matters seems to be the Weyl-Wigner-Moyal formalism, which defines a bracket structure (reducing to Poisson bracket as Planck's constant goes to zero) on certain types of "classical" observables, i.e., real-valued functions on the cotangent bundle to the configuration space of the mechanical system. I have briefly indicated in [8,16] certain potentialities for using this formalism for the purposes of nonlinear wave theory; I will now elaborate on this possibility.

4. THE MOYAL BRACKET AND PSEUDODIFFERENTIAL OPERATORS

Let q and p denote real variables. Let $\mathcal{G}$ denote all formal "Laurent" series of the form:

$$A = \sum_{n=-\infty}^{N} a_n(q)p^n, \tag{4.1}$$

where N is any integer and $q \rightarrow a_n(q)$ is a C^∞, real-valued function. Consider $\mathcal{G}$ as a real vector space. Tensor products will be defined with the real numbers as scalars. Let

$$\alpha\colon \mathcal{G} \otimes \mathcal{G} \rightarrow \mathcal{G} \otimes \mathcal{G}$$

be the linear map defined by the following formula:

$$\alpha(A_1 \otimes A_2) = \frac{\partial A_1}{\partial p} \otimes \frac{\partial A_2}{\partial q} - \frac{\partial A_1}{\partial q} \otimes \frac{\partial A_2}{\partial p} \tag{4.2}$$

Let M: $F \otimes F \rightarrow F$ be the linear map defined as follows:

$$M(A_1 \otimes A_2) = A_1 A_2. \tag{4.3}$$

(Note that products of two formal "Laurent" series of the type (4.1) can be defined by the usual formulas; it is this product which is indicated on the right hand side of (4.3).) Set

$$A_1 \# A_2 = \sum_{j=0}^{\infty} \frac{1}{j!} M\alpha^j(A_1 \otimes A_2). \tag{4.4}$$

It is readily seen that this product makes $\mathcal{G}$ into a real associative algebra. (This is the "Moyal" version of the pseudodifferential algebra of GEL'FAND, DIKKI, and MANIN [18-20].) Let $\mathcal{H}$ be the set of all $f \in \mathcal{G}$ of the form

$$f = \sum_{n=0}^{\infty} a_n(q)p^n \tag{4.5}$$

$\mathcal{H}$ then forms a subalgebra of $\mathcal{G}$; it is the "Moyal" version of what LEBEDEV and MANIN [21] call the Volterra operators.

Let $\mathcal{S}$ be the set of all $A \in \mathcal{G}$ of the form

$$A = p^2 + a(q). \tag{4.6}$$

The elements of the form (4.6) are the "Moyal" versions of the Sturm-Liouville operators. It has been shown in [7,8 (Part B)] that B: $\mathcal{S} \rightarrow \mathcal{G}$ can be chosen so that

$$B(A) = p^3 + \frac{3}{2}\, a,$$

$$[B(A),A] = \frac{1}{2}\, a\, \frac{da}{dq} + 6\, \frac{d^3a}{dq^3} \equiv K(a), \tag{4.7}$$

for $A \in \mathcal{S}$ of the form (4.6). $a \rightarrow K(a)$ is then a nonlinear, third order differential operator in functions of q. Thus, the solutions of the differential equations

$$\frac{DA}{dt} = V(A) \equiv [A,B(A)]$$

are of the form

$$A(t) = p^2 + a_t,$$

with

$$\frac{\partial a_t}{\partial t} = K(a). \tag{4.8}$$

(4.8) is the Korteweg-de Vries equation.

However, we can, in parallel with the work of GEL'FAND and DIKII [18,19], choose $C(A)$ to commute with A, for $A \in \mathcal{S}$, and such that

$$B(A) - C(A) \in \mathcal{H}. \tag{4.9}$$

In fact, it is readily seen that

$$C(A) \# C(A) = A \#' A \# A, \tag{4.10}$$

i.e.,

$$C(A) = A^{3/2},$$

if the 3/2 power is interpreted in the Moyal algebra sense. (In the work of Gel'fand and Dikii, it is defined relative to slightly different "pseudo-differential operator" algebra.)

Thus, we see that everything runs in parallel with the "Toda lattice", as treated in Sec. 2. However, the Lie algebra $\mathcal{G}$ in which this game is now played is infinite dimensional, and is not graded, but is "filtered" by powers of p with the algebraic structure (defined by the Moyal bracket) compatible with this filtration.

5. FINAL REMARKS

The theory of nonlinear waves—and its amazing relation to *linear* mathematics via Inverse Scattering—has provided a rich collection of problems in mathematics and at all levels. Most of the work has been at the level of analysis; there are also exciting ramifications into *geometry* and *Lie theory*, which go to the heart of the relation of these areas of mathematics and the theory of differential equations. Now, Lie himself was strongly motivated by this relation, but it has been obscured in work of modern times. Obviously, the underlying principle here is:

> In any problem involving differential or integral equations which is "soluble", look for the mechanism in terms of Lie groups.

In nonlinear waves, these Lie groups appear in much more subtle (and interesting) ways than in more traditional areas of Lie theory. For example, one point of view [5,7] is that the "soluble" systems (e.g., Korteweg-de Vries, sine-Gordon) can be recast to involve a Lie group (e.g., $SL(2,R)$) so that the *solutions* of the differential equations correspond to *submanifolds* of the Lie groups. We have only begun to explore the consequences of this principle!

FOOTNOTES AND REFERENCES

* This work was supported by NSF Grant MCS-78-0600 and Ames Research Center (NASA) Grant NSG-2252

[1] C. Gardner, J.M. Greene, M.D. Kruskal, R.M. Miura: "Methods for solving the Korteweg-de Vries equation," Phys. Rev. Lett. **19**, 1095-1097 (1967)

[2] P.D. Lax: "Integrals of nonlinear equations of evolution and solitary waves," Comm. Pure Appl. Math. **21**, 467-490 (1968)

[3] M.J. Ablowitz, D.J. Kaup, A.C. Newell, H. Segur: Stud. Appl. Math. **53**, 249 (1974)

[4] A. Scott, F. Chu, D. McLaughlin: "The soliton: a new concept in applied science," Proc. IEEE **61**, 1443 (1973)

[5] H. Wahlquist, F. Estabrook: "Prolongation structures of nonlinear evolution equations," J. Math. Phys. **16**, 1-7 (1975)

[6] F. Estabrook, H. Wahlquist: "Prolongation structures, connection theory, and Bäcklund transformations," in *Nonlinear Evolution Equations Solvable by the Laplace Transform*, ed. by F. Calogero (Pitman, London, 1978)

[7] R. Hermann: "The pseudopotentials of Estabrook and Wahlquist, the geometry of solitons, and the theory of connections," Phys. Rev. Lett. **36**, 835 (1976)

[8] R. Hermann: *The Geometry of Nonlinear Differential Equations, Bäcklund Transformations, and Solitons*, Parts A and B (Math Sci Press, Brookline, Massachusetts, 1976)

[9] R. Hermann, *Toda Lattices, Cosymplectic Manifolds, Bäcklund Transformations, and Kinks*, Parts A and B (Math Sci Press, Brookline, Massachusetts, 1977)

[10] R. Hermann: "Modern differential geometry in elementary particle physics," in *Proceedings of the VII GIFT Summer School*, Lecture Notes in Physics, ed. by J. Azcárraga (Springer, New York, 1978)

[11] R. Hermann: "The Lie-Cartan geometric theory of differential equations and scattering theory," in *Proceedings of the 1977 Park City (Utah) Conference on Differential Equations*, ed. by P. Bynes, M. Dekker (New York, 1979)

[12] R. Hermann: "Prolongation, Bäcklund transformations, and Lie theory as algorithms for solving and understanding nonlinear differential equations," in *Solitons in Action*, ed. by K. Lonngren, A. Scott (Academic, New York, 1978)

[13] *The 1976 Ames Research Center (NASA) Conference on the Geometric Theory of Nonlinear Waves*, ed. by R. Hermann (articles by Estabrook, Wahlquist, Hermann, Morris, Gorones, R. Gardner, and Scott) (Math Sci Press, Brookline, Massachusetts, 1977)

[14] R. Hermann: "Time-varying linear systems and the theory of nonlinear waves," in *Proceedings of the CDC Conference of the IEEE*, San Diego, 1979

[15] R. Hermann: *Cartanian Geometry, Nonlinear Waves, and Control Theory*, Part A, Interdisciplinary Mathematics, Vol. 20 (Math Sci Press, Brookline, Massachusetts, 1979)

[16] R. Hermann: "The inverse scattering technique of soliton theory, Lie algebras, the quantum mechanical Poisson-Moyal bracket, and the rotating rigid body," Phys. Rev. Lett. **37**, 1591 (1976)

[18] I.M. Gel'fand, L.A. Dikii: "Resolvents and Hamiltonian systems," Functional Anal. Appl. **11**, 93-105 (1977)

[19] I.M. Gel'fand, L.A. Dikii: "Asymptotic behavior of the resolvent of Sturm-Liouville equations and the algebra of the Korteweg-de Vries equations," Russian Math. Surveys **30**(5), 67-100 (1975)

[20] Y.U. Manin: "Algebraic aspects of nonlinear differential equations," in *Modern Problems in Mathematics*, (Viniti, Moscow, 1978), Vol. 11, pp. 5-152

[21] D.R. Lebedev, Y.U. Manin: "Gel'fand-Dikii Hamiltonian operators and coadjoint representation of Volterra group," IETP preprint, Moscow (1978)

[22] M. Adler: "On a trace functional for pseudo-differential operators and symplectic structure of the Korteweg-de Vries type equations," Inv. Math. **50**, 219-248 (1979)

SOLITONS, SOLUTIONS OF NONLINEAR EVOLUTION EQUATIONS, AND THE INVERSE SCATTERING TRANSFORM

M.J. Ablowitz

Department of Mathematics
Clarkson College of Technology
Potsdam, New York 13676

In recent years, a new method of mathematical physics has emerged. This method, which we shall refer to as the Inverse Scattering Transform (IST), allows us to solve certain physically significant nonlinear evolution equations. In one dimension, some of the interesting equations are: the Korteweg deVries (KdV), modified KdV, sine Gordon, nonlinear Schrödinger, self-induced transparency, three-wave interaction equations etc. It is certainly significant that these ideas apply also to certain discrete nonlinear wave problems; i.e. equations of differential-difference (e.g. Toda Lattice, a differential-difference nonlinear Schrödinger equation, the self-dual network, etc.) and partial-difference (e.g. numerical schemes) type. (For a detailed review of this work the reader may consult [1]). Some of the ideas involved with this new solution method have been extended to higher dimensions (see, for example, [1]-[3]). In these proceedings, I will describe certain "lump" type solutions to the so called Kadomstev-Petviashvili equation (a two dimensional generalization of the KdV equation) and Kaup will discuss certain aspects of the multidimensional three wave equations. In any event, while the full power of IST applies in one dimension, important advances are being made in multidimensional problems. It should also be mentioned that ideas originating from I.S.T. have successfully been used in lower dimensional problems, namely certain ODE's. In these proceedings, Segur will discuss how linear integral equations can be used to solve certain classical ODE's (i.e., Painlevé transcendents). These ODE's are similarity solutions of the related nonlinear evolution equations (solvable by IST) [4].

As is well known, many problems exhibiting interesting behavior are modeled by nonlinear evolution equations. In most cases, the analyst is reduced either to finding special solutions, or, essentially, to linearizing the system (perhaps about some special nonlinear solution) and exploiting the smallness of some parameter. The remarkable new development in the subject has been the discovery of a method which can exactly solve certain nonlinear evolution equations. The method uses the results of scattering and inverse scattering theory. This method is conceptually a generalization of Fourier analysis.

The method was originated by GARDNER, GREENE, KRUSKAL, and MIURA [5], (an extradordinary research contribution) and requires some understanding of scattering theory. They showed that associated with the Korteweg-deVries equation (KdV) [6]

$$u_t + 6uu_x + u_{xxx} = 0 \tag{1}$$

is a linear eigenvalue problem

$$v_{xx} + (k^2 + u(x,t))v = 0. \tag{2}$$

The latter is the well-known Schrödinger eigenvalue problem and has been extensively studied by physicists and mathematicians. The eigenvalues, and the behavior of the eigenfunctions as $|x| \to \infty$ determine the "scattering data", $S(k)$, which depend on the potential u. The direct scattering problem maps the potential into the scattering data. The inverse scattering problem reconstructs the potential from the scattering data. For appropriate boundary conditions this amounts to solving an integral equation, often referred to as the Gel'fand-Levitan integral equation.

We now outline the conceptual steps in order to obtain the solution. At the initial time, we are given $u(x,0)$, decaying sufficiently rapidly as $|x| \to \infty$. The direct scattering problem yields the scattering data at the initial instant, $S(k,0)$. For later times, even though $u(x,t)$ evolves according to a nonlinear partial differential equation, the required parts of $S(k, t)$ satisfy simple equations. One part of $S(k, t)$, the eigenvalues, do not depend on time, and for the other portion the role which was played by the dispersion relation in the linear problem is now played by the dispersion relation of the linearized problem. We are easily able to compute

$S(k,t)$. From this, one recovers the solution of the evolution equation, $u(x,t)$, by mapping back to physical space via the inverse scattering equations. Schematically, the procedure can be summarized as follows:

$$u(x,0) \to S(k,0) \xrightarrow{\omega(k)} S(k,t) \to u(x,t).$$

The procedure is conceptually analogous to Fourier analysis. Here, the scattering data plays the role of the Fouier transform, and the inverse scattering equation the inverse Fourier transform, ABLOWITZ, KAUP, NEWELL, and SEGUR [7] have termed the technique "the inverse scattering transform-Fourier analysis for nonlinear problems."

The soliton (first discovered by ZABUSKY and KRUSKAL [8]) is an important special solution of such an evolution equation. The soliton is such that it retains its identity despite nonlinear interactions. The general solution is usually quite complicated. However, the asymptotic ($t \to \infty$) character of the solution is fairly simple. For appropriate initial data, one obtains a finite number of solitons and an algebraically decaying background [1].

The method is quite powerful, but also quite specialized. Certainly, not all nonlinear evolution equations can be solved by this method, but it is remarkable that a subclass of those that are solvable are physically significant evolution equations. For example, the KdV equation arises in systems which are weakly dispersive and weakly nonlinear (quadratic nonlinearity).

The work of ZAKHAROV and SHABAT [9] showed that the method of solution for KdV was indeed no fluke. Employing methods developed by LAX [10], they showed that the nonlinear Schrödinger equation,

$$iu_t + u_{xx} + ku^2u^* = 0 \tag{3}$$

(here u^* is the complex conjugate of u) is intimately associated with a new eigenvalue problem, and, using inverse scattering, the equation (3) can be solved. This equation also has wide application. Here, $u(x,t)$ is the complex amplitude of an envelope of an almost monochromatic wavetrain propagating in a dispersive, weakly nonlinear medium.

This work led other researchers to the solution of the Sinė-Gordon equation (see also [1,7]) (a) in "light-cone" coordinates:

$$u_{xt} = \sin u \tag{4a}$$

or (b) in laboratory coordinates:

$$u_{tt} - u_{xx} + \sin u = 0, \tag{4b}$$

the modified KdV equation

$$u_t \pm 6u^2u_x + u_{xxx} = 0, \tag{5}$$

as well as others (SIT, three wave equation etc.). As briefly mentioned earlier, certain interesting discrete nonlinear equations are also solvable: those of the Toda Lattice

$$u_{ntt} = e^{-(u_n-u_{n-1})} - e^{-(u_{n+1}-u_n)}, \tag{6}$$

a differential-difference NL Schrödinger equation ,

$$iu_{nt} + u_{n+1} + u_{n-1} - 2u_n + ku_nu_n^*(u_{n+1} + u_{n-1}), \tag{7}$$

as well as partial difference schemes (see, for example, [1]).

Generally speaking, IST is effective when the following is true [10]. Consider two operators L,M; L is a spectral operator (one which is "rich" enough to employ inverse scattering)

$$L\mathrm{v} = \lambda \mathrm{v} \tag{8a}$$

and M governs the time dependence of the eigenfunctions:

$$\mathrm{v}_t = M\mathrm{v}. \tag{8b}$$

Taking ∂_t of (8a) yields

$$L_t\mathrm{v} + L\mathrm{v}_t = \lambda_t\mathrm{v} + \lambda\mathrm{v}_t. \tag{9a}$$

Using (8b) gives

$$L_t \mathrm{v} + LM\mathrm{v} = \lambda_t \mathrm{v} + ML\mathrm{v}, \tag{9b}$$

or

$$(L_t + [L,M])\mathrm{v} = \lambda_t \mathrm{v}, \tag{9c}$$

where $[L,M] \equiv LM - ML$ is the commutator of L and M. Hence, if $\lambda_t = 0$ ("isospectral flow"), then

$$L_t + [L,M] = 0. \tag{10}$$

If (10) is a nontrivial evolution equation, then it will be solvable by IST.

In this note we shall review how this method applies to the KdV equation, as well as discussing some interesting special solutions. For KdV, the operator $L = \partial_x^2 + u(x,-1)$, i.e., the Schrödinger eigenvalue equation (2) ($\lambda = k^2$). The time dependence (the operator) M is given by

$$\begin{aligned} \mathrm{v}_t &= A\mathrm{v} + B\mathrm{v}_x, \\ A &= u_x,\ B = 4k^2 - 2u. \end{aligned} \tag{11}$$

Requiring $\partial\lambda/\partial t$ (or $\partial k/\partial t$) $= 0$ insures that (10) is equivalent to the KdV equation (1). (Alternatively we satisfy the compatibility condition $\mathrm{v}_{xxt} = \mathrm{v}_{txx}$.) It should be noted that, given a suitable operator L, there is a simple deductive method to obtain the M operator associated with a nonlinear evolution equation [1].

The solution of (1) corresponding to $u \to 0$ as $|x| \to \infty$ proceeds as follows:

(i) At $t = 0$ we give $u(x,0)$ and solve the direct scattering problem (8) for eigenvalues (of which there are a finite number of discrete ones, $k = ik_n$, and a continuum for k real) and specific information about certain eigenfunctions, called the reflection coefficient $r(k,0)$, and bound-state normalization constants $C_{n,0}$ [1,5,11,12].

(ii) Equations (11) allow us to find how the scattering data evolves in time. It may be deduced that

$$k_n = \text{const.}, \rho(k,t) = \rho(k,0)e^{8ik^3 t}, C_n(t) = C_{n,0}e^{4k_n^3 t} \tag{12}$$

We refer to (12) as the scattering data at any time t:$S(k,t)$.

(iii) The theory of inverse scattering shows that, given the scattering data, we may reconstruct the potential u. The essential results are as follows. First, given (12), compute

$$F(x;t) = \frac{1}{2\pi}\int_{-\infty}^{\infty} \nu(k,t)e^{ikx}dk + \sum_{1}^{N} C_n^2(t)e^{-k_n x}. \tag{13a}$$

Then solve the integral equation ($y > x$)

$$K(x,y;t) + F(x+y;t) + \int_x^{\infty} K(x,z;t)F(z+y;t)\,dz = 0 \tag{13b}$$

for $K(x,y;t)$. The potential is reconstructed (hence the inversion) by the relation

$$u(x,t) = 2\frac{d}{dx}K(x,x;t). \tag{13c}$$

(13) constitutes the method of solution for KdV. Special soliton solutions corresponds to having no continuous spectrum. In the case $r(k,0) = 0$ the integral equation is degenerate and was solved in closed form by KAY and MOSES [13]. See also [5]. The results of such a calculation show that the N-soliton solution is given by

$$u(x,t) = 2\frac{\partial^2}{\partial x^2}\log\det(I + A), \tag{14a}$$

where

$$A_{ij} = \delta_{ij} + \frac{C_{i,0}C_{j,0}}{K_i + K_j}\exp(-(K_i + K_j)x + 4K_i^3 t + 4K_j^3 K). \tag{14b}$$

Thus, a one soliton solution has the form:

$$u = 2K_1^2 \operatorname{sech}^2 K_1(x - 4K_1^2 t - x_0). \tag{14c}$$

The N-soliton solution corresponds to waves which asymptotically ($t \to \infty$) have the form (14c). As mentioned earlier, these waves interact in such a way that they maintain their identities in the asymptotic limit. HIROTA [14] has developed a procedure by which one can develop formulas analogous to (14) without the need for inverse scattering. He has shown that the N-soliton solution has the form:

$$u(x,t) = \sum_{\mu = 0,1} \exp\left[\sum_{i=1}^{N} \mu_i \eta_i + \sum_{1 \leq i < j}^{N} A_{ij}\mu_i\mu_j\right] \tag{15a}$$

where the first sum is over all μ_i, $i = 1, \ldots, N$, $\mu_i = 0,1$, and

$$\eta_i = k_i x - k_i^3 t - \eta_0,$$

$$\exp(A_{ij}) = \left(\frac{k_i - k_j}{k_i + k_j}\right)^2, \tag{15b}$$

η_0 = arbitrary constant.

Besides all of this, there are many other special features associated with these nonlinear evolution equations, for example, (i) an infinite number of conserved quantities [15]:

$$\frac{\partial}{\partial t}T_i + \frac{\partial}{\partial x}F_i = 0, \quad i = 1, 2, \ldots,$$

for KdV, where $T_1 = u$, $T_2 = u^2$, $T_3 = \frac{1}{2}u_x^2 - u_3, \ldots\ldots$

(ii) long-time ($t \to \infty$) asymptotic states [1,16]; (iii) Bäcklund transformations [17]; (iv) complete integrability [18]; (v) periodic boundary-value problems [19]; (vi) direct methods to establish that linear integral equations solve nonlinear evolution equations [2], [20]; (vi) connection with ODE's of Painlevé type [4]; etc.

In what follows, we shall briefly discuss some other interesting solutions to the KdV equation (other nonlinear evolution equations also have such special solutions). First, we shall consider the rational solutions. A one-soliton solution given by (14c) transforms to

$$u = -2K_1^2 \operatorname{cosech}^2 K_1(x - 4K_1^2 t) \tag{16}$$

if we take $e^{K_1 x_0} = i$. Note that by taking the limit $K_1 \to 0$, u simplifies to:

$$u = -\frac{2}{x^2} = 2\frac{d^2}{dx^2}\log x. \tag{17}$$

It turns out that (17) is the first of a sequence of rational solutions given by

$$u = 2\frac{d^2}{dx^2}\log \theta_N. \tag{18a}$$

Whereas the first few θ_N are determined easily enough by long-wave limits as in (17), the higher θ_N are easily derived by limiting forms of the Bäcklund transformation. The first few are given by

$$\theta_0 = 1,\ \theta_2 = x^3 + 12t,$$

$$\theta_1 = x,\ \theta_3 = x^6 + 60x^3 t - 720t^2, \tag{18b}$$

and the general recurrence relations satisfy

$$D_x\theta_{N+1} \cdot \theta_{N-1} = (2N+1)\theta_N^2, \tag{19a}$$

$$(D_t + D_x^3)\theta_N \cdot \theta_{N+1} = 0, \tag{19b}$$

where D_x (similarly for D_t) is defined by

$$D_x a \cdot b \equiv (\partial_x - \partial_{x'})a(x)b(x')|x = x'$$

in the notation introduced by HIROTA [14]. The original work on rational solutions has appeared in [21], followed also by [22]. The work described here, using limits of soliton solutions and Bäcklund transformations has appeared in [23].

It is interesting to note that the above rational solutions may be "perturbed" in a rather general way. Specifically, it turns out that for the KdV equation we may solve for the perturbation $\mathrm{v}(x,t)$, where

$$u(x,t) = u_0(x,t) + \mathrm{v}(x,t), \tag{20}$$

and $u(x,t)$, $u_0(x,t)$ satisfy KdV (6). Here, $u_0(x,t)$ is a special solution of the KdV equation. For example, taking $u_0(x,t) = -2/x^2$ (this corresponds to scattering with a centrifugal barrier), then to obtain $v(x,t)$ we first solve the linear integral equation

$$K(x,y,t) + F(x,y,t) + \int_x^\infty K(x,z;t)F(z,y;t)\,dz = 0,$$

where

$$F(x,y;t) = \int r(k)\left[1 + \frac{i}{kx}\right]\left[1 + \frac{i}{ky}\right]e^{ik(x+y)+8ik^3t}dk + \sum_1^N C_n^2\left[1 + \frac{1}{k_n x}\right]\left[1 + \frac{1}{k_n y}\right]e^{-k_n(x+y)+8k_n^3k}. \quad (21b)$$

Then, $v(x,t)$ is obtained from

$$v(x,t) = 2\frac{d}{dx}K(x,x;t)\ . \quad (21c)$$

If we have only "discrete" spectra, then we have a combination of rational and exponential solutions:

$$v = 2\frac{d^2}{dx^2}\log(x\det(I + A))$$

$$A_{ij} = C_iC_j\left[\frac{1}{K_iK_jx} + \frac{1}{K_i + K_j}\right]\exp(-(K_i + K_j)x + 4K_i^3t + 4K_j^3t). \quad (22)$$

When $N = 1$, we have

$$u = 2\frac{d^2}{dx^2}\log x\left[1 + C_1^2\left[\frac{1}{K_1^2x} + \frac{1}{2K_1}\right]\exp(-2K_1x + 8K_1^3t)\right]. \quad (23)$$

We refer to these solutions as quasi-solitons. They may have application to the solution of KdV on $(0,\infty)$. These solutions are singular at some location $x = x_0(t) < 0$, but not for $x \geqslant 0$. The above analysis appears in [24].

Next, we turn to discuss some results pertaining to multidimensional problems. It is now well known that some of the techniques of I.S.T. apply to problems of higher dimensions. For example, it is known that there are L,M operators associated with the so called Kadomstev-Petviashvilli ("two dimensional" KdV) equation:

$$\frac{\partial}{\partial x}(u_t + 6uu_x + u_{xxx}) + \sigma u_{yy} = 0 \quad (24)$$

and a two dimensional version of the nonlinear Schrödinger equation

$$iu_k - u_{xx} + u_{yy} = \sigma u|u|^2 + u\Phi_x, \quad (25a)$$

$$\sigma\Phi_{xx} + \Phi_{yy} = -2(|u|^2)_x, \quad (25b)$$

$\sigma = \pm 1$ and others (Three-wave problems etc. [1]). Moreover, it can be established that these equations are associated with linear integral equations [2], have N-plane-wave-soliton solutions, etc.

It turns out that by taking long-wave limits of the plane wave soliton solutions certain algebraically decaying (in all directions) "lump" type solutions can be constructed [25,26]. A 1-lump solution to (24) is given by ($\sigma = -1$):

$$u = 4\frac{\left[-(x' + P_Ry')^2 + P_I^2y'^2 + \frac{3}{P_I^2}\right]}{\left[(x' + P_Ry')^2 + P_I^2y'^2 + \frac{3}{P_I^2}\right]^2}, \quad (26)$$

where

$$x' = x - (P_R^2 + P_I^2)t,$$

$$y' = y + 2P_Rt,$$

and P_R, P_I ($P_I > 0$) are constant. Alternatively, we may write u in the form

$$u = 2\frac{d^2}{dx^2}\log F_N \quad (27a)$$

1-Lump: $F_2 = \theta_1\theta_2 + B_{12},$ (27b)

2-Lumps:
$$F_4 = \theta_1\theta_2\theta_3\theta_4 + B_{12}\theta_3\theta_4 + B_{13}\theta_2\theta_4 + B_{14}\theta_2\theta_3 + B_{23}\theta_1\theta_4 + B_{24}\theta_1\theta_3 + B_{34}\theta_1\theta_2 + B_{12}B_{34} + B_{13}B_{24} + B_{14}B_{23}, \tag{27c}$$

where

$$\theta_i = x + P_i y - P_i^2 t,$$
$$B_{ij} = \frac{12}{\sigma(P_i - P_j)^2},$$
$$N = 2M,\ P_{M+i} = P_i^* \ (i = 1,2, \ldots,M).$$

These solutions are rational functions of x,y,t. As $|x|$, $|y| \to \infty$, $u \sim 0(1/x^2,\ 1/y^2)$. They produce no phase shift upon interaction. Analogous solutions may be constructed for the two-dimensional version of the nonlinear Schrödinger equation (25) (26). It should also be noted that the plane wave solutions are unstable when $\sigma = -1$, and stable when $\sigma = +1$. (Real lump-type solutions are not constructed by this method when $\sigma = +1$.) It can be expected that the complete solution to these special multidimensional problems will be found.

Finally, we briefly note that there is an important connection between nonlinear evolution equations solvable by IST and ODEs without movable initial points. This connection and its ramifications are considered in [4]. For example, consider the modified KdV equation:

$$u_t - 6u^2u_x + u_{xxx} = 0. \tag{28}$$

(28) has a similarity solution of scaling type:

$$u = \frac{1}{(3t)^{1/3}}\, w(z),\ z = \frac{x}{(3t)^{1/3}}, \tag{29}$$

where $W(z)$ satisfies

$$w'' = 2w^3 + zw + \alpha, \tag{30}$$

α = constant. (30) is a classical O.D.E. discovered by Painlevé (see [27], Chap. 14). It is one of the so called six "irreducible" Painlevé transcendents. It turns out that the methods of IST can now linearize this equation! Later in these proceedings, Segur will discuss these and other aspects in more detail.

REFERENCES

[1] M.J. Ablowitz: Stud. Appl. Math., **58**, 17 (1978)

[2] V.E. Zakharov, A.B. Shabat: Functional Anal. Appl. **8**, 226 (1974)

[3] H. Cornille: "Solutions of the nonlinear three-wave equations in three spatial dimensions," preprint, 1978, to be published in Phys. Lett; D.J. Kaup: "A method to solve the initial value problem of the three-wave equations in three dimensions," preprint (1978)

[4] M.J. Ablowitz, A. Ramani, H. Segur: Lett. Nuovo Cimento, **23**, 333, 1978; M.J. Ablowitz, A. Ramani, H. Segur: "A connection between nonlinear evolution equations and ordinary differential equations of Painlevé type I," preprint; M.J. Ablowitz, A. Ramani, H. Segur: "A connection between nonlinear evolution equations and ordinary differential equations of Painlevé type II," preprint

[5] C.S. Gardner, J.M. Greene, M.D. Kruskal, R.M. Miura: Phys. Rev. Lett. **19**, 1095 (1967); C.S. Gardner, J.M. Greene, M.D. Kruskal, R.M. Miura: Comm. Pure Appl. Math. **27**, 97 (1974)

[6] D.J. Korteweg, G. DeVries: Phil. Mag. **39**, 422 (1895)

[7] M.J. Ablowitz, D.J. Kaup, A.C. Newell, H. Segur: Stud. Appl. Math. **53**, 249 (1974)

[8] N.J. Zabusky, M.D. Kruskal: Phys. Rev. Lett. **15**, 240 (1965)

[9] V.E. Zakharov, A.B. Shabat: Sov. Phys.—JETP **34**, 62 (1972)

[10] P.D. Lax: Comm. Pure Appl. Math. **21**, 467 (1968)

[11] L. Faddeev: J. Math. Phys. **4**, 72 (1963)

[12] P. Deift, E. Trubowitz: "Inverse scattering on the line," Comm. Pure Appl. Math. **32**, 121 (1979)

[13] I. Kay, H.E. Moses: J. Appl. Phys. **27**, 1503 (1956)

[14] R. Hirota: Phys. Rev. Lett. **27**, 1192 (1971)

[15] R.M. Miura: J. Math. Phys. **9**, 1202 (1968)

[16] M.J. Ablowitz, H. Segur: Stud. Appl. Math. **57**, 13 (1977)

[17] H. Chen: Phys. Rev. Lett. **33**, 925 (1974)
[18] V.E. Zakharov, L.D. Faddeev: Functional Anal. Appl. **5**, 10 (1971)
[19] B.A. Dubrovin, V.B. Matveev, S.P. Novikov, Russian Math. Surveys **31**, 59 (1976); E. Trubowitz, H. McKean: to be published
[20] H. Cornille, J. Math. Phys. **17**, 2143, (1976)
[21] H. Airault, H.P. McKean, J. Moser: Comm. Pure Appl. Math. **30**, 1 (1977)
[22] M. Adler, J. Moser: Commun. Math. Phys. **61**, 1 (1978)
[23] M.J. Ablowitz, J. Satsuma: J. Math. Phys. **19**, 2180 (1978)
[24] M.J. Ablowitz, H. Cornille: "On solutions of the Korteweg-de Vries equation," preprint (1979), to be published in Phys. Lett.
[25] S.V. Manakov, V.E. Zakharov, L.A. Bordag, A.R. Its, V.B. Matveev: Phys. Lett. A **63**, 205 (1977).
[26] J. Satsuma, M.J. Ablowitz: J. Math. Phys. **20**, 1496 (1979)
[27] E.L. Ince: *Ordinary Differential Equations* (Dover, New York, 1944)

DETERMINING THE FINAL PROFILES FROM THE INITIAL PROFILES FOR THE FULL THREE-DIMENSIONAL THREE-WAVE RESONANT INTERACTION

D. J. Kaup*

Department of Physics
Clarkson College of Technology
Potsdam, New York 13676

Based on a result of Cornille, a complete inverse scattering solution for the full three-dimensional three-wave resonant interaction is now possible. We illustrate this by showing how one may use inverse scattering techniques to determine the final profiles from the initial nonoverlapping profiles.

In what should be considered to be a major breakthrough, CORNILLE [1] was able to lay the basic groundwork necessary for obtaining a complete inverse scattering solution to the full three-dimensional three-wave resonant interaction (3D3WRI), whose equations in characteristic coordinates are

$$\partial_i q_i = \gamma_i q^*_j q^*_k. \tag{1}$$

In (1), i,j, and k are to be taken to be cyclic in 1,2,3 (thus, there actually are three equations contained in (1)), the q's are the three envelope amplitudes, $\partial_i \equiv \partial/\partial\chi_i$ where the three χ_i's are the three characteristic coordinates, defined by

$$\partial_i \equiv -\partial_t - \mathbf{v}_i \cdot \nabla, \tag{2}$$

and the three γ's carry the signs of the coupling constants ($\gamma_i = \pm 1$ etc.), where we assume that these coupling constants have been scaled to a unit magnitude. Cornille worked out all of his results using ordinary space-time coordinates. Since he did not use characteristic coordinates, many of the natural symmetries in his results were not apparent at that time. We [2] then recast his results into characteristic coordinates form, obtained considerable simplifications, and were able to discuss a simple, but quite important class of initial value problems. This is the class of solutions when the initial profiles are nonoverlapping.

Before continuing with the method of solution, we should first briefly describe the characteristic coordinates which we shall use and the qualitative nature of the solution of (1) when viewed in these coordinates. Equation (2) defines these characteristic coordinates, provided we introduce a fourth coordinate χ_4, defined in some suitable manner. For example, if we have a time-independent problem with all three group velocities independent, then we could choose $\chi_4 = t$, since no derivatives of time would occur in (1). In general, χ_4 is defined by

$$\partial_i \chi_4 = 0, \tag{3}$$

and is simply a fourth coordinate which is independent of the three characteristic coordinates. As an explicit example, consider the frame where $\mathrm{v}_2 = 0$. Now, rotate the x,y, and z axes so that v_1 is parallel to $\hat{x}$ and v_3 lies in the x-y plane. In this case, we could take

$$\chi_4 = z, \tag{4a}$$

and the three characteristic coordinates may be taken to be given by

$$x = -\mathrm{v}_{1x}\chi_1 - \mathrm{v}_{3x}\chi_3, \tag{4b}$$

$$y = -\mathrm{v}_{3y}\chi_3, \tag{4c}$$

$$t = -\chi_1 - \chi_2 - \chi_3. \tag{4d}$$

Note that at a fixed t the sum of the characteristic coordinates is a constant, which defines a plane in the three-dimensional characteristic coordinate space.

Consider the solution of (1) in any region where q_j or q_k are zero. We have that q_i may simply be any arbitrary function of χ_j and χ_k, which we assume to be localized about $\chi_j = \chi_k = 0$. This solution is now like a tube localized about the χ_i axis. At a fixed χ_j and χ_k, its value does not change as χ_i changes, which by (4d) corresponds to changing t. The initial profile is obtained for $\chi_i \to \infty$ and the final profile is obtained when $\chi_i \to -\infty$. (Which for $q_j = 0$ or $q_k = 0$ is the same as the initial profile.)

Of course, the above considerations only apply when $q_j = 0$ or $q_k = 0$. We now allow the other two envelopes to be present and arranged as in Fig. 1. Here we have three interacting beams, with each initial profile being obtained as the corresponding characteristic coordinate approaches ∞. Note that these profiles never change until the other two profiles overlap. When this happens, the profile changes and, as the characteristic coordinate goes to minus infinity, we obtain the final profile.

We shall concentrate on the question of how one may obtain these final profiles from the initial profiles by means of the inverse scattering transform. The basic idea is essentially the same as that used in doing the same for the two-dimensional solution [3-5]. Namely, we shall map the three initial profiles into scattering data, determine the relation between the initial and final scattering data, and then reconstruct the final profiles from the scattering data.

A scattering problem for the 3D3WRI ideally suited for characteristic coordinates was first given by ABLOWITZ and HABERMAN [6], which, strictly speaking, is only a rearranged form of the original ZAKHAROV and SHABAT [7] problem. Reducing the former to its most basic form gives

$$\partial_k \psi_i = \gamma_k q^*_j \psi_k, \tag{5a}$$

$$\partial_i \psi_k = \gamma_i q_j \psi_i, \tag{5b}$$

where i, j, and k are cyclic. The general solution of (5) is determined by specifying the three functions $g_i(\chi_i) = \psi_i(\chi_i, \chi_j \to \infty, \chi_k \to \infty)$. In place of the three arbitrary functions, we may use $g_i(\chi_i) = e^{i\zeta\chi_i}$, which, upon taking the Fourier transform with respect to ζ, will give us the general solution. Thus, we define the three fundamental solutions of (1) by

$$\psi_i^n(\zeta; \chi_i, \chi_j \to \infty, \chi_k \to \infty) = \delta_i^n e^{i\zeta\chi_i}, \tag{6}$$

where $n = 1,2,3$. Without proof, we shall simply state that this solution exists for ζ real when each profile satisfies [8]

$$|q_i(\chi_i, \chi_j, \chi_k)| < V(\chi_j) V(\chi_k), \tag{7a}$$

for all values of χ_i, where

$$\int_{-\infty}^{\infty} ds\,[V(s) + V^2(s) + |s|\,V(s)] < \infty. \tag{7b}$$

Furthermore, each fundamental solution also has certain analytic properties with respect to ζ [8]. We will not need these here, but will give a simplified version of them later.

Let us return to (5) and consider the regions in our characteristic space where q_j and q_k are both zero. From Fig. 1, one sees that this is where $|\chi_i|$ is sufficiently large. In the region where $\chi_i \to \infty$, (5) gives

$$\partial_j \psi_k^n = \gamma_j Q^*_i \psi_j^n, \tag{8a}$$

$$\partial_k \psi_j^n = \gamma_k Q_i \psi_k^n, \tag{8b}$$

$$\partial_k \psi_i^n = \partial_j \psi_i^n = 0 = \partial_i \psi_k^n = \partial_i \psi_j^n, \tag{9}$$

where

$$Q_i(\chi_j, \chi_k) = \lim_{\chi_i \to \infty} q_i(\chi_i, \chi_j, \chi_k), \tag{10}$$

and is simply the initial profile. From (6), (8), and (9), we have that the ith fundamental solution is simply

$$\psi_r^i(\zeta; \chi) = \delta_r^i e^{i\zeta\chi_i}, \tag{11}$$

where $r = i,j,k$. For the other two fundamental solutions, we have that they are now independent of χ_i and are determined by the solution of (8). Thus, the problem reduces to a simpler scattering problem [9], given by

$$\partial_j H_k^n = \gamma_j Q^*_i H_j^n, \tag{12a}$$

$$\partial_k H_j^n = \gamma_k Q_i H_k^n, \tag{12b}$$

for $n = j,k$ where H is a function only of χ_j and χ_k, and

$$H_k^n(\zeta; \chi_j \to \infty, \chi_k) = \delta_k^n e^{i\zeta\chi_k}, \tag{12c}$$

$$H_j^n(\zeta;\chi_j,\chi_k \to \infty) = \delta_j^n e^{i\zeta\chi_j}. \tag{12d}$$

Again, when Q_i satisfies (7), one can show that the solution exists.

Now, the scattering problem for (12) is not very difficult to work out. Figure 2 gives a graphic representation of the $n = j$ solution for (12). The central shaded area corresponds to the initial profile of Q_i. The vertical solid lines correspond to H_j^j being propagated along its characteristics, which are parallel to the χ_k axis. Note that, unless a given characteristic intersects the initial profile, nothing happens to it and it remains "unscattered." The dashed horizontal lines represent H_k^j being generated due to interaction of H_j^j with the initial profile, Q_i, and it travels along characteristics parallel to χ_j. Thus, H_j^k is only generated by those lines of H_j^j which intersect the initial profile.

From the Neumann-series solution of (12), one may determine that $H_k^n e^{-i\zeta\chi_n}$ is analytic in the upper half ζ−plane and approaches δ_k^n as $|\zeta| \to \infty$. One can define the scattering matrix by

$$H_j^n(\zeta;\chi_j,\chi_k \to -\infty) = \int_{-\infty}^{\infty} \frac{d\lambda}{2\pi} \mu_j^n(\zeta,\lambda) e^{i\lambda\chi_j}, \tag{13a}$$

$$H_k^n(\zeta;\chi_j \to -\infty,\chi_k) = \int_{-\infty}^{\infty} \frac{d\lambda}{2\pi} \mu_k^n(\zeta,\lambda) e^{i\lambda\chi_k}, \tag{13b}$$

Referring to Fig. 2, we see that μ_k^j is simply the Fourier transform of H_k^j on the left, while μ_j^j is the Fourier transform of H_j^j at the bottom. Note that μ_j^j must contain a part proportional to the Dirac delta function $\delta(\zeta' - \zeta)$, since for large $|\chi_j|$ there are "unscattered" lines.

Directly from the conservation law

$$\partial_k[\gamma_j H_j^n(\zeta) H_j^m(\zeta')^*] - \partial_j[\gamma_k H_k^n(\zeta) H_k^m(\zeta')^*] = 0, \tag{14}$$

one may show that

$$\mu_j^j \mu_j^{j\dagger} - \gamma \mu_k^j \mu_k^{j\dagger} = I, \tag{15a}$$

$$\mu_k^k \mu_k^{k\dagger} - \gamma \mu_j^k \mu_j^{k\dagger} = I, \tag{15b}$$

$$\mu_j^k \mu_j^{j\dagger} - \gamma \mu_k^k \mu_k^{j\dagger} = 0, \tag{15c}$$

and

$$\mu_j^{j\dagger} \mu_j^j - \gamma \mu_j^{k\dagger} \mu_j^k = I, \tag{16a}$$

$$\mu_k^{k\dagger} \mu_k^k - \gamma \mu_k^{j\dagger} \mu_k^j = I, \tag{16b}$$

$$\mu_k^{k\dagger} \mu_j^k - \gamma \mu_k^{j\dagger} \mu_j^j = 0, \tag{16c}$$

where we are using a condensed operator notation (i.e., a product implies an integration, etc.), where

$$I \equiv 2\pi\delta(\zeta - \zeta') \tag{17}$$

and

$$\gamma \equiv \gamma_j \gamma_k. \tag{18}$$

We note that (16) corresponds to the orthogonality condition for the states H^n and (15) corresponds to the closure condition. Thus, these states are *complete* for ζ real, and there are never any bound states for (12).

Of course, our choice for (12c) and (12d) was arbitrary, and choosing any other corner of a square would have given equivalent results. We mention one other choice because we shall need the corresponding scattering matrix. It is

$$G_j^n(\zeta;\chi_j,\chi_k \to \infty) = \delta_j^n e^{i\zeta\chi_j}, \tag{19a}$$

$$G_k^n(\zeta;\chi_j \to -\infty,\chi_k) = \delta_k^n e^{i\zeta\chi_k}, \tag{19b}$$

and the referred to scattering matrix is given by

$$G_j^n(\zeta;\chi_j,\chi_k \to -\infty) = \int_{-\infty}^{\infty} \frac{d\lambda}{2\pi} \nu_j^n(\zeta,\lambda) e^{i\lambda\chi_j}, \tag{20a}$$

$$G_k^n(\zeta;\chi_j \to +\infty,\chi_k) = \int_{-\infty}^{\infty} \frac{d\lambda}{2\pi} \nu_k^n(\zeta,\lambda) e^{i\lambda\chi_j}. \tag{20b}$$

This scattering matrix is related to the first one by

$$\nu_k^j = -\gamma \nu_j^j \mu_j^{k\dagger}, \tag{21a}$$

$$\nu_j^k = +\gamma \mu_k^{k\dagger} \nu_j^j, \tag{21b}$$

and, most importantly,

$$\mu_j^j \nu_j^{j\dagger} = \nu_j^{j\dagger} \mu_j^j = I, \tag{22a}$$

$$\mu_k^k \nu_k^k = \nu_k^k \mu_k^k = I. \tag{22b}$$

This last relation shows that μ_j^j and μ_k^k possesses unique inverses, which is another consequence of the nonexistence of bound states for (12).

We shall now simply quote the major results for the inverse scattering problem of (12). From the above relations, one may show that the linear dispersion relations for the states H^n are given by

$$H_r^j(\zeta)e^{-i\zeta\chi_j} = \delta_r^j + \frac{i}{2\pi}\int_{-\infty}^{\infty}\frac{d\zeta'}{\zeta'-\zeta}\int_{-\infty}^{\infty}\frac{d\lambda}{2\pi}e^{i\zeta'\chi_j}\nu_k^j(\zeta',\lambda)H_r^k(\lambda), \tag{23a}$$

$$H_r^k(\zeta)e^{-i\zeta\chi_k} = \delta_r^k + \frac{i\gamma}{2\pi}\int_{-\infty}^{\infty}\frac{d\zeta'}{\zeta'-\zeta}\int_{-\infty}^{\infty}\frac{d\lambda}{2\pi}e^{-i\zeta'\chi_k}\nu_k^{j*}(\lambda,\zeta')H_r^j(\lambda), \tag{23b}$$

for ζ in the upper half ζ-plane. One may define a transformation kernel by

$$H_r^n(\zeta)e^{-i\zeta\chi_n} = \delta_r^n + \int_0^{\infty} L_r^n(s;\chi_j,\chi_k)e^{i\zeta s}ds, \tag{24}$$

and from (23) and (24) one obtains for $s > 0$,

$$L_r^n(s;\chi_j,\chi_k) + E_r^n(s+\chi_n,\chi_r) + \sum_m \int_0^{\infty} dt\ E_m^n(s+\chi_n,t+\chi_r)\,L_r^m(t;\chi_j,\chi_k) = 0, \tag{25a}$$

with the profile being recovered by

$$Q_i(\chi_j,\chi_k) = -\gamma_k L_j^k(0;\chi_j,\chi_k), \tag{25b}$$

where

$$E_j^j = 0 = E_k^k, \tag{26a}$$

$$E_k^j(u,v) = \int_{-\infty}^{\infty}\frac{d\zeta}{2\pi}\int_{-\infty}^{\infty}\frac{d\lambda}{2\pi}e^{-i\zeta u}\nu_k^j(\zeta,\lambda)e^{i\lambda v}, \tag{26b}$$

$$E_j^k(u,v) = \gamma E_k^{j*}(v,u). \tag{26c}$$

Furthermore, one can show that the solution of (25) exists and is unique when the scattering matrices ν and μ satisfy (15), (16), (21), and (22).

So, as shown above, we can solve the scattering problem associated with (8), and the solution of (8) is then

$$\psi_r^i(\zeta;\chi) = \delta_r^i e^{i\zeta\chi_i}, \tag{27a}$$

for $r = i,j,k$, and then for $n = j,k$,

$$\psi_r^n(\zeta;\chi) = H_{ir}^n(\zeta;\chi_j,\chi_k), \tag{27b}$$

$$\psi_i^n(\zeta;\chi) = 0, \tag{27c}$$

where $r = j,k$. Note the extra subscript in (27b). We shall use the convention that the first subscript indicates which profile is to be used in (12). Of course, (12), (13), and (27) remain valid as we cyclically permute $i,j,$ and k, provided we use this extra subscript to denote which profile. The same remains true of all the following relations, (14)-(26), once the extra subscript is included.

Now we have all the necessary tools for determining the scattering matrix of the final profiles. The general approach can be seen by considering Fig. 1. There it is seen that everywhere, except for the interaction region localized about the origin, we are able to use the above solution (27) to determine the fundamental solutions of (3). For example, (27) gives the solution for $\chi_i \to \infty$. The solution for $\chi_j \to \infty$ which matches onto this solution in their region of overlap (both χ_i and χ_j large) is similarly

$$\psi_r^n(\zeta;\chi) = H_{jr}^n(\zeta;\chi_k,\chi_i) \tag{28a}$$

for $n = i,k$ and $r = i,k$, and

$$\psi_r^j(\zeta;\chi) = \delta_r^j e^{i\zeta\chi_j}, \tag{28b}$$

where now $r = i,j,k$. Likewise, for $\chi_k \to \infty$, we have

$$\psi_r^n(\zeta;\chi) = H_{kr}^n(\zeta;\chi_i,\chi_j) \tag{29a}$$

for $n,r = i,j$, and

$$\psi_r^k(\zeta;\chi) = \delta_r^k e^{i\zeta\chi_k} \tag{29b}$$

for $r = i,j,k$.

The asymptotic forms of the solution to (3) as given by (27)-(29) give us one-half of the asymptotic solution. The other asymptotic forms will depend on the final profiles,

$$\tilde{Q}_i(\chi_j,\chi_k) = \lim_{\chi_i \to -\infty} q_i(\chi_i,\chi_j,\chi_k). \tag{30}$$

All quantities depending on the final profiles will be indicated by a wavy line above them. Now consider the solution as $\chi_i \to -\infty$. The general solution will be a linear combination (due to completeness) of $\delta_r^i e^{i\zeta\chi_i}$, $\tilde{H}_{ir}^j(\zeta;\chi_j,\chi_k)$ and $\tilde{H}_{ir}^k(\zeta;\chi_j,\chi_k)$. To determine what components are present, we simply require the solution to match (28) as $\chi_j \to \infty$ and (29) as $\chi_k \to \infty$. From this, one finds when $\chi_i \to -\infty$ that

$$\psi_r^i(\zeta;\chi) = \delta_r^i e^{i\zeta\chi_i}, \tag{31a}$$

$$\psi_r^n(\zeta;\chi) = \int_{-\infty}^{\infty} \frac{d\lambda}{2\pi} \mu_{kj}^n(\zeta,\lambda) \tilde{H}_{ir}^j(\lambda;\chi_j,\chi_k) + \int_{-\infty}^{\infty} \frac{d\lambda}{2\pi} \mu_{jk}^n(\zeta,\lambda) \tilde{H}_{ir}^k(\lambda;\chi_j,\chi_k), \tag{31b}$$

for $n = j,k$, where here and subsequently $r = i,j,k$. Similarly, one finds when $\chi_j \to -\infty$ that

$$\psi_r^j(\zeta;\chi) = \delta_r^j e^{i\zeta\chi_j}, \tag{32a}$$

$$\psi_r^n(\zeta;\chi) = \int_{-\infty}^{\infty} \frac{d\lambda}{2\pi} \mu_{ki}^n(\zeta,\lambda) \tilde{H}_{jr}^i(\lambda;\chi_k,\chi_i) + \int_{-\infty}^{\infty} \frac{d\lambda}{2\pi} \mu_{ik}^n(\zeta,\lambda) \tilde{H}_{jr}^k(\lambda;\chi_k,\chi_i), \tag{32b}$$

for $n = i,k$, and when $\chi_k \to -\infty$ that

$$\psi_r^k(\zeta;\chi) = \delta_r^k e^{i\zeta\chi_k}, \tag{33a}$$

$$\psi_r^n(\zeta;\chi) = \int_{-\infty}^{\infty} \frac{d\lambda}{2\pi} \mu_{ji}^n(\zeta,\lambda) H_{kr}^i(\lambda;\chi_j,\chi_j) + \int_{-\infty}^{\infty} \frac{d\lambda}{2\pi} \mu_{ij}^n(\zeta,\lambda) H_{kr}^j(\lambda;\chi_i,\chi_j), \tag{33b}$$

where now $n = i,j$.

To obtain the final relation, we now require (31)-(33) to agree in all regions where they pairwise overlap. From this we obtain only the relation (in condensed notation)

$$\mu_{kj}^n \tilde{\mu}_{ik}^j + \mu_{jk}^n \tilde{\mu}_{ik}^k - \mu_{ki}^n \tilde{\mu}_{jk}^i - \mu_{ik}^n \tilde{\mu}_{jk}^k = 0, \tag{34}$$

which is true for $n = i,j,k$, and for i,j,k cyclic. Without proof, we shall simply state that one can solve the above for $\tilde{\nu}_{ik}^j$. One finds

$$\tilde{\nu}_{ik}^j = \mu_{kj}^{j\dagger} \nu_{ik}^j \mu_{jk}^k - \gamma_i \gamma_j \mu_{kj}^{i\dagger} \mu_{jk}^i, \tag{35}$$

with which one may construct the function $\tilde{E}_{ik}^j$, then solve (25) for $\tilde{L}_{ir}^n$, and finally for $\tilde{Q}_i$.

This is our main result at this time. We are still working on further aspects of the problem, some of which I shall briefly mention. First, we have not shown that the $\tilde{\mu}$'s determined by (34) will satisfy (15) and (16). Until this is done, it is not certain that the solution of (25) for $\tilde{Q}_i$ necessarily exists. It is possible that for certain signs of the γ's, singular solutions may arise. If this were the case, the $\tilde{\mu}$'s would not satisfy (15) and (16). Also, one should be able to determine from (34) the final values of the infinity of conserved quantities in terms of their initial values. We should have this result shortly. Other aspects to be done include the total inverse scattering solution for (3), not just the asymptotic form obtained here [8]. One also wants to consider the general initial value problem when the initial envelopes are possibly overlapping [10]. Note that the result presented here can only be applied to the case when the initial envelopes are nonoverlapping, because we have assumed that we know the profiles as $t \to -\infty$. Lastly, there is the question of how all these results will mesh in with the results for the two-dimensional case [5], where solitons and bound states frequently occur.

REFERENCES

*Research supported in part by NSF Grant MCS78-03979 and ONR Contract N00014-76-C-0867

[1] H. Cornille: J. Math. Phys. **20**, 1653 (1979)
[2] D.J. Kaup: "A method for solving the separable initial value problem of the full three-dimensional three-wave resonant interaction", to appear in Stud. Appl. Math.
[3] D.J. Kaup: Stud. Appl. Math. **55**, 9 (1976)
[4] V.E. Zakharov, S.V. Manakov: Zh. Eksp. Teor. Fiz. **69**, 1654 (1975) [Sov. Phys. – JETP **42**, 842 (1976)]
[5] D.J. Kaup, A.H. Reiman, A. Bers: Rev. Mod. Phys. **51**, 275 (1979)
[6] M.J. Ablowitz, R. Haberman: J. Math. Phys. **16**, 2301 (1975)
[7] V.E. Zakharov, A.B. Shabat: Funk. Anal. Prilož. **8**, 43 (1974)
[8] D.J. Kaup: "The inverse scattering solution for the full three-dimensional three-wave resonant interaction," to appear in Physica D.
[9] Parts of this two-dimensional inverse scattering problem have also been solved by L.P. Nižnik: *Inverse Nonstationary Problem of Scattering Theory* (Naukora Dumka, Kiev, 1973), 182 pp.
[10] D.J. Kaup: "The solution of the general initial value problem for the full three-dimensional three-wave resonant interaction," to appear in the proceedings of the Soviet-American Soliton Symposium, Kiev, September 2-14, 1979

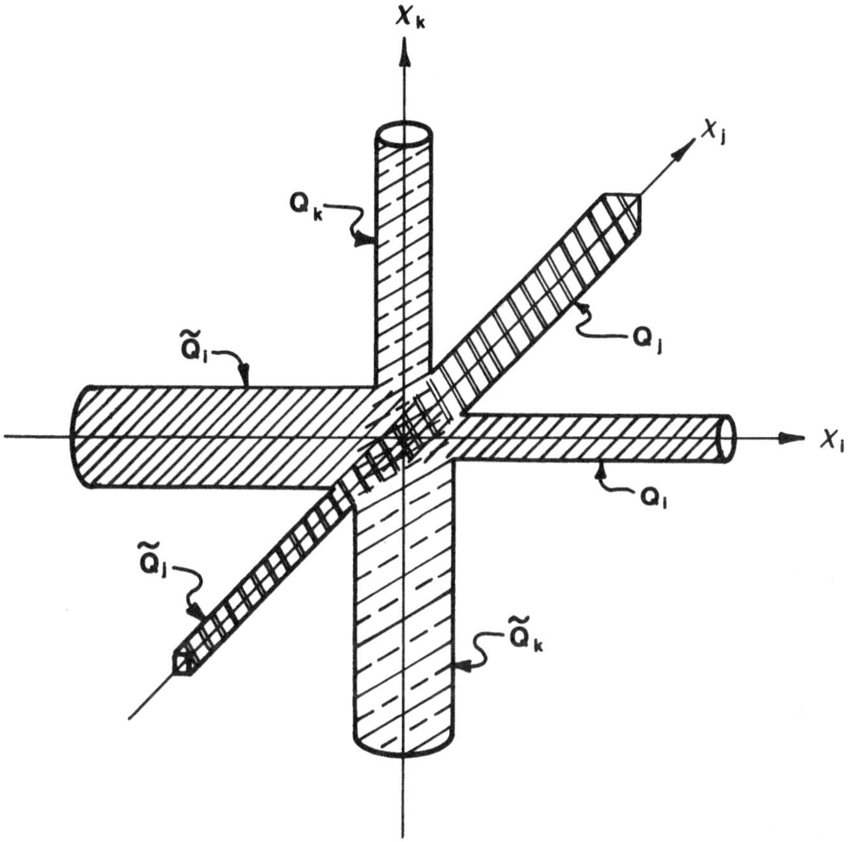

Fig. 1. A diagramatic representation of a possible three-dimensional solution in the characteristic-coordinate space.

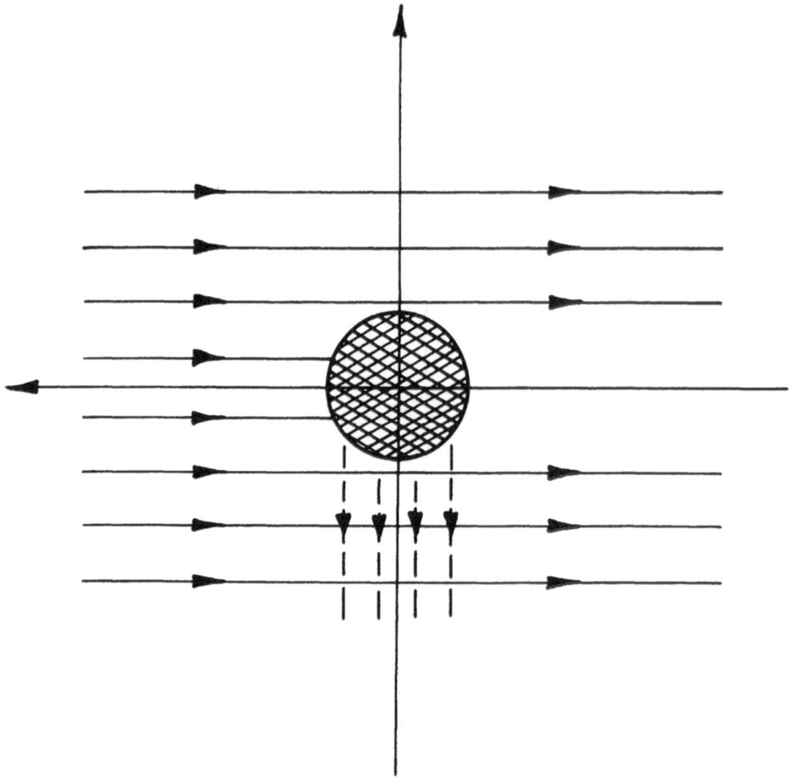

Fig. 2. Graphical representation of the $n = j$ solution of (12). The initial profile is the shaded region in the middle. The vertical solid lines at the top correspond to the initial data, $H_j^j = e^{i\zeta x_j}$. This component propagates downward to the bottom along vertical characteristics, and is, therefore, only modified in the region directly below the initial profile (scattering potential). The horizontal dashed lines represent the solution for H_k^j (the scattered wave) which is only nonzero to the direct left of the scattering potential. The μ_j^j (μ_k^j) component of the scattering matrix is simply the Fourier transform of H_j^j (H_k^j) at the bottom (left).

ORDINARY DIFFERENTIAL EQUATIONS OF PAINLEVE-TYPE AND THE INVERSE SCATTERING TRANSFORM

Harvey Segur*

Aeronautical Research Associates of Princeton, Inc.
Princeton, New Jersey 08540

The previous two talks have discussed an application of the theory of scattering and inverse scattering that has developed only within the last dozen years: solving nonlinear partial differential equations by some version of the inverse scattering transform (IST). The equations that can be solved by IST are known to be very special, and one of the major outstanding problems in the field is to characterize these equations. The question is: What structure must a partial differential equation possess in order to be solvable by IST? One of the two main points of this lecture is a conjecture, formulated by Ablowitz, Ramani, and me, on the correct characterization of these equations. The other main idea is this: inverse scattering provides an exact linearization of certain nonlinear partial differential equations; it also provides an exact linearization of certain nonlinear ordinary differential equations. The ODE's that can be solved by this method have a simple characterization: they are of Painlevé-type (I will say what that means shortly).

The material presented in this talk summarizes some of the work done by Ablowitz, Ramani, and me over the last two years. Additional information can be found in [2-5].

The general outline of the talk will be as follows. First, we must define the Painlevé property for ODE's, because everything else follows from it. Once you know what the Painlevé transcendents are, I can show you that they are related to evolution equations solvable by inverse scattering transforms. Our conjecture about characterizing these nonlinear PDE's is then almost obvious; they must reduce to ODE's of Painlevé-type. What is less obvious is how to prove the conjecture, but we can give a partial proof. This relation between ODE's of P-type and inverse scattering transforms can be exploited to obtain information about either the ordinary or the partial differential equations, and I can give you some idea of the kind of results that you get rather easily by this approach.

What is the Painlevé property? Let's start with a linear ordinary differential equation, say of second order:

$$\frac{d^2w}{dz^2} + p(z)\frac{dw}{dz} + q(z)w = 0. \tag{1}$$

For suitable $p(z)$ and $q(z)$, this equation can be viewed in the complex plane and, as everyone here knows, the singularities of the solution of (1) can be found by examing $p(z)$ and $q(z)$ (e.g., [8], Chap. 15). In particular, the general solution has two constants of integration,

$$w(z;A,B) = Aw_1(z) + Bw_2(z), \tag{2}$$

and the location in the complex plane of the singularities of $w(z)$ do *not* depend on A or B. The singularities of a linear differential equation are said to be *fixed*, because they do not depend on the constants of integration.

Nonlinear differential equations lose this property. A very simple example of a nonlinear ODE is

$$\frac{dw}{dz} + w^2 = 0; \tag{3}$$

its general solution is

$$w(z;z_o) = \frac{1}{z - z_o}. \tag{4}$$

Here, z_o is the constant of integration and it also defines the location of the singularity. This singularity is *movable*, because its location depends on the constant of integration.

So while linear differential equations have only fixed singularities, nonlinear equations can have both fixed and movable singularities. About 100 years ago, mathematicians asked the following question:

Which nonlinear ODE's admit no movable branch points or essential singularities?

Movable poles are allowed, as are fixed singularities of any kind. We will refer to this property as the Painlevé property, and equations that possess it will be said to be of *Painlevé-type*, or *P-type*.

It turns out that the only first order equations with the Painlevé property are generalized Ricatti equations:

$$\frac{dw}{dz} = p_0(z) + p_1(z)w + p_2(z)w^2. \tag{5}$$

(A complete review of the nineteenth century work in this field may be found in [8], Chaps. 12-14.)

Painlevé and his coworkers were able to answer the question comprehensively for second-order equations of the form

$$\frac{d^2w}{dz^2} = F\left(\frac{dw}{dz}, w, z\right), \tag{6}$$

where F is rational in $\frac{dw}{dz}$ and w, and analytic in z. They showed that out of all possible equations of the form (6), only 50 canonical equations have the Painlevé property of no movable branch points or essential singularities. Further, they showed that 44 of these equations can be reduced to something already known, such as elliptic functions. That left six equations that defined new transcendental functions, called the Painlevé transcendents. The first four of these are:

$$\frac{d^2w}{dz^2} = 6w^2 + z, \qquad P_I$$

$$\frac{d^2w}{dz^2} = 2w^3 + zw + \alpha, \qquad P_{II}$$

$$\frac{d^2w}{dz^2} = \frac{1}{w}\left(\frac{dw}{dz}\right)^2 - \frac{1}{z}\left(\frac{dw}{dz}\right) + \frac{1}{z}(\alpha w^2 + \beta) + \gamma w^3 + \frac{\delta}{w}, \qquad P_{III}$$

$$\frac{d^2w}{dz^2} = \frac{1}{2w}\left(\frac{dw}{dz}\right)^2 + \frac{3w^3}{2} + 4zw^2 + 2(z^2 - \alpha)w + \frac{\beta}{w}. \qquad P_{IV}$$

There are two more.

The question of which equations have the Painlevé property is appropriate at any order, but comprehensive results are available only at the first and second order.

That is all history. What does it have to do with inverse scattering? The best way to answer that is to state our *conjecture*.

Every nonlinear ODE obtained by an exact reduction of a nonlinear PDE solvable by some inverse scattering transform has the Painlevé property.

Here are some examples. The Boussinesq equation

$$u_{tt} = u_{xx} + \left(\frac{u^2}{2}\right)_{xx} + \frac{1}{4}u_{xxxx} \tag{7}$$

is a nonlinear PDE solvable by IST [12, 14]. An exact reduction to an ODE may be obtained by looking for a traveling wave solution:

$$u(x,t) = w(x - ct) = w(z).$$

Then (7) becomes

$$(1 - c^2)w'' + \left(\frac{w^2}{2}\right)'' + \frac{1}{4}w'''' = 0, \tag{8}$$

which can be integrated twice. Depending on the constants of integration, the result after rescaling is either

$$w'' + 2w^2 + a = 0 \quad \text{or} \quad w'' + 2w^2 + z = 0.$$

The first possibility defines an elliptic function, whose only singularieties are poles. The second possibility is the equation for P_I. In either case, the ODE has the Painlevé property. So the PDE solvable by inverse scattering reduces to an ODE of P-type

Another example is the modified KdV equation

$$u_t - 6u^2u_x + u_{xxx} = 0, \tag{9}$$

which can be solved by IST [13]. An exact reduction to an ODE may be obtained by looking for a self-similar solution:

$$u(x,t) = (3t)^{-2/3}w(z); \quad z = x/(3t)^{1/3}, \tag{10}$$

$$\rightarrow \qquad w''' - 6w^2w' - (zw)' = 0.$$

This can be integrated once:

$$w'' = 2w^3 + zw + \alpha. \qquad \mathrm{P_{II}}$$

Again, the ODE is of P-type.

The sine-Gordon equation

$$u_{xt} = \sin u \tag{11}$$

can be solved by IST [1]. It has a self-similar solution

$$u(x,t) = f(z), \quad z = xt. \tag{12}$$

If we set $w(z) = \exp(if)$, then

$$w'' = \frac{1}{w}(w')^2 - \frac{1}{z}(w') + \frac{1}{2z}(w^2 - 1). \qquad \mathrm{P_{III}}$$

Again, the ODE is of P-type.

The derivative nonlinear Schrödinger equation

$$iq_t = q_{xx} - 4iq^2(q^*)_x + 8|q|^4q \tag{13}$$

can be solved by IST [10]. Its similarity solution eventually reduces to P_{IV}. We have checked an enormous number of examples. In every case we checked, PDE's that can be solved by IST reduce to ODE's of P-type and PDE's that are not solvable by IST (e.g., this may be determined by observing numerically that two solitary waves do not interact like solitons) reduce to ODE's that are *not* of P-type.

So there is some kind of relation between partial differential equations solvable by IST and ordinary differential equations of P-type. This relation can be used to examine either the ODE's or the PDE's. I can demonstrate how it helps in the study of the ODE's with the modified KdV equation, (9), and P_{II}. The last step of IST, the inverse scattering part, goes like this. $F(x,t)$ satisfies a linear partial differential equation

$$F_t + F_{xxx} = 0, \tag{14}$$

subject to some boundary and initial conditions. Then, $K(x,y;t)$ satisfies a linear integral equation of the Gel'fand-Levitan-Marchenko type

$$K(x,t) = F(x+y) + \int_x^\infty\int_x^\infty K(x,z)F(z+s)F(s+y)\,dzds, \quad y \geqslant x. \tag{15}$$

Once K is known, then $q(x,t) = K(x,x;t)$ satisfies mKdV:

$$q_t - 6q^2q_x + q_{xxx} = 0.$$

In the full IST treatment, F depends on the initial data of $q(x,0)$ through the direct scattering problem. Here, we simply start with F and force everything to be self-similar:

$$\xi = x/(3t)^{1/3}, \quad \eta = y/(3t)^{1/3}, \tag{16}$$

$$F(x,t) = (3t)^{-1/3}F(\xi), K(x,y;t) = (3t)^{-1/3}K(\xi,\eta).$$

Then, (14) becomes a linear ODE,

$$F'''(\xi) - (\xi F)' = 0,$$

and a one-parameter family of solutions is

$$F(\xi+\eta) = rAi\left(\frac{\xi+\eta}{2}\right), \tag{17}$$

where $Ai(\xi)$ is the Airy function. The integral equation (15) becomes

$$K(\xi,\eta) = r\, Ai\left(\frac{\xi+\eta}{2}\right) + \frac{r^2}{4}\int_\xi^\infty\int_\xi^\infty K(\xi,\zeta) Ai\left(\frac{\zeta+\theta}{2}\right) Ai\left(\frac{\theta+\eta}{2}\right) d\zeta\, d\theta, \qquad \eta \geqslant \xi. \tag{18}$$

The Airy function decreases rapidly as its argument becomes large, so the integral term in (18) is very well-behaved. Therefore, it is relatively easy to solve (18) for $\eta \geqslant \xi$. On $\eta = \xi$, the solution of (18) satisfies the self-similar form of mKdV, *viz.* P_{II}:

$$\frac{d^2}{d\xi^2} K(\xi,\xi) = 2K^3(\xi,\xi) + \xi K(\xi,\xi). \tag{19}$$

(Two different proofs of this fact are given in [4] and [5].) The point here is that (18) is an exact linearization of P_{II}: every solution of the linear integral equation also solves P_{II}. The general solution of (19) involves two arbitrary constants; the linear integral equation gives a one parameter (r) family, which includes all of the bounded real solutions of (19).

You may recall that the most convenient way to obtain global information about solutions of linear differential equations is to replace them by linear integral equations. The same thing seems to be true for these nonlinear equations. Global information about the bounded solutions of P_{II} comes out of the integral equation (18) quite easily. Details about this global description of P_{II} can be found in [4] and [5].

Here I have concentrated on P_{II}, but clearly the idea is not so restricted. Any ODE that is an exact reduction of a PDE solvable by inverse scattering has an exact linearization in terms of a linear integral equation like (18). If your objective is to study the ODE, then the linear integral equation gives some of the information you seek quite easily.

But suppose your objective is not an ODE but a PDE. Suppose you have been staring at a particular partial differential equation for three days, and you want to know whether or not it can be solved by IST. Now we are back to the conjecture: reduce it to an ODE by looking for a traveling wave or a self-similar solution, and determine whether the ODE is of P-type. If the ODE is not of P-type, then we conjecture that the PDE cannot be solved by IST. But it is hard to place much confidence in such a test unless you have some idea of why it should work. So let me give you a partial proof of why this test actually works.

Consider a linear integral equation of the form

$$K(x,y) = F(x+y) + \int_x^\infty K(x,z) N(x;z,y)\, dz, \quad y \geqslant x, \tag{20}$$

where F vanishes rapidly for large values of the argument and N depends on F. For example, for the

$$N(x,z,y) = \int_x^\infty F(z+s) F(s+y)\, ds.$$

Other choices are also possible. We want to show that every solution of a linear integral equation like (20) must have the Painlevé property. Then if K also satisfies an ODE, the family of solutions of the ODE obtained *via* (20) necessarily has the Painlevé property as well. So the Painlevé property is not out of the blue, it is a consequence of the linear integral equation.

Very roughly, the proof goes like this (for details, see [3]):

(i) F satisfies a linear ODE, and therefore has no movable singularities at all.

(ii) If F vanishes rapidly enough, then the Fredholm theory of integral equations applies. It follows that (20) has a unique solution in the form

$$K(x,y) = F(x+y) + \int_x^\infty F(x+z) \frac{D_1(x,z,y)}{D_2(x)} dz. \tag{21}$$

where D_1 and D_2 are entire functions of their arguments. Then the singularities of K can only come from the fixed singularities of F, or the movable zeros of D_2. But D_2 is analytic, so these movable singularities must be poles.

I would like to close by giving two examples of how the conjecture may be used.

Example 1

In (1 + 1) dimensions, the nonlinear Schrodinger equation is

$$iu_t = u_{xx} + a|u|^2u \tag{22}$$

It can be solved by IST [15]. A natural generalization to (2 + 1) dimensions is

$$iu_t = \nabla^2 u + a|u|^2 u. \tag{23}$$

We claim this equation cannot be solved by IST, because (23) has a similarity solution in the form

$$u(x,y,t) = R(\sqrt{x^2+y^2},\lambda)\exp(i\lambda t),$$

and the ODE for $R(r)$ is not of P-type. So the nonlinear Schrödinger equation is solvable in (1+1) dimensions, but not in (2+1) or (3+1) dimensions.

Example 2

In (1+1) dimensions, the KdV equation [7] is

$$u_t + 6uu_x + u_{xxx} = 0 \tag{24}$$

and the modified KdV equation is

$$\mathrm{v}_t - 6\mathrm{v}^2\mathrm{v}_x + \mathrm{v}_{xxx} = 0. \tag{25}$$

Both can be solved by IST, and MIURA's [11] transformation relates them. A generalization of KdV to (2+1) dimensions is the KADOMTSEV-PETVIASHVILI [9] equation

$$(u_t + 6uu_x + u_{xxx})_x + au_{yy} = 0. \tag{26}$$

This is also of IST-type [6]. The same generalization of mKdV to (2+1) dimensions is

$$(\mathrm{v}_t - 6\mathrm{v}^2\mathrm{v}_x + \mathrm{v}_{xxx})_x + a\mathrm{v}_{yy} = 0. \tag{27}$$

We claim this equation cannot be solved by IST, because (27) has a time-independent similarity solution in the form

$$\mathrm{v}(x,y,t) = (2y)^{-1/2}V(x/(2y)^{1/2}),$$

and the equation for V is not of P-type.

To summarize, the question of determining whether a given equation is or is not solvable by IST is rather delicate. At the moment, this test for the Painlevé property is the best test we know. It is direct, it requires no exceptional cleverness from the user, and there are no known counterexamples.

FOOTNOTES AND REFERENCES

*This work was supported in part by the Office of Naval Research and by the U.S. Army Research Office

[1] M.J. Ablowitz, D.J. Kaup, A.C. Newell, H. Segur: Stud. Appl. Math. **53**, 249 (1974)
[2] M.J. Ablowitz, A. Ramani, H. Segur: Lett. Nuovo Cimento **23**, 333 (1978)
[3] M.J. Ablowitz, A. Ramani, H. Segur: "A connection between nonlinear evolution equations and ordinary differential equations of P-type. I", J. Math. Phys., to be published
[4] M.J. Ablowitz, A. Ramani, H. Segur: "A connection between nonlinear evolution equations and ordinary differential equations of P-type. II", J. Math. Phys., to be published
[5] M.J. Ablowitz, H. Segur: Phys. Rev. Lett. **38**, 1103 (1977)
[6] V. Dryuma: Sov. Phys.—JETP Lett. **19**, 387 (1974)
[7] C.J. Gardner, J.M. Greene, M.D. Kruskal, R.M. Miura: Phys. Rev. Lett. **19** 1095 (1967); Comm. Pure Appl. Math. **27**, 97 (1974)
[8] E.L. Ince: *Ordinary Differential Equations* (Dover, New York, 1944)
[9] B. Kadomstev, V. Petviashvili: Sov. Phys.-Dokl. **15**, 539 (1970)
[10] D.J. Kaup, A.C. Newell: J. Math. Phys. **19**, 798, (1978)
[11] R.M. Miura: J. Math. Phys. **9**, 1202, (1968)
[12] E. Trubowitz: invited talk at this conference, unpublished
[13] M. Wadati: J. Phys. Soc. Japan **32**, 1681, (1972)
[14] V.E. Zakharov: Sov. Phys.—JETP **65**, 219, (1973)
[15] V.E. Zakharov, A.B. Shabat: Sov. Phys.—JETP **34**, 62 (1972)

EXACT SOLUTIONS FOR THE THREE-DIMENSIONAL SCHRÖDINGER EQUATION WITH QUASI-LOCAL POTENTIALS OBTAINED FROM A THREE-DIMENSIONAL GEL'FAND-LEVITAN EQUATION. EXAMPLES OF TOTALLY REFLECTIONLESS SCATTERING

Harry E. Moses*

University of Lowell, Center for Atmospheric Research
College of Pure and Applied Science
Lowell, Massachusetts 01854

ABSTRACT

In an early paper on the inverse scattering problem for the three-dimensional Schrödinger equation using a Gel'fand-Levitan equation, Kay and Moses introduced nonlocal potentials which in the present paper are called "quasi-local." These potentials are diagonal in the radial variable, but are integral operators in the angular variables and represent a generalization of the usual local potential. In the early paper, we were unable to give explicit potentials for which the Schrödinger equation could be solved, since we did not solve the corresponding Gel'fand-Levitan equation. In the present paper, we introduce another Gel'fand-Levitan equation for which many solutions can be found. Each solution yields a quasi-local potential for which the corresponding three-dimensional Schrödinger equation can be solved in closed form. As far as the author knows, these are the first potentials, local or nonlocal, other than separable potentials, for which the Schrödinger equation can be solved in closed form. They should be useful in testing hypotheses of formal scattering theory.

In the present paper, examples of quasi-local potentials are given which support point eigenvalues and for which there is no scattering whatever. These potentials are analogues of the reflectionless potentials of the one-dimensional problem.

Finally, we indicate how the scattering operator can be found from a wave operator satisfying any boundary or initial or final value conditions and the corresponding completeness relation. This result enables us to obtain the scattering operator from spectral data for the three-dimensional problem of the present paper.

INTRODUCTION

An early method of solving the inverse scattering problem for the three-dimensional Schrödinger equation was given by the writer [1], who generalized an inverse method of JOST and KOHN [2] for the radial Schrödinger equation. This method was a nonlinear one and sought to identify a minimal (but not necessarily unique) set of scattering data from which the potential could be reconstructed. An iterative scheme was given for the potential in terms of this minimal data. We shall now describe these data.

Consider the Schrödinger equation for the continuous spectrum corresponding to the local potential $V(\mathbf{x})$:

$$[-\Delta + V(\mathbf{x})]\psi_-(\mathbf{x}|\mathbf{k}) = k^2\psi_-(\mathbf{x}|\mathbf{k}) \,, \tag{1}$$

where $k = |\mathbf{k}|$, $r = |\mathbf{x}|$, and $\psi_-(\mathbf{x}|\mathbf{k})$ is required to be the scattering eigenfunction and thus required to have the asymptotic form

$$\psi_-(\mathbf{x}|\mathbf{k}) \sim (2\pi)^{-3/2}\, e^{i\mathbf{k}\cdot\mathbf{x}} + b_k(\boldsymbol{\eta},\boldsymbol{\eta}')\, \frac{e^{ikr}}{r},\ r \to \infty \,, \tag{2}$$

where $\boldsymbol{\eta} = \mathbf{x}/r$ and $\boldsymbol{\eta}' = \mathbf{k}/k$.

In [1], it was indicated that if there were no bound states the potential would be determined by the amplitude of the spherical scattered wave. Not all of the amplitude had to be known. It would be sufficient to know $b_k(-\boldsymbol{\eta},\boldsymbol{\eta})$ for $\boldsymbol{\eta}$ restricted to a hemisphere and for all k.

Having obtained the iterative method of [1], we wished to obtain exact solutions to inverse problems. We were guided by our experience with the inverse problem for the one-dimensional Schrödinger equation, for which we could find exact solutions.

KAY and MOSES [3] set up an abstract formalism for the inverse problem in terms of spectral theory, which was a generalization of the original Gel'fand-Levitan approach of [4] for the radial equation, and it was shown in [5] that the Gel'fand-Levitan equation for the one-dimensional equation was a special case of the abstract formalism (KAY's original method [6] used a clever time-dependent approach which the present writer felt might have limitations in other situations). FADDEEV [7] showed explicitly that both the Gel'fand-Levitan approach to the inverse problem [4] for the radial equation and the treatment of MARCHENKO [8] for the radial equation could be recovered from the abstract formalism of [3].

Generally, the abstract formalism indicated that for a large area of inverse spectral theory a Gel'fand-Levitan equation could be set up which would enable one to reconstruct a self-adjoint operator from its spectral decomposition. To be somewhat more specific, one would be given the domain of an operator by giving the boundary conditions on its eigenfunctions. One would also be given the spectral decomposition (i.e., the completeness relation on the eigenfunctions). Then if one assumed certain triangularity conditions on an operator which gave the transformation from a known self-adjoint operator (i.e. one whose proper and improper eigenfunctions were known explicitly) to the desired operator, one could find the transformation explicitly. The triangularity conditions had to be compatible with the boundary conditions on the eigenfunctions of the desired operator. The equation for the triangular portion of the transformation operator was the generalized Gel'fand-Levitan equation. In the one-dimensional case, the completeness relation on the eigenfunctions of the desired operator could be expressed in terms of the reflection coefficient. Moreover, the triangularity relation assured one that the scattering potential was local.

One of the very useful applications of the Gel'fand-Levitan and Marchenko equations for the one-dimensional and radial-equation cases is that, for some of the completeness relationships, the appropriate Gel'fand-Levitan equations can be solved explicitly in closed form. The corresponding potentials can then be found explicitly. We thereby add to the small collection of completely solved problems for the one-dimensional or radial equation cases. An example of new results were the reflectionless potentials of KAY and MOSES [9] whose existence was not suspected until the Gel'fand-Levitan equation for the one-dimensional problem was set up. Indeed these solutions were particularly easy to come by. It was assumed that the spectral measure function for the continuous spectrum of the perturbed Hamiltonian was identical to that of the unperturbed Hamiltonian or kinetic energy. However, the perturbed Hamiltonian had negative point eigenvalues. The three-dimensional result of the present paper is a direct analogue of this situation.

1. EARLIER THREE-DIMENSIONAL GEL'FAND-LEVITAN EQUATIONS, COMPLETENESS RELATIONS, AND QUASI-LOCAL POTENTIALS.

In [10] and [11], KAY and MOSES wrote the first Gel'fand-Levitan equation for the three-dimensional inverse problem. The first problem which one has to contend with is to give a completeness relation compatible with the three-dimensional scattering problem and to determine how will it involve the scattering data. The second problem is how to determine the choice of the triangularity properties of the Gel'fand-Levitan kernel to obtain a multiplicative potential (i.e., diagonal in the $\mathbf{x}$-representation). Our initial Gel'fand-Levitan equation for the three-dimensional problem assumed that the Gel'fand-Levitan kernel was triangular in the radial variable. That is, let us denote by $\psi(\mathbf{x}|\mathbf{k})$ the solution of the perturbed Schrödinger equation $H\psi(\mathbf{x}|\mathbf{k}) = k^2\psi(\mathbf{x}|\mathbf{k})$, where $\psi(\mathbf{x}|\mathbf{k})$ is analogous to the Jost function of the one-dimensional or radial problem. Thus, writing

$$\psi(\mathbf{x}|\mathbf{k}) \equiv \psi(r,\theta,\phi|\mathbf{k}),\ \psi_0(\mathbf{x}|\mathbf{k}) \equiv \psi_0(r,\theta,\phi|\mathbf{k}), \tag{1.1}$$

where r,θ,ϕ are the polar coordinates of $\mathbf{x}$ (i.e., $\mathbf{x} = r(\sin\theta\cos\phi,\ \sin\theta\sin\phi,\ \cos\theta)$), we required $\psi(\mathbf{x}|\mathbf{k})$ to satisfy the boundary condition (in the sense of distributions in θ and ϕ)

$$\lim_{r\to\infty}[\psi(r,\theta,\phi|\mathbf{k}) - \psi_0(r,\theta,\phi|\mathbf{k})] = 0 \ \ for\ \pi/2 \leqslant \theta \leqslant \pi\ . \tag{1.2}$$

Furthermore, we assumed that H had no point eigenvalues.

In [10] and [11], it was shown that the boundary condition (1.2) leads to a completeness relation which involves the amplitude $b_k(\boldsymbol{\eta},\boldsymbol{\eta}')$ for $\boldsymbol{\eta},\boldsymbol{\eta}'$ on a hemisphere. Hence, if we seek a local potential $V(\mathbf{x})$, we require too much scattering data, since the earlier method of iteration indicated that only $b_k(-\boldsymbol{\eta},\boldsymbol{\eta})$ had to be known for $\boldsymbol{\eta}$ on a hemisphere. We assumed a triangularity condition on the Gel'fand-Levitan kernel $K(r,\theta,\phi|r',\theta',\phi')$ compatible with the boundary condition (1.2), namely,

$$\psi(\mathbf{x}|\mathbf{k}) = \psi_0(\mathbf{x}|\mathbf{k}) + \int_r^\infty r'^2 dr' \int_0^\pi \sin\theta' d\theta' \int_0^{2\pi} d\phi' \; K(r,\theta,\phi|r',\theta',\phi')\, \psi_0(r',\theta',\phi'|\mathbf{k}) \,. \tag{1.3}$$

thus, the Gel'fand-Levitan kernel was taken to be triangular in the radial variable.

The potential which is obtained with this triangularization is in general a nonlocal one of a particular form, namely, it is diagonal in the radial variable, but is an integral operator in terms of the angular variables. We term such potentials "quasi-local." To be explicit, a potential V is said to be quasi-local if a kernel of an integral operator $V(r;\theta,\phi|\theta',\phi')$ can be assigned to it, such that if $\psi(\mathbf{x}) \equiv \psi(r,\theta,\phi)$ is an element of the Hilbert space then

$$V\phi(\mathbf{x}) = \int_0^\pi \sin\theta' \, d\theta' \int_0^{2\pi} d\phi' V(r;\theta,\phi|\theta',\phi')\psi(r,\theta',\phi') \,. \tag{1.4}$$

The requirement that V be self-adjoint leads to

$$V(r;\theta,\phi|\theta',\phi') = V^* \, (r;\theta',\phi'|\theta,\phi) \,, \tag{1.5}$$

where the asterisk means complex conjugate. Having solved for the Gel'fand-Levitan kernel, the kernel of the quasi-local potential is given by

$$V(r;\theta,\phi|\theta',\phi') = -2 \frac{\partial}{\partial r} [r^2 K(r,\theta,\phi|r,\theta',\phi')] \,. \tag{1.6}$$

From the point of view of [10], [11], and the present paper, local potentials are particular cases of quasi-local potentials, for in the case of local potentials the kernel for the potential contains the factor $\delta(\theta-\theta')\delta(\phi-\phi')$. If a local potential gave rise to the scattering amplitude $b_k(\eta,\eta')$, the Gel'fand-Levitan equation would reproduce this local potential. If one knew in advance that the potential was a local one, however, most of the data $b_k(\eta,\eta')$ would be redundant, since from [1] only $b_k(-\eta,\eta)$ need be known on a hemisphere. This result implies a relation between $b_k(\eta,\eta')$ and $b_k(-\eta,\eta)$ or, more generally, a constraint on $b_k(\eta,\eta')$ which would assure us that the scattering came from a local potential. (A constraint was suggested in [10] and [11], based on the assumption that the iteration procedure of [1] and the iteration of the Gel'fand-Levitan equation of [10] and [11] gave the same result for the potential to first order. It remains to be proved that this constraint is valid.)

Though the data given by $b_k(\eta,\eta')$ for η,η' on a hemisphere is redundant for a local potential, it represents a set of minimal data for potentials which are quasi-local. We may, therefore, change our point of view and regard the Gel'fand-Levitan equation of [10] and [11] as giving quasi-local potentials from minimal scattering data. Therefore, we have the important result that, for every set of data $b_k(\eta,\eta')$ such that η,η' lie on a hemisphere, a unique quasi-local potential can be found which reproduces the scattering data. In exceptional cases, the quasi-local potential will be a local one.

Recently, we have been able to add point eigenvalues to the algorithm of [10] and [11]. It has been our intention to try to obtain exact quasi-local solutions to the Gel'fand-Levitan equations. One of the possible ways would appear to use the analogous methods of [9] and assume that $b_k(\eta,\eta') \equiv 0$ for η,η' on a hemisphere. Only the point eigenvalues would then contribute to the kernel of the Gel'fand-Levitan equation. We have been unable to solve such Gel'fand-Levitan equations. It appears that a reflection coefficient $b_k(\eta,\eta')$ must always exist when there are point eigenvalues and that its singularities in the complex k-plane correspond to the point eigenvalues. In this approach to the three-dimensional problem, we mirror the treatment of the one-dimensional case of [5], but thus far have not found analogues to the reflectionless potential of [9]. It should be mentioned that in a recent series of papers, of which [12] is the first, we are pressing the analogy of the three-dimensional problem to the one-dimensional problem and finding analogues to Jost functions, Green's functions, and so on.

Before we consider a Gel'fand-Levitan equation with bound states, we should mention the work of FADDEEV [13] and NEWTON [14]. Instead of using Gel'fand-Levitan kernels which are triangular in the radial variable, they require their kernels to be triangular with respect to an axis, say, the z-axis. Their potentials are diagonal with respect to the z-axis but are integral operators with respect to the x- and y-axes. Thus, their potentials are quasi-local in a somewhat similar sense to ours. The completeness relations for the eigenfunctions are more complicated than ours. The way that the completeness relations can be obtained from the scattering amplitudes is discussed in [13] and [14]. Furthermore, much of the emphasis in the last two references is on the placement of conditions on the scattering amplitudes which assure one that the potential is local and independent of the axis of triangularization. As in [10] and [11] point eigenvalues are not included and no exact solutions are given.

2. AN ALTERNATIVE GEL'FAND-LEVITAN EQUATION FOR QUASI-LOCAL POTENTIALS IN TERMS OF THE SPECTRUM OF THE PERTURBED OPERATOR

Whereas our previous algorithm for the three-dimensional problem was patterned after the one-dimensional problem as treated in [5], the treatment which has enabled us to obtain exact solutions of the three-dimensional Schrödinger equation for some quasi-local potentials represents a generalization of the original Gel'fand-Levitan method [4]. the algorithm which will now be given is obtained directly from the abstract treatment of the Gel'fand-Levitan equation [3].

Let us denote the eigenfunctions of $H_0 = -\Delta$ by

$$\psi_0(\mathbf{x}|\mathbf{k}) = (2\pi)^{-3/2}e^{i\mathbf{k}\cdot\mathbf{x}} . \tag{2.1}$$

In (2.1) the eigenfunctions of H_0 are expressed in terms of the momentum $\mathbf{k}$. It is also convenient to express the eigenfunctions of H_0 in terms of the eigenvalues of H_0 and the polar angles of $\mathbf{k}$ where $\mathbf{k} = k(\sin\theta\cos\phi, \sin\theta\sin\phi, \cos\theta)$, with $0\leqslant\theta\leqslant\pi$, $0\leqslant\phi\leqslant 2\pi$. We write

$$\psi_0(\mathbf{x}|E,\theta,\phi) = E^{1/4}[(\sin\theta)/2]^{1/2}\psi_0(\mathbf{x}|\mathbf{k}),\ E = k^2 . \tag{2.2}$$

Our purpose to find a quasi-local potential V from spectral data on H (i.e., the completeness relation for the eigenfunctions) and the boundary conditions on the eigenfunctions. The spectral data from the continuous spectrum consists in giving the function $<\theta,\phi|\omega(E)|\theta',\phi'>$ which satisfies the hermiticity and positive-definiteness conditions

$$<\theta,\phi|\omega(E)|\theta',\phi'> = <\theta',\phi'|\omega(E)|\theta,\phi>^* , \tag{2.3}$$

$$\int_0^\infty dE \int_0^\pi d\theta \int_0^{2\pi} d\phi \int_0^\pi d\theta' \int_0^{2\pi} d\phi' f^*(E,\theta,\phi)<\theta,\phi|\omega(E)|\theta',\phi'>f(E,\theta',\phi')\geqslant 0 . \tag{2.4}$$

where equality holds if and only if $f = 0$, $f(E,\theta,\phi)$ being a complex function in the Hilbert space with norm

$$||f|| = \left[\int_0^\infty dE \int_0^\pi d\theta \int_0^{2\pi} d\phi |f(E,\theta,\phi)|^2\right]^{1/2} . \tag{2.5}$$

The spectral data for the discrete spectrum of H is specified by giving the eigenvalues E_i $(i = 1,2,\ldots,n)$ and eigenfunctions of H_0, namely, *any* nonzero solutions of the differential equations

$$-\Delta\psi_{0i}(\mathbf{x}) = E_i\psi_{0i}(\mathbf{x}) . \tag{2.6}$$

Moreover, we must specify for each E_i a positive constant C_i. The point eigenvalues may be any real number (positive, negative, or zero). It is permitted that several E_i be equal (the case of degeneracy), but it is convenient to require that all the solutions of (2.6) be linearly independent.

The basic theorem which is proved by the methods of [3] is the following (we use λ and σ instead of θ and ϕ respectively for the polar angles of $\mathbf{x}$ to prevent confusion with the polar angles of $\mathbf{k}$):

Define

$$\begin{aligned}\Omega(\mathbf{x}|\mathbf{x}') \equiv \Omega(r,\lambda,\sigma|r',\lambda',\sigma') = &\int_0^\infty dE\int_0^\pi d\theta\int_0^2 \pi d\theta'\int_0^\pi\int_0^{2\pi} d\phi'\ \psi_0(\mathbf{x}|E,\theta,\phi)\\ &\times <\theta,\phi|\omega(E)|\theta',\phi'>\ \psi_0^*(\mathbf{x}'|E,\theta',\phi')\\ &+\sum_i \frac{\psi_{0i}(\mathbf{x})\psi_{0i}^*(\mathbf{x}')}{C_i} - \delta(\mathbf{x}-\mathbf{x}') .\end{aligned} \tag{2.7}$$

We then have the Gel'fand-Levitan equation for

$$\begin{aligned}K(r,\lambda,\sigma|r',\lambda',\sigma') = &-\Omega(r,\lambda,\sigma|r',\lambda',\sigma')\\ &-\int_0^r r''^2dr''\int_0^\pi \sin\lambda''\,d\lambda''\int_0^{2\pi} d\sigma'' K(r,\lambda,\sigma|r'',\lambda'',\sigma'')\Omega(r'',\lambda'',\sigma''|r',\lambda',\sigma') .\end{aligned} \tag{2.8}$$

Then, the functions $\psi(\mathbf{x}|E,\theta,\phi)$ defined by

$$\begin{aligned}\psi(\mathbf{x}|E,\theta,\phi) \equiv \psi(r,\lambda,\sigma|E,\theta,\phi) = &\ \psi_0(r,\lambda,\sigma|E,\theta,\phi)\\ &+\int_0^r r'^2\,dr'\int_0^\pi \sin\lambda'\,d\lambda'\int_0^{2\pi} d\sigma'\ K(r,\lambda,\sigma|r'\lambda',\sigma')\times\psi_0(r',\lambda',\sigma'|E,\theta,\phi) ,\end{aligned} \tag{2.9}$$

with

$$\psi_0(r,\lambda,\sigma|E,\theta,\phi) \equiv \psi_0(\mathbf{x}|E,\theta,\phi) , \tag{2.9a}$$

are eigenfunctions of the operator $H = H_0 + V$, i.e.,

$$H\psi(\mathbf{x}|E,\theta,\phi) = E\psi(\mathbf{x}|E,\theta,\phi) \ , \tag{2.10}$$

where V is a quasi-local potential whose kernel is given by

$$V(r;\lambda,\sigma|\lambda',\sigma') = 2\frac{\partial}{\partial r}[r^2K(r,\lambda,\sigma|r',\lambda',\sigma')] \ . \tag{2.11}$$

Moreover, the functions $\psi_i(\mathbf{x})$ defined by

$$\psi_i(\mathbf{x}) = \psi_{0i}(\mathbf{x}) + \int_0^r r'^2dr' \int_0^\pi \sin\lambda'\, d\lambda' \int_0^{2\pi} d\sigma' K(r,\lambda,\sigma|r',\lambda',\sigma')\, \psi_{0i}(r',\lambda',\sigma') \tag{2.12}$$

with

$$\psi_{0i}(r,\lambda,\sigma) \equiv \psi_{0i}(\mathbf{x}) \ , \tag{2.12a}$$

are proper eigenfunctions of H, i.e.,

$$H\psi_i(\mathbf{x}) = E_i\psi_i(\mathbf{x}) \ , \tag{2.13}$$

and are normalizable, with the norm being given by

$$\int |\psi_i(\mathbf{x})|^2\, d\mathbf{x} = C_i \ . \tag{2.14}$$

The eigenfunctions of H satisfy the completeness relation

$$\int_0^\infty dE \int_0^\pi d\theta \int_0^{2\pi} d\phi \int_0^\pi d\theta' \int_0^{2\pi} d\phi'\, \psi(\mathbf{x}|E,\theta,\phi) <\theta,\phi|\omega(E)|\theta'\phi'> \psi^*(\mathbf{x}'|E',\theta',\phi')$$

$$+ \sum_i \frac{\psi_i(\mathbf{x})\psi_i^*(\mathbf{X}')}{C_i} = \delta(\mathbf{x}-\mathbf{x}') \ . \tag{2.15}$$

Because of the triangularity properties of the Gel'fand-Levitan kernel, the eigenfunctions of H satisfy the boundary conditions

$$\psi(0|E,\theta,\phi) = \psi_0(0|E,\theta,\phi),\ \psi_i(0) = \psi_{0i}(0) \ ,$$

$$\nabla\psi(0|E,\theta,\phi) = \nabla\psi_0(0|E,\theta,\phi),\ \nabla\psi_i(0) = \nabla\psi_{0i}(0) \ . \tag{2.16}$$

As mentioned earlier, the proof of the theorem comes directly from an application of [3].

3. EXACT SOLUTIONS OF THE GEL'FAND-LEVITAN EQUATION AND THE SCHRÖDINGER EQUATION

We shall now give three explicit examples of solutions of the Gel'fand-Levitan equations and the corresponding solutions of the Schrödinger equation. Each of the cases correspond to the particularly simple situation in which

$$<\theta,\phi|\omega(E)|\theta',\phi'> = \delta(\theta-\theta')\,\delta(\phi-\phi') \tag{3.1}$$

and there is one nonnegative point eigenvalue. Though these are particularly simple cases, they are sufficient to show that the procedure is not an empty one. In each of the cases presented here, we have verified that the eigenfunctions really do satisfy the Schrödinger equation with the appropriate potentials and also the completeness relation (2.15), which in the present case becomes

$$\int_0^\infty dE \int_0^\pi d\theta \int_0^{2\pi} d\phi\, \psi(\mathbf{x}|E,\theta,\phi)\psi^*(\mathbf{x}'|E,\theta,\phi) + \frac{\psi_1(\mathbf{x})\psi_1^*(\mathbf{x}')}{C_1} = \delta(\mathbf{x}-\mathbf{x}') \ . \tag{3.2}$$

The potentials are not trivial and the verification, though straightforward. is somewhat arduous, In all three cases, there is no scattering! That is, the boundary condition (1.2), which can also be required for quasi-local potentials, is satisfied by the continuous spectrum eigenfunctions, but with the amplitude of the spherical wave identically zero. It should be mentioned that SÁENZ and ZACHARY [15] have shown the possibility of nonlocal potentials which do not scatter.

Simple as the present examples are, the potentials given here appear to be the first (aside from separable potentials of the form $V(\mathbf{x}|\mathbf{x}') = \sum_i f_i(\mathbf{x})f_i^*(\mathbf{x}')$ where $V(\mathbf{x}|\mathbf{x}')$ is the kernel of the nonlocal operator [16]) for which the three-dimensional Schrödinger equation can be solved in terms of elementary functions and in closed form.

Case 1

The simplest eigenfunction $\psi_{0i}(\mathbf{x})$ that we can think of is

$$\psi_{01}(\mathbf{x}) = 1 . \tag{3.3}$$

Clearly, from $-\Delta\psi_{01}(\mathbf{x}) = E_1\psi_{01}(\mathbf{x})$,

$$E_i = 0 . \tag{3.3a}$$

The kernel $\Omega(\mathbf{x}|\mathbf{x}')$ of the Gel'fand-Levitan equation is very simple indeed:

$$\Omega(\mathbf{x}|\mathbf{x}') = \frac{1}{C_1} , \tag{3.4}$$

$$K(r,\lambda,\sigma|r',\lambda',\sigma') = -\frac{1}{C_1} - \frac{1}{C_1} = \int_0^r r''^2 dr'' \int_0^\pi \sin\lambda''\, d\lambda'' \int_0^{2\pi} d\sigma''\, K(r,\lambda,\sigma|r'',\lambda'',\sigma'') . \tag{3.5}$$

From (3.5), it is clear that the Gel'fand-Levitan kernel $K(r,\lambda,\sigma|r',\lambda',\sigma')$ is independent of r',λ', σ' and, hence, we may write

$$K(r,\lambda,\sigma|r',\lambda',\sigma') \equiv F(r,\lambda,\sigma) . \tag{3.6}$$

On substituting (3.6) into (3.5), we obtain an equation for $F(r,\lambda,\sigma)$:

$$F(r,\lambda,\sigma) = -\frac{1}{C_1} - \frac{4\pi}{3C_1} r^3 F(r,\lambda,\sigma) . \tag{3.7}$$

On writing

$$r_0 = (3C_1/4\pi)^{1/3} , \tag{3.8}$$

we have the following result for the present case:

$$K(r,\lambda,\sigma|r',\lambda',\sigma') = -\frac{3}{4\pi}\frac{1}{(r^3+r_0^3)} , \tag{3.9}$$

$$V(r;\lambda,\sigma|\lambda',\sigma') = \frac{3}{2\pi}\frac{r(r^3-2r_0^3)}{(r^3+r_0^3)^2} , \tag{3.10}$$

$$\psi(\mathbf{x}|E,\theta,\phi) = E^{1/4}\,[(\sin\theta)/2]^{1/2}\,\psi(\mathbf{x}/\mathbf{k}) , \tag{3.11}$$

with

$$\psi(\mathbf{x}|\mathbf{k}) = \psi_0(\mathbf{x}|\mathbf{k}) - (2\pi)^{-3/2}\,\frac{3}{k}\,\frac{r^2}{(r^3+r_0^3)}\, j_1(kr) , \tag{3.11a}$$

$$\psi_1(\mathbf{x}) = \frac{r_0^3}{(r^3+r_0^3)} . \tag{3.12}$$

In (3.11a), $r = |\mathbf{x}|$, as usual, and $k = E^{1/2}$. In (4.11) and subsequently, $j_n(x)$ is the spherical Bessel function of order n. Also, $\mathbf{k}$ is the vector given by the polar coordinates k,θ,ϕ.

Case 2

In this case, we take

$$\psi_{01}(\mathbf{x}) = z = r\cos\lambda, \tag{3.13}$$

from which

$$E_1 = 0 , \tag{3.14}$$

as before.

On defining

$$r_0 = (15C_1/4\pi)^{1/5} , \tag{3.15}$$

we have

$$K(r,\lambda,\sigma|r',\lambda',\sigma') = -\frac{15}{4\pi}\frac{(r\cos\lambda)(r'\cos\lambda')}{(r^5+r_0^5)}, \tag{3.16}$$

$$V(r;\lambda,\sigma|\lambda',\sigma') = \frac{15}{2\pi}\frac{r^3(r^5-4r_0^5)}{(r^5+r_0^5)^2}\cos\lambda\cos\lambda' , \tag{3.17}$$

$$\psi(\mathbf{x}|\mathbf{k}) = \psi_0(\mathbf{x}|\mathbf{k}) - (2\pi)^{-3/2}\frac{15\,r^4\cos\lambda}{k(r^5+r_0^5)}j_2(kr) , \tag{3.18}$$

$$\psi_1(\mathbf{x}) = \frac{rr_0^5}{(r^5+r_0^5)}\cos\lambda . \tag{3.19}$$

Cases 1 and 2 are special cases in which the "eigenfunctions" $\psi_{0i}(\mathbf{x})$ of H_0 are expressed as a sum of spherical harmonics:

$$\psi_{0i}(\mathbf{x}) = \sum_{l,m} c_{i,lm}Y_{lm}(\lambda,\sigma)f_l(\kappa_i r) , \tag{3.20}$$

where $c_{i\,lm}$ are constants and the functions $f_l(k_i r)$ are given by

$$f_l(\kappa_i r) = \begin{cases} j_l(\kappa_i r) , & \text{for } E_i = \kappa_i^2 > 0 , \\ r^l , & \text{for } E_i = 0 , \\ j_l(i\kappa_i r), & \text{for } E_i = -\kappa_i^2 < 0 . \end{cases} \tag{3.21}$$

(The subscript i should not be confused with $i = (-1)^{1/2}$ which appears in the argument of j_l in the last of (3.21).)

In Case 1, the subscript i takes on the value 1, $E_1 = 0$, and the only non-vanishing $c_{i,lm}$ is $c_{1,00}$. The results of Case 1 could have been obtained using separation of variables and looking for a phaseless potential for the $l = 0$ radial equation as in [17]. The potentials for all of the other radial equations would be taken equal to zero. On reconstructing the solution in the three-dimensional space, the present nonlocal potential would result. A similar procedure could be used for Case 2, in which $c_{1,10}$ is the only nonvanishing coefficient. A potential would be obtained only for the $l = 1$ radial equation. However, if there is more than one term in the sum in (3.20), the method of separation of variables becomes very awkward. Even in the "simple" Cases 1 and 2, the three-dimensional approach of the present paper is far less difficult than the use of separation of variables.

It should be emphasized that the form (3.20) is not always a convenient form for representing $\psi_{0i}(\mathbf{x})$ as Case 3, which follows, shows.

Case 3

In this case we take

$$\psi_{01}(\mathbf{x}) = e^{i\mathbf{p}\cdot\mathbf{x}}, \tag{3.22}$$

where $\mathbf{p}$ is a fixed vector.

Clearly,

$$E_1 = p^2 \geqslant 0,\ p = |\mathbf{p}| . \tag{3.23}$$

Case 1 is reproduced when $p = 0$.

Let us define the unit vectors $\boldsymbol{\lambda},\boldsymbol{\lambda}'$ as being those determined by the angles λ,σ and λ',σ' respectively in terms of polar coordinates. Then, on defining r_0 as in (3.8),

$$K(r,\lambda,\sigma|r',\lambda',\sigma') = -\frac{3}{4\pi}\frac{1}{(r^3+r_0^3)}e^{i\mathbf{p}\cdot(\mathbf{x}-\mathbf{x}')} , \tag{3.24}$$

where the vectors $\mathbf{x}$ and $\mathbf{x}'$ are given by the polar coordinates (r,λ,σ) and (r',λ',σ') respectively,

$$V(r:\lambda,\sigma|\lambda',\sigma') = \frac{3}{2\pi}e^{ir\mathbf{p}\cdot(\boldsymbol{\lambda}-\boldsymbol{\lambda}')}\left[\frac{r(r^2-2r_0^3)}{(r^3+r_0^3)^2} - i\,\frac{r^2\mathbf{p}\cdot(\boldsymbol{\lambda}-\boldsymbol{\lambda}')}{r^3+r_0^3}\right] , \tag{3.25}$$

$$\psi(\mathbf{x}|\mathbf{k}) = \psi_0(\mathbf{x}|\mathbf{k}) - (2\pi)^{-3/2}\frac{3r^2e^{i\mathbf{p}\cdot\mathbf{x}}}{(r^3+r_0^3)|\mathbf{k}-\mathbf{p}|}j_1(|\mathbf{k}-\mathbf{p}|r) , \tag{3.26}$$

$$\psi_1(\mathbf{x}) = r_0^3\frac{e^{i\mathbf{p}\cdot\mathbf{x}}}{(r^3+r_0^3)} . \tag{3.27}$$

4. FINAL COMMENTS: COMPARISION POTENTIALS, VARIATIONAL PRINCIPLES, AND THE SCATTERING OPERATOR

The present section consists of additional comments relating to the Gel'fand-Levitan equation of the present paper. First of all, the Gel'fand-Levitan equation can be set up to use comparison potentials and comparison measures as discussed in [18]. The variational principle for finding the potential discussed in [19] also applies. The variational principle can also be combined with the use of a comparison potential to obtain the total potential. For the sake of brevity, we refrain from going into the details.

In the three examples of the preceding section, the scattering operator was found to be the identity because the amplitude of the outgoing spherical wave was zero. Generally, however, the scattering operator differs from the identity. The wave function which is found through the use of the Gel'fand-Levitan equation is usually *not* the outgoing wave. Hence we must give a method whereby the scattering operator can be calculated from a wave function that is not the outgoing wave and for which we know the completeness relation. The result which we shall give is very general and is a result of abstract scattering theory as discussed in [3]. For the sake of brevity, we shall assume that the reader is familiar with [3] and use the results and notation of that reference. (For simplicity, we take $\epsilon = 1$.)

Knowing a complete set of eigenfunctions for the continuous spectrum and knowing their completeness relationship is equivalent to knowing the projection of the wave operator in the continuous spectrum, denoted by $U\eta(H_0)$, and the weight operator W and, in particular, its continuous part W_c. The relation for the scattering operator S in terms of the weight and wave operators is

$$S = M_+ W_c M_-^*, \tag{4.1}$$

where the asterisk means adjoint. The operator W_c is given in terms of the H_0-representation by $<\theta,\phi|\omega(E)|\theta',\phi'>$ used in the completeness relation. The operators $M_\pm$ (which are generalizations of the Jost functions) are obtained from the known wave operators U through

$$M_\pm = U\eta(H_0) - \int \gamma_\pm(E;H_0)\, VU\eta(H_0)\delta(E-H_0)\,dE\ . \tag{4.2}$$

We should like to emphasize the generality of (4.1) and (4.2). They could be used, for example, to find the scattering operator in multichannel scattering if it is more convenient to solve the Schrödinger equation in terms of wave functions that satisfy boundary conditions other than those which lead to the outgoing wave functions.

In the case of the inverse problem of the present paper, $U = I + K$ where K is the integral operator whose kernel is the solution of the Gel'fand-Levitan equation (3.8).

FOOTNOTES AND REFERENCES

*Research sponsored by the U.S. Army Research Office under Grant DAAG 29-78-G-0003 P-14919-M

[1] H.E. Moses : Phys. Rev. **102**, 559 (1956)
[2] R. Jost, W. Kohn: Phys. Rev. **87**, 977 (1952)
[3] I. Kay, H.E. Moses: Nuovo Cimento **2**, 917 (1955); *ibid.* **3**, 66 (1956)
[4] I.M. Gel'fand, B.M. Levitan: Amer. Math. Soc. Trans. **1**, 253 (1951)
[5] I. Kay, H.E. Moses: Nuovo Cimento **3**, 276 (1956)
[6] I. Kay: New York University Institute of Mathematical Sciences Research Report EM-74 (1955)
[7] L.D. Faddeev: J. Math. Phys. **4**, 72 (1963) (translated by B. Seckler)
[8] V.A. Marchenko: Dokl. Akad, Nauk SSSR **104**, 695 (1955)
[9] I. Kay, H.E. Moses: J. Appl. Phys. **27**, 1503 (1956)
[10] I. Kay, H.E. Moses: Nuovo Cimento **22**, 689 (1961)
[11] I. Kay, H.E. Moses: Comm. Pure Appl. Math. **14**, 435 (1961)
[12] H.E. Moses: J. Math. Phys. **20 , 1151 (1979)**
[13] L.D. Faddeev: J. Sov. Math. **5**, 334 (1976)
[14] R.G. Newton: "The three-dimensional inverse scattering problem in quantum mechanics," invited lectures delivered at the 1974 Summer Seminar on Inverse Problems, Am. Math. Soc. (August 5-16, 1974)
[15] A.W. Sáenz, W.W. Zachary: J. Math. Phys. **17**, 409 (1976)
[16] R.G. Newton: *Scattering Theory of Waves and Particles* (McGraw-Hill, New York, 1966), pp. 274-276
[17] H.E. Moses, S.F. Tuan: Nuovo Cimento **13**, 197 (1959)
[18] H.E. Moses: J. Math. Phys., in press
[19] M. Kanal, H.E. Moses: J. Math. Phys. **19**, 1258 (1978)

THE JOST-KOHN ALGORITHM FOR INVERSE SCATTERING

Reese T. Prosser

Department of Mathematics
Dartmouth College
Hanover, New Hampshire 03755

INTRODUCTION

In 1952 JOST and KOHN [1] introduced a straightforward procedure for solving the inverse problem for radial (i.e., one-dimensional) potential scattering [1]. In 1956 MOSES extended this procedure to include non-radial (three dimensional) problems [2], and in 1975 I was able to show that his extension actually converges if the scattering data are sufficiently restricted [3]. Here, I want to describe briefly the algorithm, and comment briefly on its usefulness in theory and practice.

1. DESCRIPTION

The scattering of a quantum mechanical wave function $\phi(\mathbf{x},\mathbf{k})$ from a fixed potential $V(\mathbf{x})$ is governed by the time-independent Schrödinger equation

$$(\Delta+k^2)\phi(\mathbf{x},\mathbf{k}) = V(\mathbf{x})\phi(\mathbf{x},\mathbf{k}) \; . \tag{1}$$

The solution, which is to consist of an ingoing plane wave plus an outgoing scattered wave, may be expressed as

$$\phi(\mathbf{x},\mathbf{k}) = e^{i\mathbf{k}\cdot\mathbf{x}} + \int \frac{e^{i|\mathbf{k}||\mathbf{x}-\mathbf{y}|}}{4\pi|\mathbf{x}-\mathbf{y}|} V(\mathbf{y})\phi(\mathbf{y},\mathbf{k})d\mathbf{y} \; . \tag{2}$$

As $|\mathbf{x}| \rightarrow \infty$, the behavior of $\phi(\mathbf{x},\mathbf{k})$ is given by

$$\phi(\mathbf{x},\mathbf{k}) \rightarrow e^{i\mathbf{k}\cdot\mathbf{x}} + \frac{e^{i|\mathbf{k}||\mathbf{x}|}}{4\pi|\mathbf{x}|} T(\mathbf{k}',\mathbf{k}) + O\left(\frac{1}{|\mathbf{x}|}\right) . \tag{3}$$

Here $\mathbf{k}' = (|\mathbf{k}|/|\mathbf{x}|)\mathbf{x}$ and $T(\mathbf{k}',\mathbf{k})$, which contains the scattering data, is given by

$$T(\mathbf{k}',\mathbf{k}) = \int e^{-i\mathbf{k}'\cdot\mathbf{y}} \, V(\mathbf{y})\phi(\mathbf{y},\mathbf{k})d\mathbf{y} \; . \tag{4}$$

An iterative solution for $T(\mathbf{k}',\mathbf{k})$ is obtained by first solving (2) for $\phi(\mathbf{x},\mathbf{k})$ and then substituting the result in (4):

$$T(\mathbf{k}',\mathbf{k}) = \int e^{-i\mathbf{k}'\cdot\mathbf{y}} \, V(\mathbf{y})e^{i\mathbf{k}\cdot\mathbf{y}} \, d\mathbf{y} + \int\int e^{-i\mathbf{k}'\cdot\mathbf{y}} V(\mathbf{y}) \, \frac{e^{i|\mathbf{k}||\mathbf{y}-\mathbf{y}'|}}{4\pi|\mathbf{y}-\mathbf{y}'|} \, V(\mathbf{y}')e^{i\mathbf{k}\cdot\mathbf{y}'} \, d\mathbf{y}'d\mathbf{y} + \ldots . \tag{5}$$

In the momentum representation, this solution becomes

$$T(\mathbf{k}',\mathbf{k}) = V(\mathbf{k}'-\mathbf{k}) + \int V(\mathbf{k}'-\mathbf{k}'')(\mathbf{k}''^2-\mathbf{k}^2+i0)^{-1}V(\mathbf{k}''-\mathbf{k})d\mathbf{k}'' + \ldots , \tag{6}$$

or, more formally,

$$T = V + V(\Gamma V) + V\Gamma(V(\Gamma V)) + \ldots = (1 - V\Gamma)^{-1}V , \tag{7}$$

where ΓV is the kernel

$$(\Gamma V)(\mathbf{k}',\mathbf{k}) = (\mathbf{k}'^2-\mathbf{k}^2+i0)^{-1}V(\mathbf{k}'-\mathbf{k}) \; . \tag{8}$$

It is known that this iterative solution (6) for $T(\mathbf{k}',\mathbf{k})$ actually converges, provided that the potential is sufficiently weak. Specifically, one can define a class of kernels $K(\mathbf{k}',\mathbf{k})$ which includes potentials of the form $V(\mathbf{k}'-\mathbf{k})$ and a norm $|| \; ||$ for this class such that

$$||K(\Gamma M)|| \leqslant ||K|| \; ||M|| < \infty \; . \tag{9}$$

Then, if $V(\mathbf{k}'-\mathbf{k})$ is in this class and if

$$||V|| = a < 1\,, \tag{10}$$

then $T(\mathbf{k}',\mathbf{k})$ is also in this class with norm

$$||T|| \leqslant \frac{a}{1-a}, \tag{11}$$

and the series (6) converges to T in norm [4,3].

For a solution of the inverse problem, we have only to invert the series (6). This presents no problem if we already know $T(\mathbf{k}',\mathbf{k})$ for *all* values of $(\mathbf{k}',\mathbf{k})$, since then

$$V = T - T(\Gamma T) + T(\Gamma T)(\Gamma T) - \ldots = T(1+\Gamma T)^{-1}, \tag{12}$$

and the series (12) converges to V provided that

$$||T|| < 1\,. \tag{13}$$

In practice, however, we know $T(\mathbf{k}',\mathbf{k})$ only for certain values ("on-shell" values) of $T(\mathbf{k}',\mathbf{k})$, e.g., only for $\mathbf{k}' = -\mathbf{k}$ (backscatter data), or only for $\omega = \mathbf{k}/|\mathbf{k}| =$ fixed, $|\mathbf{k}'| = |\mathbf{k}|$ (fixed-aspect data). On the other hand, if the potential is local, then we should not need a six-parameter family of data $T(\mathbf{k}',\mathbf{k})$ to determine the three-parameter potential $V(\mathbf{k})$.

To see how to proceed, suppose first that we know the backscatter data $T(-\mathbf{k},\mathbf{k})$ for all values of $\mathbf{k}$. Then, following Jost and Kohn, we replace $T(\mathbf{k}',\mathbf{k})$ by $T_\epsilon(\mathbf{k}',\mathbf{k}) = \epsilon\, T(\mathbf{k}',\mathbf{k})$ and $V(\mathbf{k}'-\mathbf{k})$ by $V_\epsilon(\mathbf{k}'-\mathbf{k}) = \sum_{m=1}^{\infty} \epsilon^m\, V_m(\mathbf{k}'-\mathbf{k})$, and substitute into (6). Then we equate the coefficients of ϵ^m. The result is:

$$\begin{aligned} m = 1:&\quad T(-\mathbf{k},\mathbf{k}) = V_1(-2\mathbf{k}),\\ m = 2:&\quad 0 = V_2(-2\mathbf{k}) + \int V_1(-\mathbf{k}-\mathbf{k}'')(\mathbf{k}''^2-\mathbf{k}^2+i0)^{-1}\, V_1(\mathbf{k}''-\mathbf{k})d\mathbf{k}'', \end{aligned} \tag{14}$$

etc.

Hence, if we put

$$\begin{aligned} T_1(\mathbf{k}',\mathbf{k}) &= T(\mathbf{k}',\mathbf{k}),\\ V_1(-2\mathbf{k}) &= T_1(-\mathbf{k},\mathbf{k}),\\ T_2(\mathbf{k}'\mathbf{k}) &= \int V_1(\mathbf{k}'-\mathbf{k}''),\,(\mathbf{k}''^2-\mathbf{k}^2+i0)^{-1}V_1(\mathbf{k}''-\mathbf{k})d\mathbf{k}'',\\ V_2(-2\mathbf{k}) &= T_2(-\mathbf{k},\mathbf{k}), \end{aligned} \tag{15}$$

etc.

then,

$$V_\epsilon(-2\mathbf{k}) = \sum_{m=1}^{\infty} \epsilon^m\, V_m(-2\mathbf{k})\,. \tag{16}$$

(15) and (16) give the potential $V_\epsilon(-2\mathbf{k})$ in terms of the backscatter data $T_\epsilon(-\mathbf{k},\mathbf{k})$, and together comprise the Jost-Kohn algorithm. It is shown in [3] that the series (16) actually converges in norm for all ϵ for which

$$\epsilon||V_1|| < 3 - 2\sqrt{2} = 0.172\ldots \tag{17}$$

and that the sum V_ϵ is a potential which will yield the given backscatter data $T_\epsilon(-\mathbf{k},\mathbf{k})$.

2. COMMENTS

(1) This algorithm shows clearly that the inverse scattering problem is solvable in a neighborhood of the origin in the space of backscatter data, and that the solution depends analytically on the data.

(2) The algorithm is simply described and easily adapted for numerical computation. For this purpose, it is useful to note that the data enter only at the first step—in the computation of V_1. The remaining steps are all independent of the data.

(3) The algorithm seems remarkably stable and insensitive to the details of the problem. Versions can be developed for any dimension, for fixed aspect or other scattering data, and for energy-dependent potentials, as in the case of scattering from a variable index of refraction (see [3]).

(4) The algorithm requires only a three-parameter family of data in three dimensions, in contrast to the various versions of the Gel'fand-Levitan algorithm available in three dimensions, which all require a full five-parameter family of data [5].

(5) If only approximate data are known, then the algorithm will give an approximate potential. This potential will be in error in norm by an amount determined by the error in norm of the data. Specifically, if the difference between the true and approximate backscatter data differ by δ in norm, so that $||V_1 - V'_1|| < \delta$, then, if the defining series converge, the resulting true and approximate potentials will differ by at most 2δ in norm, so that $||V_\epsilon - V'_\epsilon|| < 2\delta$.

(6) If only partial data are known, then V_1 can still be determined from the partial data together with reasonable assumptions about the model and standard interpretation and extrapolation techniques. Then V_ϵ can be determined from V_1 as before. Thus, partial data will still give an approximate solution.

(7) Moses has shown that in one dimension the first two terms V_1 and V_2 of this algorithm coincide with the first two terms of the potential obtained from an iterative solution of the Marchenko equation. He conjectures that these two algorithms agree term by term. I know of no evidence to the contrary.

(8) The primary weakness of this algorithm lies in its restriction to weak potentials and weak scattering data. It requires that the Born series (6) converge, and this rules out many interesting applications. There seems to be no way, for example, to accommodate bound states, or discontinuous indices of refraction, in the manner of the Gel'fand-Levitan algorithm. It is therefore of interest to try to relax these restrictions to weak potentials.

(9) For instance, it may be possible to improve the situation by replacing the Born series by a suitable version of the Fredholm theory, giving $T(\mathbf{k}',\mathbf{k})$ as a ratio of two series converging for all potentials, or equivalently, by replacing the Born series by a suitable assortment of Padé approximants. Bound states then appear as zeros in the denominator of the ratio for $T(\mathbf{k}',\mathbf{k})$.

(10) It may also be possible to construct a reference potential V' in such a way that the difference ΔT between the measured backscatter data $T(-\mathbf{k},\mathbf{k})$ and the reference data $T'(-\mathbf{k},\mathbf{k})$ is small. The algorithm could then be used to construct the difference ΔV between the true potential V and the reference potential V'. Bound states could then be included in the reference potential.

(11) Finally, it may be possible to sum the series (16), giving V in terms of V_1 via an integral equation of Gel'fand-Levitan type. These possibilities are under current consideration.

REFERENCES

[1] R. Jost, W. Kohn: Phys. Rev. **87**, 977 (1952)
[2] H. Moses: Phys. Rev. **102**, 559 (1956)
[3] R. T. Prosser: J. Math. Phys. **17**, 1773 (1976)
[4] K.O. Friedrichs: *Perturbations of Spectra in Hilbert Space* (American Mathematical Society, Providence, Rhode Island, 1965)
[5] I. Kay, H. Moses: Comm. Pure Appl. Math. **14**, 435 (1961)

APPLICATION OF NONLINEAR TECHNIQUES TO THE INVERSE PROBLEM

V.H. Weston

Department of Mathematics
Purdue University
West Lafayette, Indiana 47907

ABSTRACT

The inverse problem for the reduced wave equation $\Delta u + k^2 n^2(x) u = 0$, where n is real and continuous and $n^2 - 1$ has compact support in $\mathbf{R}^3$, is examined for the case where the scattering data consists of a set of measurements of the near or far field produced by a prescribed incident wave. The inverse problem is formulated in terms of a system of functional equations, a quadratic nonlinear integral equation, plus an additional inequality or constraint. The general nonlinear theory of the complete system is examined.

INTRODUCTION

We will examine the inverse scattering problem for the scalar wave equation

$$\Delta u + k^2 n^2(x) u = 0$$

where the index of refraction $n(x)$ is identically equal to one outside some compact region. The scattering object is characterized by the compact domain where $n(x) \not\equiv 1$. Here, $n(x)$ will be assumed to be real and continuous, with the support of $n^2(x) - 1$ being contained in some sphere D of radius R with center at the origin.

The emphasis in this paper is on the limited but important problem where a finite set of scattering measurements are made on either the scattered field u^s or the total field $u = u^i + u^s$, for a fixed frequency or wave number k, and fixed incident field u^i. Once this problem is well understood, one can then treat the problem where sets of measurements are made for a finite set of frequencies or incident fields.

As a preliminary, a few remarks on the direct scattering problem will be made. In LEIS [1], it is shown that given the above assumption on $n(x)$, and if the incident wave is continuous in the region D (all sources are external to the scatterer), then the direct scattering problem has a unique solution. The results hold independently of frequency. The direct scattering problem can then be transformed to the integral equation

$$u(x) = u^i(x) + \frac{k^2}{4\pi} \int_D \frac{e^{ik|x-y|}}{|x-y|} \mathrm{v}(y) u(y)\, dy, \tag{1}$$

where

$$\mathrm{v}(x) = n^2(x) - 1.$$

(The condition on continuity of $n(x)$ or $\mathrm{v}(x)$ can be relaxed somewhat. However, the existence and uniqueness of the solution will depend upon the size of the spectral radius of the integral operator in the above equation and this in turn will depend upon k.)

When $|x| \to \infty$, the far field behaviour in the direction $x/|x|$ is given by

$$u^s(x) \sim \frac{e^{ik|x|}}{|x|} g(k^s),$$

where $k^s = (k/|x|)x$ and the complex scattering amplitude $g(k^s)$ has the form

$$g(k^s) = \frac{k^2}{4\pi} \int_D e^{-ik^s \cdot y} v(y) u(y) dy$$

1. FORMULATION OF THE INVERSE PROBLEM

We want to determine $v(x)$ from a set of measured values of $u(x)$ or $u^s(x)$ at points outside the scatterer (the domain D).

Measurements of u^s made in the near-field of the scatterer at the N points $\{x_l\}$ yield the relations

$$\frac{k^2}{4\pi} \int_D \frac{e^{ik|x_l - y|}}{|x_l - y|} v(y) u(y) dy = b_l, \quad l = 1, 2, \ldots, N,$$

where the complex numbers b_l are known or measured quantities.

For measurements in the far field at a set of N scattered directions represented by the spherical polar variables (θ_l^s, ϕ_l^s), or by the vector k^s in the same direction with length k, the measured quantities are the complex scattering amplitudes. In this case we have the relations

$$\frac{k^2}{4\pi} \int_D e^{-ik_l^s \cdot y} v(y) u(y) dy = b_l, \quad l = 1, 2, \ldots, N.$$

In either case the results of the measurements yield a set of N functional equations of the form

$$\int_D h_l(y) v(y) u(y) dy = b_l, \quad l = 1, 2, \ldots, N, \tag{2}$$

where h_l are known functions and $v(y) u(y)$ is unknown [2].

We are of course assuming that we can measure phase and amplitude, and for present purposes are neglecting the effect of errors in the data. The more complicated problem where only the amplitude is measured leads directly to a nonlinear equation involving the data.

We will work in terms of the unknown quantity in (2); hence, we will set

$$v(x) u(x) = w(x). \tag{3}$$

Then (1) can be represented in the form

$$w(x) = v(x) [u^i(x) + \mathbf{K} w],$$

where $\mathbf{K}$ is the integral operator with kernel

$$\frac{k^2}{4\pi} \frac{e^{ik|x-y|}}{|x-y|}.$$

We see that $v(x)$ can be recovered from knowledge of $w(x)$ as follows:

$$v = w / [u^i + \mathbf{K} w], \tag{4}$$

or

$$v = \frac{\bar{w} [u^i + \mathbf{K} w]}{|u^i + \mathbf{K} w|^2}.$$

But we required that $v(x)$ be real, continuous on D, and vanish on the boundary of D, ∂D. Thus, we have the condition

$$\operatorname{Im} \bar{w} [u^i + \mathbf{K} w] = 0. \tag{5}$$

The inverse problem consists of finding the complex function w with real and imaginary parts which are real and continuous on D, vanish on ∂D, and are such that they satisfy the nonlinear integral equation (5) and the functional equations (2). In addition, we shall impose a mild constraint that for some $\delta_1 > 0$,

$$|u^i + \mathbf{K} w| > \delta_1 |w|. \tag{6}$$

This last constraint insures that when $v(x)$ is determined from (4) it will be bounded.

Thus, the inverse problem reduces to solving the system (2), (5) and (6).

Remark. When an infinite set of measurements are made over some cone of scattering directions in the far-field, or at all points in a rectangle in the near field (as in Holography), then the set of N equations (2) can be replaced by a Fredholm equation of the first kind. In turn, the above set of N equations (2) can be thought of as resulting from the decomposition of Fredholm equation of the first kind with a degenerate kernel into the corresponding algebraic system.

2. PLANE WAVE INCIDENCE AND FAR FIELD MEASUREMENTS

To simplify the further analysis, we shall consider now the case where the incident field is a plane wave propagating in the direction (θ^i, ϕ^i) as indicated by the vector k^i of length k. The formulation of the inverse problem is simplified by decomposing $w\bar{u}^i$ into real and imaginary parts $\phi(x)$ and $\psi(x)$ respectively by setting

$$\bar{u}^i(x) w(x) = \phi(x) + i\psi(x). \tag{7}$$

(Note that when the incident field is not a plane wave we can use the same procedure provided that $|1/u^i|$ is bounded for all $x \in D$.)

Introduce the integral operator Π as follows:

$$\Pi u = \frac{k^2}{4\pi} \int_D \frac{\exp[ik|x-y| - ik^i \cdot (x-y)]}{|x-y|} u(y) dy \tag{8}$$

and decompose its kernel into real and imaginary parts, each yielding the corresponding integral Π_R and Π_I, such that

$$\Pi = \Pi_R + i\Pi_I.$$

Equation (5) reduces to the following equation involving real quantities only

$$\psi = \psi(\Pi_I\psi - \Pi_R\phi) + \phi(\Pi_R\psi + \Pi_I\phi) \tag{9}$$

The functional equations (2) now have the explicit form

$$\left.\begin{aligned} \frac{k^2}{4\pi}\int_D \cos\alpha_l(y)\phi(y)\,dy &= \frac{k^2}{4\pi}\int_D \sin\alpha_l(y)\psi(y)\,dy + b_l^1, \\ \frac{k^2}{4\pi}\int_D \sin\alpha_l(y)\phi(y)\,dy &= -\frac{k^2}{4\pi}\int_D \cos\alpha_l(y)\psi(y)\,dy + b_l^2, \end{aligned}\right\} \tag{10}$$

where $\alpha_l(y) = (k^i - k_l^s)\cdot y$ and $b_l = b_l^1 + ib_l^2$.

Inequality (6) reduces to either

$$|1 + \Pi_R\phi - \Pi_I\psi| > \delta_1|\phi| \tag{11}$$

or

$$(1 + \Pi_R\phi - \Pi_I\psi)^2 + (\Pi_R\psi + \Pi_I\phi)^2 > \delta_1(\phi^2 + \psi^2). \tag{11a}$$

The inverse problem now reduces to finding real continuous functions ϕ, ψ that vanish on D and satisfy the relations (9), (10), and (11). Once these are known, $v(x)$ is determined from

$$v(x) = \phi(x)/[1 + \Pi_R\phi - \Pi_I\psi], \tag{12}$$

which is the reduced form of (4).

3. ANALYSIS OF THE SYSTEM OF EQUATIONS

The system of linear functional equations (10) will be inverted first. We shall assume that the functions $\cos\alpha_l(x)$, $\sin\alpha_l(x)$, $l = 1,2, \ldots, N$ are linearly independent and hence span a $2N$ dimensional space X. Pick out a suitable choice of basis vectors Φ_m, $m = 1,2, \ldots, 2N$, which must be continuous and vanish on the boundary ∂D and are such that $\det\{a_{ij}\} \neq 0$, where the matrix elements a_{ij} are given by

$$a_{ij} = \frac{k^2}{4\pi} \int_D \cos \alpha_i(x)\Phi_j(x)\,dx, \quad i = 1,2, \ldots, N,$$
$$= \frac{k^2}{4\pi} \int_D \sin \alpha_{i-N}(x)\Phi_j(x)\,dx, \quad i = N+1, \ldots, 2N.$$

Then expand

$$\phi = \sum_{m=1}^{2N} c_m \Phi_m + \phi^{\perp}, \tag{13}$$

where $\phi^{\perp}$ is a continuous function which vanishes on D and is perpendicular to the space X, treated as a subspace of the real Hilbert space with inner product

$$(u, v) = \int_D u(x)v(x)\,dx.$$

Then the linear functional equations becomes an algebraic system. Solve for c_n and substitute back into expression (13) to obtain the form

$$\phi(x) = \phi_0(x) + \mathbf{K}_0\psi + \phi^{\perp}, \tag{14}$$

where $\phi_0(x)$ is the linear combination of Φ_m whose coefficients depend upon the measured quantities b_m^1 and b_m^2. $\mathbf{K}_0$ is an integral operator with degenerate kernel of the form

$$k_0(x,y) = \sum_{i,j=1}^{2N} \beta_{ij}\Phi_i(x)\Psi_j(y), \tag{15}$$

where the coefficients depend upon a_{ij} only.

As $N \to \infty$ or as the scattered directions become close together the matrix becomes ill-conditioned and one has to use alternative techniques to solve the system (HILGERS [3], NASHED [4]). One either uses the generalized inverse or, what amounts to the same thing, if (10) are expressed in the general form

$$(h_m^1, \phi) = (h_m^2, \psi) + B_m, \quad m = 1,2, \ldots, 2N,$$

then one minimizes the following

$$\min\left\{\sum_{m=1}^{2N} [(h_m^1,\phi) - (h_m^2,\psi) - B_m]^2 + \tilde{\alpha}(\phi,\phi)\right\}$$

for some $\tilde{\alpha} > 0$.

Because of the undetermined nature of $\phi^{\perp}$, we have nonuniqueness. To obtain uniqueness, additional conditions have to be imposed. Two such conditions are given as follows:

(1) Find a solution ϕ that minimizes (ϕ,ϕ). This yields $\phi^{\perp} \equiv 0$, if the proper basis is chosen.

(2) If we have an *a priori* estimate for n, namely $n^*(x)$, then we can compute the corresponding value of $\phi^*(x)$ and seek the solution ϕ that minimizes $(\phi - \phi^*, \phi - \phi^*)$. This yields the value (or a proper basis)

$$\phi^{\perp} = \phi^* - \mathbf{P}\phi^* = \phi^* - \sum_{m=1}^{2N} c^*_m \Phi_m$$

where $\mathbf{P}$ is the projection operator on the space span $\{\Phi_m\}_{m=1}^{2N}$. If the basis Φ_m is a real orthonormal set, then $c^*_m = (\phi^*, \Phi_m)$.

One may want to work with a nonorthonormal basis, especially if the finite-element method is employed [5].

We now substitute expression (14), where $\phi^{\perp}$ is now prescribed, into (9) and obtain a nonlinear integral equation of the Lyapunov-Schmidt [6] type which has the general form

$$\psi = S(\psi). \tag{16}$$

Since the equation is in particular, a quadratic equation, we can make use of the result of RALL [7] which states that it will have a unique solution in a convex set for which the integral operator $(I - S'(\psi))$ is nonsingular, where $S'(\psi)$ is the Fréchet derivative of $S(\psi)$. Here we are interested in solutions $\psi \in C_0(D)$, the space of continuous functions on D, vanishing on ∂D.

The usual iteration techniques that may be used to solve (16) are

(i) method of successive substitutions,

$$\psi_{n+1} = S(\psi_n);$$

(ii) Newton's method,

$$\psi_{n+1} = \psi_n - (I - S'(\psi_n))^{-1}(\psi_n - S(\psi_n));$$

(iii) modified Newton's method

$$\psi_{n+1} = \psi_n - (I - S'(\psi_0))^{-1}(\psi_n - S(\psi_n)).$$

In each case, an appropriate initial approximation ψ_0 is required.

Estimates for uniform convergence of the iteration processes are obtained using a majorizing method [8]. A rough estimate of conditions needed for convergence of the successive approximation scheme starting from the initial value $\psi_0 \equiv 0$ is given by [9]

$$||\phi_1|| l [1 + ||K_0||] < 0.22,$$

where the norms are the uniform norms, $\phi_1 = \phi_0 + \phi^{\perp}$, and $l = \text{Max}[||\Pi_R||, ||\Pi_I||]$. Using, further, the rough estimate $l < (1/2)\ k^2R^2$, where R is the radius of the region D, we see that, as expected, the successive approximation method starting from $\psi_0 = 0$ is a low-frequency approximation. It depends upon $||K_0||$, which in turn depends upon the size of $\det\{a_{ij}\}$. If the matrix is ill-conditioned, then $||K_0||$ can be quite large if normal inversion techniques are used to solve $\{c_i\}$ or ϕ.

A similar estimate, although slightly less restrictive, can be made for the modified Newton method, with $\psi_0 = 0$. For either method, it can be shown that the solution ψ obtained from the iteration processes satisfies the inequality $||\psi|| \leqslant 4l||\phi_1||^2$, in which case it can be shown

$$||\Pi_R\phi - \Pi_I\psi|| \leqslant 1/2 < 1,$$

which implies that the constraint (11) is automatically satisfied. Hence, the solution $v(x)$ obtained from equations (12) and (14) with the computed values of ψ is bounded.

For the non-low-frequency region, the initial approximation ψ_0 in the iterative procedures will have to have a value other than zero. If we have an *a priori* estimate or rough guess for $n(x)$, say $n^*(x)$, then we can compute the corresponding value ψ^* and employ this for the intial approximation, i.e., $\psi_0 = \psi^*$. The modified Newton process converges if ψ^* is sufficiently close to the solution.

If one does not have a good *a priori* estimate for ψ^*, an approach that can be used to obtain a good initial estimate is the following. Split $S(\psi)$ into two parts, a homogeneous portion (with regard to ψ) $S_1(\psi)$ and a nonhomogeneous portion, in particular the term $\phi_1\Pi_I\phi_1$, where $\phi_1 = \phi_0 + \phi^{\perp}$. Introduce a parameter λ so that equation (16) corresponds to the equation

$$\psi = S_1(\psi) + \lambda\phi_1\Pi_1\phi_1,$$

with $\lambda = 1$. Starting from $\lambda = 0$ with the solution $\psi(x,0) = 0$, one solves this equation for $\psi(x,\lambda)$ for a set of increasing values of the parameter λ. Each solution $\psi(x,\lambda + \Delta\lambda)$ is found by using the previously determined solution $\psi(x,\lambda)$ as the initial approximation ψ_0 in the Newton process [10]. This process may lead to possible bifurcation points λ_0, points for which the operator $[I - S'(\psi,\lambda_0)]$ is singular, and one can end up with more than one solution as $\lambda \rightarrow 1$.

There still remains the question of whether or not in general these solutions satisfy constraint (11). If they do not, then one can modify the solution by adding a function $\tilde{\phi}$ to $\phi^{\perp}$ (Recall that $\phi^{\perp}$ was uniquely specified only after requiring the additional condition of a minimum L_2 norm for ϕ or $\phi - \phi^*$; thus this constraint has to be relaxed.) For some cases, $\tilde{\phi}$ need only be a small perturbation to yield the desired result. However, the general question of the best way to choose $\tilde{\phi}$ needs to be investigated in detail.

Note that the Galerkin technique cannot be directly applied to (16), since the operator $S(\psi)$ is not a completely continuous nonlinear operator [11]. By applying the constraint

$$||\Pi_R\phi - \Pi_I\psi|| < 1, \tag{17}$$

(16) can be written in the form

$$\psi = \phi(\Pi_R\psi + \Pi_I\phi) \sum_{n=0}^{\infty} (\Pi_I\psi - \Pi_R\phi)^n. \tag{18}$$

The operator given by the right-hand side of (18) with $\phi = \phi_0 + \phi^\perp + \mathbf{K}_0\psi$, ($\phi_0$, $\phi^\perp$ fixed) is a completely continuous operator; hence, the Galerkin technique will yield a good approximation. The above constraint (17) does not restrict k to the low-frequency region, since it only implies that $\mathrm{Re}\{u\bar{u}'\} \neq 0$. Thus, (8) can be applied to large frequencies for almost transparent material where $n(x)$ fluctuates about unity, i.e.,

$$k\left|\int^x (n-1)\,dx\right| < \pi/2.$$

4. NOTE ON STABILITY

Changes in the measured data $\{b_i\}_{i=1}^N$ register corresponding changes in ϕ directly through the component $\phi_0(x)$. As pointed out earlier, to reduce the effects of the ill-conditioning of the matrix $\{a_{ij}\}$ for large N, the generalized inverse or an equivalent method is emplooyed to find $\phi_0(x)$ and reduce the size of the change $\delta\phi_0$. To see how $\delta\phi_0$ effects $v(x)$ and hence $n(x)$ we note from equation (12) that

$$\delta v(1 + \Pi_R\phi - \Phi_I\psi) = (I - v\Pi_R)\delta\phi_0 + [\mathbf{K}_0 - \mathrm{v}(\Pi_R\mathbf{K}_0 - \Pi_I)]\delta\psi_0$$

and

$$\delta\psi_0 = [I - S'(\psi)]^{-1}\delta\phi_0(\Pi_I\phi + \Pi_R\psi) + [I - S'(\psi)]^{-1}[-\psi\Pi_R + \phi\Pi_I]\delta\phi_0.$$

Note that, in the pointwise sense, the change $\delta\mathrm{v}$ is large when $(1 + \Pi_R\phi - \Pi_I\psi)$ is small, corresponding to points where v has a maximum, which is to be expected. For these points, the relative change $(1/n^2)\,\delta n^2 = [1/(\mathrm{v}+1)]\,\delta\mathrm{v}$ is more important. Apart from this, it is seen that a problem only occurs in certain critical cases, namely when $I - S'(\psi)$ is nonsingular. This corresponds to a bifurcation point and, hence, small changes in $\delta\phi_0$ can produce nonunique changes in $\delta\psi$ (more than one branch of solution).

5. COMMENTS

It should be pointed out that at high frequencies, if n is sufficiently smooth, procedures based upon ray-tracing (which is a nonlinear process) may be more practical.

Measurements made for a set of different incident waves will lead to a system of nonlinear equations. This more complicated system should be studied only after a thorough analysis of the single incident wave case.

The details in the present paper and additional analysis will appear elsewhere.

FOOTNOTES AND REFERENCES

[1] R. Leis: *Vorlesungen über partielle Differentialgleichungen zweiter Ordunung* (Bibliographisches Institut, Mannheim, Germany, 1967)
[2] As a functional of v, these are nonlinear functional equations, since u depends on v. Otherwise, they are a linear functional of the product vu.
[3] J.W. Hilgers: "Non-iterative methods for solving operator equations of the first kind," MRC. Report 1413 (1974)
[4] M.Z. Nashed: *Generalized Inverses and Applications* (Academic, New York, 1976)
[5] G. Strang, G. Fix: *An Analysis of the Finite Element Method* (Prentice-Hall, Englewood Cliffs, New Jersey, 1973)
[6] M.M. Vainberg, V.A. Trenogin: *Theory of Branching of Solutions of Non-linear Equations* (Noordhoff, Groninger, 1974)
[7] L.B. Rall: SIAM Rev. **11**, 386 (1969)
[8] M.M. Vainberg: *Variational Methods for the Study of Nonlinear Operators* (Holden-Day, San Francisco, 1964)
[9] Unless otherwise specified the norm is the uniform (max) norm. Better estimates can be obtained using the $L_2(D)$ norm
[10] D.W. Decker: "Topics in Bifurcation Theory," thesis, Cal Tech (1978)
[11] M.A. Krasnosel'skii: *Topological Methods in the Theory of Nonlinear Integral Equations* (Pergamon, New York, 1964)

N-DIMENSIONAL FAST FOURIER TRANSFORM TOMOGRAPHY FOR INCOMPLETE INFORMATION AND ITS APPLICATION TO INVERSE SCATTERING THEORY

Norbert N. Bojarski*

16 Pine Valley Lane
Newport Beach, California 92660

INTRODUCTION

The n-dimensional tomography problem is solved in closed form by means of the Fast Fourier Transform algorithm, thus requiring of the order of $N\log_2 N$ complex arithmetic add-multiply operations, where N is the number of data points specifying the problem; vis-a-vis the conventional Radon transform solution which requires of the order of N^2 operations. The extention from two-dimensional to three-dimensional tomography is achieved in a simple and natural fashion.

For incomplete input information, this solution yields simply and directly a Fredholm Integral Equation of the Second Kind, which is again solvable in $N\log_2 N$ operations by means of the Fast Fourier Transform algorithm.

Some numerico-experimental results are presented, and the connection between these solutions and the Physical Optics Inverse Scattering Identity of this author, as well as the Radon transform, are discussed in some detail.

1. A FOURIER TRANSFORM SOLUTION OF THE TOMOGRAPHY PROBLEM

Consider a two-dimensional function $f(\mathbf{x})$ and its two dimensional Fourier Transform $F(\mathbf{k})$, represented in the unprimed coordinate system $\mathbf{x}$ and $\mathbf{k}$ by

$$F(\mathbf{k}) = \iint e^{i\mathbf{k}\cdot\mathbf{x}} f(\mathbf{x})\, d^2x \quad . \tag{1}$$

In the primed coordinate system $\mathbf{x}'$ and $\mathbf{k}'$, formed by rotation by an angle ϕ about the origin (see Fig. 1), (1) is

$$F(k'_1,k'_2) = \iint e^{i(k'_1x'_1+k'_2x'_2)} f(x'_1,x'_2)\, dx'_1\, dx'_2, \tag{2}$$

which along the k'_1 axis only, i.e., for $k'_2 = 0$, yields

$$F(k'_1,0) = \iint e^{ik'_1x'_1} f(x'_1,x'_2)\, dx'_1\, dx'_2 \tag{3}$$

$$= \int e^{ik'_1x'_1} \int f(x'_1,x'_2)\, dx'_2\, dx'_1. \tag{4}$$

Next, let the function $g(x')$ be defined as

$$g(x'_1) = \int f(x'_1,x'_2)\, dx'_2, \tag{5}$$

which reduces (4) to

$$F(k'_1,0) = \int e^{ik'_1x'_1} g(x'_1)\, dx'_1 \tag{6}$$

Since (2) through (6) are invariant under rotation, i.e., valid for all angles ϕ, it follows that in the unprimed coordinate systems x and k (see Fig. 2)

$$F(k,\phi) = \int e^{ikr}\, g(r,\phi)\, dr \quad , \tag{7}$$

where, consistent with (5), again in the unprimed coordinate system, in which r' is orthogonal to r for any given value of the angle ϕ (see Fig. 2)

$$g(r,\phi) = \int f(r,r',\phi)\, dr' \ . \tag{8}$$

Taking the two-dimensional inverse Fourier transform of (7) yields

$$f(\mathbf{x}) = \frac{1}{(2\pi)^2} \int\int e^{-i\mathbf{k}\cdot\mathbf{x}} \int e^{ikr}\, g(r,\phi)\, dr\, d^2k \ , \tag{9}$$

where

$$\mathbf{k} = k \begin{pmatrix} \cos\phi \\ \sin\phi \end{pmatrix} \ , \tag{10}$$

$$d^2k = k\, dk\, d\phi \ . \tag{11}$$

Equation (9) is thus the closed form Fourier transform solution of the Tomography problem, i.e., a solution for the function f, given the function g defined by (8).

The three-dimensional generalization of the problem and its solution can thus be accomplished by mere inspection of (8) and (9), i.e.,

$$g(r,\phi,\theta) = \int\int f(r,r',r'',\phi,\theta)\, dr'\, dr'' \tag{12}$$

$$f(\mathbf{x}) = \frac{1}{(2\pi)^3} \int\int\int e^{-i\mathbf{k}\cdot\mathbf{x}} \int e^{ikr} g(r,\phi,\theta)\, dr\, d^3k \ , \tag{13}$$

where r' and r'' are orthogonal to each other and to r for any given value of the angles ϕ and θ. In terms of the standard spherical coordinates,

$$\mathbf{k} = k \begin{pmatrix} \cos\theta\, \cos\phi \\ \cos\theta\, \sin\phi \\ \sin\theta \end{pmatrix} \ , \tag{14}$$

and

$$d^3k = k^2\, dk\, \sin\theta\, d\theta\, d\phi \ . \tag{15}$$

2. AN INTEGRAL EQUATION FOR INCOMPLETE INFORMATION TOMOGRAPHY

Solution (9) requires knowledge of $g(r,\phi)$ for all angles ϕ, i.e., $\phi \in [0,\pi]$. It should be noted that knowledge of $g(r,\theta)$ is not needed for all $\phi \in (0, 2\pi)$ since $g(r,\phi) = g(r,\phi+\pi)$; (see (8) and Fig. 2). In most practical applications $g(r,\phi)$ is known only for some incomplete subset $[\phi_1,\phi_2]$ of the domain $[0,\pi]$. It is to this problem of solving (8) for $f(x)$, given $g(r,\phi)$ for the incomplete subset domain $\phi \in [\phi_1,\phi_2]$, that this section is addressed. Let a characteristic k-space aperture information function $A(\mathbf{k})$ and its complement $A_c(\mathbf{k})$ be defined as

$$A(\mathbf{k}) = \begin{cases} 1, & \forall\, k \in [\phi_1,\phi_2] \\ 0, & \forall\, k \notin [\phi_1,\phi_2] \end{cases} \ , \tag{16}$$

and

$$A_c(\mathbf{k}) = 1 - A(\mathbf{k}) \ , \tag{17}$$

respectively, and let $a(x)$ and $a_c(x)$ be the two dimensional Fourier transforms of $A(k)$ and $A_c(k)$ respectively, i.e.,

$$a(\mathbf{x}) \Longleftrightarrow A(\mathbf{k}) \ , \tag{18}$$

and

$$a_c(\mathbf{x}) \Longleftrightarrow A_c(\mathbf{k}) \ . \tag{19}$$

Thus, by (13) and (15)

$$a_c(\mathbf{x}) = \delta(\mathbf{x}) - a(\mathbf{x}) \ . \tag{20}$$

Convolving (9) by $a(\mathbf{x})$ thus yields with the aid of the Fourier transform convolution theorem

$$a(\mathbf{x}) * f(\mathbf{x}) = a(\mathbf{x})\frac{1}{(2\pi)^2}\int\int e^{-i\mathbf{k}\cdot\mathbf{x}}\int e^{ikr} g(r,\phi)\, dr\, d^2k \tag{21}$$

$$= \frac{1}{(2\pi)^2}\int\int e^{-i\mathbf{k}\cdot\mathbf{x}}A(\mathbf{k})\int e^{ikr} g(r,\phi)\, dr\, d^2k\ , \tag{22}$$

which by (16) reduces to

$$a(\mathbf{x}) * f(\mathbf{x}) = \frac{1}{(2\pi)^2}\int\int_A e^{-i\mathbf{k}\cdot\mathbf{x}}\int e^{ikr} g(r,\phi)\, dr\, d^2k\ . \tag{23}$$

Since the right hand side term of (23) requires an integration over $\phi \in [\phi_1, \phi_2]$ only, for which $g(r,\phi)$ is known, it follows that this right hand side term of (23) can always be evaluated. It thus becomes convenient to define this right hand side term of (23) as the known Ansatz $h(x)$, i.e.,

$$h(\mathbf{x}) = \frac{1}{(2\pi)^2}\int\int_A e^{-i\mathbf{k}\cdot\mathbf{x}}\int e^{ikr} g(r,\phi)\, dr\, d^2k \tag{24}$$

Equation (23), with the aid of (20) and (24), yields

$$[\delta(\mathbf{x}) - a_c(\mathbf{x})] * f(\mathbf{x}) = h(\mathbf{x})\ , \tag{25}$$

which reduces to

$$f(\mathbf{x}) - a_c(\mathbf{x}) * f(\mathbf{x}) = h(\mathbf{x})\ . \tag{26}$$

Written out explicitly, (26) is a Fredholm Integral Equation of the Second Kind for the unknown function $f(\mathbf{x})$, in terms of the known kernel $a_c(\mathbf{x})$ and known Ansatz for $h(\mathbf{x})$, i.e.,

$$f(\mathbf{x}) - \int\int a_c(\mathbf{x} - \mathbf{x}')\, f(\mathbf{x}')\, d^2x' = h(\mathbf{x})\ . \tag{27}$$

Next, an analytic expression for the kernel $a_c(\mathbf{x})$ shall be obtained. By (19)

$$a_c(\mathbf{x}) = \frac{1}{(2\pi)^2}\int\int e^{-i\mathbf{k}\cdot\mathbf{x}}\, A_c(\mathbf{k})\, k^2k\ , \tag{28}$$

which with the aid of (16, (17), and referring to Fig. 3, reduces to

$$a_c(\mathbf{x}) = \frac{1}{(2\pi)^2}\int_{-\infty}^{\infty}\int_{\beta k_2}^{\alpha k_2} e^{-i\mathbf{k}\cdot\mathbf{x}}\, \mathrm{sgn}(k_2)\, dk_1\, dk_2\ , \tag{29}$$

where the term sgn (k_2) is necessitated by the fact that the direction of integration along the k_1 axis from βk_2 to αk_2 is reversed for $k < 0$.

With the aid of the one dimensional Fourier transform statement

$$\mathrm{sgn}(k) \Longleftrightarrow \frac{2}{ix} \tag{30}$$

and a modest amount of straightforward algebra, (25) yields

$$a_c(\mathbf{x}) = \frac{(\alpha - \beta)}{2\pi^2(x_2 + \alpha x_1)(x_2 + \beta x_1)}\ . \tag{31}$$

For the case $\beta = -\alpha$, which involves no loss of generality by virtue of the freedom of choice of the cartesian coordinate system, (31) reduces to

$$a_c(\mathbf{x}) = \frac{\alpha}{\pi^2(x_2 - \alpha x_1)(x_2 + \alpha x_1)}\ , \tag{32}$$

which, in the polar coordinate system of Fig. 4, reduces to

$$a_c(r,\phi) = \frac{\cos^2\phi_0}{\pi^2 r^2 \cos^2(\phi - \phi_0)\, \cos^2(\phi + \phi_0)}\ . \tag{33}$$

3. THE CONNECTION TO THE RADON TRANSFORM

The two-dimensional solution (9) is in a mixed coordinate system, i.e., the known $g(r,\phi)$ is in the polar coordinate system (r,ϕ) and the unknown $f(x)$ is in the cartesian coordinate system (x_1,x_2). Transforming (9) into a single polar coordinate system, i.e.,

$$\begin{pmatrix} x_1 \\ x_2 \end{pmatrix} = r' \begin{pmatrix} \cos \phi' \\ \sin \phi' \end{pmatrix} , \tag{34}$$

$$\begin{pmatrix} k_1 \\ k_2 \end{pmatrix} = k \begin{pmatrix} \cos \phi \\ \sin \phi \end{pmatrix} , \tag{35}$$

$$d^2k = k\ dk\ d\phi, \tag{36}$$

yields

$$f(r',\phi') = \frac{1}{(2\pi)^2} \int_0^{2\pi} \int_0^{\infty} e^{-ikr'\cos(\phi-\phi')} \int e^{ikr} g(r,\phi)\ dr\ k\ dk\ d\phi \tag{37}$$

$$= \frac{1}{(2\pi)^2} \int_0^{2\pi} \int_{-\infty}^{\infty} \int_0^{\infty} e^{-ik[r'\cos(\phi-\phi')-r]} k\ dk\ g(r,\phi)\ dr\ d\phi\ . \tag{38}$$

The innermost integral over k can be executed analytically, and with a modest amount of algebra (38) reduces to the Radon-transform solution of (8). A similar transformation of the three-dimensional solution (13) into a single spherical coordinate system yields the generalized three-dimensional Radon-transform solution of (12). As shall be shown in the next section, from a numerical-evaluation-efficiency perspective, this analytic evaluation of the innermost integral is precisely not the thing to do.

4. NUMERICAL IMPLEMENTATION

If $g(r,\phi)$ is given for N data points in (r,ϕ), then a numerical execution of the Radon transform solution of (8) requires of the order of N^2 arithmetic operations. In contrast, however, solution (9) can be executed with the aid of the Fast Fourier Transform algorithm in the order of $N \log_2 N$ arithmetic operations, in the mixed polar-cartesian coordinate system of (9). Specifically, the innermost integral over r, say

$$F(k,\phi) = \int e^{ikr}\ g(r,\phi)\ dr\ , \tag{39}$$

is executed in $N_\phi(1/2N_r \log_2 N_r)$ operations with the aid of the one-dimensional FFT algorithm N_ϕ times, where N_r and N_ϕ are the number of data points in r and ϕ respectively, and $N = N_\phi N_r$. Next, a two-dimensional cartesian array $F(k_1,k_2)$ of size N is filled with zeros, and then the two-dimensional array $F(k,\phi)$ is mapped into the two-dimensional cartesian array $F(k_1,k_2)$ by a straightforward rotation. The mismatch between the polar and cartesian grid points is ignored, and taken merely to the nearest point. Such a mapping is necessitated by the fact that the FFT algorithm requires equally spaced grid points in each dimension. Such a mapping can be executed in N-logic operation. The outer integral over k, say

$$f(x_1,x_2) = \frac{1}{(2\pi)^2} \int\int e^{i(k_1x_1+k_2x_2)}\ F(k_1,k_2)\ dk_1\ dk_2\ , \tag{40}$$

is then executed in $1/2N\log_2 N$ operations with the aid of the two-dimensional (inverse) FFT algorithm. The previously mentioned mismatch between the polar and cartesian grid points introduces errors in the k-space. The FFT operation into the x-space thus constitutes a global smoothing of these errors, which decrease with increasing N. Any more sophisticated, and therefore more computer-time-consuming interpolation schemes for the polar to cartesian-grid-point mapping is thus totally unwarranted.

5. NUMERICO-EXPERIMENTAL RESULTS

Solution (9) was computer implemented for $N = 32 \times 32$ for a unity density disc of radius 7.5, and a unity density disc of radius 11.5 with a concentric hole of radius 5.5, with the results shown graphically in Figs. 5 and 7 respectively. Figures 6 and 8 are replicas of Figs. 5 and 7 respectively, with the correct reference circle superimposed. In these figures the size of the solid and hollow squares are proportional to the positive and negative values of the reconstructed density functions $f(x)$ respectively. It appears that a raster of $N = 32 \times 32$ is marginally adequate in size to accomplish the previously mentioned polar to cartesian grid point mismatch error smoothing.

This solution was also computer implemented for a raster of $N = 64 \times 64$ for a unity density disc of radius 24.5, and a unity density disc of radius 24.5 with a concentric hole of radius 16.5, with the results shown graphically in Figs. 9 and 10 respectively. It appears that a raster of $N = 64 \times 64$ is most adequate in size for the grids mismatch error smoothing.

The method-of-successive-approximation solution

$$f_{n+1}(\mathbf{x}) = \int\int a_c(\mathbf{x}-\mathbf{x}')\, f_n(\mathbf{x}')\, d^2x' + h(\mathbf{x}) \ , \tag{41}$$

of the integral equation (27) for incomplete information was also computer implemented. The initial approximation $f_0(\mathbf{x}) = 0$, which yields analytically $f_1(\mathbf{x}) = h(\mathbf{x})$, was taken as the starting point of the numerical iteration. The *Ansatz* $h(\mathbf{x})$ was evaluated as per (24) by the previously described method for the complete information case in $N \log_2 N$ operations, and the convolution implied by (41) was executed with the aid of the two-dimensional FFT algorithm in $N \log_2 N$ operations. In Figs. 11, 13, and 15 are shown the *Ansatz* $h(\mathbf{x})$, the first iteration $f_2(\mathbf{x})$, and the sixteenth iteration $f_{18}(\mathbf{x})$ respectively, for a unity density disc of radius 7.5 in a $N = 32 \times 32$ field. The information aperture was taken as 45 degrees (out of the full 180 degrees field). Figures 12, 14, and 16 are replicas of Figs. 11, 13, and 15 respectively, with the correct reference circle and information aperture superimposed.

6. CONCLUDING REMARKS

The FFT implementation of the two-dimensional solution (9) is idealy suited to x-ray and ultrasonic tomography applied to medical radiology and nondestructive material testing. A firm-wired array processor FFT implementation of this solution will yield real-time displays for rasters as high as 1024 × 1024.

The disc-to-disc FFT implementation of the two-dimensional solution (41) for the incomplete information case is well suited for ocean tomography for data up to the order of 10^7.

The three-dimensional solution (13) deserves the following discussion. If $g(r,\phi,\theta)$ as per (12) is taken as the projected area function of r, for the complete set of the spherical angles (ϕ,θ), as yielded by the One-Dimensional Physical Optics Inverse Scattering solution of this author [1] and KENNOUGH and MOFFATT [2], then solution (13) is a method for the reconstruction of the full three-dimensional scatterer. Furthermore, if $G(k,\phi,\theta)$, the one-dimensional Fourier transform of $g(r,\phi,\theta)$, is properly recognized as the directly measured complex field cross section of a scatterer divided by k^2, the solution (13) is nothing more then a restatement of the Physical Optics Inverse Scattering Identity [3] of this author.

FOOTNOTES AND REFERENCES

*Work prepared for and supported by the Office of Naval Research under contract N00014-76-C-0082

[1] N.N. Bojarski: "One dimensional physical optics inverse scattering," RCA-Moorestown Report NNB-64-9, April 1964

[2] E.M. Kennough, D.L. Moffatt: "Transient and impulse response approximation," Proc. IEEE **53**, 893-901 (1965)

[3] N.N. Bojarski: "Three dimensional electromagnetic short pulse inverse scattering," Syracuse University Research Corporation, Special Projects Laboratory Report SPL 67-3 (February, 1967) (AD-845 126)

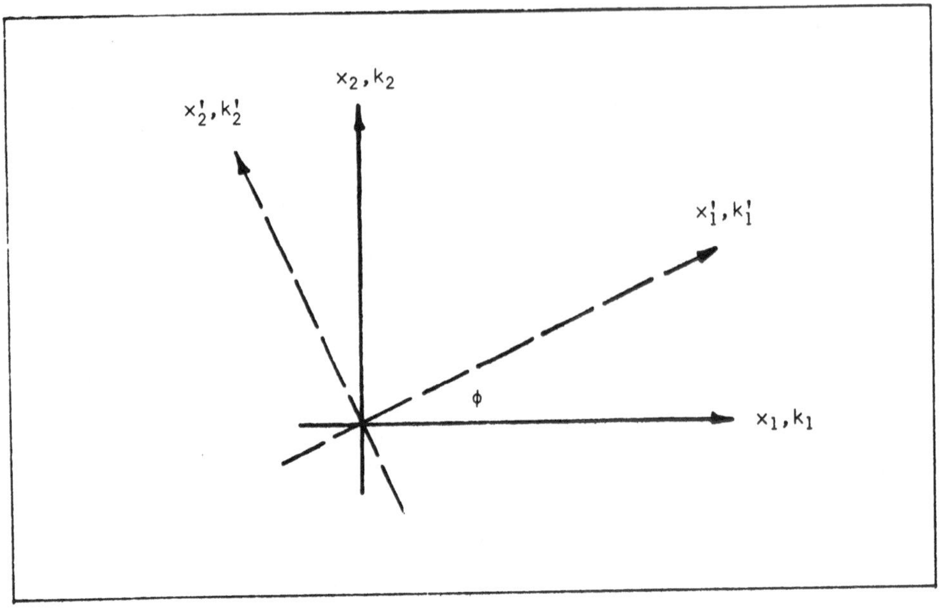

Fig. 1. Unprimed and primed coordinate systems related by a rotation of angle ϕ.

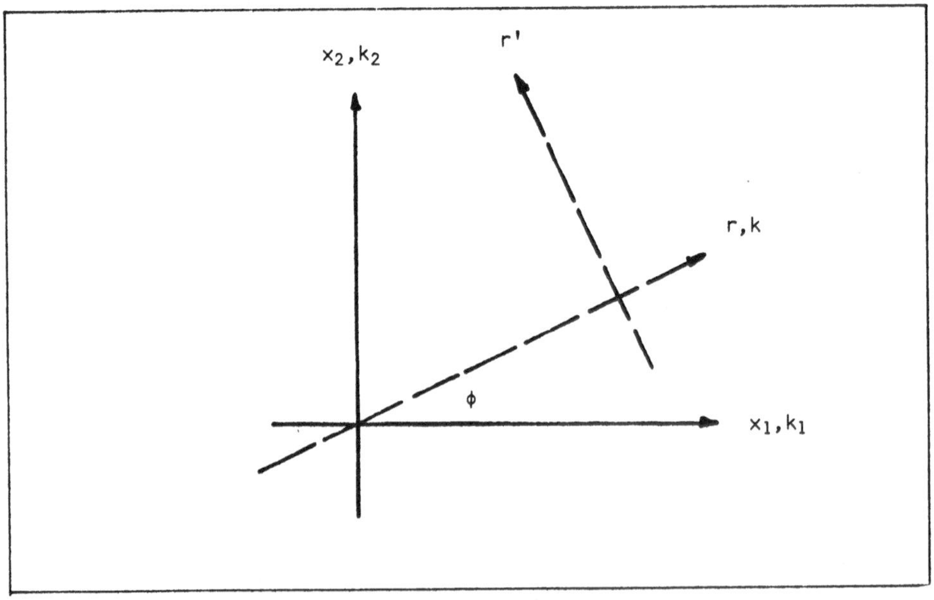

Fig. 2. Unprimed coordinate system using radial notation, and the coordinate radius in the primed system.

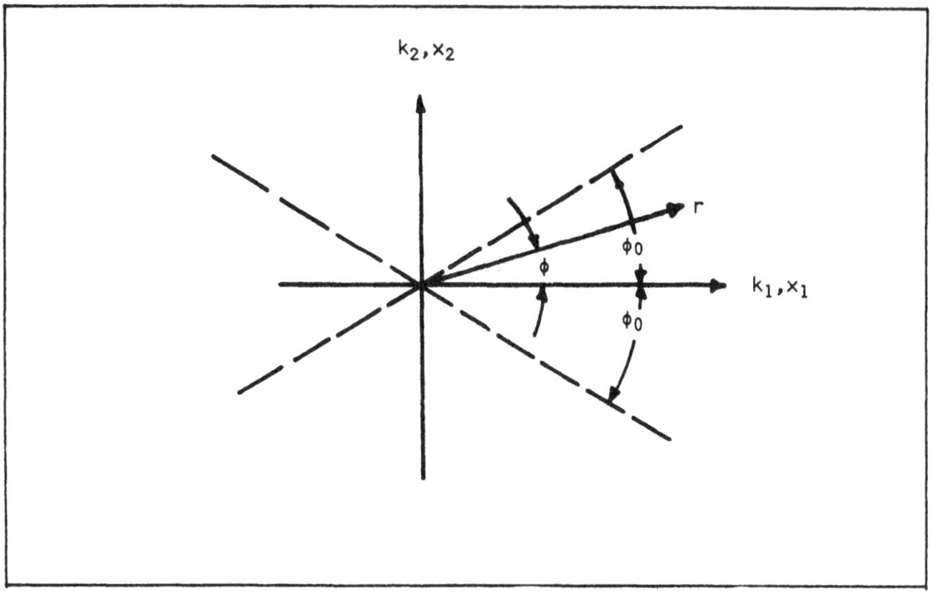

Fig. 3. Coordinate variables referring to Eq. (29).

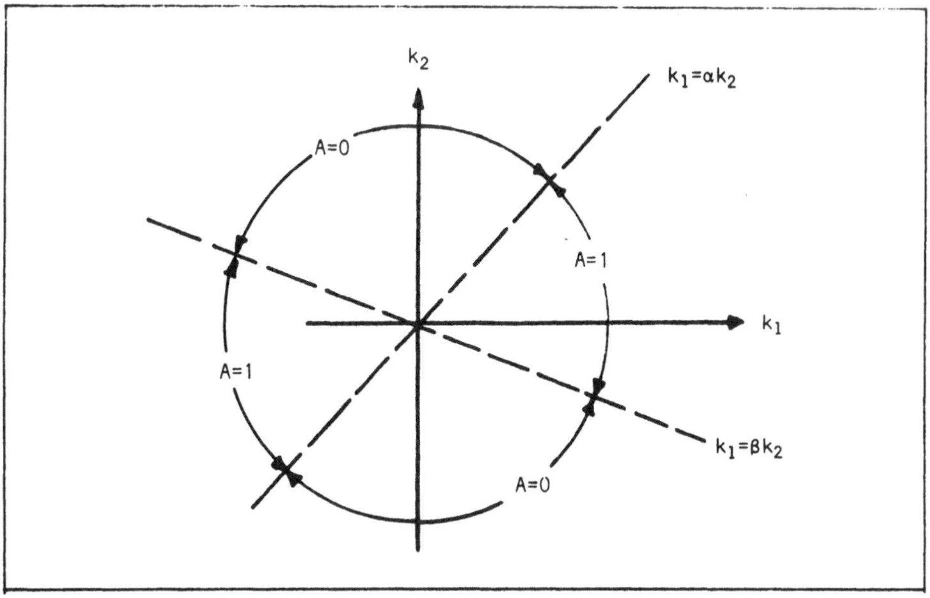

Fig. 4. Polar coordinate system referring to Eq. (33).

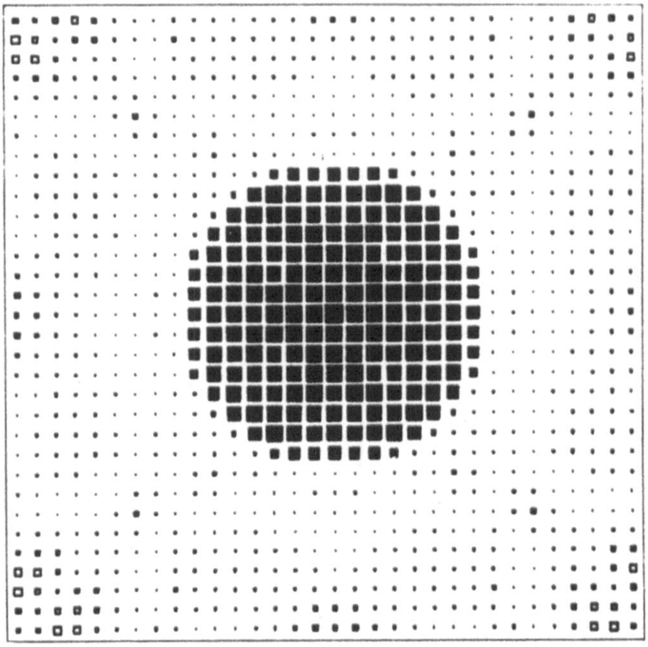

Fig. 5. Reconstructed unity density disc of radius 7.5 in a 32× 32 field.

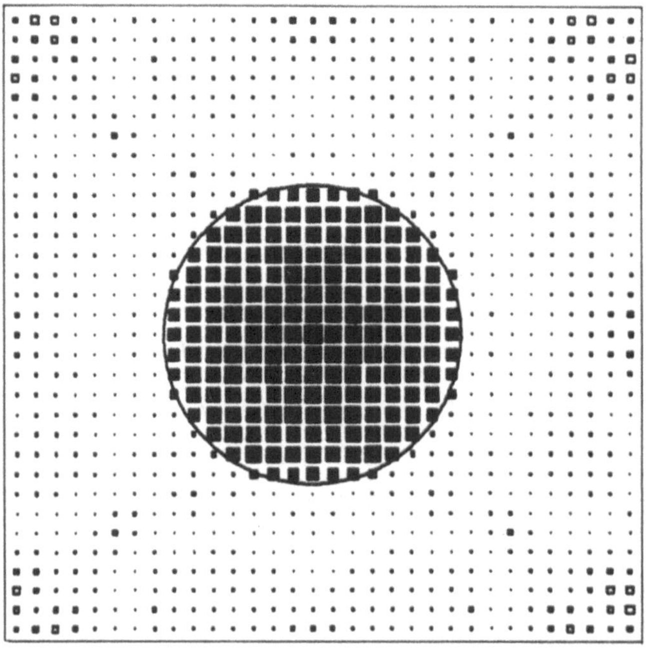

Fig. 6. Reconstructed unity density disc of radius 7.5 in a 32× 32 field with reference circle.

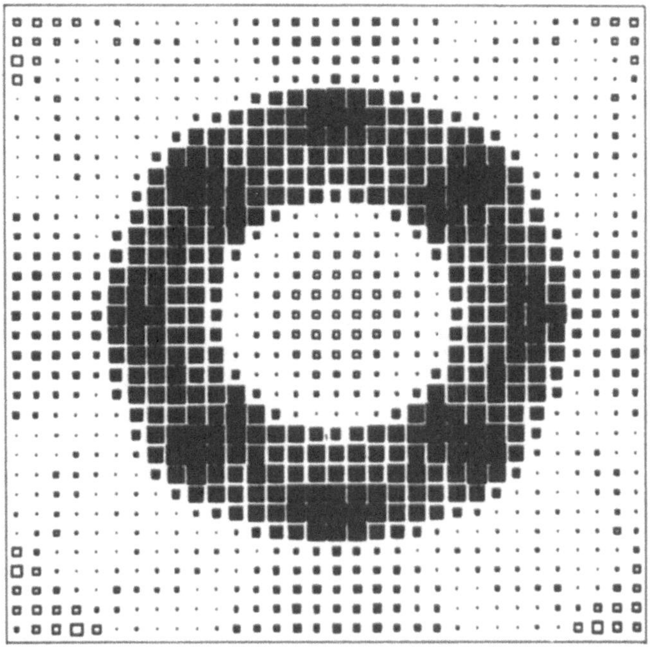

Fig. 7. Reconstructed unity density disc of radius 11.5 with concentric hole of radius 5.5 in a 32×32 field.

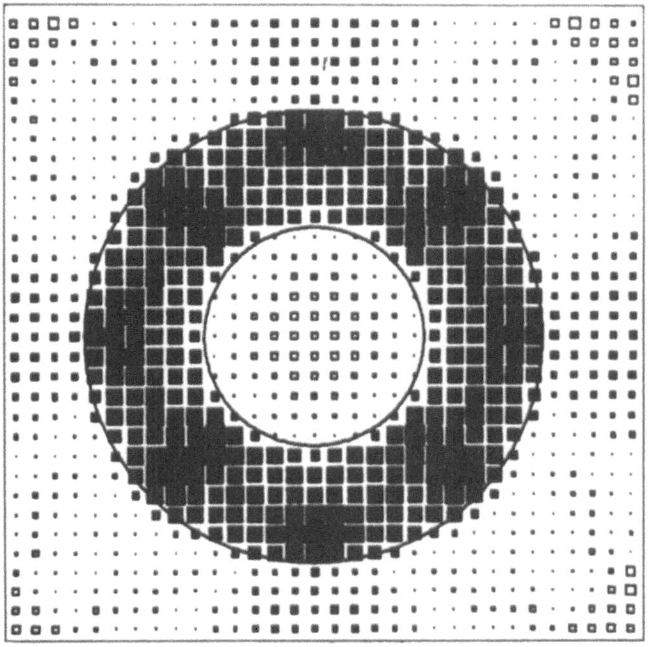

Fig. 8. Reconstructed density disc of radius 11.5 concentric hole of radius 5.5 in a 32×32 field with reference circles.

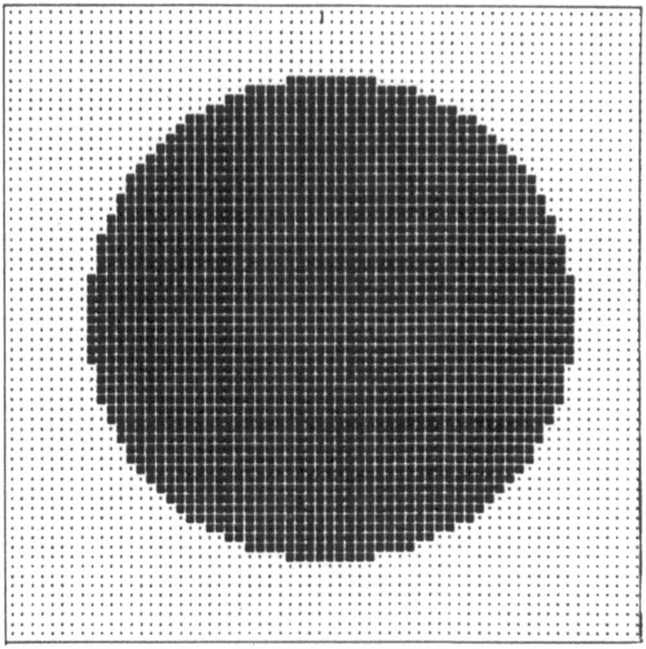

Fig. 9. Reconstructed unity density disc of radius 24.5 in a 64×64 field.

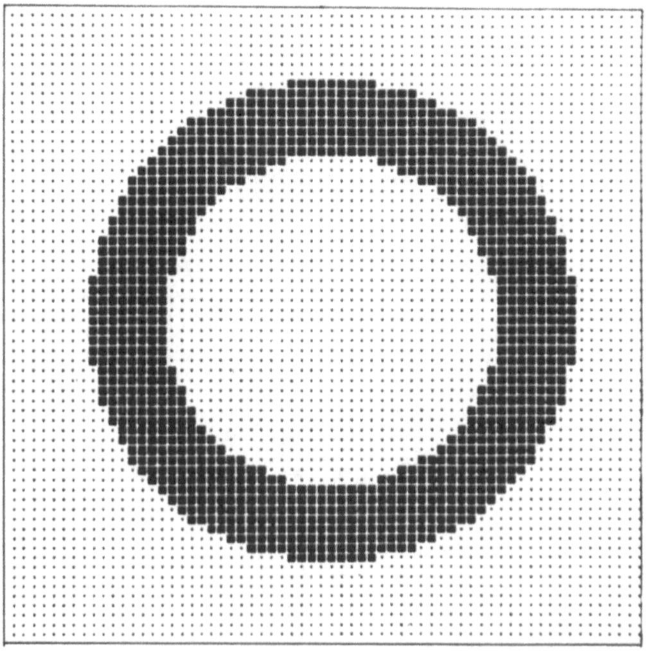

Fig. 10. Reconstructed unity density disc of radius 24.5 with concentric hole of radius 16.5 in 64×64 field.

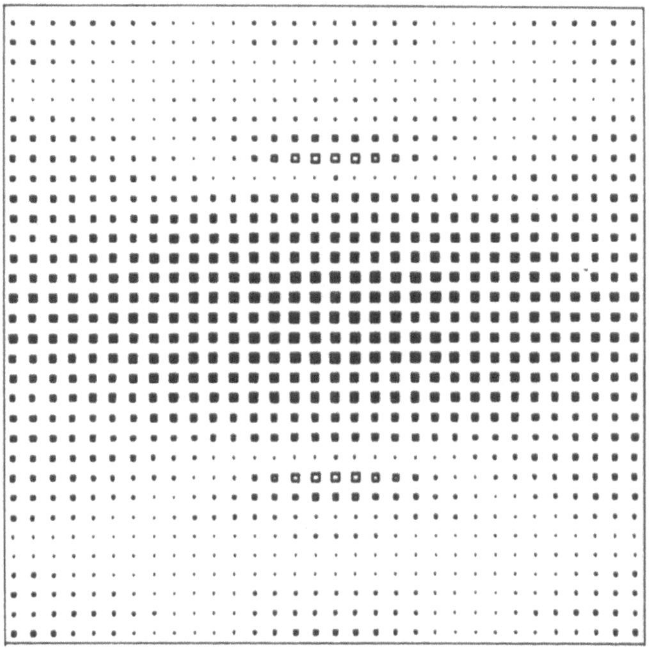

Fig. 11. Ansatz of reconstructed unity density disc of radius 7.5 in 32× 32 field and 45 degrees information field.

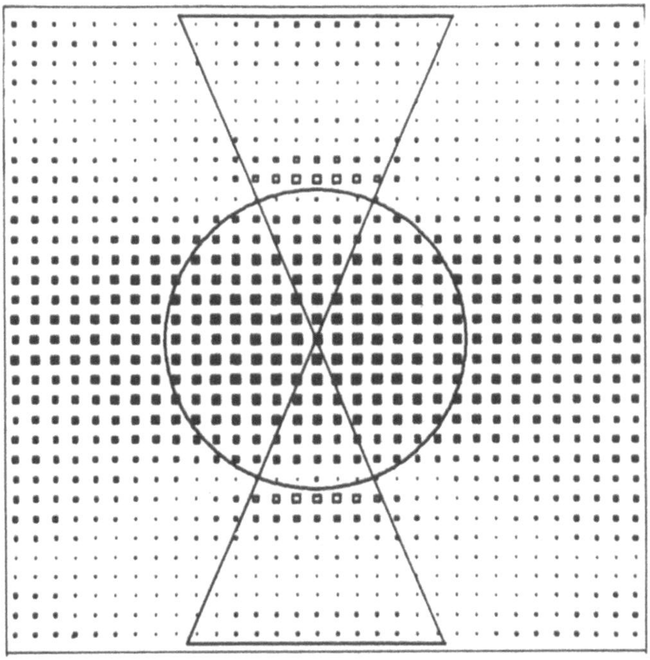

Fig. 12. Ansatz of reconstructed unity density disc of radius 7.5 in 32× 32 field and 45 degrees information field with reference circle information aperture.

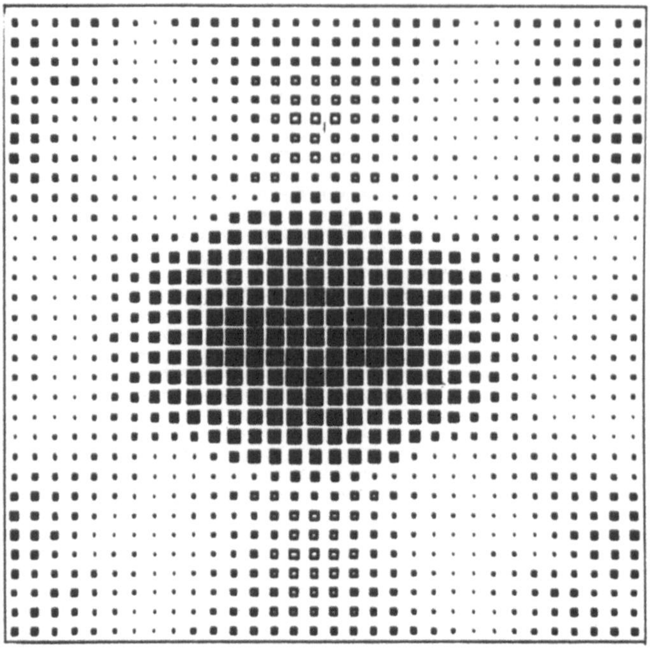

Fig. 13. First iteration of reconstructed unity density disc of radius 7.5 in 32×32 field and 45 degrees information field.

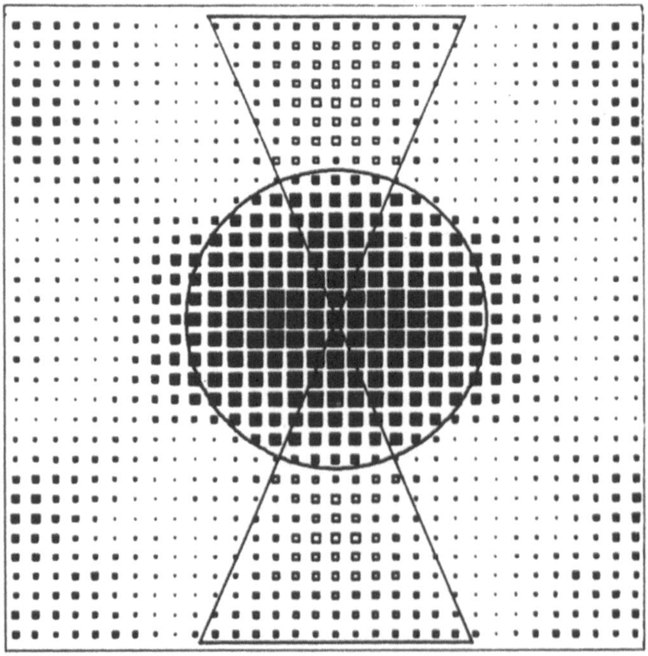

Fig. 14. First iteration of reconstructed unity density disc of radius 7.5 in 32×32 field and 45 degrees information field with reference circle and information aperture.

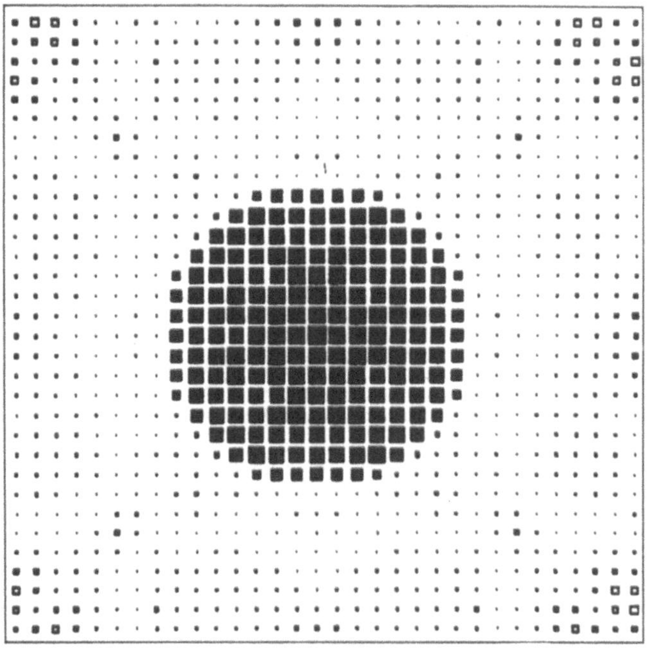

Fig. 15. Sixteenth iteration of reconstructed unity density disc of radius 7.5 in 32× 32 field and 45 degrees information field.

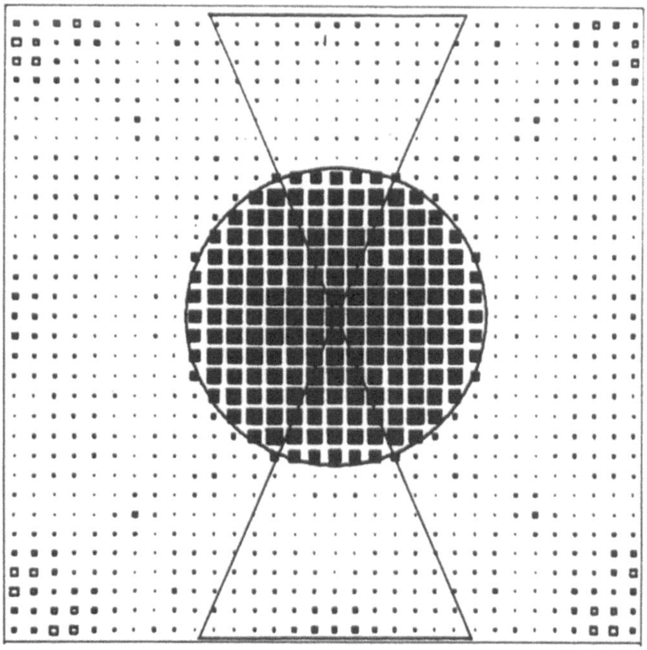

Fig. 16. Sixteenth iteration of reconstructed unity density disc of radius 7.5 in 32× 32 field and 45 degrees information field with reference circle and information aperture.

TWO ACOUSTICAL INVERSE PROBLEMS IN SPEECH AND HEARING

Man Mohan Sondhi

Acoustics Research Department
Bell Laboratories
Murray Hill, New Jersey 07974

INTRODUCTION

I would like to describe to you two acoustical inverse problems that arise in the physical description of speech production and of hearing.

The first problem is that of estimating the (time-varying) shape of the human vocal tract during normal speech production. The use of acoustical measurements for this purpose dates back to the late 1960's. I will summarize the progress made on this problem, with particular emphasis on time-domain methods.

The second problem concerns the estimation of the mechanical properties of a certain membrane - the basilar membrane - in the inner ear, which plays a crucial role in hearing. To the best of my knowledge, the application of inverse-scattering theory to this problem is being suggested here for the first time.

To the present audience, probably the most interesting aspect of this work is the fact that inverse scattering theory has application in such a remote field as anatomical measurements. I will therefore spend a considerable part of the paper in giving you our motivation for studying these problems, and also in giving you an idea of the type of physical measurements that need to be made in order to apply the theory.

As we shall presently see, both problems are formally equivalent to the problem of synthesis of nonuniform distributed transmission lines. The end result of our analysis will therefore be an integral equation which should be familiar to most of you. However, the manner in which we derive the equation, using the physical concepts of causality, charge (or mass) conservation, and passivity, is novel and of independent interest. Another novel feature is the inclusion of certain types of dissipative loss and other perturbations into the transmission line equations.

1. THE VOCAL-TRACT PROBLEM

Figure 1 shows a sketch of the side view of a human vocal tract. During nasal speech sounds — e.g., *m, n, ng* — there is an opening into the nasal passage. We will ignore such situations. For our purposes, therefore, the vocal tract is just a bent tube with a nonuniform cross section. To produce a speech sound, we adjust the tongue, jaw, and lips so as to produce the shape appropriate to the speech sound. Then we excite the tube with a sound source. For vowels, the source is the stream of quasi-periodic pulses produced by the vocal cords (glottis). For fricative sounds — e.g., *s, sh, f, th* — the sound source is the turbulent air flow through a constriction. Some sounds — e.g., *v, zh, z* — require a combination of these sources.

For the purposes of studying acoustic wave propagation, the tract is assumed to be straightened out as shown in Fig. 1b, and wave motion is assumed to be planar. These assumptions are justified for frequencies below about 4 kHz. Under these assumptions, the "shape" of the tract at any instant of time is adequately specified by the cross-sectional *area* $C(x)$ as a function of the distance x from the lips. During speech production, of course, C is also a function of time. However, the time variation of C is so slow that the tract can be regarded as assuming a succession of stationary shapes. We will, therefore, talk of $C(x)$ as if it were time-independent.

(a) Ideal Stationary Tract

Assume first that the tube has rigid walls and is filled with a perfect gas with density ρ_0 and sound velocity c. Then, choosing units such that $C(0) = c = \rho_0 = 1$, the pressure $p(x,t)$ and the volume velocity $u(x,t)$ in the tract satisfy the equations

$$\frac{\partial p}{\partial x} = -\frac{1}{C(x)}\frac{\partial u}{\partial t}, \tag{1a}$$

$$\frac{\partial u}{\partial x} = -C(x)\frac{\partial p}{\partial t}. \tag{1b}$$

If the function $C(x)$ is known and the excitation is known, then the speech wave (i.e., the volume velocity at the lips) can be computed by standard methods. Thus, for example, suppose that the excitation is a volume velocity $-g(t)$ at the vocal cords. Then what is needed is a solution of the partial differential equations (1a, 1b) subject to the boundary conditions:

$u(L,t) = -g(t)$ at the glottis,

$p(0,t) = 0$ at the lips.

(The second condition assumes a pressure release boundary at the lips. A more accurate boundary condition would be a termination representing the radiation load.)

This is the *direct* problem of speech production. However, in order to carry out the solution, the area function $C(x)$ must first be estimated. The traditional method of estimating the area function is to make an x-ray movie of the side view of the tract during normal speech production. Combining this information with estimates of the cross-dimensions from castings of the stationary tract, gives $C(x)$ at any instant during a speech sound.

Besides the fact that exposure to x-rays is highly undesirable, the method is extremely laborious and not very accurate, because of the uncertainty in the cross dimensions.

An alternative to this approach is to solve the *inverse acoustical problem*; i.e., to estimate $C(x)$ from the behavior of the solutions of (1a) and (1b) at the boundaries. The obvious first question is: Can $C(x)$ be estimated from the speech wave? The answer is *no, even if the excitation function is known!* (It *is* possible to derive $C(x)$ from the speech wave but only if one is willing to make a number of drastic, unjustified assumptions. See [1] for a discussion of the issues involved.)

However, the area function can be uniquely determined, provided one is willing to make appropriate acoustical measurements at the lips. The first such suggestion [2,3] was based on a frequency domain measurement of the driving point impedance at the lips. (The driving point impedance $Z(s)$ is the Laplace transform of the pressure developed at a point when an impulse of volume velocity is applied at the same point. What is measured is $Z(s)$ for $s = i\omega$.) The poles and zeros of this impedance constitute two infinite sequences of interlaced real eigenvalues (corresponding to the tract being, respectively, closed and open at the lips). Given *all* these eigenvalues, it is well known [4] that $C(x)$ in Eqs. (1a) and (1b) can be uniquely recovered.

It is, of course, not possible to measure the infinite sets of eigenvalues. Even if it *were* possible to measure them, they would be useless, because, as mentioned above, the assumption of planar wave motion is invalid at high frequencies. One can get around this difficulty if the length of the tract and the boundary condition at the glottis are both known. Then the eigenvalues above about 4 kHz are assumed to be those of the uniform tract of the same length with the appropriate boundary condition.

An alternative frequency domain approach [5] recovers $C(x)$ from the poles and residues of the impedance (with a similar assumption about the high frequency information).

The trouble with these frequency domain methods is that neither the length nor the boundary condition at the glottis is known accurately; and each of these strongly affects the reconstructed area function.

The time-domain method, which we will now discuss, recovers $C(x)$ without a knowledge of the length of the tract or of the boundary condition at the glottis. This method is based on a measurement of the impulse response at the lips. (See subsection (c) of this section for a brief description of the measurement procedure.) The impulse response $H(t)$ is the pressure developed at the lips due to an impulse of volume velocity at the lips. (The function $Z(s)$ above is just the Laplace transform of $H(t)$.) Thus,

$$p_0(t) = (H * u_0)(t) \tag{2}$$

where the subscript 0 refers to the lips and $*$ denotes convolution. Now, if the vocal tract were an infinitely long uniform tube, then $H(t) = \delta(t)$. (The strength of the impulse is 1 because of our choice of units.)

Clearly, then, for an arbitrary tract $H(t) = \delta(t) + h(t)$. It can be shown that $h(t)$ is continuous if $C(x)$ is continuously differentiable.

Two rather different, highly physically motivated arguments give $C(x)$, $x \leqslant T$, in terms of $h(t)$, $t \leqslant 2T$.

i) If a volume velocity $f(t)$, $t \geqslant 0$, is applied to the lips, then at $t = a$, the disturbance in the tube travels a distance a (since $c = 1$ in our units). Therefore, from conservation of mass,

$$\int_0^a f(t)\,dt = \int_0^a \rho(x,a)C(x)\,dx, \tag{3}$$

where $\rho(x,t)$ is the density perturbation in the acoustic wave. However, $p = \rho c^2$, so since $c = 1$,

$$\int_0^a f(t)\,dt = \int_0^a p(x,a)C(x)\,dx. \tag{3a}$$

As shown in [6], if $f(t)$ satisfies the integral equation

$$f(t) + \frac{1}{2}\int_0^{2a} h(|t-\tau|)f(\tau)\,d\tau = 1, \quad 0 \leqslant t \leqslant 2a, \tag{4}$$

then $p(x,a) = 1$, $x \leqslant a$. Thus, if $f(t)$ satisfies (4), then

$$\int_0^a f(t)\,dt = \int_0^a C(x)\,dx = V(a). \tag{5}$$

Thus, given $h(t)$, $0 \leqslant t \leqslant 2a$, one solves (4) for $f(t)$. It can be shown that the kernel $\delta(t-\tau) + \frac{1}{2}h(|t-\tau|)$ is positive definite, provided $C(x)$ is finite and bounded away from 0. Positive definiteness guarantees that (4) has a unique solution. Once $f(t)$ has been found, (5) gives the *volume* $V(a)$ up to the distance a. Repeating for $0 \leqslant a \leqslant T$, one gets $V(a)$ and, hence, by differentiation, $C(a)$ as a function of a. The computation can be further simplified because it can be shown that

$$f^2(0) = C(a). \tag{5a}$$

ii) The second physical argument is intuitively more appealing in the electrical analog of (1a) and (1b). Note that if pressure is identified with voltage and volume velocity with current, then (1a) and (1b) become the telegrapher's equations for a transmission line with capacitance per unit length given by $C(x)$ and inductance per unit length by $C^{-1}(x)$.

Suppose we connect a *negative* capacitance $-\Gamma$ to the transmission line at $x = 0$, as shown in Fig. 2. The combined system is in general no longer passive. However, after connection of the capacitance, the system stays passive for a time interval whose duration depends on Γ. During this interval, an arbitrary current $i(t)$ flowing into the circuit *delivers energy to it.*

To see this, note that the input impedance at the terminals in Fig. 2 is $Z(s) - \frac{1}{\Gamma s}$. Therefore, a current $i(t)$ flowing for an interval $0 \leqslant t \leqslant 2a$ and carrying a total charge Q delivers an energy $E(2a)$ given by

$$E(2a) = \int_0^{2a} i^2(t)\,dt + \frac{1}{2}\int_0^{2a}\int_0^{2a} h(|t-\tau|)i(t)i(\tau)\,dt\,d\tau - \frac{Q^2}{2\Gamma}. \tag{6}$$

The $i(t)$ which minimizes this turns out to be precisely the function $f(t)$ which is a solution of (4). And the minimum value of energy corresponding to this current is

$$E_{\min}(2a) = \frac{Q^2}{2}\left[\frac{1}{\int_0^a f(t)\,dt} - \frac{1}{\Gamma}\right]. \tag{7}$$

Thus, the system is stable if

$$\Gamma \geqslant \int_0^a f(t)\,dt. \tag{8}$$

On the other hand, it is intuitively clear (and proven rigorously in [7]) that the system is stable up to $t = 2a$ as long as the series connection of $-\Gamma$ and the cumulative capacitance $V(a)$ of the line up to $x = a$ positive. That is, as long as

$$\frac{1}{V(a)} - \frac{1}{\Gamma} \geqslant 0. \tag{9}$$

The inequalities (8) and (9) again give (5).

(b) Tract with losses and yielding walls

When the tract has dissipative losses or if the walls are not rigid, then (1a) and (1b) no longer apply. To introduce these effects, it will be convenient to take Laplace transforms. Let $P(x,s)$, $U(x,s)$ represent the Laplace transforms of $p(x,t)$, $u(x,t)$, respectively. Then, for the tract with losses and yielding walls, P and U satisfy

$$\frac{\partial P}{\partial x} = -\left[\frac{s}{C(x)} + z(x,s)\right]U, \tag{10a}$$

$$\frac{\partial U}{\partial x} = -\,[sC(x) + y(x,s)]P. \tag{10b}$$

Physically, $z(x,s)$ represents viscous losses in the air, per unit length of the tube, and $y(x,s)$ represents the wall admittance per unit length.

It is easy to see that in this case input measurements alone cannot recover $C(x)$. This is because the impulse response in this case is again of the form $\delta(t) + h(t)$ and the corresponding kernel is positive definite. There is, therefore, no way of telling whether the impulse response is that of an ideal tube or of a tube with losses and/or yielding walls.

Suppose, however, that $z(x,s)$ and $y(x,s)$ are known, or their functional forms are known. Is it possible then to recover $C(x)$ from the measured $H(t)$? In the following three cases we know how.

(i) If

$$z(x,s) = \frac{1}{C(x)}\alpha(s), \tag{11a}$$

$$y(x,s) = C(x)\beta(s), \tag{11b}$$

and $\alpha(s)$ and $\beta(s)$ are known, then [8] $C(x)$ can be uniquely recovered. Although these functional forms for $z(x,s)$ and $y(x,s)$ appear to be rather contrived, it can be shown [9] that the vocal tract is accurately modeled this way, with $\alpha(s) = 0$ and $\beta(s) = \frac{b}{s+\epsilon}$, and the quantities b and ϵ can be estimated from a variety of experimental data.

For $z(x,s)$ and $y(x,s)$ given by (11a) and (11b), it is fairly straightforward to derive a relation between the measured input impedance $Z(s)$ and the input impedance $\hat{Z}(s)$ *of the same tract but with* $\alpha = \beta = 0$. This relation turns out to be

$$\hat{Z}(s) = \sqrt{\frac{s+\alpha(s)}{s+\beta(s)}}\; Z(\sqrt{[s+\alpha(s)]\,[s+\beta(s)]}). \tag{12}$$

It is important to realize that (12) implies a causal, invertible relation between the corresponding impulse responses, $H(t)$ and $\hat{H}(t)$. Thus, T seconds of $H(t)$ yields $\hat{H}(t)$ for T seconds. Once $\hat{H}(t)$ is known, $C(x)$ is computed as in Sec. 1(a).

(ii) If $z(x,s) = 0$ and $y(x,s) = \frac{1}{sm(x)}$ (i.e., a lossless tract whose walls have mass $m(x)$ per unit length), then the problem is solvable for any given $m(x)$. In this case [8], if $D(x)$ is the area function obtained by the method of Sec. 1(a) from the measured $H(t)$, then the true area function $C(x)$ satisfies

$$\frac{\sqrt{C}''}{\sqrt{C}} + \frac{1}{m(x)\sqrt{C}} = \frac{\sqrt{D}''}{\sqrt{D}} \tag{13}$$

with $C(0) = D(0)$, and $C'(0) = D'(0)$.

(iii) If $z(x,s) = F(x,C(x))$ and $y(x,s) = G(x,C(x))$, with the functions $F(.,.)$ and $G(.,.)$ known, then $C(x)$ can be determined [10]. Note that this is a very general distribution of resistive and conductive losses. The algorithm is too complicated to summarize here. It is to be described (along with generalizations and experimental results) in a forthcoming dissertation [11].

(c) Measurements

Figure 3 shows a sketch of the impedance tube used for the measurement of impulse response. The sound source produces short duration impulses of volume velocity and the wedge speeds up the return to quiescent conditions. The person holds the flexible coupler to the lips and moves the vocal tract as in normal speech without phonating. The apparatus is an adaptation of the arrangement first used for frequency-domain measurements [2,3]. In one respect, the time-domain measurement is more exacting because it is harder to generate impulsive excitation than sinusoidal. On the other hand, the wedge need not be reflectionless; it need only ensure that sound energy from one pulse has died out adequately before another pulse is applied. This makes the wedge design much simpler.

The hard-walled impedance tube transmits acoustic waves undistorted in either direction. Therefore, the microphone measurements can be translated to give the corresponding measurements at $x = 0$. The point $x = 0$ is chosen to be inside the tube (rather than exactly at the lips); therefore, as assumed in the derivations above, $C(0)$ is known.

For a typical vocal tract of length about 17 cm and the velocity of sound about 34 km/sec, the round trip travel time from lips to glottis is about a millisecond. Therefore, a measurement of $h(t)$ for about 1.5 msec is more than adequate to recover the shape of the entire tract; and it is quite feasible to design the wedge to allow measurements once every 20 msec or so. Thus, it is possible to make a 50 frame/sec movie displaying the variations of $C(x)$ during normal speech. I made such a movie for a number of spoken sentences [12] and demonstrated the feasibility of this procedure. However, at the time the conclusion was that the experiment needed to be performed with much greater accuracy than I had been able to attain. More recently, several groups have started making such measurements [11,13,14,15]. It is to be hoped that these efforts will lead to accurate estimates of dynamically varying vocal tract shapes.

2. THE BASILAR MEMBRANE PROBLEM

We turn now to a problem that arises at the other end of the "speech chain" — the inner ear. The successive panels of Fig. 4 show the organ of hearing in progressively increasing detail. The motion of the ear drum in response to an impinging sound wave is transmitted, via a mechanical linkage of three small bones, to a snail-shaped organ called the cochlea. If the cochlea could be unrolled, it would be a fluid-filled conical tube, longitudinally partitioned into three chambers. The basilar membrane, with which the present discussion is concerned, is one of the partitions. The other dividing membrane, called Reissner's membrane, will be ignored here; it appears to be acoustically transparent, although it serves the important function of electrically isolating the fluids on its two sides.

The motion of the stapes (which is the last of the bones in the linkage) pushes fluid in and out at the oval window. The fluid motion produces motion of the basilar membrane perpendicular to its resting position. This motion is sensed by hair-like endings of nerve cells (see Fig. 4); and it is the changes in the firing patterns of these nerve cells that are ultimately interpreted as the sensation of sound by the brain. An accurate description of the motion of the basilar membrane is, therefore, a necessary first step toward the understanding of the mechanism of hearing.

It has been known for quite some time that if a sinusoidal velocity is imparted to the stapes, then the displacement of the basilar membrane shows resonance behavior. The displacement has maximum amplitude at a certain point along the membrane. This point of maximum displacement moves closer to the stapes as the frequency of the excitation increases. Accurate measurement of this motion is a rather difficult task. Some idea of the difficulty is conveyed by the dimensions involved. In human cochleas, the entire length of the basilar membrane is about 35 mm, the cross dimension of the chamber on either side of it is 1-3 mm, and the displacement of the basilar membrane due to a typical sound wave is less than a micron. Further, as mentioned above, this structure is rolled up into a spiral and embedded in bone, which makes it highly inaccessible. In spite of these difficulties, several ingenious experiments were conducted during the 1940s and 1950s on cochleas excised from cadavers. Many measurements were made and models proposed to explain the resonance phenomena. (See [16] for a review of this work.) However, it is only in recent years that accurate measurements of its motion have been made *in vivo*, using sophisticated measuring aids such as capacitance probes, the Mossbauer effect and laser interferometry. These refined measurements have provided the impetus for more refined models of cochlear mechanics [17, 18,19]. Our discussion will be based on the formulation of [19].

The "unrolled" cochlea is shown schematically in Fig. 5a. To a good approximation, the outer walls of the conical tube can be assumed rigid and the fluid filling it may be assumed incompressible. The cross-sectional area varies by a factor of about 4 or 5 along the length of the tube. As we shall see presently, this is a very small variation in comparison with the range of variation of the mechanical properties of the basilar membrane. Therefore, we will idealize the conical tube to a rectangular tube, as shown in Fig. 5b. Also, since the major components of fluid motion are in the x- and y-directions in Fig. 5b, we will ignore variations of quantities in the z-direction (i.e., perpendicular to the plane of the paper). Finally, we will assume perfect symmetry. That is, at two points located symmetrically with respect to the membrane, we will assume the y-components of velocity to be identical and the x-components to be equal in magnitude and opposite in direction. With these idealizations, we might just as well replace the partitioned chamber of Fig. 5b by the single rectangular box shown in Fig. 5c. That figure also shows the coordinate system, as well as the length L and the height H of the box.

It is generally assumed that the membrane is locally reacting, i.e., its velocity at any point depends only on the pressure across it at that point. As long as the motion is linear, therefore, the membrane is represented by a wall with an impedance $Z(x,s)$. This impedance is well approximated by a damped spring-mass oscillator, i.e.,

$$Z(x,s) = sm(x) + \frac{K(x)}{s} + R(x). \tag{14}$$

Estimates are that K varies by a factor of over 100,000:1 from one end to the other, and R by a factor of over 300:1. On the other hand, m is more or less constant.

With these preliminaries, let us sketch the derivation of the equation of basilar membrane motion. Since the fluid is incompressible, the pressure $P(x,y,s)$ satisfies Laplace's equation

$$\nabla^2 P = 0, \tag{15}$$

subject to the boundary conditions

$$\frac{\partial P}{\partial y} = \begin{cases} -\rho s V, & y = 0, \\ 0 & y = H, \end{cases}$$

$$\frac{\partial P}{\partial x} = -\rho s U_0, \quad x = \pm L.$$

Here, ρ is the fluid density, U_0 is the velocity applied to the stapes, and $V(x,s)$ is the velocity of the basilar membrane. This problem has a unique solution for P in terms of U_0 and V. The solution is facilitated by a conformal map of the rectangle to the upper half-plane by means of an elliptic function. The result, specialized to $y = 0$, is

$$P(x,0,s) = -\rho s U_0 x - \int_{-L}^{L} F(|x-x'|)\, V(x',s)\, dx'. \tag{16}$$

The function $F(x)$ is expressed in terms of elliptic functions [19]. However, for the dimensions of the cochlea, it turns out that, to a very good approximation,

$$F(x) = \frac{\rho s}{2H}(|x| - L) + \frac{\rho s H}{3}\delta(x).$$

Substituting for $F(x)$ in (16) gives

$$P(x,0,s) = -\rho s U_0 x - \frac{\rho s}{2H}\int_{-L}^{L} |x-x'|\, V(x',s)\, dx' + \frac{\rho s H}{3} V(x,s).$$

However, $P(x,0,s) = -Z(x,s)\, V(x,s)$. Thus,

$$\hat{Z}(x,s)\, V(x,s) = \rho s U_0 x + \frac{\rho s}{2H}\int_{-L}^{L} |x-x'|\; V(x',s)\, dx', \tag{17}$$

where we have defined $\hat{Z} = Z + \rho s H/3$. Equation (17) is an integral equation for V, whose solution relates V to the input velocity U_0. If we define $\hat{P} = \hat{Z}V$ and differentiate (17) twice with respect to x, we get

$$\frac{d^2\hat{P}}{dx^2} = \frac{\rho s}{H\hat{Z}}\, \hat{P}. \tag{18}$$

This is recognized as a canonical form of the transmission line equation, with somewhat unusual series and shunt elements. Fortunately, it turns out that $\hat{Z}$ is well approximated by $\hat{Z} \approx K(x)/s$ up to reasonably high frequencies at the stapes. This is because $K(x)$ is greatest near the stapes, and also the stapes is far away from the resonance point for most frequencies of interest. With this approximation, (18) becomes

$$\frac{d^2\hat{P}}{dx^2} = \frac{\rho s^2}{HK(x)}\, \hat{P}. \tag{19}$$

It is well known that this equation is equivalent to the second order equation for pressure for the lossless vocal tract. To see this, note that eliminating U from (10a) and (10b) in the lossless case ($z = y = 0$) gives

$$\frac{d}{dx} C \frac{d}{dx} P = s^2 CP. \tag{20}$$

Transform x to the independent variable ξ defined by

$$\xi = \int_0^x \frac{d\lambda}{C(\lambda)}.$$

Then, if $P(x) = Q(\xi)$, (20) becomes

$$\frac{d^2}{d\xi^2} Q = s^2 C^2(x(\xi)) Q, \tag{21}$$

which has exactly the same form as the basilar membrane equation (19). Thus, the theory of Sec. 1 above can be applied to give $K(x)$ in terms of impulse response measurement at the stapes.

REFERENCES

[1] M.M. Sondhi: "Estimation of vocal tract areas: the need for acoustical measurements," *Proceedings of the Symposium on Articulary Modelling and Phonetics*, Grenoble, France, 1977, pp. 77-88. (To be reprinted in IEEE Trans. on Acoustical Speech and Signal Processes, June 1979)

[2] M.R. Schroeder: "Determination of the geometry of the human vocal tract," J. Acoust. Soc. Am. **41**, 1002-1010 (1967)

[3] P. Mermelstein: "Determination of the vocal-tract shape from measured formant frequencies," J. Acoust. Soc. Am. **41**, 1283-1294 (1967)

[4] G. Borg: "Eine Umkehrung der Sturm-Liouvilleschen Eigenwertaufgabe," Acta Math. **78**, 1-97 (1954)

[5] B. Gopinath, M.M. Sondhi: "Determination of the shape of the human vocal tract from acoustical measurements," Bell. Sys. Tech. J. **49**, 1195-1214 (1970)

[6] M.M. Sondhi and B. Gopinath: "Determination of vocal-tract shape from impulse response at the lips," J. Acoust. Soc. Am. **49**, 1867-1973 (1971)

[7] B. Gopinath, M.M. Sondhi: "Inversion of the telegraph equation and the synthesis of nonuniform lines," Proc. IEEE **59**, 383-392 (1971)

[8] M.M. Sondhi, B. Gopinath: "Determination of the shape of a lossy vocal tract," *Proceedings of the International Congress on Acoustics*, Budapest, Hungary, 1971, paper 23CIO

[9] M.M. Sondhi: "Model for wave propagation in lossy vocal tract," J. Acoust. Soc. Am. **55**, 1070-1075 (1974)

[10] J.R. Resnick, J.M. Heinz: "Vocal tract area function determination in the presence of spatially-varying losses," J. Acoust. Soc. Am. **62**, S28 (1977) (abstract of talk)

[11] J.R. Resnick: forthcoming dissertation, Johns Hopkins University (1979)

[12] M.M. Sondhi: "Experimental determination of the area function of a lossy dynamically varying vocal tract," J. Acqust. Soc. Am. **53**, 294 (1973) (abstract of talk)

[13] R. Descout, B. Tousignant, M. Lecours: "Vocal-tract area measurements: two time-domain methods," *Record of IEEE International Conference on Acoustical Speech and Signal Processes*, Philadelphia, April 12-14, 1976, pp. 75-78

[14] B. Tousignant, J-P. Lefevre, M. Lecours: "Speech synthesis from vocal tract area function acoustical measurements," *Record of IEEE International Conference on Acoustical Speech and Signal Processes*, Washington, D.C., April 2-4, 1979, pp. 921-924

[15] J. Blauert's group at the University, Bochum, Germany

[16] M.R. Schroeder: "Models of hearing," Proc. IEEE **63**, 1332-1350 (1975)

[17] M.B. Lesser, D.A. Berkley: "Fluid mechanics of the cochlea. Part I," J. Fluid Mech. **51**, 497-512 (1972)

[18] J.B. Allen: "Two-dimensional cochlear fluid motion: new results," J. Acoust. Soc. Am. **61**, 110-119 (1977)

[19] M.M. Sondhi: "Method of computing motion in a two-dimensional cochlear model," J. Acoust. Soc. Am. **63**, 1468-1477 (1978)

[20] J.L. Flanagan: *Speech Analysis Synthesis and Perception* (Springer, New York, 1972), 2nd. ed.

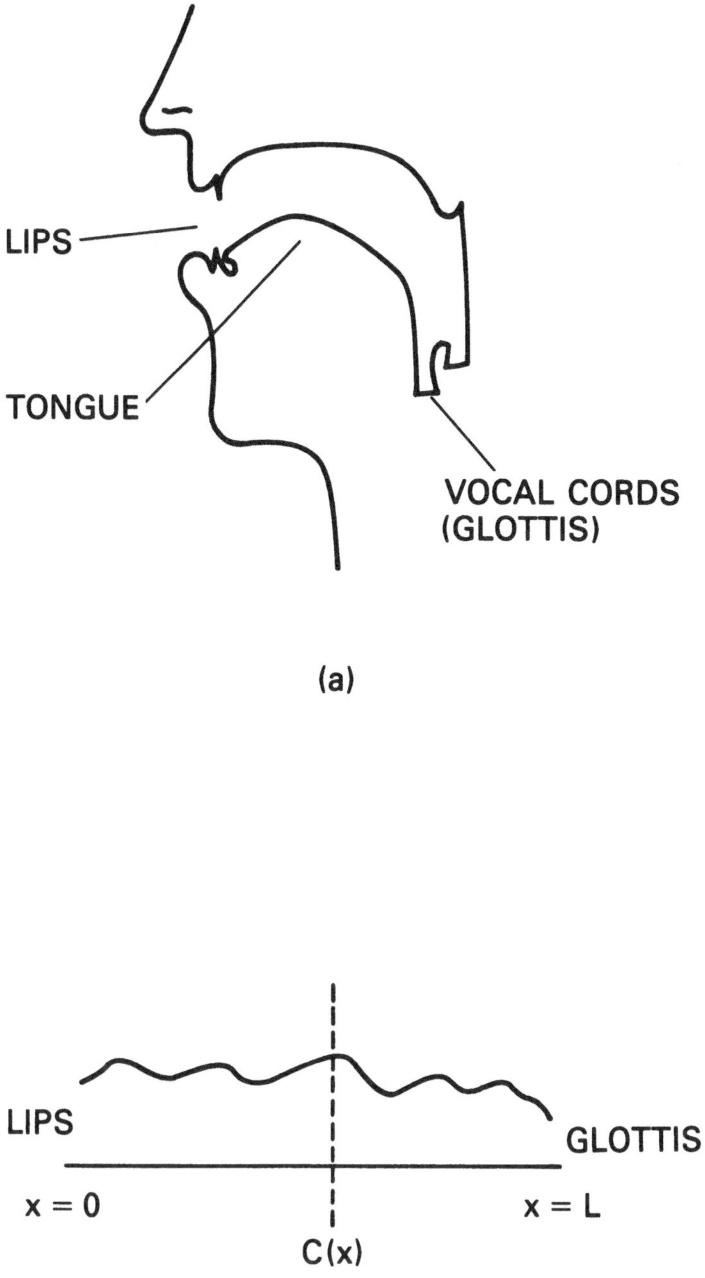

Fig. 1. (a) Side view of the vocal tract; (b) idealization to a straight tube with variable cross-sectional area.

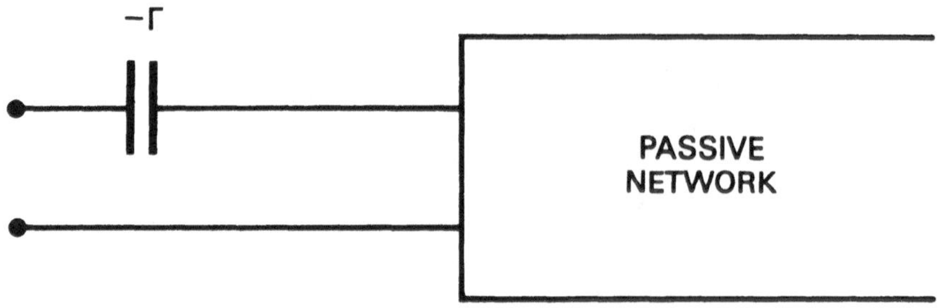

Fig. 2. The passive network (transmission line) seen through a negative capacitance.

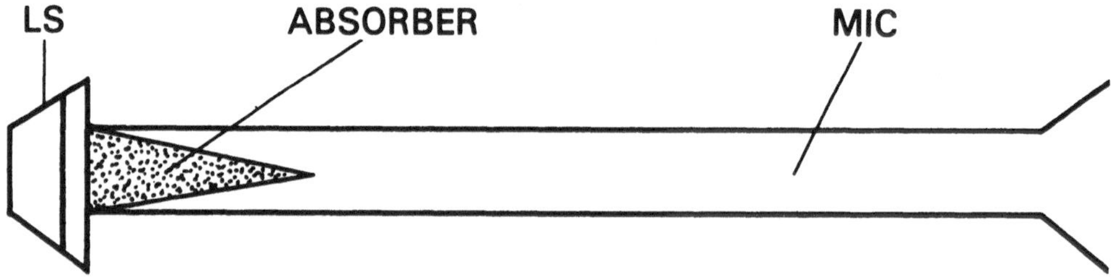

Fig. 3. One form of the impedance tube used to measure impulse response.

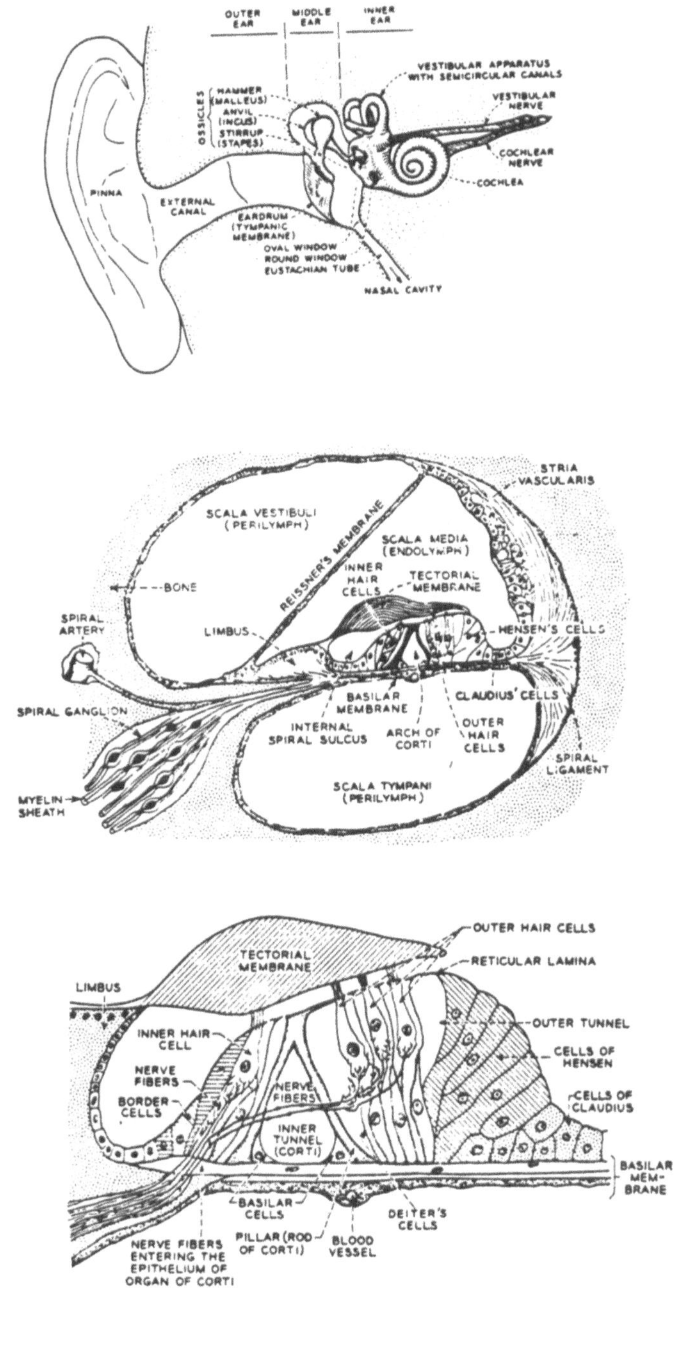

Fig. 4. (a), (b), (c): Diagrams of the hearing organ in successively increasing detail (after Flanagan [20]).

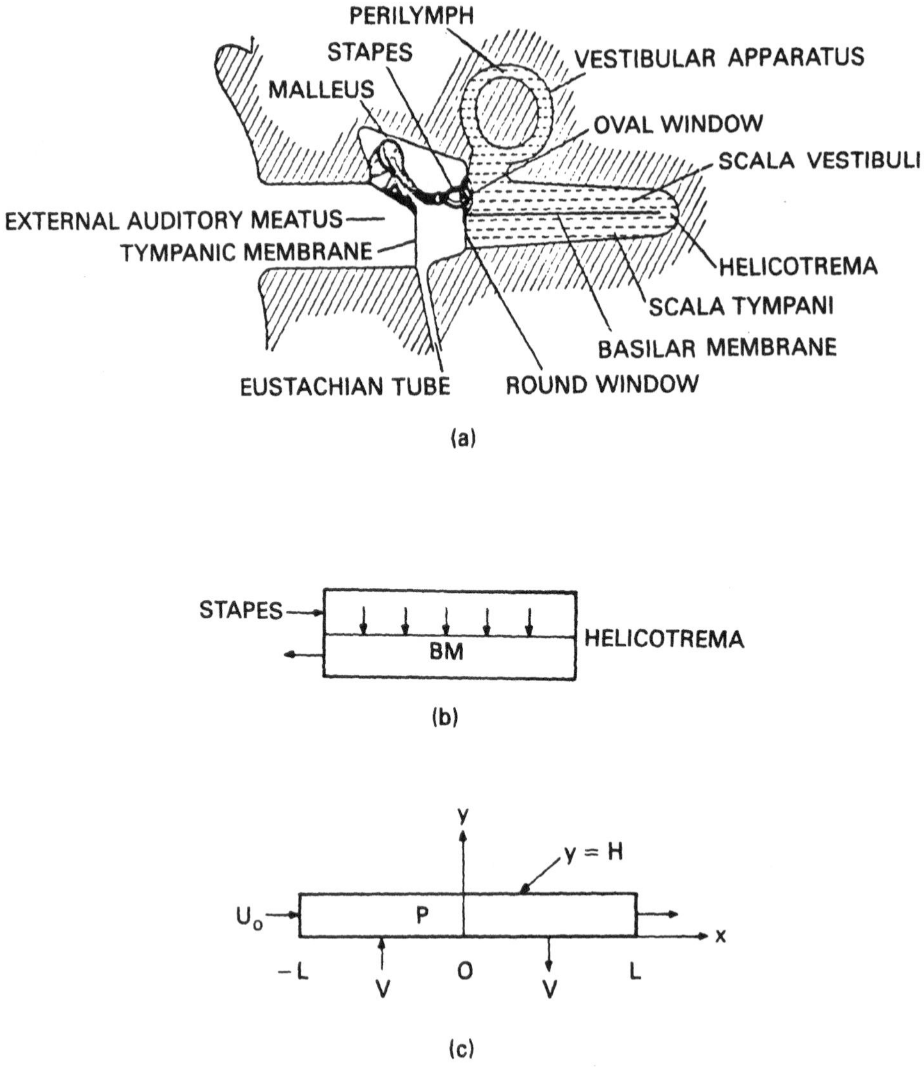

Fig. 5. (a) The "unrolled" cochlea; (b) idealization of the cochlea to a rectangular chamber with a yielding partition; (c) equivalent single-chamber representation.

DEVELOPMENT OF PHYSICAL OPTICS INVERSE SCATTERING TECHNIQUES USING RADON PROJECTION THEORY

Wolfgang-M. Boerner

Communications Laboratory, Department of Information Engineering
University of Illinois at Chicago Circle
Chicago, Illinois 60680

INTRODUCTION

Recent advances in broadband antenna theory and technology have made possible monostatic and bistatic polarization radar systems with sufficient isolation between the co- and cross-polarized channels for both linear and circular polarization measurement over a wide band of harmonic frequencies. Since most of the applicable microwave imaging and target identification schemes are based on high frequency approximations, those techniques based on Physical Optics Far Field Inverse Scattering will be restudied applying Radon's theory of image reconstruction from projections. In particular, there still exist various unresolved questions relating to the band-limited, aspect-direction-limited, sparse-data case which can best be treated in the Fourier-Radon projection space.

1. THE PHYSICAL OPTICS APPROXIMATION

The development of electromagnetic inverse scattering theories applicable to practical radar imaging of isolated, perfectly conducting, smooth, and closed-shape scatterers derived from the Physical Optics (PO) approximation

$$\mathbf{J}(\mathbf{r},t)_{PO} = 2\hat{\mathbf{n}} \times \mathbf{H}_i(\mathbf{r},t)\ , \tag{1}$$

with first order correction terms [1]

$$\mathbf{J}(\mathbf{r},t)_{PO_1} = \frac{1}{2\pi} \iint_{S_{ill}} \hat{\mathbf{n}} \times \left\{ \left[\frac{1}{R^2} + \frac{1}{c_0 R} \frac{\partial}{\partial t} \right] \mathbf{J}_{PO}(\mathbf{r},t) \times \hat{\mathbf{a}}_r \right\} dS, \tag{2}$$

is not complete. A succinct introduction in the historical development sequence of various Physical Optics Far Field Inverse Scattering (POFFIS) identities is presented here with the specific objective to isolate still existing open problems [2, Chap. 5].

2. THE RAMP RESPONSE IDENTITY

One of the first POFFIS identities was discovered in the fifties at ESL-OSU and was documented first in [3]. It showed that the target ramp response to $F_R(t')$ is proportional to the target silhouette area $A(r')$ along the radar line of incidence $r' = ct'/2$, where

$$F_R(t') = \frac{1}{\pi c^2} A(r')\big|_{r'=ct'/2}, \quad \text{for } 0 \leqslant r' \leqslant r_0'. \tag{3}$$

Here, $r' = +O$ identifies the tip of the target for which $A(+O) = 0$, $r = r_0'$ the shadow boundary for which $A(r_0') = A_{\max}$, and $r' = l$ the tip in the shadow region, respectively; and (3) is not valid in the umbra region $r' > r_0'$. This temporal relation, which was stated in the literature much later also for the radar case [4] and for sonar applications [5], was first utilized in radar target imaging by KENNAUGH and MOFFATT [6, 1965] and in Fourier Transform Imaging by KELLER [7]. This identity, applied as formulated, can only be used to recover the weighted shape of axially symmetric targets within the illuminated region for incidence along the invariant axis of a rotationally symmetric target. It also requires low-frequency-regime data for reconstructing the target and response [6], thus apparently violating the inherent PO assumption. A rather intuitive attempt using a three-orthogonal look-angle approach by YOUNG [8] is not unique, but may yet be useful in developing approximate target-portrayal discrimination techniques. It also should be noted here that the formulation of (3), in essence, provided a means to introduce the concept of a target's *unique* set of natural frequencies [9]. Yet, for true three-dimensional image reconstruction of three-dimensional arbitrary, nonsymmetrical-shaped targets, a generalization of (3) must be sought or another entirely different approach may be required.

developing approximate target-portrayal discrimination techniques. It also should be noted here that the formulation of (3), in essence, provided a means to introduce the concept of a target's *unique* set of natural frequencies [9]. Yet, for true three-dimensional image reconstruction of three-dimensional arbitrary, nonsymmetrical-shaped targets, a generalization of (3) must be sought or another entirely different approach may be required.

3. THE SPATIAL FREQUENCY κ-SPACE IDENTITY

A spatial frequency domain POFFIS identity was developed in the early sixties and seems to have been documented first by BOJARSKI [10] relating the monstatic scalar complex field cross-section $\rho(\kappa)$ in Fourier space $\kappa = -2k\hat{\mathbf{a}}_{inc}$

$$\rho(\kappa) = \frac{j}{2\sqrt{\pi}} \iint_{\kappa\cdot\hat{\mathbf{a}}_n>0} \kappa\cdot\hat{\mathbf{a}}_n \exp(-j\mathbf{x}\cdot\kappa)dS, \tag{4}$$

to the characteristic target-shape function

$$\gamma(\mathbf{x}) = (2\pi)^{-3} \iiint \Gamma(\kappa)\exp(j\kappa\cdot\mathbf{x})d^3\kappa, \tag{5}$$

where

$$\Gamma(\kappa) = 2\sqrt{\pi}\,\frac{\rho(\kappa)+\rho^*(-\kappa)}{|\kappa|^2} = \iiint \exp(-j\kappa\cdot\mathbf{x})d^3\mathbf{x} = \iiint \gamma(\mathbf{x})\exp(-j\kappa\cdot\mathbf{x})d^3\mathbf{x}, \tag{6}$$

and where use was made of joining the two complementary sets of monostatic-field cross sections $\rho(\kappa)$ and $\rho^*(-\kappa)$ for κ and $(-\kappa)$ across an idealized smooth shadow boundary [2, Chap. 5.2]. This κ to $\mathbf{x}$-space Fourier transform identity states that if the backscattered field could be measured in amplitude and relative phase at all frequencies $\omega = kc$ and at all aspects $\kappa/|\kappa|$, then $\Gamma(\kappa)$ would be known for all κ, and (4)-(6) would yield a self-consistent solution of the inverse diffraction problem for a perfectly conducting scatterer in the PO-limit [11]. A bistatic extension of this scalar POFFIS identity was attempted by ROSENBAUM-RAZ [12] which is inconsistent with the monostatic-bistatic equivalence theorem of KELL [13].

4. THE RADON PROJECTION SPACE (ξ, q) FORMULATION

The construction of $\Gamma(\kappa)$ from two complementary monostatic scalar-field cross sections according to (6) suggests from the formal derivation of (3) given by BOERNER [2, Chap. 5.1] that a strict transform relation between (3) and (4)-(6) must exist. Such a relation was established by DAS and BOERNER [14, 1978] with the aid of the projection transformation theory of RADON [15]. Similar to constructing the complete augmented cross section $\Gamma(\kappa)$ from two complementary monostatic cross sections $\rho(\kappa)$ and $\rho^*(-\kappa)$ respectively (view angles $\kappa/|\kappa|$ and $(-\kappa)/|\kappa|$ are considered complementary), the silhouette area functions $A(r')$ and $A(-r')$ for two complementary view angles need to be curve-fitted across the shadow boundary at $r' = r_0'$ so that the complete projection $A_c(r')$ is thus obtained from $A(r')|r'_0$ (extracted from $F_R(t',\kappa/|\kappa|)$), and $A(-r')|r_0'$, (extracted from $F_R(t',-\kappa/|\kappa|)$), and thus extending across the length of intersection from $r' = 0$ across the shadow boundary at $r' = r_0'$ to the other specular point in the shadow region at $r' = l$, so that

$$A_c(r') = A(r')|_0^{r_0} + A(-r')|_l^{r_0}. \tag{7}$$

Truncation of $F_R(t')$ at $r' = r'_0$, which is still unresolved for the inverse problem, and curve fitting the data across the shadow boundary will introduce an error that has not been analyzed properly in the literature. It should be noted that for the rotationally symmetric target case with nose-on incidence MOFFATT and YOUNG [16] obtained rather satisfactory results by extending $A(r')$ from r_0' to l, i.e., utilizing $F_R(t')$ up to its first zero crossing. Major additional analyses are required [17].

In Radon projection space (ξ: unit vector in direction of incidence $\kappa/|\kappa|$; q: Euclidean distance along projection line r') the Radon transform $\hat{\gamma}(\xi,q)$ of the characteristic silhouette function $\gamma(\bar{x})$ in image space, as defined in (5), is given by [2, Chap. 5.3]

$$\gamma(\xi,q) = \iiint_\infty \gamma(\mathbf{x})\delta(q-(\xi\cdot\mathbf{x}))d^3\mathbf{x}, \tag{8}$$

which relates to its Fourier transform $\tilde{\gamma}(\alpha\xi)$ in α space [2, Chap. 2.3.3] as

$$\hat{\gamma}(\xi,q) = (2\pi)^{1/2} \int_{-\infty}^{\infty} \tilde{\gamma}(\alpha\xi)e^{j\alpha q}d\alpha, \tag{9}$$

where α is a scalar variable denoting reciprocal space radius. For the particular aspect direction $\kappa = (0,0,\kappa')$, i.e., projection for a particular (ξ,q) along r', $\alpha = \kappa'$, the relation between $\hat{\gamma}(\xi,q)$ and $A_c(r',\kappa/|\kappa|)$, as well as that between $\gamma(\kappa'\xi)$ and $\Gamma(\kappa)$, needs to be established. For targets with rotational symmetry, the relation was established in [14] and for the general nonsymmetrical case in [17], where it is shown [2, Chap. 5.3] that

$$\hat{\gamma}(\xi,q) = A_c(r',\kappa/|\kappa|) \tag{10}$$

and

$$\tilde{\gamma}(\kappa'\xi) = \Gamma(\kappa) \text{ for } \kappa = (0,0,\kappa'). \tag{11}$$

Thus, we have established that the complemented Kennaugh-Cosgriff identity (3) and the Bojarski-Lewis identity (4)-(6) establish a Radon-Fourier transform pair [18]. This relation requires further exhaustive analysis, particularly with respect to the complementation and curve-fitting procedures adopted in constructing $A_c(r')$ and $\Gamma(\kappa)$ in (3)-(11).

5. THE LIMITED-APERTURE PROBLEM AND ITS SOLUTION

Since in practice data are given or usually known only over a subset of κ-space, the Fourier transform of (5) can no longer be used to obtain $\Gamma(\kappa)$ directly. It was LEWIS [11] who showed if D is the limited region in which $\Gamma(\kappa)$ is known, one may choose $K(\kappa)$ being zero outside Q and nonzero inside so that $K(\kappa)\Gamma(\kappa) \equiv \tilde{\gamma}(\kappa)$ and

$$f(\mathbf{x}) = (2\pi)^{-3} \iiint_{-\infty}^{\infty} F(\kappa)\exp(j\kappa\cdot\mathbf{x})d^3\kappa. \tag{12}$$

Lewis shows that if (12) can be solved, the size *ka* and the shape of the target can be recovered, at least within D. However, LEWIS [11, 1969], and then more rigorously PERRY [19,20], clearly demonstrated the ill-posed nature of (12), although TABBARA [21,22] was able to reconstruct with striking accuracy the shape and the electrical size *ka* of a conducting sphere using only low-frequency data, which in itself is violating the underlying PO assumption (1). It should be noted here that this perplexing result may yet be due to projecting a given solution, though assumed to be unknown, back onto itself by virtue of using its *spherical* wave expansion [23] which ultimately converges into its *spherical* dominant momentum *ka* [24].

A unique solution for the band and aspect-limited ($\kappa_- < |\kappa| < \kappa_+, \Omega < \kappa|\kappa|$), ill-posed case of (12) was given by MAGER and BLEISTEIN [25] making use of the directional derivative concept introduced by MAJDA [26]. The Fourier transform $\tilde{\Delta}(\kappa,\hat{\mathbf{p}})$ of the directional derivative $\Delta(\mathbf{x},\hat{\mathbf{p}})$ of (3) across the bounding surface S of B in direction $\hat{\mathbf{p}}$, with $\hat{\mathbf{n}}$ the local outward unit normal to S, may be expressed as

$$\tilde{\Delta}(\kappa,\hat{\mathbf{p}}) = \iint_S \hat{\mathbf{p}}\cdot\hat{\mathbf{n}} \exp(-j\kappa\cdot\xi)dS(\xi). \tag{13}$$

Using a multidimensional stationary-phase evaluation [25], the reconstructed directional derivative $\Delta_K(\mathbf{x},\hat{\mathbf{p}})$ over the limited $K(\kappa)$ aperture defined in (12) becomes

$$\mathrm{Re}\{\Delta_K(\mathbf{x},\hat{\mathbf{p}}(\xi_s))\} \cong \frac{\hat{n}(\xi_s)\cdot\hat{p}(\xi_s)}{2\pi D^{1/2}} \sin\{|\kappa(\mathbf{x}-\xi_s)|\}|_{\kappa_-}^{\kappa_+}. \tag{14}$$

This identity states that the real part of $\Delta_K(\mathbf{x},\hat{\mathbf{p}})$ behaves as a band-limited delta function with a central lobe peaking on the target surface S in regions with surface normals $\hat{n}$ falling within the family of aperture directions, with the height of the central lobe being proportional to $(\kappa_+ - \kappa_-)$ and the width to $2\pi/\kappa_+$, with $\kappa_+ >> \kappa_-$. Thus, the specular target regions are identifiable with a resolution that depends directly on the high-frequency character of the data and on the bandwidth of information available, i.e., the ill-posedness of (12) can be avoided to obtain a solution for the band- and aspect direction-limited case. However, the case of aspect-sparse information still needs to be considered in depth utilizing the Radon projection theory [27].

6. POLARIZATIONAL CORRECTION

The POFFIS method suffers one critical deficiency in that $A(r')$ can be recovered with sufficient accuracy from $F_R(t')$ only up to the shadow boundary at $r' = r_0'$ for which $A(r' = r_0') = A_{\max}$ becomes the peak transverse cross section in the evaluation of (3). The formulation of $\Gamma(\kappa)$ from $\rho(\kappa)$ in (5) and $\hat{\gamma}(\hat{\xi},q)$ from $A(r')$ in (7) requires matching of RCS response data across the shadow boundary using complementary data sets along one radar line when viewed in the two opposite directions. A systematic approach to optimal matching across the shadow boundary for general nonsymmetrical shapes is nonexistent and mechanisms [6,8] developed for axially symmetric cases are very heuristic and require thorough additional analysis. Diffraction processes in the penumbra region are highly polarization-dependent and so are those for regions of nonidentical principal radii of curvature. Since the PO approximation of (1) is polarization-independent, depolarization effects of nonrotationally symmetric bodies will strongly affect the accuracy of POFFIS shape reconstruction. Therefore, it will be necessary to develop a vector extension of scalar POFFIS utilizing available data of co/cross polarized components of the target RCS polarization scattering matrix [28] and of first-order vector corrections of the temporal PO approximation (1) which were derived in [29]. Using the notation established in [30], (1) may be extended to $\mathbf{J}_{ex} - \mathbf{J}_{PO}+\mathbf{J}_{PO_1}$, where the related far-field response for (1) is given by (3) or by

$$\mathbf{H}_{PO} = \frac{1}{2\pi}\frac{d^2A(r')}{dr'^2}(H_{i_1}\hat{\mathbf{a}}_1 + H_{i_2}\hat{\mathbf{a}}_2)\frac{1}{r}, \tag{15}$$

and that for (2) becomes

$$\mathbf{H}_{PO_1} = \frac{1}{2\pi}\frac{dA(r')}{dr'}(H_{i_1}\hat{\mathbf{a}}_1 - H_{i_2}\hat{\mathbf{a}}_2)\frac{(K_1-K_2)}{r}, \tag{16}$$

where $\hat{\mathbf{a}}_1$, $\hat{\mathbf{a}}_2$ are orthogonal unit vectors along the incident magnetic fields and K_1,K_2 denote the associated principal curvatures along $\hat{\mathbf{a}}_1$ and $\hat{\mathbf{a}}_2$. Whereas (15) results by integration in the ramp-response identity (3), (16) becomes a step-response identity [1],

$$\mathbf{F}_u(t') = \frac{1}{2\pi c}A(r')(H_{i_1}\hat{\mathbf{a}}_1 - H_{i_2}\hat{\mathbf{a}}_2)\frac{(K_1-K_2)}{r}. \tag{17}$$

From inspection of (15), (16), and (3), we find that the correction terms are certainly nonnegligible if the principal curvatures differ appreciably, in which case the POFFIS identities (3)-(11) become highly inadequate. The antisymmetric nature of (15) and (16) also shows that, even if the cross-polarized components are neglected in applying POFFIS, different results in shape reconstruction are obtained if incidence is chosen purely along $\hat{\mathbf{a}}_1$ or purely along $\hat{\mathbf{a}}_2$ for nonsymmetrical target shapes, as was demonstrated beyond doubt in [6,8,30]. Thus, instead of simplistically generalizing a scalar-wave approach to a vector wave solution [31], we need to revisit the highly complex diffraction processes of skew incidence on nonsymmetrical targets, and relevant relations are reviewed next. Similarly, we need to reevaluate the performance abilities of most recently developed broadband radar systems with dual polarization facility for extracting useful coherent target information from incoherent clutter-perturbed data.

7. PENUMBRA DIFFRACTION AND HF INVERSE SCATTERING

In analyzing the validity of various scattering theories applicable to profile reconstruction of closed, perfectly conducting shapes, derived from the physical optics or the geometrical optics approximations, it is necessary to establish the field properties in the transition region about the shadow boundary and deep into the umbra region [2, Chap. 4.4]. FOCK [32], in analyzing the exact integral equation for the induced surface currents, showed that for HF scattering the current distribution in the transition region along the shadow boundary depends only on the local curvature, and, in particular, on the curvature ρ_0 of the GO shadow boundary, where the width d_F of the penumbra region becomes

$$d_F = (\lambda\rho_0^2/\pi)^{1/3}. \tag{18}$$

This parameter will play an important role in smoothing the two silhouette area functions $A(r')$ across the shadow boundary $r' = r_0'$ within the penumbra region $(r_0' - d_F/2) < r' < (r_0' + d_F/2)$. Fock's theory, which introduces special diffraction functions along local coordinates parallel and normal to the shadow boundary [33-35], will prove very useful in analyzing diffraction effects, including creeping-wave contributions, which must also be taken into consideration in correcting POFFIS.

8. DATA UTILIZATION OF ADVANCED BROADBAND POLARIZATION RADAR SYSTEMS

Although we have found that POFFIS suffers many limitations, its near-future applicability to practical target imagery cannot be ruled out, since modern advanced broadband radar systems with dual polarization facility will provide all the input data for POFFIS and its correction, to the degree of accuracy required. Since in radar tactical environments the useful coherent target signal has to be recovered from incoherent clutter perturbed data, it is essential that we analyze the properties of the polarization scattering matrix and its relation to the Stokes vectors and the Mueller matrix target operators [2, Chap. 8]. Namely, it is possible to derive a polarization processing algorithm which selects optimum polarization for target parameter discrimination in the presence of background clutter [36], so that in consideration of POFFIS useful coherent target information (amplitude, phase, Doppler-range, and polarization) can be recovered to a degree of accuracy dictated by the amount of a priori known background clutter statistics [38-41].

Following STEINBACH [39], the target scattering matrix $[S(t)]$ possessing relative, but not absolute phase may be defined as

$$[S] = \begin{bmatrix} S_{xx} & S_{xy} \\ S_{yx} & S_{yy} \end{bmatrix}, \; S_{xy} = S_{yx}, \; \phi_{xy} = \phi_{yx} = 0, \tag{19}$$

where $\hat{x}$ may be chosen along the horizontal and $\hat{y}$ along the vertical direction [2, Chap. 7.3], and for the monostatic case the complex field-cross section $\rho(\kappa)$ of (4) is the Fourier transform in $\bar{\kappa}$-space of either $S_{xx}(t)$ or $S_{yy}(t)$. The time averages (<>) of the square moduli $|S_{ij}(t)|^2$, the inner products $S_{ij}S^*_{kl}$, their real or imaginary parts, and any specifically defined combinations can be related to the components M_{ij} of the 4×4 Mueller polarization matrix [M] [40] or the average Stokes scattering operator, as [M] is denoted frequently [41]. It can be shown that, of the sixteen M_{ij}, nine are independent and all of them can be determined uniquely from the values of $<S_{xx}(t)S^*_{yy}(t)>$, $<|S_{xx}(t)|>$, $<|S_{yy}(t)|^2>$, $<|S_{xy}(t)|^2>$, $<S_{xx}(t)S^*_{xy}(t)>$, and $<S_{yy}(t)S^*_{xy}(t)>$ which have been reconstructed from measurements [42]. In case the target is embedded in background clutter, the objective is to separate the useful coherent clutter components, and can be achieved by using clustering properties of a characteristic set of null polarizations [28, 1970] on the polarization chart which is the projection plane of Poincaré's polarization sphere [43]. Namely, it can be shown that co- and cross-polarized null pairs of target and background clutter have highly different clustering properties, and this should make the extraction of useful target signals a feasible task. Here it is to be noted that the target scattering matrix [S] constitutes a complete description of the reflecting properties of a target for given frequency and target orientation. Although the elements of the matrix depend on the manner in which the measurements are made, the intrinsic properties of the matrix are functions of only the target and not the measurement technique, i.e. it is always possible to transform the matrix for one polarization pair to that for any other, and there are an infinite number of such transformations. If in (19) we let S_{ij} refer to the i th polarization transmitted and the j th polarization received, the infinity of transformations gives rise to two chacteristic ones, the cross-polarization null pair for which $S_{ij}=S_{ji}=0$, and the co-polarization null pair for which $S_{ii}=S_{jj}=0$. It can now be shown that the two cross-pol nulls and the two co-pol nulls must lie on one great circle path on the Poincaré sphere, where the cross-pol nulls are orthogonal, i.e., lie on the opposite ends of the great circle and bisect the great circle in between the co-pol nulls, which will in general not be orthogonal [41]. Thus if we know the location of one crosspol null and one co-pol null or only the two co-pol nulls, the locations of the other two nulls is determined. The original radar cross section matrix [S] required five numbers to describe it: relative phase and amplitude of S_{ii}, S_{ij} and the amplitude of S_{ij}, where the absolute magnitude $p = (\Sigma_i \Sigma_j |S_{ij}|^2)^{1/2}$ is invariant with respect to the [S] matrix transformation and defines the radius of the Poincaré sphere, being a function of target reflectivity. The properties are unique and their utilization in radar target detection and imaging is by far not exhausted [2, Chaps. 7,8].

9. SUMMARY AND RECOMMENDATIONS

It has been established that considerable improvements of and deeper insight into the POFFIS technique can be obtained by utilizing properties of the Radon-projection-transform theory [2, Chap. 2]. Although the relation between the ramp-response identity (3) and the Bojarski-Lewis identity (4)-(6) has been established [44, Eq. (4)], extensive studies are still required to fully exhaust all new insights that can be gained by the use of Radon's theory. Particular emphasis needs to be placed on developing reconstruction algorithms which can be applied directly to the three-dimensional nonsymmetric target case. Furthermore, the Radon transform approach is also very well suited for analyzing the aspect direction-limited and the sparse-data case [27, 45, 46], as well as the question of uniqueness, self-consistency, and accuracy, by utilizing the determinacy theorems of LUDWIG [47] which have been well presented in [27].

The polarizational correction of POFFIS and other HF imaging techniques needs to be advanced along the direction presented in CHAUDHURI and BOERNER [30] employing first-order PO corrections in (3), and (15)-(17) and properties of penumbra diffraction (18). The particular unique properties of the RCS polarization scattering matrix (19) need to be investigated in depth, and the additional polarization information should be integrated into applicable HF radar imaging techniques and HF inverse scattering theories.

ACKNOWLEDGMENTS

This research was initiated when the author was still with the Applied Electromagnetics Laboratory, Department of Electrical Engineering, University of Manitoba, Winnipeg, Canada, and it was supported in part by the Natural Sciences and Engineering Research Council of Canada under Grant A7240, NATO Research Grant 1405, and a Humboldt Fellowship. The invitation by the Workshop Organizing Committee to participate in the Conference on MMAST is acknowledged with sincere appreciation.

FOOTNOTES AND REFERENCES

[1] C.L. Bennett: "Inverse Scattering," and "Time domain solutions via integral equations — surfaces and composite bodies," Invited Papers, NATO Institute, Norwich University, England, July, 1979

[2] W-M. Boerner: "Polarization utilization in electromagnetic inverse scattering," UICC, Communications Laboratory Report 78-3 October 1978, UICC, SEO-1104

[3] E.M. Kennaugh, R.L. Cosgriff: "The use of impulse response in electromagnetic scattering problems," *IRE National Convention Record,* Part I, 72-77 (1958)

[4] Yu.N. Barabanenkov, AA. Tolkachev, N.A. Aytkhozhin, O.K. Lesota: "Scattering of electromagnetic delta pulses by ideally conducting bodies of finite dimensions," Radioteknika i Elektronika (USSR) **8**, 1061-1063 (1963)

[5] A. Freedman: "The portrayal of body shape by a sonar or radar system," Radio Electr. Eng. **25**, 51-64 (1963)

[6] E.M. Kennaugh, D.L. Moffatt: "Transient and impulse response approximations," IEEE Proc. **53** (8), 893-901 (1965)

[7] J.B. Keller: "On the use of a short-pulse broad-band radar for target identification," report, RCA, Moorestown, New Jersey, (February 17, 1965)

[8] J.D. Young: "Target imagining from multiple-frequency radar returns," ESL-OSU, Columbus, Ohio, Technical Report 2768-6, (June, 1971) (AD-728235). See also "Radar imaging from ramp response signatures," IEEE Trans. **AP-24**, 276-282 (1976)

[9] D.L. Moffatt, R.K. Mains: "Detection and discrimination of radar targets," IEEE Trans. **AP-23**, 358-367 (1975)

[10] N.N. Bojarski: "Three-dimensional electromagnetic short pulse inverse scattering," Syracuse University Research Corporation, Syracuse, New York, (February 1967)

[11] R.M. Lewis: "Physical optics inverse diffraction," IEEE Trans. **AP-17**, 308-314 (1969)

[12] S. Rosenbaum-Raz: "On scatter reconstruction from far-field data (a bistatic generalization of Bojarski-Lewis' physical optics inverse scattering theory)," IEEE Trans. **AP-24**, 66-70 (1976)

[13] R.E. Kell: "On the derivation of bistatic RCS from monostatic measurements," Proc. IEEE **53**, 983-988 (1965)

[14] Y. Das and W.-M. Boerner: "On radar target shape estimation using algorithms for reconstruction from projections," IEEE Trans. **AP-26**, 274-279 (1978). See also Y. Das: "Application of concepts of image reconstruction from projections and Radon transform theory to radar target identification," Ph.D. thesis, University of Manitoba 1977; Y. Das, W.-M. Boerner: "Applications of algorithms for 3-D image reconstruction from 2-D projections to electromagnetic inverse scattering," USNC/URSI Annual Meeting, Boulder, Colorado, October 20-23, 1975, Session III-7-7, p. 184

[15] J. Radon: "Uber die Bestimmung von Funktionen durch ihre Integralwerte längs gewisser Mannigfaltigkeiten (On the determination of functions from their integrals along certain manifolds)," Ber. sachs. Akad. Wiss. (Leipzig) Math. Phys. Klasse **69**, 262-271 (1917)

[16] D.L. Moffatt, J.D. Young: "A chronological history of radar target imagery at the Ohio State University," *Proceedings of the International IEEE-APS Symposium*, Seattle, 1979, Vol. I. pp. 244-247. See also ESL-ESU, Class Notes, Radar Target Identification, Vols. I and II, Department of Electrical Engineering, ElectroScience Laboratory, Ohio State University, (September, 1976)

[17] W.-M. Boerner, C.-M. Ho: "Development of physical optics far field inverse scattering (POFFIS) and its limitations," *Proceedings of the International IEEE-APS Symposium*, Seattle, 1979, Vol. I, pp. 240-243

[18] Contrary to some misleading information, after extensive historical surveys we are now certain that Prof. R. Lewis did not establish the above relationship as expressed in (7)-(11). See [2, p. 56, Chap. 5.3].

[19] W.L. Perry: "On the Bojarski-Lewis inverse scattering theory," IEEE Trans. **AP-22**, 826-829 (1974)

[20] W.L. Perry: "Approximate solution of inverse problems with piecewise continuous solutions," Radio Science **12**(5), 637-642 (1977)

[21] W. Tabbara: "On an inverse scattering method," IEEE Trans. **AP-21**, 245-247 (1973)

[22] W. Tabbara: "On the feasibility of an inverse scattering method," IEEE Trans. **AP-23**, 446-448 (1975)

[23] W.-M. Boerner, F.H. Vandenberghe: "Determination of the electrical radius *ka* of a spherical scatterer from the scattered field," Can. J. Phys. **49**, 1507-1535 (1971); "Determination of the electrical radius *ka* of a circular cylindrical scatterer from the scattered field;" *ibid.* **49**, 804-819 (1971); "On the inverse problem of electromagnetic scattering by a perfectly conducting prolate spheroid," **50**, 754-759 (1972); "On the inverse problem of electromagnetic scattering by a perfectly conducting elliptic cylinder," **50**, 1987-1992 (1972)

[24] W-M. Boerner, O.A. Aboul-Atta: "Properties of a determinant associated with inverse scattering in spherical coordinates," Utilitas Mathematica **3**, 163-273 (1973)

[25] R.D. Mager, N. Bleistein: "An approach to the limited aperture problem of physical optics far field inverse scattering," University of Denver Research Institute Report MS-R-7704 (1976) and IEEE Trans. **AP-26**, 695-699 (1978)

[26] A. Majda: "A representation for the scattering operator and the inverse problem for arbitrary bodies," Comm. Pure. Appl. Math **30**, 165-194 (1977)
[27] K.T. Smith, D.C. Solomon, S.L. Wagner: "Practical and mathematical aspects of the problem of reconstructing objects from radiographs," Bull. Am. Math. Soc. **83**(6), 1227-1270 (1977)
[28] J.R. Huynen: "Radar target sorting based upon polarization signature analysis," Lockheed Missiles and Space Division, Report 28-82-16, (May 1960) (AD318597), and *Phenomenological Theory of Radar Targets* (Drukkerij Bronder-offset N.V., Rotterdam, 1970), dissertation (obtainable from author); see also *Proceedings of the National Conference on Electromagnetic Scattering*, UICC, June, 1976, pp. 91-94; J. R. Huynen: "Radar target phenomenology," in *Electromagnetic Scattering*, ed. by P.L.E. Uslenghi (Academic, New York, 1968), pp. 653-712
[29] C.L. Bennett, A.M. Auckenthaler, R.S. Smith, J.D. DeLorenzo: "Space time integral equation approach to the large body scattering problem," Sperry Research Center, Sudbury, Massachusetts, SCRCR-Cr-73-1 (1973)
[30] S.K. Chaudhuri, W-M. Boerner: "A monostatic inverse scattering model based on polarization utilization," Applied Physics (Springer) **11**, 337-350 (1976); "Polarization utilization in profile inversion of a perfectly conducting prolate spheriod," IEEE Trans. **AP-25**, 505-511 (1977)
[31] N.N. Bojarski: "Inverse scattering," Company Report N00019-73-C-312/F, prepared for NASD (AD-775-235/5) (1974)
[32] V.A. Fock: *Electromagnetic Diffraction and Propagation Problems* (Pergamon, New York, 1965). See also J. Phys. USSR **9**, 255-266 (1945); *ibid.* **10**, 130-136, 339-409 (1946)
[33] R.F. Goodrich: "Fock theory—an appraisal and exposition," IRE Trans. **AP-7**, 528-536 (1959)
[34] P.H. Pathak, R.G. Kouyoumjian: "An analysis of the radiation from apertures in curved surfaces by GTD," Proc. IEEE **62**, 1438-1447 (1974)
[35] V.A. Borovikov, B.Ye. Kinber: "Some problems in the asymptotic theory of diffraction," Proc. IEEE **62**, 1416-1437 (1974)
[36] G. Ioannidis, D.E. Hammers: "Adaptive antenna polarization schemes for clutter suppression and target identification," RADC-TR-79-4, (February 1979) and IEEE Trans. **AP-27**, 357-363 (1979)
[37] M.W. Long: *Reflectivity of Land and Sea* (Lexington Books, Heath, Lexington, Massachusetts, 1975)
[38] D.E. Hammers, A.J. MacKinnon: "Radar target recognition, an operator-theoretical systems approach," *International Symposium on Operator Theory and Networks*, Montreal, 1975
[39] K.H. Steinbach: "Nonconventional aspects of radar target classification by polarization properties," USAMERDC, Fort Belvoir, VA, Report 2065, (June, 1973) (AD-763155). See also "On the polarization transform power of radar target" in *Atmospheric Effects on Radar Target Identification and Imaging*, ed. by E. Jeske (Reidel, Dordrecht-Holland, 1976), pp. 65-82
[40] P.T. Gough, W-M. Boerner: "Depolarization of specular scatter as an aid to identifying a rough dielectric surface from an identical rough metallic surface," J. Opt. Soc. Am. **69**, (July 1979) in press
[41] E. Kennaugh: "Polarization properties of target reflections," Griffis AFB Report No. 389-8 (April 1951); final report (March 1952) (AD-002-494). See also *Proceedings of the Workshop on Radar Backscatter from Terrain*, January 1979, U.S. Army ETL, Fort Belvoir, Virginia, ed. by J.A. Styles, J.C. Holtzman; RSL Technical Report 374-2 (DAAG 29-78-C-0019)
[42] A. Ishimaru: *Wave Propagation and Scattering in Random Media; I: Single Scattering and Transport Theory; II: Multiple Scattering, Turbulence, Rough Surfaces, and Remote Sensing* (Academic, New York, 1978)
[43] G.A. Deschamps: "Geometrical representation of the polarization of a plane electromagnetic wave," Proc. IRE **39**(5), 543-548 (1951)
[44] W-M. Boerner, Y. Das: "Application of the Radon transform theory to electromagnetic inverse scattering," ISAP 1978, Sendai, August 29-31, 1978
[45] B.E. Oppenheim: "More accurate algorithms for iterative 3-dimensional reconstruction," IEEE Trans. **NS-21**, 72-83 (1974)
[46] R.N. Bracewell, S.J. Wernecke: "Image reconstruction over a finite field of view," J. Opt. Soc. Am. **65**, 1342-1346 (1975)
[47] D. Ludwig: "The Radon transform on Euclidean space," Comm. Pure Appl. Math. **19**, 49-81 (1966)

AN ITERATIVE PROCEDURE FOR SOLVING INVERSE SCATTERING PROBLEMS ARISING FROM ACTIVE REMOTE SENSING

Yung Ming Chen

Department of Applied Mathematics and Statistics
State University of New York
Stony Brook, New York 11794

INTRODUCTION

Remote sensing techniques have been used in many branches of science and engineering very recently. In general, remote sensing techniques can be divided into two types, passive and active techniques. Passive techniques are basically listening devices and mathematically they lead to inverse eigenvalue problems of differential equations. Active techniques are basically transmitting and receiving devices and mathematically they lead to noneigenvalue inverse problems of differential equations.

Here, an iterative algorithm for solving the inverse problems of determining the velocity coefficient of a wave equation from the partial information of the solution on the boundary surface is introduced and its convergence is discussed. But first a literature survey of methods for solving this type of inverse problems is given here. The above-mentioned iterative numerical algorithm was first introduced by TSIEN and CHEN [1] for solving an idealized velocity inverse problem in fluid dynamics; then it was further developed to have the capability of handling information with measurement errors and information from a large range of frequencies [2]. Later, it was used to solve an inverse problem in electromagnetic wave propagation [3] and it compared favorably with the spectral domain method [4]. Independently, COHEN and BLEISTEIN [5] have developed a perturbation method for solving this type of inverse problem where their first order solution is similar to the first order iteration of ours [1,2,3]. However, this method is not suitable for numerical computation as it stands. In Russia, NIGUL [6] and NIGUL and ENGELBRECHT [7] have presented a different perturbation method which can take care of dissipative and nonlinear effects in the wave equation; however, it is limited to layered media and is not suitable for numerical computation.

1. NUMERICAL ALGORITHM

Consider the simple initial-boundary value problem for the wave equation,

$$c^2(x)\partial^2 u(x,t)/\partial x^2 - \partial^2 u(x,t)/\partial t^2 = 0, \quad 0 < x < 1, \quad t > 0 \; ,$$
$$u(x,0) = \partial u(x,0)/\partial t = 0, \quad u(0,t) = f(t), \text{ and } u(1,t) = 0 \; . \tag{1}$$

An inverse problem of (1) is to determine $c^2(x)$ or $c(x)$ from the given $\partial u(0,t)/\partial x$. By a Fourier sine transform, (1) is reduced to

$$c^2(x)\; d^2u(x,\omega)/dx^2 + \omega^2 u(x,\omega) = 0, \quad 0 < x < 1 \; ,$$
$$u(0,\omega) = f(\omega)\; , \quad u(1,\omega) = 0 \; , \tag{2}$$

and now the corresponding inverse problem is to determine $c(x)$ from $du(0,\omega)/dx$.

The iterative algorithm is defined by

$$u_{n+1}(x,\omega) = u_n(x,\omega) + \delta u_n(x,\omega), \quad c^2_{n+1}(x) = c^2_n(x) + \delta c^2_n(x),$$
$$n = 0,1,2,3, \; \ldots \; , \tag{3}$$

where $|\delta u_n(x,\omega)| < |u_n(x,\omega)|$, $|\delta c^2_n(x)| < |c^2_n(x)|$, $\delta c^2_n(0) = 0$, and $c_0^2(x)$ is the initial guess. Upon substituting (3) into (2) and neglecting second order and higher order terms, one obtains

$$c^2_n(x)\, d^2u_n(x,\omega)/dx^2 + \omega^2 u_n(x,\omega) = 0, \quad 0 < x < 1 \; ,$$

$$u_n(0,\omega) = f(\omega), \quad u_n(1,\omega) = 0 \; , \tag{4}$$

and

$$c_n^2(x)\, d^2\delta u_n(x,\omega)/dx^2 + \omega^2\, \delta u_n(x,\omega) = -\delta c_n^2(x)\, d^2 u_n(x,\omega)/dx^2, \quad 0 < x < 1 \quad ,$$

$$\delta u_n(0,\omega) = \delta u_n(1,\omega) = 0 \tag{5}$$

By using the method of Green's functions and replacing $du_{n+1}(0,\omega)/dx$ by the given data $du(0,\omega)/dx$, (5) is reduced to a first kind Fredholm integral equation,

$$\int_0^1 [1/c_n^2(x')\, du_n(x',\omega)/dx']^2\, \delta c_n^2(x')\, dx' = du(0,\omega)/dx - du_n(0,\omega)/dx \quad . \tag{6}$$

Equations (3), (4) and (6) form the basic structure of the iterative numerical method.

Equation (6) can be solved by using any one of the following techniques; the regularization method of TIHONOV [8], the linear inversion technique of BACKUS and GILBERT [9], the MOORE-PENROSE pseudoinverse method [10], etc. The choice of frequencies ω's can be achieved by adopting any one of the following criteria; low noise-computational efficiency criterion [2], minimum error criterion [11], or well-conditioned matrix criterion [12].

2. ASSOCIATED INVERSE MATRIX PROBLEM

Consider the linear system

$$\underline{A} x^i = f^i \quad , \tag{7}$$

where $\underline{A}$ is a $k \times k$ nonsingular symmetric matrix, f^i, $i = 1, \ldots, p \leqslant k$ are linearly independent input vectors and x^i, $i = 1, \ldots, p$ their corresponding response vectors. The associated inverse matrix problem is to determine $\underline{A}$ from $\{f^i\}$ and some components of $\{x^i\}$. For simplicity, only the ideal case where $\underline{X} = (x^1, x^2, \ldots, x^k)$ and $\underline{F} = (f^1, f^2, \ldots, f^k)$ are symmetric and completely known is considered here.

Let

$$\underline{A}_{n+1} = \underline{A}_n + \delta\underline{A}_n \text{ and } x^i_{n+1} = x^i_n + \delta x^i_n \,, \quad n = 0, 1, 2, \ldots, \tag{8}$$

where $||\delta\underline{A}_n|| < ||\underline{A}_n||$ and $||\delta x^i_n|| < ||x^i_n||$. Then the corresponding iterative algorithm is defined by

$$\underline{A}_n\, x^i_n = f^i \tag{9}$$

and

$$\underline{A}_n^{-1}\, \delta\underline{A}_n x^i_n = -\delta x^i_n \quad . \tag{10}$$

After a considerable amount of algebraic manipulation, (9) and (10) can be written as

$$\delta\underline{A}_n = \underline{A}_n(\underline{I} - \underline{F}^{-1}\underline{A}_n\underline{X}) \quad . \tag{11}$$

Theorem: *The iterative algorithm of (8) and (11) converges quadratically when* $||\underline{I} - \underline{F}^{-1}\underline{A}_0\underline{X}|| < 1$.

Proof. Repeated application of (11) leads to

$$\delta\underline{A}_n = \underline{A}_0\{I + (\underline{I} - \underline{F}^{-1}\underline{A}_0\underline{X})\}\{\underline{I} + (\underline{I} - \underline{F}^{-1}\underline{A}_0\underline{X})^2\}\ldots\{\underline{I} + (\underline{I} - \underline{A}_0\underline{X})^{2^{n-1}}\}$$
$$\cdot\, (\underline{I} - \underline{F}^{-1}\underline{A}_0\underline{X})^{2^n} \tag{12}$$

Therefore, a necessary condition for convergence here is $||\underline{I} - \underline{F}^{-1}\underline{A}_0\underline{X}|| < 1$. Moreover, $R_n \equiv ||\delta\underline{A}_n||/||\delta\underline{A}_{n-1}||^\alpha \sim ||\underline{A}_0||^{-\alpha+1}\, ||\underline{I} - \underline{F}^{-1}\underline{A}_0\underline{X}||^{2^{n-1}(2-\alpha)}$ Hence for R_n to be a non-zero constant as $n \to \infty$, $\alpha = 2$. This means that the iterative algorithm converges quadratically. Q.E.D.

REFERENCES

[1] D.S. Tsien, Y.M. Chen (1974): "A numerical method for nonlinear inverse problems in fluid dynamics," in *Computational Methods in Nonlinear Mechanics,* Proceedings of the International Conference on Computational Methods in Nonlinear Mechanics, University of Texas at Austin (1974), pp. 935-943

[2] Y.M. Chen, D.S. Tsien: "A numerical algorithm for remote sensing of density profiles of a simple ocean model by acoustic pulses," J. Comp. Phys. **25**, 366-385 (1977)

[3] D.S. Tsien, Y.M. Chen: "A pulse-spectrum technique for remote sensing of stratified media," Radio Science **13**, 775-783 (1978)

[4] D.H. Schaubert, R. Mittra: "A spectral domain method for remotely probing stratified media," IEEE Trans. **AP-25**, 261-265 (1977)

[5] J.K. Cohen, N. Bleistein: "An inverse method for determining small variations in propagation speed," SIAM J. Appl. Math. **32**, 784-799 (1977)

[6] U. Nigul:. "Influence of nonlinear effects on one-dimensional echo signals from elastic targets," Soviet Phys. Acoust. **21**, 93-95 (1975)

[7] U. Nigul, J. Engelbrecht: "On dissipative, diffractional, and nonlinear effects in acoustodiagnostics of layered media," in *Modern Problems in Elastic Wave Propagation*, ed. by J. Miklowitz, J. D. Achenbach (Wiley-Interscience, New York, 1978), pp. 265-282

[8] A.N. Tihonov, : "Solution of incorrectly formulated problems and the regularization method," Sov. Math. Dokl. **4**, 1035-1038 (1963)

[9] G. Backus, F. Gilbert : "Numerical applications of a formalism for geophysical inverse problems," Geophys. J. Roy. Astrn. Soc. **13**, 247-276 (1967)

[10] A. Albert: *Regression and the Moore-Penrose Pseudoinverse* (Academic, New York, 1972)

[11] E. Tsimis: "On the inverse problem by means of the integral equation of the first kind," Ph.D. thesis, Dept. of Applied Math. and Stat., State University of New York at Stony Brook (1977)

[12] F. Hagin: "On the construction of well-conditioned systems for Fredholm I problems by mesh adapting," J. Comput. Phys., to be published

ONE-DIMENSIONAL VELOCITY INVERSION FOR ACOUSTIC WAVES: NUMERICAL RESULTS

Samuel Gray
Naval Research Laboratory, Washington, District of Columbia 20375
and Department of Mathematics and Engineering Mechanics,
General Motors Institute, Flint, Michigan 48502

Norman Bleistein
Department of Mathematics
University of Denver, Denver, Colorado 80208

We consider the inverse problem of determining small variations in propagation speed from remote observations of signals which pass through an inhomogeneous medium. Under the conditions (1) that the variations can be written as a small perturbation from a known reference value and (2) that the medium of interest varies in one direction only, an integral equation has been developed for the variations which can be solved in closed form. Here, a technique is presented to obtain and process synthetic data from a scattering profile of arbitrary shape. The results of numerical testing show that, as long as a velocity variation is indeed "small", both its size and its shape can be reproduced with negligible error by this method.

INTRODUCTION

A problem of interest in seismic exploration is to determine a sound wave's propagation velocity at points below the earth's surface, or ocean bottom, given reflection data at points near the surface of the earth or ocean. Mathematically, this is the inverse scattering problem of determining the coefficients of a wave equation, given only a knowledge of the waveform used to probe the medium and a limited knowledge of the solution. The basic underlying assumption is that the unknown coefficient can be written as a small perturbation from a known reference value. A further assumption for the one-dimensional case is that the coefficient varies in one direction only (e.g., with depth).

This problem was treated by COHEN and BLEISTEIN [1], who showed that, when the probes used are plane waves of all frequencies, both the size and the shape of the unknown coefficient can be determined by taking a Fourier transform of the scattering data. Here, a method is developed to test this result on synthetically produced scattering data; the method used to produce the data is independent of the method used to process it. Also, the analogous time-domain inversion result is presented and tested.

1. ANALYTICAL RESULTS

We consider a medium for which the index of refraction is known up to small perturbations, and varies in one direction only. The objective is to determine these perturbations from observations of a scattered field generated by probing the medium with an impulsive signal (or, equivalently, with plane waves of all frequencies). Thus, in the frequency domain, the wave field $u(k,z)$ is a solution of the following reduced wave equation (the prime denotes differentiation with respect to z):

$$u'' + k^2(1 + \alpha(z))\, u = 0. \tag{1}$$

Here, $k = \omega/c$ with c a known constant and ω the frequency. We also assume that the function $\alpha(z)$ is nonzero only in a finite interval, say $0 < z < H$, and that $\alpha(z)$ is $0(\epsilon)$, where ϵ is a small parameter. Our objective is to find $\alpha(z)$ from observations of u at $z = 0$.

Let us suppose that we probe the region of inhomogeneity with a pulse from the left. The Fourier transform in time of the incident wave is then given by

$$u_I(k, z) = \exp(ikz). \tag{2}$$

Then the total solution u of (1.1) will be of the form

$$u = u_I + u_S, \tag{3}$$

with u_S a wave which satisfies

$$u'_s + iku_S = 0\ ,\ \ z \leqslant 0,$$

$$u'_s - iku_S = 0\ ,\ \ z \geqslant H. \tag{4}$$

In COHEN and BLEISTEIN [1], the following integral equation for α is obtained:

$$\int_{-\infty}^{\infty} \alpha(z) \exp(2ikz)\, dz = \theta(k), \tag{5}$$

with error $O(\epsilon^2)$. Here,

$$\theta(k) = \frac{u_S'(k,0) - iku_S(k,0)}{k^2} = -\frac{2iu_S(k,0)}{k}\ . \tag{6}$$

Then, to the same order of accuracy,

$$\alpha(z) = \frac{1}{\pi} \int_{-\infty}^{\infty} \theta(k) \exp(-2ikz)\, dk. \tag{7}$$

In the time domain, instead of (1-4), the differential equation, incident wave, total wave decomposition, and boundary conditions are as follows:

$$(\partial_z^2 - c^{-2}(1 + \alpha(z))\partial_t^2)\, U = 0, \tag{8}$$

$$U_I(t,z) = \delta(t - z/c), \tag{9}$$

$$U(t,z) = U_I(t,z) + U_S(t,z), \tag{10}$$

$$\partial_z U_S - c^{-1}\partial_t U_S = 0\ ,\ \ z \leqslant 0\ ,$$

$$\partial_z U_S + c^{-1}\partial_t U_S = 0\ ,\ \ z \geqslant H. \tag{11}$$

Here ∂_z (∂_t) denotes a partial derivative with respect to $z(t)$. In this case, the integral equation for α is

$$\int_0^{\infty} dz\ \alpha(z) \int_0^{\tau} dt\ U_I(t,z)\, U_I(\tau - t,\ z) = -2c \int_0^{\tau} U_S(t,0)\, dt. \tag{12}$$

The inversion of this equation yields

$$\alpha(z) = -4 \int_0^{2z/c} U_S(t,0)\, dt. \tag{13}$$

Formulas (12) and (13) are derived in the Appendix.

2. NUMERICAL METHOD AND RESULTS

These results were tested on several examples where the scattering data was given as a function of frequency, and on one example where the scattering data was given as a function of time.

To generate the scattering data in (6), we first solve (1-4) with a known profile $\alpha(z)$. Then $\theta(k)$ in (6) is generated and substituted into (7). The integration is carried out by using a Fast Fourier Transform (FFT) routine. The system (1-4) is solved numerically for each nonzero k in the discrete set required for the FFT routine as follows: For a given k, $u(k,z)$ is expressed in terms of solutions to the initial-value problem related to (1-4). Thus, the system to be solved numerically is

$$u_j'' + k^2(1 + \alpha)\, u_j = 0\ ,\ \ j = 1,2,$$

$$u_1''(k,0) = 1,\ \ u_1'(k,0) = 0,$$

$$u_2(k,0) = 0,\ \ u_2'(k,0) = k, \tag{14}$$

so that

$$u(k,\ z) = c_1 u_1(k,\ z) + c_2 u_2(k,\ z). \tag{15}$$

Here, c_1 and c_2 are constants to be determined. From (6), (2), and (14),

$$\theta(k) = \frac{u'(k,0) - iku(k,0)}{k^{2\cdot}} = \frac{c_2 - ic_1}{k}. \tag{16}$$

From (2) and (4),

$$u' + iku = 2ik \quad , \quad z = 0$$

$$u' - iku = 0 \quad , \quad z = H. \tag{17}$$

Equations (17) are two linear equations for c_1 and c_2 which can be solved in terms of $u_1(k,H)$, $u_1'(k,H)$, $u_2(k,H)$ and $u_2'(k,H)$. This yields $\theta(k)$ (observations at $z = 0$) as a function of the wave field at $z = H$. The wave field at $z = H$ is obtained by solving (14) using a fourth order Runge-Kutta scheme (see, for example, DORN & McCRACKEN [2], p. 373). Lacking information for $k = 0$ is equivalent to determining $\alpha(z)$ up to an additive constant. However, it is assumed that $\alpha(0)$ can be observed as well as the wave field at $z = 0$, and this value adjusts the constant. In all examples, we have taken $\alpha(0) = 0$.

Examples of the results of the processing are shown in Figs. 1-5. The reference velocity c has been set equal to unity, so that the length scale is dimensionless. In each example, the agreement between the true $\alpha(z)$ and the reconstructed $\alpha(z)$ is excellent.

A final example was treated to test (13). The data for this example was furnished by L. YOST [3] of Marathon Oil Company, and was processed (including the deconvolution of the source wavelet) without prior knowledge of the velocity profile. Here, the velocity profile is given by

$$\frac{c}{\sqrt{1+\alpha(z)}} = \begin{cases} 10,000 \text{ ft/sec.}, & 000 \text{ ft} < z < 100 \text{ ft} \\ 11,000 \text{ ft/sec.}, & 100 \text{ ft} < z < 200 \text{ ft..} \\ 10,000 \text{ ft/sec.}, & 200 \text{ ft} < z \end{cases}$$

In Fig. 6, the velocity is plotted against depth z. Taking $c = 10,000$ ft/sec yields $|\alpha(z)|$ as large as 0.17; however, the agreement between the true and reconstructed values for the velocity is still very good.

3. CONCLUSIONS

The computer processing described here has been shown to illustrate the validity of an analytical result in inverse scattering theory. In particular, we note the independence of the method used to obtain the scattering data from the method used to process it. (Compare GJEVIK, NILSEN AND HÖYEN [4], where the same equation is used both to generate the synthetic scattering data which solves the "direct" problem and to process that scattering data in solving the "inverse" problem.)

More importantly, we emphasize the success of this method in reproducing the *exact* shape of a profile, given scattering data of all frequencies. (Compare further work by BLEISTEIN, COHEN AND GRAY (e.g. [5,6], in which the general shape of a scattering profile is assumed *a priori* in processing frequency bandlimited data.)

APPENDIX

We will derive (1.12-13) using methods developed in Bleistein and Cohen. Firstly, we define the unperturbed operator L_0 by

$$L_0 U = (\partial_z^2 - c^{-2}\partial_t^2)\, U. \tag{A.1}$$

Then

$$L_0\, U_I = 0$$

$$L_0\, U_S = \alpha c^{-2} \partial_t^2\, U. \tag{A.2}$$

Next, the "adjoint source" V is introduced which satisfies the unperturbed adjoint problem:

$$L_0 V = 0; \tag{A.3}$$

$$\partial_z V + c^{-1}\partial_t V = -2c^{-1} H(\tau - t) \quad , \quad z = 0 \tag{A.4}$$

$$\partial_z V - c^{-1}\partial_t V = 0 \quad , \quad z = H.$$

By comparing the problems for U_I and V, one can see that

$$\partial_t^2 V(t,z) = U_I(\tau - t,z). \tag{A.5}$$

Also, U_S and V satisfy

$$U_S \equiv 0 \ , \ t < 0; \tag{A.6}$$

$$V \equiv 0 \ , \ t > \tau. \tag{A.7}$$

Now, by (A.2-3),

$$\begin{aligned} - \alpha c^{-2} V \partial_t^2 U &= U_S L_0 V - V L_0 U_S \\ &= \partial_z [U_S \partial_z V - V \partial_z U_S] - c^{-2} \partial_t [U_S \partial_t \ V - V \partial_t U_S]. \end{aligned} \tag{A.8}$$

Integrating this equation with respect to t from 0- to $\tau+$ and using (A.6-7), one finds

$$- \alpha c^{-2} \int_0^\tau [U \partial_t^2 V] \, dt = \int_0^\tau \partial_z [U_S \partial_z V - V \partial_z \, U_S] \, dt. \tag{A.9}$$

By integrating over space and using (A.5) and the boundary conditions in space and time (1.11) and (A.4,6,7), it follows that

$$- \int_0^H dz \, \alpha c^{-2} \int_0^\tau dt \; U(t,z) \, U_I(\tau - t,z) = 2c^{-1} \int_0^\tau U_S(t,0) \, dt. \tag{A.10}$$

Since the support of α is contained in the interval $(0,H)$, we may replace the limits of integration in (A.10) by $(0, \infty)$, yielding

$$- \int_0^\infty dz \, \alpha c^{-2} \int_0^\tau dt \; U(t,z) \, U_I(\tau - t,z) = 2c^{-1} \int_0^\tau U_S(t,0) \, dt. \tag{A.11}$$

The right side is a function of the field observations and, hence, known. The left side has two unknowns, namely, α and U_S. However, U_S appears only in U and, therefore, only through the product αU_S. However, from (A.2), it is seen that U_S is itself of the order of α. Thus, for small α, it is expected that αU can be reasonably approximated by αU_I. In this case, (A.11) becomes an integral equation for α alone, namely,

$$- \int_0^\infty dz \, \alpha c^{-2} \int_0^\tau dt \, \delta(t - z/c) \delta(\tau - t - z/c) = 2c^{-1} \int_0^\tau U_S(t,0) \, dt. \tag{A.12}$$

The integral on the left can be evaluated, yielding

$$\alpha(z) = -4 \int_0^{2z/c} U_S(t,0) \, dt. \tag{A.13}$$

FOOTNOTES AND REFERENCES

[1] J.K. Cohen, N. Bleistein: "An Inverse Method for Determining Small Variations in Propagation Speed", SIAM J. Appl. Math. **32**, 784-799 (1977)

[2] N. Bleistein, J.K. Cohen: "Inverse methods for reflector mapping and sound speed profiling", in *Ocean Acoustics*, ed. by J.A. De Santo, Topics in Current Physics, Vol. 8 (Springer, New York, 1979)

[3] W.S. Dorn, D.D. McCracken: *Numerical Methods with Fortran IV Case Studies* (Wiley, New York, 1972)

[4] L. Yost: Marathon Oil Company, Denver, Colorado, private communication.

[5] B. Gjevik, A. Nilsen, J. Höyen: "An attempt at the inversion of reflection data", Geophysical Prospecting **24**, 592-505 (1976)

[6] S. Gray, N. Bleistein, J.K. Cohen, "Direct inversion for strongly depth dependent velocity profiles", Report MS-R-7902, University of Denver (1978)

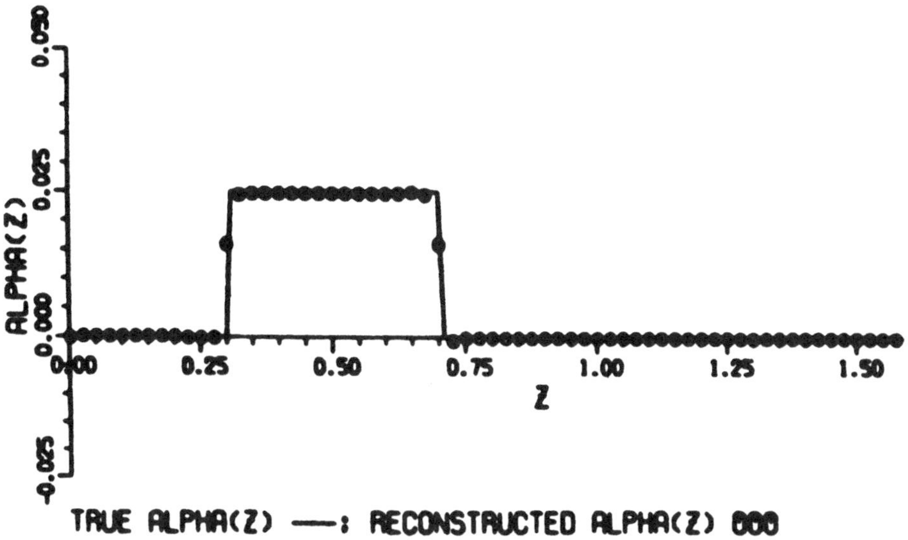

0.050
0.025
0.000
-0.025
ALPHA(Z)
0.00
0.25
0.50
0.75
1.00
1.25
1.50
Z
TRUE ALPHA(Z) ——: RECONSTRUCTED ALPHA(Z) OOO

Fig. 1

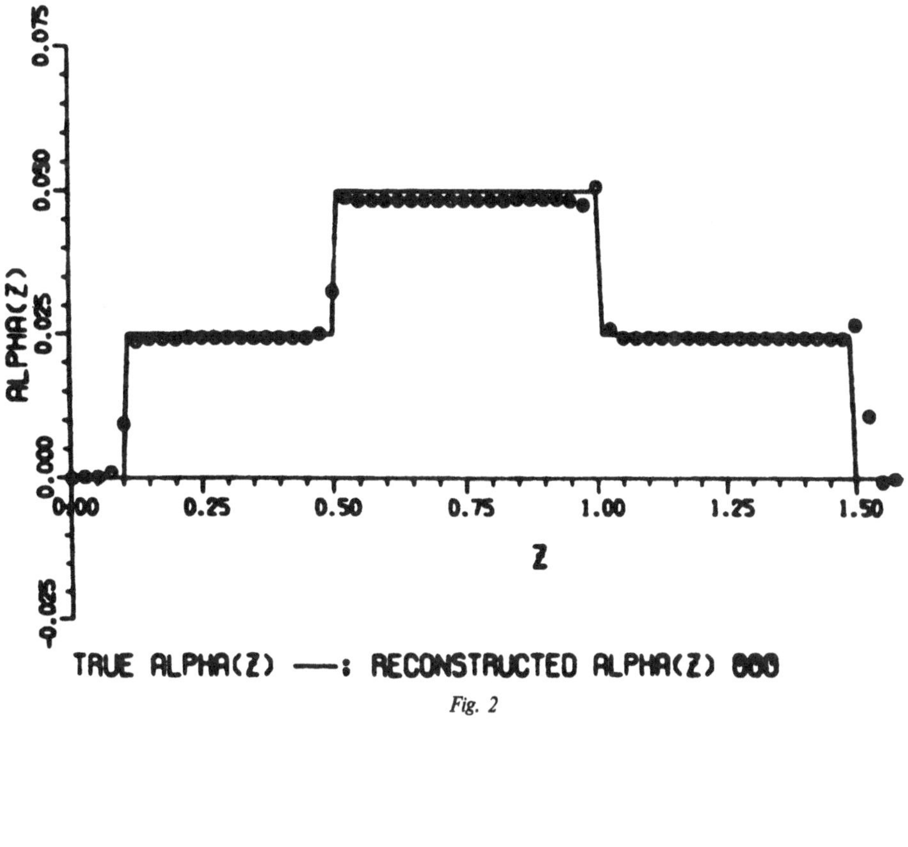

0.075
0.050
0.025
0.000
-0.025
ALPHA(Z)
0.00
0.25
0.50
0.75
1.00
1.25
1.50
Z
TRUE ALPHA(Z) ——: RECONSTRUCTED ALPHA(Z) OOO

Fig. 2

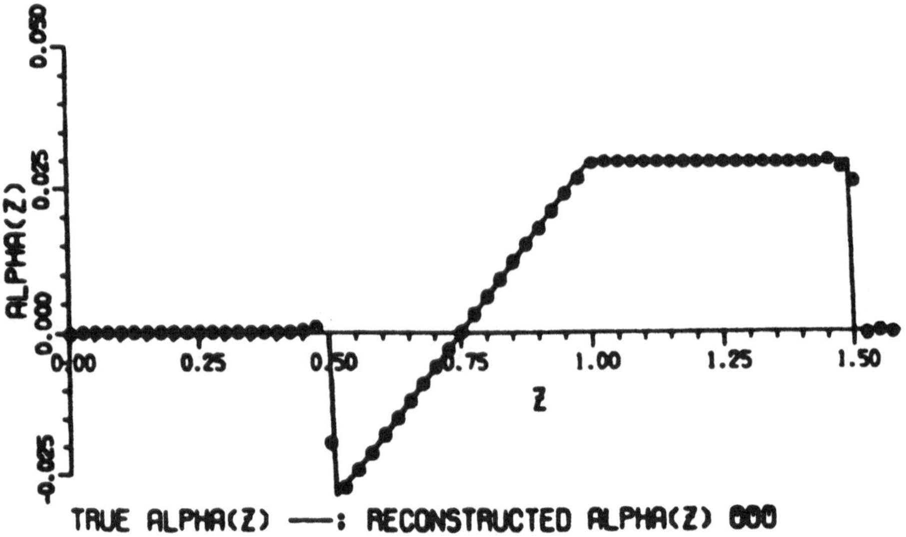
0.050
0.025
0.000
-0.025
ALPHA(Z)
0.00
0.25
0.50
0.75
1.00
1.25
1.50
Z
TRUE ALPHA(Z) —: RECONSTRUCTED ALPHA(Z) OOO

Fig. 3

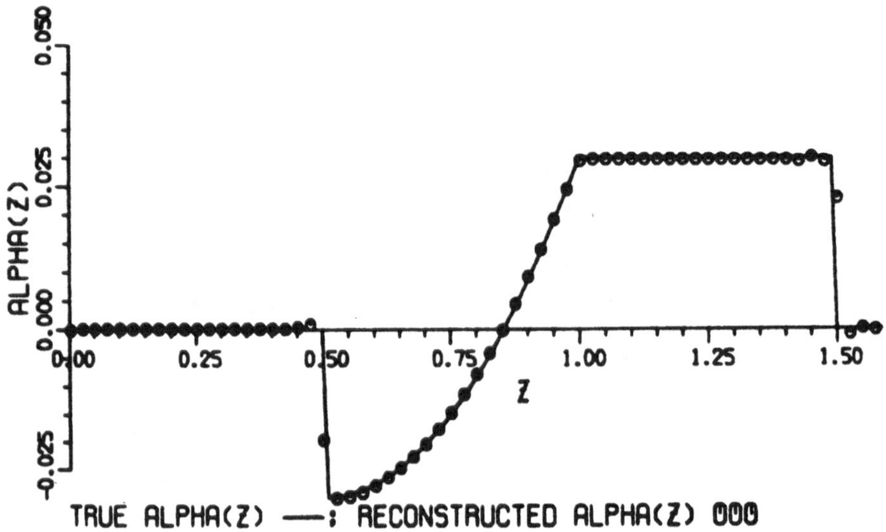
0.050
0.025
0.000
-0.025
ALPHA(Z)
0.00
0.25
0.50
0.75
1.00
1.25
1.50
Z
TRUE ALPHA(Z) —: RECONSTRUCTED ALPHA(Z) OOO

Fig. 4

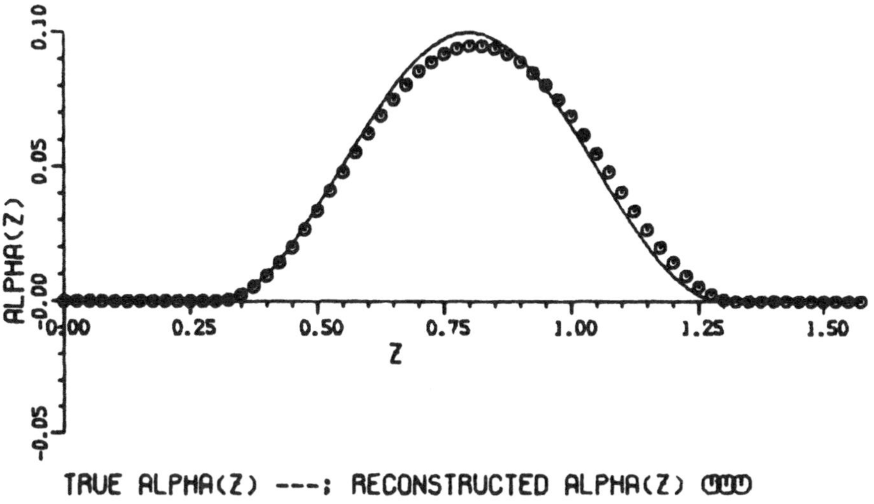

TRUE ALPHA(Z) ---; RECONSTRUCTED ALPHA(Z) ⊙⊙⊙

Fig. 5

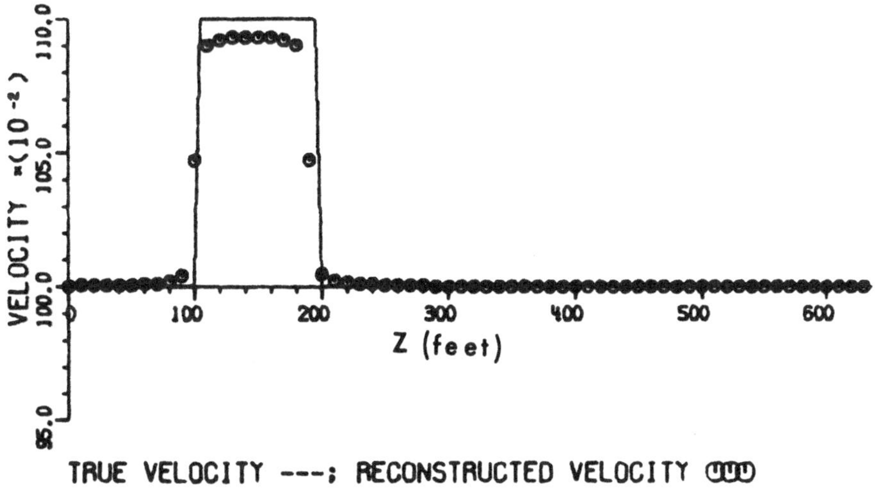

TRUE VELOCITY ---; RECONSTRUCTED VELOCITY ⊙⊙⊙

Fig. 6

INVERSE SCATTERING THEORY: EXACT AND APPROXIMATE SOLUTIONS

Arthur K. Jordan

Naval Research Laboratory
Washington, District of Columbia 20375

INTRODUCTION

The customary procedure for constructing a theory for an electromagnetic scattering phenomenon is first to assume some specific model for the scattering object and then to calculate the resultant scattered fields. This procedure is known as *direct* scattering theory. The scattered fields thus predicted are compared with the experimental scattering data and the specific model is altered until theory and experiment agree according to some acceptable criterion.

Inverse scattering theory assumes only the general physical properties of the scattering object and then determines analytically the specific model of that target using only the knowledge of the incident fields and the scattering data. This procedure inverts the customary analysis of the cause-and-effect relationship and so is known as *inverse* scattering theory. This general problem can be simplified if the scattering object is assumed to be an inhomogeneous region whose index of refraction has only a one-dimensional spatial variation. This procedure is known as *profile reconstruction.*

A general theory of profile reconstruction, due to GEL'FAND and LEVITAN [1], provides exact solutions for the profiles of refractive index by using an analytic representation of the scattering data. Exact solutions of this inverse scattering problem will be presented and several approximate solutions will be discussed.

The general physical model which we consider is the scattering of electromagnetic waves from a stratified ionized region, as shown in Fig. 1. This model has been used to study ionospheric radio wave propagation [2]. The effects of electron collisions and static magnetic fields have been neglected, so that the relative permittivity of the inhomogeneous region is

$$\frac{\epsilon(k,x)}{\epsilon_o} = 1 - \frac{1}{k^2} q(x), x \geqslant 0, \tag{1}$$

where ϵ_o is the permittivity of free space and $k = \omega/c$, the wave number in free space, and where ω = radian frequency and c = velocity of light. The profile function $q(x)$ is proportional to the electron density.

The time-harmonic amplitude $u(\kappa,x)$ of the horizontally polarized electromagnetic field $\mathbf{E} = \hat{\mathbf{a}}_y E_y$ satisfies the differential equation in the variable x

$$\frac{d^2}{dx^2} u(\kappa,x) + \left[\kappa^2 - q(x)\right] u(\kappa,x) = 0, \tag{2}$$

where the spectral variable is the wavenumber in the x-direction, $\kappa = k\cos\theta$, $u(\kappa,x)$ is the Fourier transform of $E_y(ct,x)$, and t the time. The case of a vertically polarized field will be discussed below.

The profile function $q(x)$ in (2) will be assumed to be real, bounded, and piecewise continuous in $0 \leqslant x < \infty$, with $q(x) \equiv 0$ for $x < 0$. Thus, it is possible to obtain a solution of (2) which satisfies the asymptotic conditions

$$u(\kappa,x) = \begin{cases} r(\kappa) e^{-i\kappa x} + e^{i\kappa x}, & x \to -\infty, \\ T(\kappa) e^{i\kappa x}, & x \to +\infty. \end{cases} \tag{4}$$

The reflection coefficient $r(\kappa)$ and transmission coefficient $T(\kappa)$ are assumed to be represented by analytic functions of κ in the complex κ-plane. From the conservation of energy,

$$|r(\kappa)|^2 + |T(\kappa)|^2 = 1, \tag{5}$$

where

$$|r(\kappa)|^2 = r(\kappa)\cdot\overline{r(\kappa)} = r(\kappa)\cdot r(-\kappa),$$
$$|T(\kappa)|^2 = T(\kappa)\cdot\overline{T(\kappa)} = T(\kappa)\cdot T(-\kappa),$$

for real κ , and where $\overline{r(\kappa)}$ means complex conjugate of $r(\kappa)$.

The inverse scattering problem can now be stated: Given the reflection coefficient $r(\kappa)$ as an analytic function of the wave number κ, find the self-consistent profile function $q(x)$. (Here, self-consistent means that no further information is needed to find a unique profile.) We will direct our attention to the analytic relationship between the reflection coefficient and the profile function. The problem of the appropriate data processing to obtain the reflection coefficient $r(\kappa)$ warrants a separate investigation.

1. INVERSE SCATTERING PROBLEM

The time-harmonic wave amplitude in the free-space region is

$$u_o(\kappa,x) = e^{i\kappa x} + r(\kappa)e^{-i\kappa x},\ x \leqslant 0, \tag{6}$$

where only the dependence on the x spatial coordinate is shown. The corresponding time-dependent electric field is

$$E_{yo}(x,ct) = \delta(x-ct) + R(x+ct),\ x \leqslant 0, \tag{7}$$

where $\delta(x-ct)$ = incident δ-function impulse, and $R(x+ct)$ = reflected transient.

The time-dependent field in the inhomogeneous region satisfies the differential equation

$$\frac{\partial^2 E_y}{\partial x^2} - \frac{1}{c^2}\frac{\partial^2 E_y}{\partial t^2} - q(x)E_y = 0, x \geqslant 0. \tag{8}$$

The retarded electric field can be represented in terms of the electric field defined by (7) with the transformation [1,3]

$$E_y(x,ct) = E_{yo}(x,ct) + \int_{-x}^{x} K(x,\xi)E_{yo}(\xi,ct)\,d\xi, \tag{9}$$

where the function $K(x,\xi)$ also satisfies the differential equation (8) with the boundary conditions

$$K(x,-x) = 0, \tag{10}$$

$$\frac{dK(x,x)}{dx} = \frac{1}{2}q(x). \tag{11}$$

For a wave moving toward the right, the retarded field satisfies the condition

$$E_y(x,ct) = 0,\ ct < x, \tag{12}$$

so that (9) together with (7) provides the integral equation

$$R(x+ct) + K(x,ct) + \int_{-ct}^{x} K(x,\xi)R(\xi+ct)\,d\xi = 0. \tag{13}$$

If this equation can be solved for $K(x,\xi)$, then (11) gives the solution to the inverse scattering problem.

2. EXACT SOLUTIONS

In principle, the inverse scattering problem is solved if the Gel'fand-Levitan integral equation (13) can be solved for $K(x,ct)$ so that the profile function $q(x)$ is found from condition (11). The general solution of (13) is not a simple matter. However, there are two useful exact solutions:

i. $R(x+ct)$ is separable, $R(x+ct) = R_1(x)R_1(ct)$.

If $R(x+ct)$ is separable, direct substitution shows that the solution to (13) is

$$K(x,ct) = \frac{-R_1(x)R_1(ct)}{1+\int_{-x}^{x} R_1^2(\xi)d\xi}.$$

ii. $r(\kappa)$ is a rational function of κ (or $R(x)$ is a sum of exponentials).

The integral equation (13) can be solved exactly when $r(\kappa)$ is a rational function of κ, considered as a complex variable [3]. In this case, the reflected field is

$$R(x+ct) = \frac{1}{2\pi}\int_{-\infty}^{\infty} r(\kappa)e^{-i\kappa(x-ct)}d\kappa - i\sum_{n=1}^{N} r_n e^{i\kappa_n(x-ct)}, \tag{14}$$

where the integral represents the continuous spectrum of $r(\kappa)$ and the discrete spectrum is represented by the sum over the poles, κ_n, if any, on the positive imaginary axis with the residues, r_n. If $q(x) \geqslant 0$, as in the present model of ionospheric scattering, then $r(\kappa)$ has no poles on the positive imaginary axis. If the poles of $r(\kappa)$ lie on the unit circle, then $r(\kappa)$ is the nth-order Butterworth approximation and the integral equation (13) can be solved rather easily [4].

Examples of Exact Solutions

The reconstruction method can be demonstrated with a third-order rational approximation to $r(\kappa)$, e.g., $r(\kappa)$ has three poles in the complex κ-plane [5]. The resultant profile function $q(x)$ resembles electron density profiles that have been analyzed by direct methods [2]. We consider the reflection coefficient

$$r(\kappa) = \frac{\kappa_1\kappa_2\kappa_3}{(\kappa-\kappa_1)(\kappa-\kappa_2)(\kappa-\kappa_3)}, \tag{15}$$

where $\kappa_2 = -\bar{\kappa}_1 = c_1 - ic_2$ and $\kappa_3 = -ia$. The normalization has been chosen so that $r(0) = -1$. Conservation of energy requires that

$$|r(\kappa)|^2 \leqslant 1. \tag{16}$$

This defines the regions for the allowed pole locations, shown in Fig. 2 by the shaded portion. The reflected energy density $|r(\kappa)|^2$ is shown in Fig. 3 as a function of κ for the configuration of poles shown in Fig. 2, assuming that $a = 1.0$, $c_1 = 0.50$, $c_2 = 0.499$. This configuration is also shown in Fig. 4.3.

An example of a discrete as well as a continuous spectrum is furnished by a reflection coefficient with the pole configuration shown in Fig. 4.5. The symmetric poles on the unit circle in the lower half-plane correspond to the two symmetric poles for the third-order Butterworth approximation, these poles representing the continuous part of the spectral function for the differential equation (2); the pole on the positive imaginary axis represents the discrete part of the spectral function:

$$r(\kappa) = \frac{-i}{\kappa^3 + i},$$

$$\kappa_1 = \frac{1}{2}(\sqrt{3} - i),$$

$$\kappa_2 = -\bar{\kappa}_1,$$

$$\kappa_3 = i.$$

A solution of this inverse scattering problem has been presented by S. Ahn at this conference. An alternate, but completely equivalent method, will be summarized here.

The characteristic function $R(x)$ is found from (14) to be

$$R(x) = -\frac{1+i\sqrt{3}}{6}e^{-\frac{x}{2}(1+i\sqrt{3})} - \frac{1-i\sqrt{3}}{6}e^{-\frac{x}{2}(1-i\sqrt{3})} + \frac{1}{3}e^x, \tag{17}$$

where the first two terms represent the continuous part of the spectrum and the last term represents the discrete part of the spectrum.

We will use this example to demonstrate an alternate, but equivalent, technique [3,4] for solving the integral equation (13). It is possible to construct a differential operator $f(p)$, $p \rightarrow \frac{d}{dy}$, such that $f(p)R(x) = 0$. For a three-pole reflection coefficient,

$$f(p) = p^3 + i(\kappa_1 + \kappa_2 + \kappa_3)\, p^2 - (\kappa_1\kappa_2 + \kappa_1\kappa_3 + \kappa_2\kappa_3)p - i\kappa_1\kappa_2\kappa_3, \tag{18}$$

so that, in the present case, $f(p) = p^3 - 1$. The differential operator is applied to (13) to obtain (here we fix x and consider p to represent differentiation with respect to $y = ct \leqslant x$)

$$f(p)K(x,y) + K(x,-y) = 0, \tag{19}$$

and by symmetry

$$f(-p)K(x,-y) + K(x,y) = 0. \tag{20}$$

The boundary conditions on $K(x,y)$ are

$$K(x,y)\,|_{y=-x} = 0, \tag{21}$$

$$K'(x,y)|_{y=-x} = R'(x)|_{x=0} = 0, \tag{22}$$

$$K''(x,y)|_{y=-x} = R''(x)|_{x=0} = -1. \tag{23}$$

Eliminating $K(x,-y)$ between Eqs. (19) and (20) yields $p^6K(x,y) = 0$, so that

$$K(x,y) = C_5(x)y^5 + C_4(x)y^4 + C_3(x)y^3 + C_2(x)y^2 + C_1(x)y + C_0(x). \tag{24}$$

From (19)-(23), we obtain

$$K(x,y) = -\frac{x}{8x^3 + 12}\,(y^4 + 12y) + \frac{x^3 - 3}{4x^3 + 6}\,y^2 - \frac{x^5 + 6x^2}{2(4x^3 + 6)}, \tag{25}$$

so that the profile function is found from (17) to be

$$q(x) = \frac{24x(2x^3 - 3)}{(2x^3 + 3)^2}, \quad x \geqslant 0. \tag{26}$$

As $x \rightarrow \infty$, $q(x) = 1/x^2$, which is the same asymptotic behavior as the profile function which was derived from the second-order Butterworth approximation [4]. There is a "potential well" closer to $x = 0$ with one "bound state" or "characteristic mode" with the value $q_1 = -1$, corresponding to the pole on the positive imaginary axis at $\kappa_1 = +\,i$. There is a simple check on the number M of bound states which was obtained by BARGMANN [6] for direct quantum scattering theory:

$$M \leqslant \int_{-\infty}^{\infty} |x|\cdot|q_-(x)|\,dx \leqslant M + 1, \tag{27}$$

where $q_-(x)$ is the portion of the profile function where $q(x) < 0$. After integrating by parts between the limits $0 \leqslant x \leqslant (3/2)^{1/3}$, we can evaluate this integral to obtain $M \leqslant 3-2\ln 2$; so that $M = 1$. (Positive and negative values of the profile function can be interpreted physically in terms of the scattering of vertically polarized waves by an inhomogeneous dielectric region, discussed below.)

A reflection coefficient with a zero at $\kappa=0$ is shown in Fig. 4.4. $r(\kappa)$ also has the second-order Butterworth poles

$$r(\kappa) = \frac{\kappa}{\kappa^2 + i\sqrt{2}\kappa - 1}.$$

The reconstruction method yields the profile function

$$q(x) = \frac{q_N(x)}{q_D(x)}, \quad x \geqslant 0,$$

where

$$\frac{q_N(x)}{4(\sqrt{2}+1)} = 2 - \frac{e^x}{\sqrt{3}}[(4-\sqrt{2})\sin\sqrt{3}x + \sqrt{6}\cos\sqrt{3}x] - \frac{e^{-x}}{\sqrt{2}+1}\left[\frac{4-3\sqrt{2}}{\sqrt{3}}\sin\sqrt{3}x + \sqrt{2}\cos\sqrt{3}x\right]$$

$$q_D(x) = \left[(\sqrt{2}+1)e^x - (\sqrt{2}-1)e^{-x} - \frac{2\sqrt{2}}{\sqrt{3}}\sin\sqrt{3}x\right]^2.$$

Since $r(\kappa)|_{\kappa=0} = 0$, there is a potential well for small x. However, there are no bound states.

If the electromagnetic field is vertically polarized so that $\mathbf{H} = \hat{\mathbf{a}}_y H_y$, then the time-harmonic amplitude $v(k,x)$ satisfies the differential equation

$$\frac{d}{dx}\left\{\frac{1}{\epsilon}\frac{dv(k,x)}{dx}\right\} + \left\{k^2 - \frac{k^2\sin^2\theta}{\epsilon}\right\}v(k,x) = 0. \tag{28}$$

This can be expressed in a form similar to (2) by using the local wave impedance $W(x)$ with the following transformation of variables

$$W(x) = \sqrt{\frac{\mu_o}{\epsilon(x)}},$$

$$\phi(k,x) = v(k,x)\sqrt{W(x)},$$

$$x = \int_0^{\xi}\sqrt{\mu_0\epsilon(\xi)}\,d\xi,$$

$$q(x) = \left\{\frac{W'(x)}{2W(x)}\right\}^2 - \left\{\frac{W'(x)}{2W(x)}\right\}'.$$

Some values of $W'(x)/(2W(x))$ can cause a negative $q(x)$. $W(x)$ can be found from

$$W(x) = \frac{W(0)}{[1+F(x,0)]^2}, \tag{29}$$

where $F(x,0) = F(x,s)|_{s=0}$, $W(0) = W(x)|_{x=0}$, and $F(x,s) = \int_{-x}^{x} K(x,\zeta)\, e^{-s\zeta}d\zeta$; details can be found in [5].

The general form of the profile function $q(x)$ is related to the pole-zero configuration of the reflection coefficient $r(\kappa)$. The continuous spectrum is represented by poles in the lower half-plane. The "smoothness" of $q(x)$ is determined by the number of poles and zeros of $r(\kappa)$. If $r(\kappa)$ has M poles and no zeros, then the $(M-2)$th derivative of $q(x)$, and all lower derivatives will be continuous at $x = 0$. This means that if $r(\kappa)$ has one pole, the corresponding $q(x)$ will be discontinuous at $x = 0$. If $r(\kappa)$ has two poles, then $q(x)$ will be finite at $x = 0$, but will have an infinite slope. If $r(\kappa)$ has three poles, both $q(x)$ and $q'(x)$ are continuous at $x = 0$ but there is an "angle discontinuity". If $r(\kappa)$ has a zero at $\kappa = 0$, then $q(x)$ will have a potential well, since waves with small energy penetrate the medium and are not reflected immediately. If a discrete spectrum is present, it can be represented by a pole on the positive imaginary axis.

3. APPROXIMATE SOLUTIONS

On the basis of the preceding discussion, several approximate solutions of the Gel'fand-Levitan equation can be suggested:

i. Neglect higher-order poles. For example in the three-pole reflection coefficient, if $k_3 = -ia \to -i\infty$, then $q_3(x) \to q_2(x)$ as $x \to \infty$, where $q_3(x)$ means the potential function obtained from a 3-pole $r(\kappa)$ and $q_2(x)$ means the potential function obtained from the corresponding 2-pole reflection coefficient as $\kappa_3 \to -i\infty$.

ii. Approximate $R(x+ct)$ by a separable function. This corresponds to a 1-pole $r(\kappa)$.

iii. Use the second iterative solution of the Gel'fand-Levitan equation. This perturbation solution was suggested by MOSES [7]. If $r(\kappa) \to \epsilon r(\kappa)$, $\epsilon << 1$ (and $R(x) \to \epsilon R(x)$), then an approximate solution of the inverse scattering problem is

$$q_\epsilon(x) = \epsilon\left\{-2\frac{d}{dx}R(2x)\right\} + \epsilon^2R^2(2x).$$

For example, in the 2-pole Butterworth case,

$$|r(\kappa)|^2 = \frac{1}{1+\kappa^4} \Rightarrow q(x) = \frac{4}{(1+\sqrt{2}x)^2}, \quad x \geqslant 0.$$

Using the approximate formulation,

$$q_\epsilon(x) = \epsilon[4e^{-\sqrt{2}x}(\cos\sqrt{2}x - \sin\sqrt{2}x)] + \epsilon^2[2e^{-\sqrt{2}x}\sin^2\sqrt{2}x], \ x \geqslant 0.$$

A more appropriate comparison could be obtained by using the potential function obtained by the method of solution demonstrated in (17)-(25).

iv. For the case of vertical polarization, if the reflection coefficient is small, then the "taper function" $\frac{W'(x)}{2W(x)}$ is related to the Fourier transform of the reflection coefficient,

$$\frac{W'(x)}{2W(x)} = \int_{-\infty}^{\infty} r(\kappa) e^{2i\kappa x} dx.$$

It is apparent that for this approximation to be valid, $r(\kappa)$ should have more than 4 poles.

ACKNOWLEDGMENT

The author gratefully acknowledges several informative discussions with I. Kay. This work was supported by the Office of Naval Research and the Naval Air Systems Command.

REFERENCES

[1] I.M. Gel'fand, B.M. Levitan: "On the determination of a differential equation from its spectral function," Trans. Amer. Math. Soc., Ser. 2, **1**, 253-304 (1955)

[2] K. G. Budden: *Radio waves in the ionosphere* (Cambridge University Press, 1961)

[3] I. Kay: "The inverse scattering problem when the reflection coefficient is a rational function," Comm. Pure Appl. Math. **13**, 371-393 (1961)

[4] A.K. Jordan, H.N. Kritikos: "An application of one-dimensional inverse-scattering theory for inhomogeneous regions," IEEE Trans. **AP-21**, 909-911 (1973)

[5] S. Ahn, A. K. Jordan: "Profile inversion of simple plasmas and nonuniform regions: three-pole reflection coefficient," IEEE Trans. **AP-24**, 879-882 (1976)

[6] V. Bargmann: "On the number of bound states in a central field of force," Proc. Nat. Acad. Sci. (USA) **38**, 961-966 (1952)

[7] H. Moses: "Calculation of the scattering potential from the reflection coefficient", Phys. Rev. **102**, 559-567 (1956)

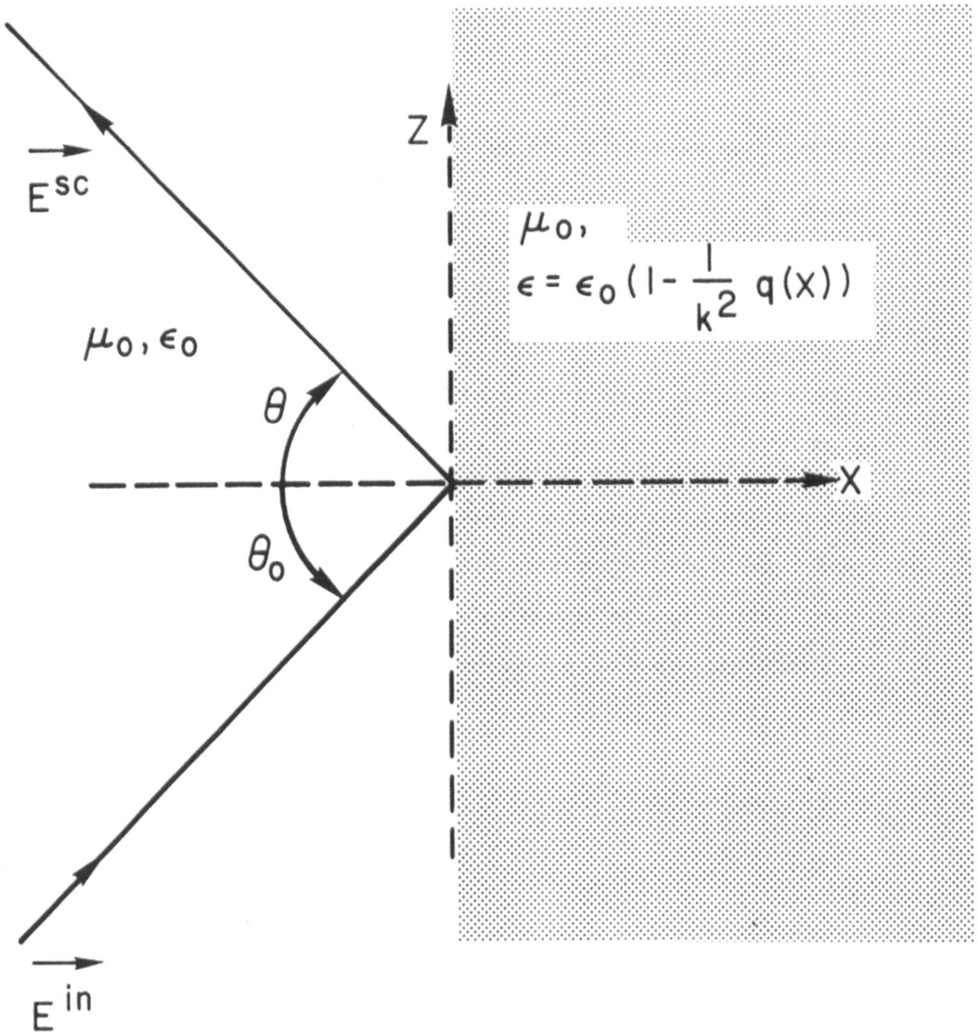

Fig. 1. Physical model for time-harmonic variation of electromagnetic wave scattering by an inhomogeneous ionized region in $x \geqslant 0$. Incidence angle $= \theta_0$ and scattering angle $= \theta$.

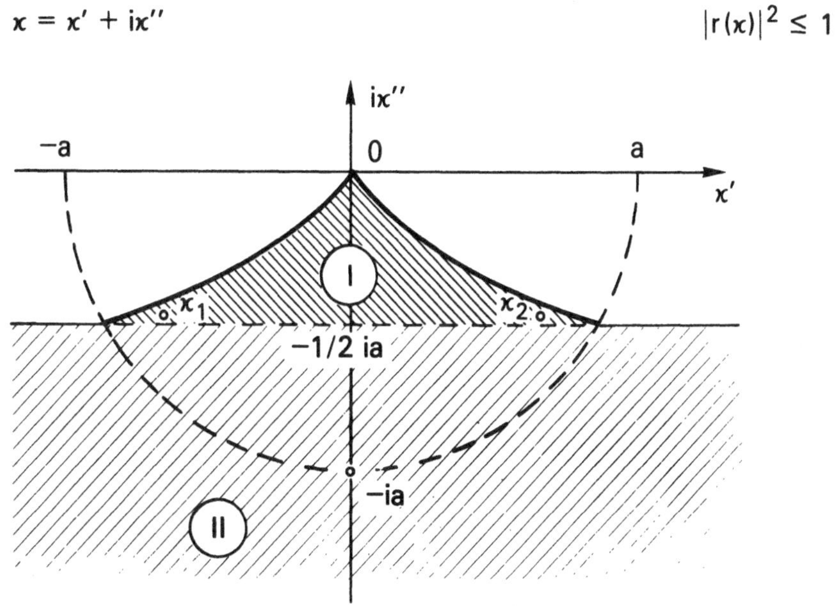

Fig. 2. Pole location in complex κ-plane for r(κ) of Equation (15). The shaded region also includes the positive imaginary axis.

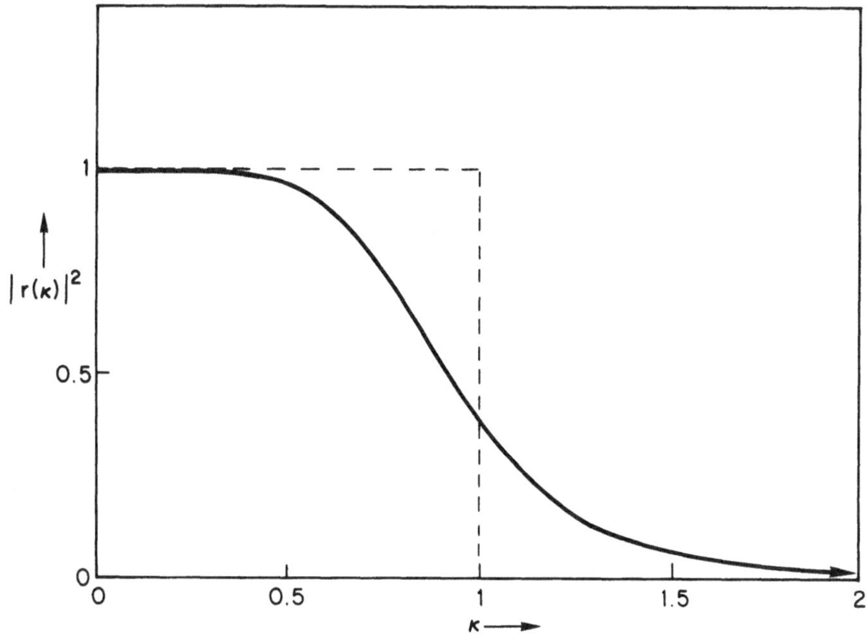

Fig. 3. Reflected energy density $|r(\kappa)|^2$ for Example 4.3 in Fig. 4.

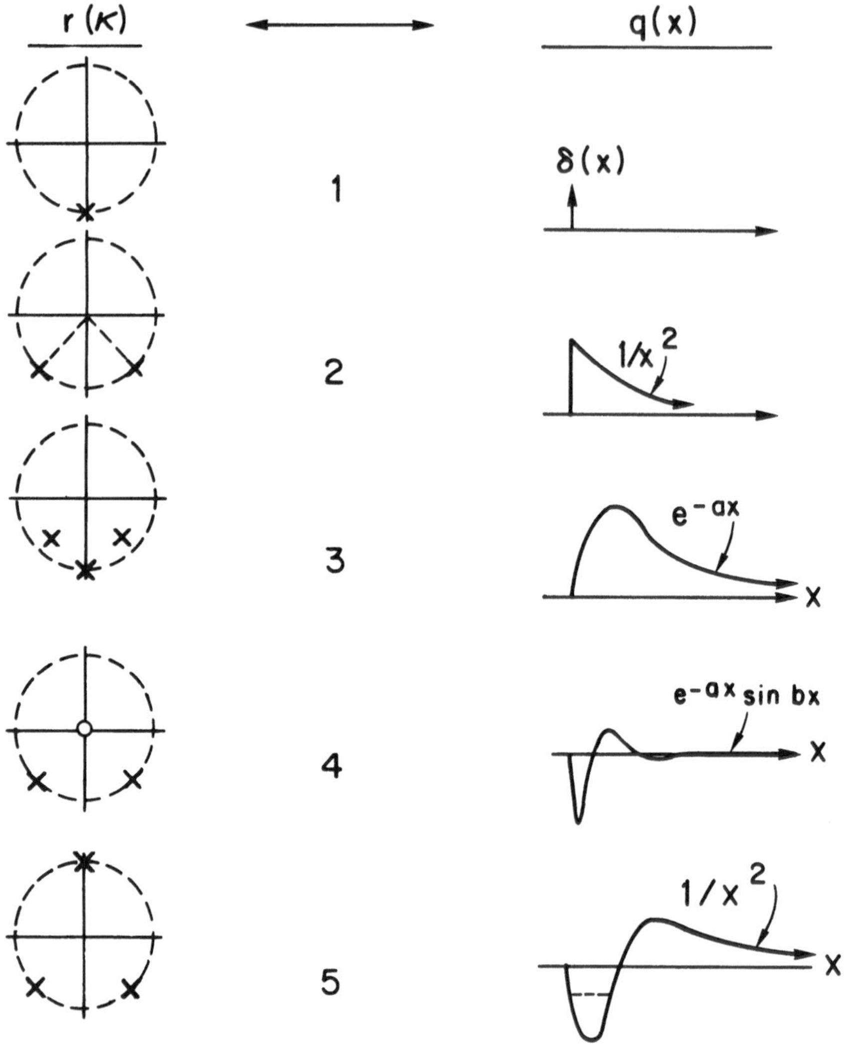

Fig. 4. Comparison of pole configuration of five examples of r (κ) and their corresponding profile functions q (x).

FUNCTIONAL EQUATION OF INVERSE SCATTERING FOR THE SCHRÖDINGER EQUATION WITH CONTINUOUS AND DISCRETE SPECTRA (Short Note)

Saeyoung Ahn

Naval Research Laboratory
Washington, D.C. 20375

The inverse scattering problem for the one-dimensional Schrödinger equation in the space $L^2(-\infty, \infty)$ that will be discussed in the note was first considered by KAY [1] in 1955. The problem is well defined, e.g., as in the superb review by FADDEYEV [2]: Find the interaction potential $q(x)$ of the particle energy operator $\hat{L}$, given the reflection coefficient $r(k)$, such that the particle-wave function satisfies [1]

$$U(x,k) \rightarrow e^{ikx} + r(k)e^{-ikx}, \; x \rightarrow -\infty, \tag{1}$$
$$U(x, k) \rightarrow t(k)e^{ikx}, \; x \rightarrow +\infty.$$

In the usual *direct* scattering problem, the reduced energy operator $\hat{L}$ in one dimension,

$$\hat{L} \equiv -\frac{d^2}{dx^2} + q(x), \tag{2}$$

determines the wave number k by the eigenvalue equation,

$$\hat{L}U(x) = k^2 U(x). \tag{3}$$

The inverse solution of (3) was obtained by Marchenko, and by Jost and Kohn; but more thoroughly by Gel'fand and Levitan by means of the so-called GEL'FAND-LEVITAN integral equation [3], i.e.

$$R(x + ct) + K(x,ct) + \int_{-\infty}^{x} K(x,y)R(y + ct)dt = 0, \tag{4}$$

where

$$q(x) = \frac{1}{2}\frac{dK(x,x)}{dx}, \tag{5}$$

and R is the Fourier transform of $r(k)$. KAY [4] found a formal solution $K(x,ct)$ of (4) when $r(k)$ is a rational function of K with no bound state included. It is possible [1,5] to find solutions of (4) for the reflection coefficient $r(k)$ of rational form with a few bound states and continuous spectra. Solutions for non-rational or meromorphic functions of $r(k)$ cannot be obtained analytically in closed form by the usual inverse method.

Some solutions were obtained for nonrational $r(k)$'s with a branch cut via the functional equation method [6] in the plasma inverse scattering problem, where the potential function is assumed to be positive and, consequently, excludes the bound-state problem.

I will go back to the fundamental equation (3) and will derive a form of the Gel'fand-Levitan integral equation (4) to solve the inverse scattering problem for a reflection coefficient $r(k)$ of *meromorphic type with bound states.* A new functional equation is then obtained by Laplace transformation of this new form of the generalized Gel'fand-Levitan integral equation by a procedure similar to that described in [6].

The question of how to find $r(k)$ directly or indirectly from the time response data of $R(x + ct)$ is not pursued here. I assume that $r(k)$ is known *a priori.*

CAUSALITY

When the incident wave $\delta(x - ct)$ corresponding to the particle wave $U(x,k)$ in (1) travels in the potential $q(x)$, where

$$q(x) = 0 \text{ for } x < 0 \text{ and real for } x \geqslant 0, \tag{6}$$

the reflected and bound-state waves are found by making use of the causality condition in the Fourier transform, i.e.,

$$R(x+ct) = \frac{1}{2\pi}\int r(k)e^{-ik(x+ct)}dk,\ x+ct \geqslant 0, x \leqslant 0 \tag{7}$$

$$B(x+ct) = -\frac{1}{2\pi}\int_C r(k)e^{-ik(x+ct)}dk, x+ct \geqslant 0,\ x \leqslant 0, \tag{8}$$

where C is the counterclockwise contour around the bound-state poles on the positive imaginary axis. The reflected wave $R(x+ct)$ is caused by $\delta(x-ct)$ and vanishes when $x+ct<0$. Since the bound state wave $B(x+ct)$ is also activated by the incident wave $\delta(x\ \ ct)$, its domain of nonvanishing values is defined to be the same as that of $R(x+ct)$. Due to causality, the total particle wave in the left half-plane is

$$U_t(x,ct) = U_0(x,ct) \equiv \delta(x-ct) + R(x+ct) + B(x+ct),\ x \leqslant 0,\ x+ct \geqslant 0, \tag{9}$$

and becomes a causal function in the same manner, i.e., $U_t(x,ct) = 0$ when $x+ct \leqslant 0$ and $x \leqslant 0$.

The retarded particle wave can be written with the aid of the MARCHENKO kernel [7], i.e.,

$$U_t(x,ct) = U_0(x,ct) + \int_{y_0}^{x} K(x,y)\,U_0(y,ct)\,dy,\ x \geqslant 0, \tag{10}$$

where $y_0 = \max(-ct,-x)$. $K(x,ct)$ defined by (10) has the property that

$$K(x,ct) = 0,\ \text{when } ct > x > 0, \tag{11a}$$

$$K(x,ct) = 0,\ \text{when } -ct > x > 0, \tag{11b}$$

$$K(x,ct) = 0, \text{when } x < 0. \tag{11c}$$

That is, $K(x,ct)$ becomes causal in the $x-t$ space and also satisfies the wave equation

$$\hat{D}(x,t)K(x,ct) = q(x)\delta(x-ct), \tag{12}$$

where we defined the transient wave operator $\hat{D}$ with $q(x)$ in (6)

$$\hat{D}(x,t) \equiv \frac{\partial^2}{\partial x^2} - \frac{1}{c^2}\frac{\partial^2}{\partial t^2} - q(x).$$

$K(x,ct)$ is thus a spacelike solution of the wave equation.

The causality property of $K(x,ct)$ in (11a) can be derived from that of $U_t(x,ct)$, namely, from the general form of (10), with the upper and lower bounds of the integral extended to $\pm\infty$. This latter solution is obtained, since a linear integral transformation exists between U_o and U_t by (10).

From the property that $U_t(x,ct) = 0$ when $x < ct$, $U_t(x,ct)$ should depend only on $U_0(z,ct)$, where $z \leqslant x$, and therefore we find the property (11a). In free space ($x \leqslant 0$), $U_t(x,ct) = U_o(x,ct)$, which results in the property (11c). The property that $U_t(x,ct) = U_o(x,ct) = 0$ when $0 \leqslant x \leqslant -ct$ leads us to (11b) and a lower bound $-x$ of the integral in (10). Another lower bound, $-ct$, is due to the property (7) of the retarded wave $R(x+ct)$ in the spacelike region (and $x > 0$).

Thanks to the causality and Marchenko kernel function, the retarded particle wave has a clear physical meaning. The observable wave at x and t inside the forward light-cone ($x < ct$) is the sum of the initial wave U_0 and the integral of $K(x,ct)\,U_o(ct,ct)$ over the accessible space-time $-x < ct < x$ in Fig. 1.

The physical meaning of $K(x,ct)$ becomes evident from (12), which can be rewritten

$$\hat{D}(x,t)\,U_s(x,ct) = 0, \tag{13}$$

where

$$U_s(x,t) \equiv \delta(x-ct) + K(x,ct). \tag{14}$$

Due to the properties (11a,b,c), $K(x,ct)$ is a *spacelike solution* of the transient wave operator $\hat{D}$ and, due to (11c), $K(x,ct)$ becomes a *reflectionless* wave.

In summary, given a timelike solution $U_t(x,ct) = U_0 = \delta(x-ct) + R(x+ct) + B(x+ct),\ x<0,$ *for* $q(x)$ *defined as in (6), there exists correspondingly a unique space-like solution* $U_s(x,ct) = \delta(x-ct) + K(x,ct)$, *to the transient wave operator* $\hat{D}(x,t)$ *of (12). Moreover,* $U_s(x,ct)$ *is reflectionless.*

Since $U_s(x,ct)$ has no reflection, nor is it reachable by an observer inside the forward light-cone due to its space-like nature, the space-like solution $U_s(x,ct)$ is an unobservable wave.

Now let me introduce an entire transient wave $\hat{U}(x,ct)$ as in [6], that is, the difference of two corresponding time-like and space-like solutions,

$$\hat{U}(x,ct) = U_s(x,ct) - U_t(x,ct), \tag{15}$$

with U_s as in (14) and $U_t(x,ct)$ as in (9) and (10). Then $\hat{U}(x,ct)$ is bounded everywhere in the x,t space and vanishes when $x + ct < 0$.

By substituting (15) into (10), we finally obtain a new version of the Gel'fand-Levitan integral equation by defining

$$\hat{R}(x + ct) = R(x + ct) + B(x + ct). \tag{16}$$

Then we get

$$\hat{R}(x + ct) + \hat{U}(x,ct) + \int_{\max[-x,-ct]}^{x} \hat{U}(x,y)\hat{R}(y + ct)\,dy = 0. \tag{17}$$

Equation (17) is a generalization to the case of the bound-state potential. The same version of (17) was previously obtained for the positive potential with no bound states in the plasma inverse problem [6].

LAPLACE TRANSFORM AND FUNCTIONAL EQUATION

The Laplace transform of (16) gives rise to a functional equation of the form

$$\tilde{R}(s)[\exp(sx) + \tilde{K}(x,-s)] + \tilde{K}(x,s) + \tilde{U}(x,s) = 0, \tag{18}$$

where

$$\tilde{R}(s) = \int_0^\infty [R(y) + B(y)]e^{-sy}dy, \tag{19}$$

$$\tilde{K}(x,s) = \int_{-x}^{x} U(x,y)e^{-sy}dy \tag{20}$$

and

$$\tilde{U}(x,s) = \int_x^\infty U(x,y)e^{-sy}dy. \tag{21}$$

From (20) and (21), we recover

$$K(x,x) = U(x,x) = \lim_{s\to\infty} se^{sx}\tilde{U}(x,s) = \lim_{s\to-\infty} se^{sx}\tilde{K}(x,s). \tag{22}$$

The power of the generalized functional equation can be exhibited for the inverse problem corresponding to the step-potential case, the so-called branch-cut problem with no bound state, i.e., $B(y) = 0$, so that

$$\tilde{R}(s) = -(\sqrt{s^2 + k_0^2} - s)^2/k_0^2. \tag{23}$$

When (23) holds, the functional equation has the form

$$\psi^2(s) - (s^2 + k_0^2)\phi^2(s) = 1, \tag{24}$$

where

$$\psi(s) \equiv \frac{1}{2}\{\tilde{K}(s) + \tilde{K}(-s) + e^{sx} + e^{sx}\}, \tag{25}$$

$$\phi(s) \equiv \frac{1}{2s}\{\tilde{K}(s) - \tilde{K}(-s) + e^{-sx} - e^{sx}\}. \tag{26}$$

The functional equation (24) has been solved by PENROSE and LEBOWITZ [8] using MUSKHELISHVILI's method [9]. A solution of (24) is $\psi(s) = 2\cosh(x\sqrt{s^2 + k_0^2})$ and $\phi(s) = -2s\sinh(x\sqrt{s^2 + k_0^2})/\sqrt{s^2 + k_0^2}$. By (22) and (13), we find the step-potential immediately.

Another example of interest is the case where $r(k)$ possesses two poles in the continuous spectrum and one bound-state pole [5],

$$r(k) = -\frac{i}{k^3 + i}, \tag{27}$$

namely, $\tilde{R}(s) = (s^3 - 1)^{-1}$. Equation (18) can be solved [9] to find that

$$\tilde{K}(x,s) = f_1 e^{sx} + f_2 e^{-sx},$$

$$f_1(s) = 6(x^2 s^{-4} - x s^{-5})(2x^3 + 3)^{-1} - s^{-3},$$

$$f_2(s) = f_1(-s) + 6x\,(1 + xs)s^{-2}(2x^3 + 3)^{-1},$$

and

$$\tilde{U}(x,s) = [6xs\,(1 + xs)\,(2x^3 + 3)^{-1} + 1](1 - s^3)^{-1} e^{-sx},$$

which yields, by (22) and (5),

$$q(x) = 24x(2x^3 - 3)/(2x^3 + 3)^2. \tag{28}$$

The true merit of solving the inverse problem by the functional equation (18) is demonstrated with the meromorphic reflection coefficient with an infinite number of poles for continuous spectra in the lower half-plane, coupled with a finite number of bound-state poles on the imaginary axis in the upper half-plane. For instance, it is very instructive to show how to solve the well-known potential well with a finite number of bound states via the functional equation (18). In this case,

$$\tilde{R}(s) = \frac{k_o^2(e^{2a\sigma} - 1)}{(\sigma - s)^2 - (\sigma + s)^2 e^{2a\sigma}}, \tag{29}$$

where k_0^2 and a are the depth and width of the potential well $q(x)$, and $\sigma \equiv \sqrt{s^2 - k_0^2}$. This is found [10] to have the similar functional form of (24), i.e. $\psi^2(s) - (s^2 - k_0^2)\,\phi^2(s) = 1$.

In this short note, I sketched the derivation of (17) and (18) to solve the inverse problem with a meromorphic reflection coefficient of an *infinite* number of poles coupled with a finite number of bound-state poles. This procedure could be easily modified for the potential

$$q(x) \to k_0^2, \text{ as } x \to \infty, \tag{30}$$

which results in $r(k)$'s with a possible branch cut via $\sqrt{k^2 + k_0^2}$.

I did not intend to give a review in this note, and many important and relevant papers were not quoted. They can be found in reviews by DYSON [11] and FADDEYEV [2]. A reader seriously interested in the inverse scattering problem should refer to the textbooks in [12] as well as to other papers presented in this conference. The conference proceedings edited by COLIN [13] has many useful papers and summaries, and the essays in honor of V. Bargmann (see [11]) also contain valuable papers.

REFERENCES

[1] I. Kay: "The inverse scattering problem", New York University Report EM-74, (AD 141205); AFCRC-TN-55-360 (February, 1955)

[2] L.D. Faddeyev: "The inverse problem in the quantum theory of scattering," J. Math. Phys. **4**, 72-104 (1963). See also "Properties of the S-matrix of the one-dimensional Schrödinger equation", Amer. Math. Soc., Ser. 2, **65**, 139-166 (1964)

[3] I.M. Gel'fand, B.M. Levitan: "On the determination of a differential equation by its spectral function," Dokl. Akad. Nauk. SSSR (N.S.) **77**, 557-560 (1951) (Russian); Izv. Akad. Nauk. SSSR Ser. Mat., **15**, 309-360 (1951) (Russian). English translation in Amer. Math. Soc. Transl., Ser. 2, **1**, 253-304 (1955)

[4] I. Kay: "The inverse scattering problem when the reflection coefficient is a rational function", Comm. Pure and Appl. Math. **13**, 371-393 (1960)

[5] A.K. Jordan, S. Ahn: "Inverse Scattering theory and profile reconstruction," Proc. IEEE, to appear (1979)

[6] H. Szu, C. E. Carroll, C. C. Yang, S. Ahn: "A new functional equation in the plasma inverse problem and its analytic properties," J. Math. Phys. **17**, 1236-1247 (1976)

[7] V.A. Marchenko: "Concerning the theory of a differential operator of a second order," Doklady Akad. Nauk. SSSR (N.S.) **72**, 457-460 (1950). See also K. Chadan and P. C. Sabatier: *Inverse problems in quantum scattering theory* (Springer, New York, 1977)

[8] O. Penrose, L. Lebowitz: "A functional equation in the theory of fluids," J. Math. Phys. **13**, 604 (1972)

[9] N. I. Muskhelisvili: *Singular Integral Equations* (Noordhoff, Groningen, 1953)

[10] S. Ahn: "The inverse scattering theory of the Schrödinger equation for meromorphic reflection coefficients with bound states" (unpublished).

[11] F. Dyson: "Old and new approaches to the inverse scattering problem," in *Studies in Mathematical Physics*, Essays in honor of V. Bargmann, ed. by E. H. Lieb, B. Simon, and A.S. Wightman (Princeton University Press, 1976).

[12] R.G. Newton: *Scattering Theory of Waves and Particles* (McGraw-Hill, New York, 1966); Z. S. Agranovich, V. A. Marchenko, *The Inverse Problem of Scattering Theory* (Gordon and Breach, New York, 1963); K. Chadan, P. C. Sabatier: *Inverse Problems in Quantum Scattering Theory* (Springer, Heidelberg, 1977)

[13] "The Mathematics of Profile Inversion," NASA Technical Memorandum Memo X-62,150 (August, 1972), ed. by L. Colin, available from National Technical Information Service, NTIS No. N73-11585, Springfield, Virginia

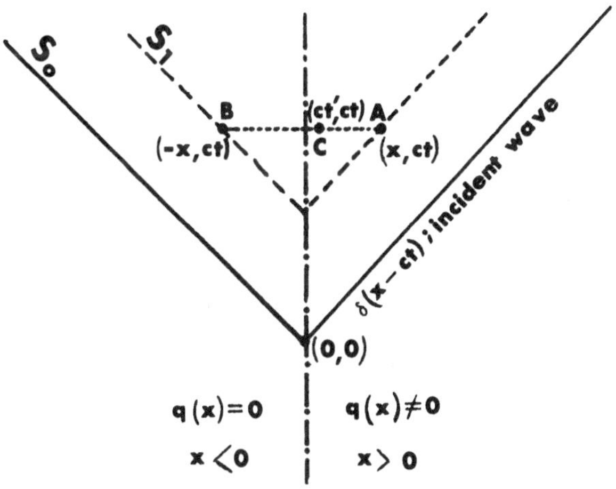

Fig. 1. The retarded wave $U_t(x,ct)$ inside the light-cone $S_0(ct>|x|)$ is the sum over the incident wave $U_0(x,ct)$ and the integration of $K(x,ct')$ $U_0(ct',ct)$ over the event C between two events A and B. The information on $q(x)$ is contained in KU_0 along the event line $[B,A]$ inside the light-cone S_1.

Selected Issues from

Lecture Notes in Mathematics

Vol. 662: Akin, The Metric Theory of Banach Manifolds. XIX, 306 pages. 1978.

Vol. 665: Journées d'Analyse Non Linéaire. Proceedings, 1977. Edité par P. Bénilan et J. Robert. VIII, 256 pages. 1978.

Vol. 667: J. Gilewicz, Approximants de Padé. XIV, 511 pages. 1978.

Vol. 668: The Structure of Attractors in Dynamical Systems. Proceedings, 1977. Edited by J. C. Martin, N. G. Markley and W. Perrizo. VI, 264 pages. 1978.

Vol. 675: J. Galambos and S. Kotz, Characterizations of Probability Distributions. VIII, 169 pages. 1978.

Vol. 676: Differential Geometrical Methods in Mathematical Physics II, Proceedings, 1977. Edited by K. Bleuler, H. R. Petry and A. Reetz. VI, 626 pages. 1978.

Vol. 678: D. Dacunha-Castelle, H. Heyer et B. Roynette. Ecole d'Eté de Probabilités de Saint-Flour. VII-1977. Edité par P. L. Hennequin. IX, 379 pages. 1978.

Vol. 679: Numerical Treatment of Differential Equations in Applications, Proceedings, 1977. Edited by R. Ansorge and W. Törnig. IX, 163 pages. 1978.

Vol. 681: Séminaire de Théorie du Potentiel Paris, No. 3, Directeurs: M. Brelot, G. Choquet et J. Deny. Rédacteurs: F. Hirsch et G. Mokobodzki. VII, 294 pages. 1978.

Vol. 682: G. D. James, The Representation Theory of the Symmetric Groups. V, 156 pages. 1978.

Vol. 684: E. E. Rosinger, Distributions and Nonlinear Partial Differential Equations. XI, 146 pages. 1978.

Vol. 690: W. J. J. Rey, Robust Statistical Methods. VI, 128 pages. 1978.

Vol. 691: G. Viennot, Algèbres de Lie Libres et Monoïdes Libres. III, 124 pages. 1978.

Vol. 693: Hilbert Space Operators, Proceedings, 1977. Edited by J. M. Bachar Jr. and D. W. Hadwin. VIII, 184 pages. 1978.

Vol. 696: P. J. Feinsilver, Special Functions, Probability Semigroups, and Hamiltonian Flows. VI, 112 pages. 1978.

Vol. 702: Yuri N. Bibikov, Local Theory of Nonlinear Analytic Ordinary Differential Equations. IX, 147 pages. 1979.

Vol. 704: Computing Methods in Applied Sciences and Engineering, 1977, I. Proceedings, 1977. Edited by R. Glowinski and J. L. Lions. VI, 391 pages. 1979.

Vol. 710: Séminaire Bourbaki vol. 1977/78, Exposés 507–524. IV, 328 pages. 1979.

Vol. 711: Asymptotic Analysis. Edited by F. Verhulst. V, 240 pages. 1979.

Vol. 712: Equations Différentielles et Systèmes de Pfaff dans le Champ Complexe. Edité par R. Gérard et J.-P. Ramis. V, 364 pages. 1979.

Vol. 716: M. A. Scheunert, The Theory of Lie Superalgebras. X, 271 pages. 1979.

Vol. 720: E. Dubinsky, The Structure of Nuclear Fréchet Spaces. V, 187 pages. 1979.

Vol. 724: D. Griffeath, Additive and Cancellative Interacting Particle Systems. V, 108 pages. 1979.

Vol. 725: Algèbres d'Opérateurs. Proceedings, 1978. Edité par P. de la Harpe. VII, 309 pages. 1979.

Vol. 726: Y.-C. Wong, Schwartz Spaces, Nuclear Spaces and Tensor Products. VI, 418 pages. 1979.

Vol. 727: Y. Saito, Spectral Representations for Schrödinger Operators With Long-Range Potentials. V, 149 pages. 1979.

Vol. 728: Non-Commutative Harmonic Analysis. Proceedings, 1978. Edited by J. Carmona and M. Vergne. V, 244 pages. 1979.

Vol. 729: Ergodic Theory. Proceedings 1978. Edited by M. Denker and K. Jacobs. XII, 209 pages. 1979.

Vol. 730: Functional Differential Equations and Approximation of Fixed Points. Proceedings, 1978. Edited by H.-O. Peitgen and H.-O. Walther. XV, 503 pages. 1979.

Vol. 731: Y. Nakagami and M. Takesaki, Duality for Crossed Products of von Neumann Algebras. IX, 139 pages. 1979.

Vol. 733: F. Bloom, Modern Differential Geometric Techniques in the Theory of Continuous Distributions of Dislocations. XII, 206 pages. 1979.

Vol. 735: B. Aupetit, Propriétés Spectrales des Algèbres de Banach. XII, 192 pages. 1979.

Vol. 738: P. E. Conner, Differentiable Periodic Maps. 2nd edition, IV, 181 pages. 1979.

Vol. 742: K. Clancey, Seminormal Operators. VII, 125 pages. 1979.

Vol. 755: Global Analysis. Proceedings, 1978. Edited by M. Grmela and J. E. Marsden. VII, 377 pages. 1979.

Vol. 756: H. O. Cordes, Elliptic Pseudo-Differential Operators – An Abstract Theory. IX, 331 pages. 1979.

Vol. 760: H.-O. Georgii, Canonical Gibbs Measures. VIII, 190 pages. 1979.

Vol. 762: D. H. Sattinger, Group Theoretic Methods in Bifurcation Theory. V, 241 pages. 1979.

Vol. 765: Padé Approximation and its Applications. Proceedings, 1979. Edited by L. Wuytack. VI, 392 pages. 1979.

Vol. 766: T. tom Dieck, Transformation Groups and Representation Theory. VIII, 309 pages. 1979.

Vol. 771: Approximation Methods for Navier-Stokes Problems. Proceedings, 1979. Edited by R. Rautmann. XVI, 581 pages. 1980.

Vol. 773: Numerical Analysis. Proceedings, 1979. Edited by G. A. Watson. X, 184 pages. 1980.

Vol. 775: Geometric Methods in Mathematical Physics. Proceedings, 1979. Edited by G. Kaiser and J. E. Marsden. VII, 257 pages. 1980.

Lecture Notes in Physics

Vol. 91: Computing Methods in Applied Sciences and Engineering, 1977, II. Proceedings, 1977. Edited by R. Glowinski and J. L. Lions. VI, 359 pages. 1979.

Vol. 92: Nuclear Interactions. Proceedings, 1978. Edited by B. A. Robson. XXIV, 507 pages. 1979.

Vol. 93: Stochastic Behavior in Classical and Quantum Hamiltonian Systems. Proceedings, 1977. Edited by G. Casati and J. Ford. VI, 375 pages. 1979.

Vol. 94: Group Theoretical Methods in Physics. Proceedings, 1978. Edited by W. Beiglböck, A. Böhm and E. Takasugi. XIII, 540 pages. 1979.

Vol. 95: Quasi One-Dimensional Conductors I. Proceedings, 1978. Edited by S. Barišić, A. Bjeliš, J. R. Cooper and B. Leontič. X, 371 pages. 1979.

Vol. 96: Quasi One-Dimensional Conductors II. Proceedings 1978. Edited by S. Barišić, A. Bjeliš, J. R. Cooper and B. Leontić. XII, 461 pages. 1979.

Vol. 97: Hughston, Twistors and Particles. VIII, 153 pages. 1979.

Vol. 98: Nonlinear Problems in Theoretical Physics. Proceedings, 1978. Edited by A. F. Rañada. X, 216 pages. 1979.

Vol. 99: M. Drieschner, Voraussage – Wahrscheinlichkeit – Objekt. XI, 308 Seiten. 1979.

Vol. 100: Einstein Symposion Berlin. Proceedings 1979. Edited by H. Nelkowski et al. VIII, 550 pages. 1979.

Vol. 101: A. Martin-Löf, Statistical Mechanics and the Foundations of Thermodynamics. V, 120 pages. 1979.

Vol. 102: H. Hora, Nonlinear Plasma Dynamics at Laser Irradiation. VIII, 242 pages. 1979.

Vol. 103: P. A. Martin, Modèles en Mécanique Statistique des Processus Irréversibles. IV, 134 pages. 1979.

Vol. 104: Dynamical Critical Phenomena and Related Topics. Proceedings, 1979. Edited by Ch. P. Enz. XII, 390 pages. 1979.

Vol. 105: Dynamics and Instability of Fluid Interfaces. Proceedings, 1978. Edited by T. S. Sørensen. V, 315 pages. 1979.

Vol. 106: Feynman Path Integrals, Proceedings, 1978. Edited by S. Albeverio et al. XI, 451 pages. 1979.

Vol. 107: J. Kijowski, W. M. Tulczyjew, A Symplectic Framework for Field Theories. IV, 257 pages. 1979.

Vol. 108: Nuclear Physics with Electromagnetic Interactions. Proceedings, 1979. Edited by H. Arenhövel and D. Drechsel. IX, 509 pages. 1979.

Vol. 109: Physics of the Expanding Universe. Proceedings, 1978. Edited by M. Demiański. V, 210 pages. 1979.

Vol. 110: D. A. Park, Classical Dynamics and Its Quantum Analogues. VIII, 339 pages. 1979.

Vol. 111: H.-J. Schmidt, Axiomatic Characterization of Physical Geometry. V, 163 pages. 1979.

Vol. 112: Imaging Processes and Coherence in Physics. Proceedings, 1979. Edited by M. Schlenker et al. XIX, 577 pages. 1980.

Vol. 113: Recent Advances in the Quantum Theory of Polymers. Proceedings 1979. Edited by J.-M. André et al. V, 306 pages. 1980.

Vol. 114: Stellar Turbulence. Proceedings, 1979. Edited by D. F. Gray and J. L. Linsky. IX, 308 pages. 1980.

Vol. 115: Modern Trends in the Theory of Condensed Matter. Proceedings, 1979. Edited by A. Pekalski and J. A. Przystawa. IX, 597 pages. 1980.

Vol. 116: Mathematical Problems in Theoretical Physics. Proceedings, 1979. Edited by K. Osterwalder. VIII, 412 pages. 1980.

Vol. 117: Deep-Inelastic and Fusion Reactions with Heavy Ions. Proceedings, 1979. Edited by W. von Oertzen. XIII, 394 pages. 1980.

Vol. 118: Quantum Chromodynamics. Proceedings, 1979. Edited by J. L. Alonso and R. Tarrach. IX, 424 pages. 1980.

Vol. 119: Nuclear Spectroscopy. Proceedings, 1979. Edited by G. F. Bertsch and D. Kurath. VII, 250 pages. 1980.

Vol. 120: Nonlinear Evolution Equations and Dynamical Systems. Proceedings, 1979. Edited by M. Boiti, F. Pempinelli and G. Soliani. VI, 368 pages. 1980.

Vol. 121: F. W. Wiegel, Fluid Flow Through Porous Macromolecular Systems. V, 102 pages. 1980.

Vol. 122: New Developments in Semiconductor Physics. Proceedings, 1979. Edited by F. Beleznay et al. V, 276 pages. 1980.

Vol. 123: D. H. Mayer, The Ruelle-Araki Transfer Operator in Classical Statistical Mechanics. VIII, 154 pages. 1980.

Vol. 124: Gravitational Radiation, Collapsed Objects and Exact Solutions. Proceedings, 1979. Edited by C. Edwards. VI, 487 pages. 1980.

Vol. 125: Nonradial and Nonlinear Stellar Pulsation. Proceedings, 1980. Edited by H. A. Hill and W. A. Dziembowski. VIII, 497 pages. 1980.

Vol. 126: Complex Analysis, Microlocal Calculus and Relativistic Quantum Theory. Proceedings, 1979. Edited by D. Iagolnitzer. VIII, 502 pages. 1980.

Vol. 127: E. Sanchez-Palencia, Non-Homogeneous Media and Vibration Theory. IX, 398 pages. 1980.

Vol. 128: Neutron Spin Echo. Proceedings, 1979. Edited by F. Mezei. VI, 253 pages. 1980.

Vol. 129: Geometrical and Topological Methods in Gauge Theories. Proceedings, 1979. Edited by J. Harnad and S. Shnider. VIII, 155 pages. 1980.

Vol. 130: Mathematical Methods and Applications of Scattering Theory. Proceedings, 1979. Edited by J. A. DeSanto, A. W. Sáenz and W. W. Zachary. XIII, 331 pages. 1980.